Bilinear Integrable Systems: From Classical to Quantum, Continuous to Discrete

NATO Science Series

A Series presenting the results of scientific meetings supported under the NATO Science Programme.

The Series is published by IOS Press, Amsterdam, and Springer (formerly Kluwer Academic Publishers) in conjunction with the NATO Public Diplomacy Division.

Sub-Series

I. Life and Behavioural Sciences	IOS Press
II. Mathematics, Physics and Chemistry	Springer (formerly Kluwer Academic Publishers)
III. Computer and Systems Science	IOS Press
IV. Earth and Environmental Sciences	Springer (formerly Kluwer Academic Publishers)

The NATO Science Series continues the series of books published formerly as the NATO ASI Series.

The NATO Science Programme offers support for collaboration in civil science between scientists of countries of the Euro-Atlantic Partnership Council. The types of scientific meeting generally supported are "Advanced Study Institutes" and "Advanced Research Workshops", and the NATO Science Series collects together the results of these meetings. The meetings are co-organized by scientists from NATO countries and scientists from NATO's Partner countries — countries of the CIS and Central and Eastern Europe.

Advanced Study Institutes are high-level tutorial courses offering in-depth study of latest advances in a field.
Advanced Research Workshops are expert meetings aimed at critical assessment of a field, and identification of directions for future action.

As a consequence of the restructuring of the NATO Science Programme in 1999, the NATO Science Series was re-organized to the four sub-series noted above. Please consult the following web sites for information on previous volumes published in the Series.

http://www.nato.int/science
http://www.springer.com
http://www.iospress.nl

Bilinear Integrable Systems: From Classical to Quantum, Continuous to Discrete

edited by

Ludwig Faddeev

Russian Academy of Sciences,
St. Petersburg, Russia

Pierre Van Moerbeke

Université Catholique de Louvain,
Louvain-la-Neuve, Belgium

and

Franklin Lambert

Vrije Universiteit Brussel,
Belgium

Springer

Published in cooperation with NATO Public Diplomacy Division

Proceedings of the NATO Advanced Research Workshop on
Bilinear Integrable Systems: From Classical to Quantum, Continuous to Discrete
St. Petersburg, Russia
15–19 September 2002

A C.I.P. Catalogue record for this book is available from the Library of Congress.

ISBN-10 1-4020-3502-0 (PB)
ISBN-13 978-1-4020-3502-9 (PB)
ISBN-10 1-4020-3501-2 (HB)
ISBN-10 1-4020-3503-9 (e-book)
ISBN-13 978-1-4020-3501-2 (HB)
ISBN-13 978-1-4020-3503-6 (e-book)

Published by Springer,
P.O. Box 17, 3300 AA Dordrecht, The Netherlands.

www.springer.com

TABLE OF CONTENTS

Preface   ix

List of Participants   xiii

ARATYN, HENRIK & VAN DE LEUR, JOHAN: **The CKP hierarchy and the WDVV prepotential**   1

ARIK, M.: **Quantum invariance groups of particle algebras**   13

ATHORNE, CHRIS: **Algebraic Hirota maps**   17

BAJNOK, Z., PALLA, L., & TAKÁCS, G.: **Boundary states in SUSY sine-Gordon model**   35

BOBENKO, ALEXANDER I.: **Geometry of discrete integrability. The consistency approach**   43

DOKTOROV, E.V. & ROTHOS, V.M.: **Homoclinic orbits and dressing method**   55

ENOLSKII, V. & GRAVA, T.: **Riemann-Hilbert problem and algebraic curves**   65

GERDJIKOV, V.S.: **Analytic and algebraic aspects of Toda field theories and their real Hamiltonian forms**   77

GRAMMATICOS, B., RAMANI, A. & CARSTEA, A. S.: **Bilinear avatars of the discrete Painlevé II equation**   85

HAINE, LUC: **Orthogonal polynomials satisfying Q-difference equations**   97

HIROTA, RYOGO: **Discretization of coupled soliton equations**   113

HOROZOV, E.: **An adelic W-algebra and rank one bispectral operators**   123

KAKEI, SABURO: **Toroidal Lie algebra and bilinear identity of the self-dual Yang-Mills hierarchy**   137

LAMBERT, F. & SPRINGAEL, J.: **From soliton equations to their zero curvature formulation**   147

LEBLE, SERGEY: **Covariant forms of Lax one-field operators: from Abelian to noncommutative**   161

MARSHAKOV, A. & ZABRODIN, A.: **On the dirichlet boundary problem and Hirota equations**   175

MATVEEV, VLADIMIR B.: **Functional-difference deformations of Darboux-Pöshl-Teller potentials**   191

MIR-KASIMOV, R.M.: **Maxwell equations for quantum space-time**   209

NAUDTS, JAN: **A solvable model of interacting photons**   219

OHTA, Y.: **Discretization of a sine-Gordon type equation**   225

POGREBKOV, A.K.: **Hierarchy of quantum explicitly solvable and integrable models**   231

PUTTOCK, S.E. & NIJHOFF, F.W.: **A two-parameter elliptic extension of the lattice KdV system**   245

ROTHOS, VASSILIS M. & FECKAN, MICHAL: **Travelling waves in a perturbed discrete sine-Gordon equation**   253

SASAKI, RYU: **Quantum VS classical Calogero–Moser systems**   259

TAKAHASHI, DAISUKE & IWAO, MASATAKA: **Geometrical dynamics of an integrable piecewise-linear mapping**   291

TAKHTAJAN, LEON A.: **Free bosons and dispersionless limit of Hirota tau-function**   301

TAMIZHMANI, K.M., GRAMMATICOS, B., RAMANI, A., OHTA, Y., & TAMIZHMANI, T.: **Similarity reductions of Hirota bilinear equations and Painlevé equations**   313

TOKIHIRO, TETSUJI: **On fundamental cycle of periodic Box-Ball systems**   325

VAN MOERBEKE, P.: **Combinatorics and integrable geometry**   335

VERHOEVEN, C., MUSETTE, M. & CONTE, R.: **On reductions of some KdV-type systems and their link to the quartic Hénon-Heiles hamiltonian**   363

WILLOX, R. & HIETARINTA, J.: **On the bilinear forms of Painlevé's 4th equation**   375

PREFACE

On April 29, 1814 Napoleon landed on the island of Elba, surrounded with a personal army of 1200 men. The allies, Russia, Prussia, England and Austria, had forced him into exile after a number of very costly defeats; he was deprived of all his titles, but could keep the title of "Emperor of Elba". History tells us that each morning he took long walks in the sun, reviewed his army each midday and discussed world matters with newly appointed advisors, following the same pattern everyday, to the great surprise of Campbell, the British officer who was to keep an eye on him. All this made everyone believe he was settled there for good. Napoleon once said: Elba is beautiful, but a bit small. Elba was definitely a source of inspiration; indeed, the early morning, March 6, 1815, Metternich, the chancellor of Austria was woken up by one of his aides with the stunning news that Napoleon had left Elba with his 1200 men and was marching to Paris with little resistance; A few days later he took up his throne again in the Tuileries. In spite of his insatiable hunger for battles and expansion, he is remembered as an important statesman. He was a pioneer in setting up much of the legal, administrative and political machinery in large parts of continental Europe.

We gathered here in a lovely and quaint fishing port, Marciana Marina on the island of Elba, to celebrate one of the pioneers of integrable systems, Hirota Sensei, and this at the occasion of his seventieth birthday. Trained as a physicist in his home university Kyushu University, Professor Hirota earned his PhD in '61 at Northwestern University with Professor Siegert in the field of "Quantum Statistical mechanics". He wrote a widely appreciated Doctoral dissertation on "Functional Integral representation of the grand partition function". As a young researcher, he entered the RCA Company in Tokyo to do research on semi-conductor plasmas. He then joined the Faculty of Science and Engineering of Ritsumeikan University in Kyoto and then later Hiroshima National University and Waseda University, until his recent retirement.

We are also celebrating another birthday, namely the birth, some thirty years ago, of multisoliton solutions for the KdV equation, the representation of integrable equations as bilinear equation and Hirota's D-operation. All this happened in the period 1971 through 1974.

Professor Hirota was led to model the Toda lattice as a non-linear network of ladder-type LC circuits. The self-dual case led to equations very reminiscent of the Sine-Gordon equation, with much the same features (existence of one soliton, soliton-soliton interaction, etc)

Meanwhile, At RCA, Hirota Sensei was looking for applications of solitons to multi-channel communication systems. As an important requirement, they needed to be stable in the presence of a ripple. Taking a 2-soliton interaction, letting one of them become very small, led to the stability of a 1-soliton solutions. What about the stability of two solitons? Professor Hirota argued as follows: If one wants to use the same method, one should look for three-soliton solutions and again let one soliton become very small. In the beginning, most naïve guesses turned out to be wrong. Finally the answer came from an ingenious use of the Bäcklund transformation and a superposition principle, for the sine-Gordon equation. In this way, Professor Hirota expressed the three-soliton solution, in terms of sums of exponentials with phases linear in x and t. These same kind of methods could then be applied to the non-linear self-dual network equation, the Toda equation and finally to the KdV equation.

In his celebrated 1971-paper: "Exact solutions of the KdV equation for multiple collisions of solitons", Hirota gave the multisoliton solution to the KdV equation in terms of the second logarithmic derivative of a determinant of exponentials and showed most importantly that the determinant satisfies a bilinear equation of order 4. So Hirota's bilinear equation was born.

The story goes that Professor Scott who was visiting Japan in the summer 1971 remarked: why do you want to replace the KdV equation by a much more complicated equation, namely the bilinear equation, which after all is 4th order? This seemingly negative comment had striking consequences. Having written bilinear equations for all those integrable PDE's, Professor Hirota became very concerned with finding simple ways to express them, which he did in a paper in 1974, where he introduced the operation, known these days as Hirota symbol or Hirota D-operator. This amazing intuition turned out to have profound consequences. Beyond being an ingenious device, it had a lasting impact onto the field. It gave rise to the famous tau-function theory, which by now has become a classic chapter of mathematical physics. One might say that the Hirota symbol has become one of those tools that everyone is using without referring to it in the bibliography, just like Schwarz's inequality or Stokes' theorem.

Hirota's career is specked with striking and stunning discoveries, often based on simple, but ingenious observations. They unleashed a great tide of energy and activity; all hell broke loose. In the 70's, one miracle came after the other, the field literally exploded in the most fascinating directions that we all know and worship. This week here in Elba will be a tribute to his work!

This NATO-sponsored workshop here in Elba was dominated by an enormous wealth of subjects around integrability, ranging from geometric to analytic questions, from Lie groups, quantum groups and W-algebras to combinatorics and quantum field theory. We would like to thank the participants for having delivered these interesting lectures. Also many thanks to those who have contributed to this volume.

The organizing committee consisted of Professors Franklin Lambert, Frank Nijhoff, Ludwig Faddeev and Pierre van Moerbeke. Last but not least, we would like to express our gratitude to Professor Franklin Lambert. It was his idea to organize the conference on this theme, he picked this wonderful spot, he was the real engine behind this enterprise, he did an enormous amount of work. Thank you Franklin!

Ludwig D. Faddeev
Pierre van Moerbeke

LIST OF PARTICIPANTS

M. Arik
Bogazici University
Physics Department
80815 Bebek
Istanbul, Turkiye
arikm@boun.edu.tr

C. Athorne
Department of Mathematics
University of Glasgow
G12 8QW
UK
ca@maths.gla.ac.uk

E. Belokolos
Department of Theoretical
    Physics
Institute of Magnetism
National Academy of Sciences of
    Ukraine
36-b, prosp. Vernadsky
03142 Kiev-142, Ukraine
bel@im.imag.kiev.ua

A. Bobenko
Department of Mathematics
Technical University of Berlin
Strasse des 17.Juni 136
10623 Berlin, Germany
bobenko@math.tu-berlin.de

R. Conte
Service de Physique de l'Etat
    Condensé
CEA-Saclay
F-91191 Gif-sur-Yvette Cedex
France
conte@spec.saclay.cea.fr

E. Doktorov
B.I. Stepanov Institute of Physics
68F. Skaryna Ave
220072 Minsk
Republic of Belarus
doktorov@dragon.bas-net.by

B. Dubrovin
Mathematics Physics Sector
International School for Advanced
    Studies (SISSA)
Via Beirut, 2-4
I-34013 Trieste, Italy
dubrovin@sissa.it

V. Enolskii
Institute of Magnetism
National Academy of Sciences of
    Ukraine
36-b, prosp. Vernadsky
03142 Kiev-142, Ukraine
V.Z.Enolskii@ma.hw.ac.uk

L. Faddeev
Steklov Mathematical Institute
Russian Academy of Sciences
27, Fontanka
St. Petersburg 191011, Russia
faddeev@pdmi.ras.ru

V.S. Gerdjikov
Department of Theoretical Physics
Institute for Nuclear Research and
    Nuclear Energy
Bulgarian Academy of Sciences
72 Tsarigradsko chaussee
1784 Sofia, Bulgaria
gerijkov@inrne.bas.bg

C. Gilson
Department of Mathematics
University of Glasgow
Glasgow G12 8QW
UK
crg@maths.gla.ac.uk

B. Grammaticos
GMPIB, Univ. Paris VII
Tour 24-14, 5e et., Case 7021
75251 Paris, France
grammati@paris7.jussieu.fr

F. Gungor
Department of Mathematics
Faculty of Science and Letters
Istanbul Technical University
80626 Maslak-Istanbul
Turkey
gungorf@itu.edu.tr

L. Haine
University of Louvain
Department of Mathematics
2 Chemin du Cyclotron
1348 Louvain-la-Neuve
Belgium
haine@math.ucl.ac.be

J. Hietarinta
Department of Physics
University of Turku
FIN-20014 Turku
Finland
Jarno.Hietarinta@utu.fi

R. Hirota
Department of Information and
    Computer Sciences
Waseda University, School of
    Science and Engineering
Shinjuku, Tokyo, Japan
roy@spnl.speednet.ne.jp

E. Horozov
University of Sofia
Faculty of Mathematics and
    Informatics
5, James Bourchier blvd.
1126 Sofia, Bulgaria
horozov@fmi.uni-sofia.bg

K. Kajiwara
Graduate School of Mathematics
Kyushu University
6-10-1 Hakozaki, Fukuoka
    812-8581
Japan
kaji@math.kyushu-u.ac.jp

S. Kakei
Department of Mathematics
Faculty of Science
Rikkyo University
Nishi-Ikebukuro, Toshima-ku
Tokyo 171-8501, Japan
kakei@rkmath.rikkyo.ac.jp

R. Kashaev
Section de mathématiques
Université de Genève
2-4, rue du Lièvre
C.P.240, 1211 Genève 24 (Suisse)
Rinat.Kashaev@math.unige.ch

B. Konopeltchenko
Dipartimento di Fisica
Universitá degli Studi di Lecce
Via Arnesano
73100 Lecce, Italy
Boris.Konopeltchenko@le.infn.it

V. Kuznetsov
Department of Applied Mathematics
University of Leeds
Woodhouse Lane, Leeds LS2 9JT
England, UK
vadim@maths.leeds.ac.uk

F. Lambert
Vrije Universiteit Brussel
Theoretische Natuurkunde (TENA)
Pleinlaan 2, B-1050 Brussels
Belgium
tenasecr@vub.ac.be

S. Leble
Faculty of Technical Phys. and
  Applied Math.
Gdansk University of Technology
ul. G. Narutowicza 11/12
80-952 Gdan'sk-Wrzeszcz, Poland
leble@mif.pg.gda.pl

J. Loris
Vrije Universiteit Brussel
Theoretische Natuurkunde (TENA)
Pleinlaan 2, B-1050 Brussels
Belgium
igloris@vub.ac.be

F. Magri
Department of Mathematics and
  Applications
University of Milano-Bicocca
Via Bicocca degli Arcimboldi
8-20126 Milano, Italy
magri@matapp.unimib.it

A. Marshakov
Russian Federation State Scientific
  Center
Institute for Theoretical and
Experimental Physics
B. Cheremushkinskaja, 25
117259 Moscow, Russia
mars@gate.itep.ru

V. Matveev
Université de Bourgogne
U.F.R. Sciences et Techniques
B.P. 47870
F-21078 Dijon Cedex, France
matveev@u-bourgogne.fr

R. Mir-Kasimov
Bogoliubov Laboratory of
Theoretical Physics
JINR
141980 Dubna, Moscow region
Russia
mirkr@thsunl.jinr.ru

M. Musette
Vrije Universiteit Brussel
Theoretische Natuurkunde (TENA)
Pleinlaan 2, B-1050 Brussels
Belgium
mmusette@vub.ac.be

J. Naudts
Departement Fysica
Universiteit Antwerpen UIA
Universiteitsplein 1
B-2610 Antwerpen, Belgium
naudts@uia.ua.ac.be

F. Nijhoff
Department of Mathematics
University of Leeds
Leeds LS2 9JT
UK
frank@amsta.leeds.ac.uk

J.J.C. Nimmo
Department of Mathematics
University of Glasgow
Glasgow G12 8QW
UK
j.nimmo@maths.gla.ac.uk

Y. Ohta
Department of Applied Mathematics
Faculty of Engineering, Hiroshima
  University
1-4-1 Kagamiyama,
  Higashi-Hiroshima 730-8527
Japan
ohta@kurims.kyoto-u.ac.jp

L. Palla
Department of Theoretical Physics
Eötvös Loránd University
1518 Budapest, Pf. 32
Hungary
palla@ludens.elte.hu

A.K. Pogrebkov
V.A. Steklov Mathematical Institute
Russian Academy of Sciences
GSP-1, ul. Gubkina 8
117966 Moscow
Russian Federation
pogreb@mi.ras.ru

R. Sasaki
Yukawa Institute for Theoretical
   Physics
Kyoto University
Kyoto, 606-8502
Japan
ryu@yukawa.kyoto-u.ac.jp

Y.B. Suris
Institut fuer Mathematik
Technische Universitaet Berlin
Str. des 17.Juni 136
D-10623 Berlin, Germany
suris@sfb288.math.tu-berlin.de

D. Takahashi
Department of Mathematical
   Sciences
Waseda University
3-4-1, Ohkubo, Shinjuku-ku
Tokyo 169-8555, Japan
tdaisuke@mub.biglobe.ne.jp

L. Takhtajan
Department of Mathematics
SUNY at Stony Brook
Stony Brook
NY 11794-3651
USA
leontak@math.sunysb.edu

K.M. Tamizhmani
Department of Mathematics
Pondicherry University
Kalapet, Pondicherry-605 014
India
tamizh@yahoo.com

T. Tokihiro
Graduate School of Mathematical
   Sciences
University of Tokyo
3-8-1 Komaba, Meguro-ku
Tokyo, Japan
toki@poisson.ms.tokyo-u.ac.jp

N. Ustinov
Theoretical Physics Department
Al. Nevsky str. 14
Kaliningrad 236041
Russia
n_ustinov@mail.ru

Johan van de Leur
Department of Mathematics
University of Utrecht
P.O. Box 80010
3508 TA Utrecht
The Netherlands
vdleur@math.uu.nl

P. van Moerbeke
Université Catholique de Louvain
Département de Mathématique
Chemin du Cyclotron, 2
1348 Louvain-la-Neuve, Belgium
vanmoerbeke@math.ucl.ac.be

P. Vanhaecke
Université de Poitiers, Dépt. de
   Mathématiques
Boulevard Marie et Pierre Curie,
   BP 30179
86962 Futuroscope Chasseneuil
   Cedex

France
Pol.Vanhaecke@mathlabo.
univ-poitiers.fr

A. Volkov
Vrije Universiteit Brussel
Theoretische Natuurkunde
Pleinlaan 2, B-1000 Brussel
Belgium
avolkov@vub.ac.be

Ralph Willox
Graduate School of Mathematical
Sciences
University of Tokyo
3-8-1 Komaba, Meguro-ku
Tokyo, Japan
willox@poisson.ms.u-tokyo.ac.jp

A. Zabrodin
Institute of theoretical and
experimental physics
B. Cheremushkinskaya 25
Moscow 117259
Russia
zabrodin@heron.itep.ru

V. Zakharov
Landau Institute for Theoretical
Physics
Kosygina 2
Moscow 117 940
Russia
zakharov@hedgehog.math.
arizona.edu

THE CKP HIERARCHY AND THE WDVV PREPOTENTIAL

Henrik Aratyn
Department of Physics, University of Illinois at Chicago,
845 W. Taylor St., Chicago, IL 60607-7059

Johan Van de Leur
Mathematical Institute,
University of Utrecht,
P.O. Box 80010, 3508 TA Utrecht,
The Netherlands

1 THE WDVV PREPOTENTIAL

In terms of the so-called flat coordinates $x^1, x^2, \ldots, x^n$ a solution to the Witten–Dijkgraaf–Verlinde–Verlinde (WDVV) equations [1, 2] is given by a prepotential $F(x^1, x^2, \ldots, x^n)$ which satisfies the associativity relations:

$$\sum_{\delta, \gamma = 1}^{n} \frac{\partial^3 F(x)}{\partial x^\alpha \partial x^\beta \partial x^\delta} \eta^{\delta \gamma} \frac{\partial^3 F(x)}{\partial x^\gamma \partial x^\omega \partial x^\rho} = \sum_{\delta, \gamma = 1}^{n} \frac{\partial^3 F(x)}{\partial x^\alpha \partial x^\omega \partial x^\delta} \eta^{\delta \gamma} \frac{\partial^3 F(x)}{\partial x^\gamma \partial x^\beta \partial x^\rho} \tag{1}$$

together with a quasi-homogeneity condition:

$$\sum_{\alpha=1}^{n} (1 + \mu_1 - \mu_\alpha) x^\alpha \frac{\partial F}{\partial x^\alpha} = (3 - d)F + \text{quadratic terms.} \tag{2}$$

where μ_i, $i = 1, \ldots, n$ and d are constants.
Furthermore, expression

$$\frac{\partial^3 F(x)}{\partial x^\alpha \partial x^\beta \partial x^1} = \eta_{\alpha\beta} \tag{3}$$

defines a constant non degenerate metric: $g = \sum_{\alpha, \beta=1}^{n} \eta_{\alpha\beta} \, dx^\alpha \, dx^\beta$.
As shown by Dubrovin (e.g., in reference [3]) there is an alternative description of the metric in terms of a special class of orthogonal curvilinear

1

L. Faddeev et al. (eds.),
Bilinear Integrable Systems: From Classical to Quantum, Continuous to Discrete, 1–11.
© 2006 *Springer. Printed in the Netherlands.*

coordinates $u_1, \ldots, u_n$

$$g = \sum_{\alpha\beta=1}^{n} \eta_{\alpha\beta} \, dx^\alpha \, dx^\beta = \sum_{i=1}^{n} h_i^2(u)(du_i)^2 \tag{4}$$

called canonical coordinates. These coordinates allow to reformulate the problem in terms of the Darboux–Egoroff metric systems and corresponding Darboux–Egoroff equations and their solutions. In the Darboux–Egoroff metric the Lamé coefficients $h_i^2(u)$ are gradients of some potential and this ensures that the so-called "rotation coefficients"

$$\beta_{ij} = \frac{1}{h_j} \frac{\partial h_i}{\partial u_j}, \quad i \neq j, \quad 1 \leq i, j \leq n, \tag{5}$$

are symmetric $\beta_{ij} = \beta_{ji}$. The Darboux–Egoroff equations for the rotation coefficients are:

$$\frac{\partial}{\partial u_k} \beta_{ij} = \beta_{ik} \beta_{kj}, \quad \text{distinct } i, j, k \tag{6}$$

$$\sum_{k=1}^{n} \frac{\partial}{\partial u_k} \beta_{ij} = 0, \quad i \neq j. \tag{7}$$

In addition to these equations one also assumes the conformal condition:

$$\sum_{k=1}^{n} u_k \frac{\partial}{\partial u_k} \beta_{ij} = -\beta_{ij}. \tag{8}$$

The Darboux–Egoroff equations (6)–(7) appear as compatibility equations of a linear system:

$$\frac{\partial \Phi_{ij}(u, z)}{\partial u_k} = \beta_{ik}(u) \Phi_{kj}(u, z), \quad i \neq k \tag{9}$$

$$\sum_{k=1}^{n} \frac{\partial \Phi_{ij}(u, z)}{\partial u_k} = z \Phi_{ij}(u, z) \tag{10}$$

Define the $n \times n$ matrices $\Phi = (\Phi_{ij})_{1 \leq i, j \leq n}$, $B = (\beta_{ij})_{1 \leq i, j \leq n}$ and $V_i = [B, E_{ii}]$, where $(E_{ij})_{kl} = \delta_{ik}\delta_{jl}$. Then the linear system (9)–(10) acquires the following form:

$$\frac{\partial \Phi(u, z)}{\partial u_i} = (z E_{ii} + V_i(u)) \, \Phi(u, z), \quad i = 1, \ldots, n, \tag{11}$$

$$\sum_{k=1}^{n} \frac{\partial \Phi(u, z)}{\partial u_k} = z \Phi(u, z). \tag{12}$$

Let, furthemore $\Phi(u, z)$ have a power series expansion

$$\Phi(u, z) = \sum_{j=0}^{\infty} z^j \Phi^{(j)}(u) = \Phi^{(0)}(u) + z \Phi^{(1)}(u) + z^2 \Phi^{(2)}(u) + \cdots \tag{13}$$

and satisfy the "twisting" condition;

$$\Phi(u, z)\eta^{-1}\Phi^T(u, -z) = I \tag{14}$$

Equation (11) implies $\sum_{k=1}^n u_k \partial \Phi(u, z)/\partial u_k = (zU + [B, U])\Phi(u, z)$, with $U = \sum_{k=1}^n u_k E_{kk}$. For a matrix $[B, U]$ which is diagonalizable the conformal condition (8) leads to

$$\sum_{k=1}^n u_k \frac{\partial \Phi(u, z)}{\partial u_k} = z\frac{\partial \Phi(u, z)}{\partial z} + \Phi(u, z)\mu, \quad \mu = \mathrm{diag}(\mu_1, \ldots, \mu_n) \tag{15}$$

where μ is a constant diagonal matrix obtained by a similarity transformation from the matrix $[B, U]$. The constant diagonal elements u_i entered the quasi-homogeneity condition (2).

Define

$$\phi_\alpha(u, z) \equiv \sum_{\beta=1}^n \Phi_{\beta 1}^{(0)}(u)\Phi_{\beta\alpha}(u, z)$$
$$= \phi_\alpha^{(0)}(u) + z\phi_\alpha^{(1)}(u) + z^2\phi_\alpha^{(2)}(u) + z^3\phi_\alpha^{(3)}(u) + \cdots \tag{16}$$

then, in terms of the flat coordinates $x^1, \ldots, x^n$

$$\phi_\alpha^{(1)}(u) = \sum_{\beta=1}^n \eta_{\alpha\beta}x^\beta(u) \tag{17}$$

and the prepotential is given by a closed expression (see e.g., [4] or [5]):

$$F = -\frac{1}{2}\phi_1^{(3)}(u) + \frac{1}{2}\sum_{\delta=1}^n x^\delta(u)\phi_\delta^{(2)}(u). \tag{18}$$

2 THE CKP HIERARCHY

The CKP hierarchy [6] can be obtained as a reduction of the KP hierarchy,

$$\frac{\partial}{\partial t_n}L = [(L^n)_+, L], \quad \text{for } L = L(t, \partial) = \partial_x + \ell^{(-1)}\partial_x^{-1} + \ell^{(-2)}\partial_x^{-2} + \cdots, \tag{19}$$

where $x = t_1$, by assuming the extra condition

$$L^* = -L. \tag{20}$$

By taking the adjoint, i.e., $*$ of (19), one sees that $\frac{\partial L}{\partial t_n} = 0$ for n even. Date, Jimbo, Kashiwara, and Miwa [6], [7] construct such L's from certain special KP wave functions $\psi(t, z) = P(t, z)e^{\sum_i t_i z^i}$ (recall $L(t, \partial) = P(t, \partial)\partial P(t, \partial)^{-1}$), where one then puts all even times t_n equal to 0. Recall that a KP wave function satisfies

$$L\psi(t, z) = z\psi(t, z), \qquad \frac{\partial \psi(t, z)}{\partial t_n} = (L^n)_+ \psi(t, z), \qquad (21)$$

and

$$\text{Res } \psi(t, z)\psi^*(s, z) = 0. \qquad (22)$$

The special wave functions which lead to an L that satisfies (20) satisfy

$$\psi^*(t, z) = \psi(\tilde{t}, -z), \quad \text{where} \quad \tilde{t}_i = (-)^{i+1} t_i. \qquad (23)$$

We call such a ψ a CKP wave function. Note that this implies that $L(t, \partial)^* = -L(\tilde{t}, \partial)$ and that

$$\text{Res } \psi(t, z)\psi(\tilde{s}, -z) = 0.$$

One can put all even times equal to 0, but we will not do that here.

The CKP wave functions correspond to very special points in the Sato Grassmannian, which consists of all linear spaces

$$W \subset H_+ \oplus H_- = \mathbb{C}[z] \oplus z^{-1}\mathbb{C}[[z^{-1}]],$$

such that the projection on H_+ has finite index. Namely, W corresponds to a CKP wave function if for any $f(z), g(z) \in W$ one has $\text{Res} f(z)g(-z) = 0$. The corresponding CKP tau functions satisfy $\tau(\tilde{t}) = \tau(t)$.

We will now generalize this to the multi-component case and show that a CKP reduction of the multi-component KP hierarchy gives the Darboux–Egoroff system. The n component KP hierarchy [8, 9] consists of the equations in $t_j^{(i)}$, $1 \le i \le n$, $j = 1, 2, \ldots$

$$\frac{\partial}{\partial t_j^{(i)}} L = \left[\left(L^j C_i \right)_+, L \right], \qquad \frac{\partial}{\partial t_j^{(i)}} C_k = \left[\left(L^j C_i \right)_+, C_k \right], \qquad (24)$$

for the $n \times n$-matrix pseudo-differential operators

$$L = \partial_x + L^{(-1)}\partial_x^{-1} + L^{(-2)}\partial_x^{-2} + \cdots,$$
$$C_i = E_{ii} + c_i^{(-1)}|\partial_x^{-1}| + C_i^{(-2)}|\partial_x^{-2}| + \cdots, \qquad (25)$$

$1 \le i \le n$, where $x = t_1^{(1)} + t_1^{(2)} + \cdots + t_1^{(n)}$. The corresponding wave function has the form

$$\Psi(t, z) = P(t, z) \exp \left(\sum_{i=1}^{n} \sum_{j=1}^{\infty} t_j^{(i)} z^j E_{ii} \right),$$

where $P(t, z) = I + P^{(-1)}(t)z^{-1} + \cdots,$

and satisfies

$$L\Psi(t, z) = z\Psi(t, z), \quad C_i\Psi(t, z) = \Psi(t, z)E_{ii},$$
$$\frac{\partial \Psi(t, z)}{\partial t_j^{(i)}} = (L^j C_i)_+ \Psi(t, z) \tag{26}$$

and

$$\text{Res } \Psi(t, z)\Psi^*(s, z)^T = 0.$$

From this we deduce that $L = P(t, \partial_x)\, \partial_x\, P(t, \partial_x)^{-1}$ and $C_i = P(t, \partial_x)E_{ii}$ $P(t, \partial_x)^{-1}$. Using this, the simplest equations in (26) are

$$\frac{\partial \Psi(t, z)}{\partial t_1^{(i)}} = (zE_{ii} + V_i(t))\Psi(t, z), \tag{27}$$

where $V_i(t) = [B(t), E_{ii}]$ with $B(t) = P^{(-1)}(t)$. In terms of the matrix coefficients β_{ij} of B we obtain (6) for $u_i = t_1^{(i)}$.

The Sato Grassmannian becomes vector valued, i.e.,

$$H_+ \oplus H_- = (\mathbb{C}[z])^n \oplus z^{-1}(\mathbb{C}[[z^{-1}]])^n.$$

The same restriction as in the 1-component case (23), viz.,

$$\Psi(t, z) = \Psi^*(\tilde{t}, -z), \quad \text{where} \quad \tilde{t}_n^i = (-)^{n+1}t_n^{(i)}.$$

leads to $L^*(\tilde{t}) = -L(t)$, $C_i^*(\tilde{t}) = C_i(t)$ and

$$\text{Res } \Psi(t, z)\Psi(\tilde{s}, -z)^T = 0, \tag{28}$$

which we call the multi component CKP hierarchy. But more importantly, it also gives the restriction

$$\beta_{ij}(t) = \beta_{ji}(\tilde{t}). \tag{29}$$

Such CKP wave functions correspond to points W in the Grassmannian for which

$$\text{Res } f(z)^T g(-z) = \text{Res } \sum_{i=1}^{n} f_i(z)g_i(-z) = 0$$

for any $f(z) = (f_1(z), f_2(z), \ldots, f_n(z))^T$, $g(z) = (g_1(z), g_2(z), \ldots, g_n(z))^T$ $\in W$.

If we finally assume that $L = \partial_x$, then Ψ, W also satisfy

$$\frac{\partial \Psi(t, z)}{\partial x} = \sum_{i=1}^{n} \frac{\partial \Psi(t, z)}{\partial t_1^{(i)}} = z \Psi(t, z), \quad zW \subset W \tag{30}$$

and thus β_{ij} satisfies (7) for $u_i = t_1^{(i)}$. Now differentiating (28) n times to x for $n = 0, 1, 2, \ldots$ and applying (30) leads to

$$\Psi(t, z)\Psi(\tilde{t}, -z)^T = I.$$

These special points in the Grassmannian can all be constructed as follows [10]. Let $G(z)$ be an element in $GL_n(\mathbb{C}[z, z^{-1}])$ that satisfies

$$G(z)G(-z)^T = 1, \tag{31}$$

then $W = G(z)H_+$. Clearly, any two $f(z), g(z) \in W$ can be written as $f(z) = G(z)a(z)$, $g(z) = G(z)b(z)$ with $a(z), b(z) \in H_+$, then $zf(z) = zG(z)a(z) = G(z)za(z) \in W$, since $za(z) \in H_+$. Moreover,

$$\operatorname{Res} f(z)^T g(-z) = \operatorname{Res} a(z)^T G(z)^T G(-z)b(-z) = \operatorname{Res} a(z)^T b(-z) = 0.$$

If we define $M(t, z) = \Psi(t, z)G(z)$, then one can prove [10, 11] that

$$M(t, z) = M^{(0)}(t) + M^{(1)}(t)z + M^{(2)}(t)z^2 + \cdots$$

We want to change $M(t, z)$ a bit more. However, we only want to do that for very special elements in this twisted loop group, i.e., to certain points of the Grassmannian that have a basis of homogeneous elements in z. Let $n = 2m$ or $n = 2m + 1$, choose non-negative integers μ_i for $1 \leq i \leq m$ and define $\mu_{n+1-j} = -\mu_j$ and let $\mu_{m+1} = 0$ if n is odd. Then take $G(z)$ of the form

$$G(z) = N(z)S^{-1} = Nz^{-\mu}S^{-1}, \quad \text{where} \quad \mu = \operatorname{diag}(\mu_1, \mu_2, \ldots, \mu_n)$$

and $N = (n_{ij})_{1 \leq i, j \leq n}$ a constant matrix that satisfies

$$N^T N = \sum_{j=1}^{n} (-1)^{\mu_j} E_{j, n+1-j} \tag{32}$$

and

$$S = \delta_{n, 2m+1} E_{m+1, m+1} + \sum_{j=1}^{m} \frac{1}{\sqrt{2}}$$
$$\times (E_{jj} + i E_{n+1-j, j} + E_{j, n+1-j} - i E_{n+1-j, n+1-j}).$$

Then [11]

$$\sum_{i=1}^{n} \sum_{j=1}^{\infty} j t_j^{(i)} \frac{\partial \Psi(t, z)}{\partial t_j^{(i)}} = z \frac{\partial \Psi(t, z)}{\partial z},$$

from which one deduces that

$$\sum_{i=1}^{n}\sum_{j=1}^{\infty} jt_j^{(i)} \frac{\partial \beta_{ij}}{\partial t_j^{(i)}} = -\beta_{ij}. \tag{33}$$

Define $\eta = (\eta_{ij})_{1 \le i,j \le n} = S^T S = \sum_{i=1}^{n} E_{i,n+1-i}$ and denote by $\Phi(t,z) = M(t,z)S = \Psi(t,z)G(z)S = \Psi(t,z)N(z)$, then $\Phi(t,z)$ satisfies the following relations:

$$\Phi(t,z) = \Phi^{(0)}(t) + \Phi^{(1)}(t)z + \Phi^{(2)}(t)z^2 + \cdots$$

$$\Phi(t,z)\eta^{-1}\Phi(t,-z)^T = I$$

$$\frac{\partial \Phi(t,z)}{\partial t_1^{(i)}} = (z E_{ii} + V_i(t))\Phi(t,z)$$

$$\sum_{i=1}^{n} \frac{\partial \Phi(t,z)}{\partial t_1^{(i)}} = z\Phi(t,z),$$

$$\sum_{i=1}^{n}\sum_{j=1}^{\infty} jt_j^{(i)} \frac{\partial \Phi(t,z)}{\partial t_j^{(i)}} = z\frac{\partial \Phi(t,z)}{\partial z} + \Phi(t,z)\mu.$$

We next put $t_j^{(i)} = 0$ for all i and all $j > 1$ and $u_i = t_1^{(i)}$, then we obtain the situtation of Section 1. Define $\phi_\alpha(u,z)$ as in (16), then $\phi_\alpha^{(1)}(u) = \sum_{\gamma=1}^{n} \eta_{\alpha\gamma} x^\gamma(u)$ and the function $F(u)$ given by (18) satisfies the WDVV equations.

3   AN EXAMPLE

We will now give an example of this construction, viz., the case that $n = 3$ (for simplicity) and $\mu_1 = -\mu_3 = 2$ and $\mu_2 = 0$. Hence, the point of the Grassmannian is given by

$$N(z)H_+ = N \begin{pmatrix} z^{-2} & 0 & 0 \\ 0 & 1 & 0 \\ 0 & 0 & z^2 \end{pmatrix} H_+.$$

More precise, let $n_i = (n_{1i}, n_{2i}, n_{3i})^T$ and $e_1 = (1,0,0)^T$, $e_2 = (0,1,0)^T$ and $e_3 = (0,0,1)^T$, then this point of the Grassmannian has as basis

$$n_1 z^{-2}, n_1 z^{-1}, n_1, n_2, n_1 z, n_2 z, e_1 z^2, e_2 z^2, e_3 z^2, e_1 z^3, e_2 z^3, \ldots.$$

Using this one can calculate in a similar way as in [12] (using the bosonfermion correspondence or vertex operator constructions) the wave function:

$$\Psi(t, z) = P(t, z) \exp\left(\sum_{i=1}^{n}\sum_{j=1}^{\infty} t_j^{(i)} z^j E_{ii}\right),$$

$$\text{where } P_{jj}(t, z) = \frac{\hat{\tau}\left(t_\ell^{(k)} - \delta_{kj}(\ell z^\ell)^{-1}\right)}{\hat{\tau}(t)},$$

$$P_{ij}(t, z) = z^{-1}\frac{\hat{\tau}_{ij}\left(t_\ell^{(k)} - \delta_{kj}(\ell z^\ell)^{-1}\right)}{\hat{\tau}(t)}.$$

and where

$$\hat{\tau}(t) = \det\begin{pmatrix} n_{11}S_2(t^{(1)}) & n_{11}S_1(t^{(1)}) & n_{11} & 0 & n_{12} & 0 \\ n_{21}S_2(t^{(2)}) & n_{21}S_1(t^{(2)}) & n_{21} & 0 & n_{22} & 0 \\ n_{31}S_2(t^{(3)}) & n_{31}S_1(t^{(3)}) & n_{31} & 0 & n_{32} & 0 \\ n_{11}S_3(t^{(1)}) & n_{11}S_2(t^{(1)}) & n_{11}S_1(t^{(1)}) & n_{11} & n_{12}S_1(t^{(1)}) & n_{12} \\ n_{21}S_3(t^{(2)}) & n_{21}S_2(t^{(2)}) & n_{21}S_1(t^{(2)}) & n_{21} & n_{22}S_1(t^{(2)}) & n_{22} \\ n_{31}S_3(t^{(3)}) & n_{31}S_2(t^{(3)}) & n_{31}S_1(t^{(3)}) & n_{31} & n_{32}S_1(t^{(3)}) & n_{32} \end{pmatrix}.$$

The functions $S_i(x)$ are the elementary Schur polynomials:

$$S_1(x) = x_1, \quad S_2(x) = \frac{x_1^2}{2} + x_2, \quad S_3(x) = \frac{x_1^3}{6} + x_2 x_1 + x_3.$$

The tau function $\tilde{\tau}_{ij}(t)$ is up to the sign $\text{sign}(i - j)$ equal to the above determinant where we replace the jth row by

$$\left(n_{i1}S_1(t^{(i)}) \quad n_{i1} \quad 0 \quad 0 \quad 0 \quad 0\right).$$

Next we put all higher times $t_j^{(i)}$ for $j > 1$ equal to 0 and write u_i for $t_1^{(i)}$. Then using the orthogonality-like condition (32) of the matrix N to reduce long expressions, the wave function becomes:

$$\Psi(u, z) = \left(I + \frac{1}{\tau(u)}\sum_{i,j=1}^{3}\left[\left(-w_1^{(3)} + w_1^{(2)}(u_i + u_j) - w_1^{(1)}u_i u_j\right)z^{-1}\right.\right.$$
$$\left.\left. + \left(w_1^{(1)}u_i - w_1^{(2)}\right)z^{-2}\right]n_{i1}n_{j1}E_{ij}\right)e^{zU},$$

where, for convenience of notation, we have introduced some new "variables"

$$w_i^{(k)} = \frac{1}{k}\sum_{\ell=1}^{3} u_\ell^k n_{\ell i} n_{\ell 1},$$

and where

$$\tau(u) = w_1^{(3)} w_1^{(1)} - w_1^{(2)} w_1^{(2)}.$$

Note that in this way we also have determined the rotation coefficients

$$\beta_{ij} = \frac{1}{\tau(u)} \left(-w_1^{(3)} + w_1^{(2)}(u_i + u_j) - w_1^{(1)} u_i u_j \right) n_{i1} n_{j1},$$

which is a new solution of order 3 of the Darboux–Egoroff equations.

Recall that $\eta = \sum_{i=1}^{3} E_{i,4-i}$. It is now straightforward but tedious to determine the flat coordinates x^α and the $\phi_\alpha^{(j)}$ for $j > 1$. One finds that for $\ell > 0$ and $p = 1, 2, 3$:

$$\phi_p^{(\ell - \mu_p)} = \frac{\tau w_p^{(\ell+2)} + \tau_1 w_p^{(\ell+1)} + \tau_2 w_p^{(\ell)}}{2(\ell - 1)! \tau} \tag{34}$$

and

$$\phi_p^{(-2-\mu_p)} = \delta_{p3}, \quad \phi_p^{(-1-\mu_p)} = -\delta_{p3}\frac{\tau_1}{2\tau}, \quad \phi_p^{(-\mu_p)} = \delta_{p3}\frac{\tau_2}{2\tau}, \tag{35}$$

where

$$\tau_1 = w_1^{(2)} w_1^{(3)} - w_1^{(1)} w_1^{(4)}$$

$$\tau_2 = w_1^{(2)} w_1^{(4)} - (w_1^{(3)})^2.$$

Note that (34) also holds for $p = 1$ and $\ell = 1, 2$, it is easy to verify that $\phi_1^{(-1)} = \phi_1^{(0)} = 0$. Using (17), one has the following flat coordinates:

$$x^1 = -\frac{\tau_1}{2\tau},$$

$$x^2 = \frac{1}{2\tau} \left(\tau w_2^{(3)} + \tau_1 w_2^{(2)} + \tau_2 w_2^{(1)} \right),$$

$$x^3 = \frac{1}{4\tau} \left(\tau w_1^{(5)} + \tau_1 w_1^{(4)} + \tau_2 w_1^{(3)} \right). \tag{36}$$

From all this it is straightforward to determine $F(u)$, given by (18):

$$F = \frac{\tau_2}{16\tau^2} \left(\tau w_1^{(5)} + \tau_1 w_1^{(4)} + \tau_2 w_1^{(3)} \right)$$

$$- \frac{\tau_1}{48\tau^2} \left(\tau w_1^{(6)} + \tau_1 w_1^{(5)} + \tau_2 w_1^{(4)} \right)$$

$$- \frac{\tau}{96\tau^2} \left(\tau w_1^{(7)} + \tau_1 w_1^{(6)} + \tau_2 w_1^{(5)} \right)$$

$$+ \frac{1}{8\tau^2} \left(\tau w_2^{(3)} + \tau_1 w_2^{(2)} + \tau_2 w_2^{(1)} \right) \left(\tau w_2^{(4)} + \tau_1 w_2^{(3)} + \tau_2 w_2^{(2)} \right).$$

We shall not determine the explicit form of this prepotential in terms of the canonical coordinates here, because it is quite long. However, there is a problem even in this "simple" example. We do not know how to express the canonical coordinates u_i in terms of the flat ones, the x^α's and thus cannot express F in terms of the flat coordinates. Hence we cannot determine the desired form of F. This can be solved in the simplest example, see [12], viz. the case that $\mu_1 = -\mu_n = 1$ and all other $\mu_i = 0$. This gives a rational prepotential F (in terms of the flat coordinates).

ACKNOWLEDGMENT

H. A. was partially supported by NSF (PHY-9820663).

REFERENCES

1. Witten, E. (1990) On the structure of the topological phase of two-dimensional gravity, *Nucl. Phys.* **B340**, pp. 281–332.
2. Dijkgraaf, R., Verlinde, E., and Verlinde, H. (1991) Topological strings in $d < 1$, *Nucl. Phys.* **B325**, p. 59.
3. Dubrovin, B. (1996) Geometry on 2D topological field theories, in: *Integrable Systems and Quantum Groups (Montecatini Terme, 1993)*, Lecture Notes in Mathematics 1620, Springer/Berlin, pp. 120–348.
4. Akhmetshin A. A., Krichever, I. M., and Volvovski, Y. S. (1999) A generating formula for solutions of associativity equations (Russian), *Uspekhi Mat. Nauk* **54** (2, 326), pp. 167–168. English version: hep-th/9904028.
5. Aratyn, H., Gomes, J. F., van de Leur, J. W., and Zimerman, A. H. (2003) WDVV equations, Darboux–Egoroff metric and the dressing method, in: *the Unesp 2002 workshop on Integrable Theories, Solitons and Duality*, Conference Proceedings of JHEP (electronic journal, see http://jhep.sissa.it/) or math-ph/0210038.
6. Date, E., Jimbo, M., Kashiwara, M., and Miwa, T. (1981) Transformation groups for soliton equations. 6. KP hierarchies of orthogonal and symplectic type, *J. Phys. Soc. Japan* **50** pp. 3813–1818.
7. Jimbo, M. and Miwa, T. (1983) Solitons and infinite dimensional Lie algebras, *Publ. RIMS, Kyoto Univ.* **19**, pp. 943–1001.
8. Aratyn, H., Nissimov, E., and Pacheva, S. (2001) Multi-component matrix KP hierarchies as symmetry-enhanced scalar KP hierarchies and their Darboux–Bäcklund solutions, in: *Bäcklund and Darboux transformations. The geometry of solitons (Halifax, NS, 1999)*, CRM Proceedings Lecture Notes **29**, Amer. Math. Soc., Providence, RI, pp. 109–120.
9. Kac, V. G. and van de Leur, J. W. (1993) The n-component KP hierarchy and representation theory, in *Important developments in soliton theory*, eds. Fokas, A.S. and Zakharov, V. E., Springer Series in Nonlinear Dynamics, pp. 302–343.
10. van de Leur, J. W. (2001) Twisted GL_n loop group orbit and solutions of WDVV equations, *Internat. Math. Res. Notices* **2001**(11), pp. 551–573.

11. Aratyn, H. and van de Leur, J. (2003) Integrable structures behind WDVV equations, *Teor. Math. Phys.* **134**(1), pp. 14–26. [arXiv:hep-th/0111243].
12. van de Leur, J. W. and Martini, R. (1999) The construction of Frobenius Manifolds from KP tau-functions, *Commun. Math. Phys.* **205**, pp. 587–616.
13. Dubrovin, B. (1993) Integrable systems and classification of 2-dimensional topological field theories, in: *Integrable Systems*, Proceedings of Luminy 1991 conference dedicated to the memory of J.-L. Verdier, eds. Babelon, O., Cartier, O., Kosmann-Schwarzbach, Y., Birkhäuser, pp. 313–359.

QUANTUM INVARIANCE GROUPS
OF PARTICLE ALGEBRAS

M. Arik
Boğaziçi University, Department of Physics, Bebek 80815, İstanbul, Turkey

Abstract   The boson and fermion algebras as well as their various generalizations posess invariance under quantum groups. One example is $FIO\,(2d, \mathcal{R})$, the fermionic inhomogenous orthogonal quantum group which is the inhomogenous quantum invariance group of the d-dimensional fermion algebra. Another is $BISp(2d, \mathcal{R})$, the bosonic inhomogenous symplectic quantum group which is the inhomogenous quantum invariance group of the d-dimensional boson algebra. Complexification, sub(quantum) groups and quantum group manifolds of these quantum groups will also be discussed.

I am honored to present this talk in this conference dedicated to celebration of professor Hirota's monumental work in Bilinear Integrable Systems. This talk is not directly related to his work in content. My hope is that it is in the same spirit. The historical road from classical mechanics to quantum field theory is most effectively summarized by the following steps. The first is the replacement of the Poisson bracket

$$\{p_i, q_i\} = \delta_{ij} \tag{1}$$

by the commutator

$$i[p_i, q_i] = \delta_{ij}\,\hbar. \tag{2}$$

The second is the passage, via a harmonic oscillator hamiltonian, to creation and annihilation operators which satisfy

$$[a_i, a_j] = 0$$
$$[a_i, a_j^*] = \delta_{ij} \tag{3}$$

so that the number operator

$$N = \sum_i a_i^{\,*} a_i \tag{4}$$

has integer eigenvalues. The third entails the replacement of the discrete indices $i, j, \ldots$ by the continuous momentum indices $\vec{p}_1, \vec{p}_2, \ldots$, and interpreting $a^*(\vec{p})$, $a(\vec{p})$ as creation and annihilation operators of a particle of momentum

L. Faddeev et al. (eds.),
Bilinear Integrable Systems: From Classical to Quantum, Continuous to Discrete, 13–16.
© 2006 *Springer. Printed in the Netherlands.*

$\vec{p}$ thus generalizing (3) to

$$[a(\vec{p}_1), a(\vec{p}_2)] = 0$$
$$[a(\vec{p}_1), a^*(\vec{p}_2)] = \delta(\vec{p}_1 - \vec{p}_2). \tag{5}$$

This simple procedure can be applied to bosons posessing integer spin and additional quantum numbers. Fermions, however, require the replacement of the commutator in (3) by the anticommutator. For the sake of clarity we will use the discrete form (3) although all our results can be generalized to the continuous form (5).

We start with the bosonic and fermionic particle algebras

$$[c_i, c_j]_\mp = c_i c_j \mp c_j c_i = 0$$
$$[c_i, c_j^*]_\mp = c_i c_j^* \mp c_j^* c_i = \delta_{ij}, i, j = 1, 2, \ldots, d \tag{6}$$

and look for inhomogenous linear transformations

$$c_i \rightarrow \alpha_{ij} \otimes c_j + \beta_{ij} \otimes c_j^* + \gamma_i \otimes 1 \tag{7}$$

which leave the commutation relations (6) invariant. α_{ij}, β_{ij}, γ_i are assumed to belong to a possibly noncommutative *-Hopf algebra where the coproduct, counit, and coinverse are respectively given by the matrix product, the identity matrix and the matrix inverse. The $(2d + 1) \times (2d + 1)$ matrix corresponding to the transformation (7) is given by

$$M = \begin{pmatrix} \alpha & \beta & \gamma \\ \beta^* & \alpha^* & \gamma^* \\ 0 & 0 & 1 \end{pmatrix}. \tag{8}$$

Here α, β, α^*, β^*, are $d \times d$ square submatrices, γ and γ^* are $d \times 1$ columns. * entails hermitean conjugation without taking the transpose of the submatrix. If the Hopf algebra $\mathcal{A}$ is taken to be commutative then the answer is well known. For bosons one obtains the inhomogenous symplectic group $ISp\,(2d, \mathcal{R})$ which is also the linear invariance group of the classical Poisson bracket (1). For fermions one obtains the orthogonal group $O(2n, \mathcal{R})$, the inhomogenous parameters γ_i (and their hermitean conjugates) being constrained to be zero. Nonzero γ_i require $\mathcal{A}$ to be noncommutative and give rise to a matrix quantum group [1]. What is somewhat surprising, however, is that when the condition that $\mathcal{A}$ is commutative is relaxed, for the bosonic case one also obtains a quantum group which contains $ISp\,(2d, \mathcal{R})$ as a subgroup. Thus the bosonic inhomogenous symplectic group $BISp(2d, \mathcal{R})$ (upper signs) and the fermionic inhomogenous orthogonal group $FIO\,(2d, \mathcal{R})$ (lower signs) are defined by [2, 3]

$$[\gamma_i, \gamma_j^*]_\mp = \delta_{ij} - \alpha_{ik}\alpha_{jk}^* \pm \beta_{ik}\beta_{jk}^*$$
$$[\gamma_i, \gamma_j]_\mp = \pm\beta_{ik}\alpha_{jk} - \alpha_{ik}\beta_{jk}$$
$$[A_{ij}, \Gamma_k]_\mp = 0$$
$$[A_{ij}, A_{k\ell}]_- = 0 \tag{9}$$

where $A_{ij} = (\alpha_{ij}, \alpha_{ij}^*, \beta_{ij}, \beta_{ij}^*)$, $\Gamma_i = (\gamma_i, \gamma_i^*)$ and summation over repeated indices is implied. Putting both sides of the first two equations equal to zero gives $ISp\,(2d, \mathcal{R})$ and $GrIO(2d, \mathcal{R})$, the second one differing from the well known $IO(2d, \mathcal{R})$ by the anticommutation relations satisfied by the inhomogenous parameters γ_i, γ_i^*. Putting $\beta_{ij} = 0$ in (9) gives the quantum groups $BIU\,(d)$ and $FIU\,(d)$ which are quantum group generalizations of the well known inhomogenous group $IU\,(d)$. For these quantum groups M can be reduced to a $(d+1) \times (d+1)$ matrix

$$M = \begin{pmatrix} \alpha & \gamma \\ 0 & 1 \end{pmatrix}. \tag{10}$$

Comparing with the Cartan classification of semisimple Lie groups one finds that $BIU\,(d)$ and $FIU\,(d)$ are type A, $BISp\,(2d, \mathcal{R})$ is of type C and $FIO\,(2d, \mathcal{R})$ is of type D. $FIO\,(2d+1, \mathcal{R})$ which corresponds to type B can be obtained by performing a similarity transformation on M in (8) to put it into real form where each matrix element is hermitean. Then,

$$M = \begin{pmatrix} A & \Gamma \\ 0 & 1 \end{pmatrix} \tag{11}$$

where for the fermionic case

$$[\Gamma_i, \Gamma_j]_+ = \delta_{ij} - A_{ik}A_{jk}$$
$$[A_{ij}, \Gamma_k]_+ = 0$$
$$[A_{ij}, A_{k\ell}]_- = 0.$$

These relations define $FIO\,(2d, \mathcal{R})$ for $i, j, k = 1, 2, \ldots, 2d$ and $FIO\,(2d+1, \mathcal{R})$ for $i, j, k = 1, 2, \ldots, 2d + 1$. Whether there exist generalizations corresponding to exceptional types is an open question.

Finally taking $\alpha_{ij}^*, \beta_{ij}^*, \gamma_i^*$ in (9) as independent elements of $\mathcal{A}$ rather than as hermitean conjugates of $\alpha_{ij}, \beta_{ij}, \gamma_i$ one obtains the "complex" quantum groups $BISp\,(2d, \mathcal{C})$ and $FIO\,(2d, \mathcal{C})$. Similarly deleting the condition that A_{ij}, Γ_j in (11) are hermitean, one obtains $FIO\,(n, \mathcal{C})$ for even or odd n. Quantum subgroups of these can be obtained by considerations similar to the real case.

The quantum groups considered in this talk are the simplest generalizations of Lie groups to quantum groups in the sense that the noncommutativity of the underlying Hopf algebra is introduced only via the inhomogenous parameters Γ_i. When these inhomogenous parameters are set to zero one obtains a Lie group which is the homogenous part of the inhomogenous quantum group. Another related feature is that when the Killing–Cartan metric of the (inhomogenous) quantum group is constructed considering the Cartan–Maurer 1-form $M^{-1}dM$, the inhomogenous parameters drop out and one obtains [4] a Riemannian manifold related to a Lie group manifold. This Riemannian manifold is given by the $GL\,(2d, \mathcal{R})$ manifold for $BISp\,(2d, \mathcal{R})$, a region of the $GL\,(n, \mathcal{R})$ manifold for $FIO(n, \mathcal{R})$ and a region of the $GL(d, \mathcal{C})$ manifold for $BIU(d)$ and $FIU(d)$. The regions mentioned are specified by the condition that the C^*-norm of the matrix A in (11) is less than unity. These regions have the structure of a semi-group. I would also like to remark that the quantum groups $BISp(2d, \mathcal{R})$ and $FIO(2d, \mathcal{R})$ can be regarded as deformations of their respective particle algebras, i.e., setting the homogenous parameters A_{ij} equal to zero gives back the respective particle algebras. In fact, the representations of $\mathcal{A}$ are most easily constructed by writing the elements of $\mathcal{A}$ in terms of the related particle algebra. Hopefully, their further consideration will give a better and more consistent approach to interacting field theory.

REFERENCES

1. Fadeev, L. D., Reshetikhin, N. Y., and Takhtajan, L. A. (1987) Quantization of Lie groups and Lie algebras, *Math. J.*, **1**, p. 193.
2. Arik, M., Gün, S., and Yildiz, A. (2003) Invariance quantum group of the fermionic oscillator, *Europ. Phys. J. C.* **27**, pp. 453–455.
3. Arik, M. and Kayserilioğlu, U. (2003) Quantum invariance groups of bosons and fermions, hep-th/0304185.
4. Arik, M. and Baykal, A. (2003) Riemannian metric of the invariance group of the fermion algebra, *Gen. Relat. Gravit.* **35**, pp. 885–890.

ALGEBRAIC HIROTA MAPS

Chris Athorne
Department of Mathematics, University Gardens, Glasgow, G14 9LZ

Abstract   We give definitions of Hirota maps acting as intertwining operators for representations of $SL_n(\mathbb{C})$. We show how these reduce to the conventional (generalized) Hirota derivatives in the limit of the dimension of the representation becoming infinite and we discuss an application to the theory of $\wp$-functions associated with hyperelliptic curves of genus 2.

1   INTRODUCTION

The Hirota derivative has been with us since the early days of soliton theory. Over the intervening years it has developed and influenced the subject to a degree extraordinary for so simple an idea: the replacement of the Leibnitz rule for derivations with a skew rule. The beauty of the bilinear forms of soliton equations coupled with this hint of perversity lends to the Hirota derivative an irresistible lure of mystery.

It is, of course, not a *derivative* at all, properly speaking. So the issue of its precise nature is crucial. There are two approaches. Firstly, the operator parts of the bilinear equations comprising soliton hierarchies have a natural role in the theory of Kac–Moody Lie algebras [1] as actions on Schur polynomials. Secondly, Hirota derivatives themselves can be associated with a simple invariance property [2] which essentially characterizes them unambiguously as well as allowing their extension to multilinear products.

The current paper generalizes the latter approach in two ways. Firstly we construct a Hirota-like operator (Hirota map) acting on representations of finite-dimensional Lie algebras almost as an intertwining operator. This allows us to construct highest weight vectors of irreducible representations. Such irreducibles are associated with polynomial functions but in the limit of infinite-dimensional representations the Hirota *map* becomes the Hirota derivative acting on analytic functions.

Secondly, we are able to push this procedure through for $SL_n(\mathbb{C})$, constructing Hirota maps, analyzing their actions on representations and their infinite-dimensional limits, recovering old and new Hirota derivatives. The Hirota derivatives of [2] can be regarded as intertwining for the Hiesenberg algebra. The new class of Hirota derivatives are partially intertwining either

17

L. Faddeev et al. (eds.),
Bilinear Integrable Systems: From Classical to Quantum, Continuous to Discrete, 17–33.
© 2006 *Springer. Printed in the Netherlands.*

for $SL_n(\mathbb{C})$ or the n-dimensional Heisenberg algebra, but not, in general, both.

As evidence for the potential usefulness of these operators we cite their occurrence in the theory of genus 2 hyperelliptic functions.

2 BASIC DEFINITIONS

We will deal mostly with $SL_n(\mathbb{C})$-modules and the action of the algebra $\mathfrak{sl}_n(\mathbb{C})$. By $\mathfrak{h}$ we will mean a Cartan subalgebra with basis $\{h_1, \ldots, h_{n-1}\}$ (to be made explicit shortly) and by e_{ij}, for $i \neq j$, nilpotent elements of the algebra associated with the roots. The roots α_{ij} are elements of $\mathfrak{h}^*$ and the e_{ij} are eigenvectors under the adjoint action of $\mathfrak{h}$:

$$[h, e_{jk}] = < h, \alpha_{jk} > e_{jk}, \quad \forall h \in \mathfrak{h}. \tag{1}$$

where $< \cdot, \cdot >$ is the natural pairing between $\mathfrak{h}$ and its dual.

Any (finite-dimensional) $SL_n(\mathbb{C})$ (or $\mathfrak{sl}_n(\mathbb{C})$) module Γ decomposes into a finite number of irreducible modules each associated with a *highest weight*, ω:

$$\Gamma = \bigoplus_\omega \Gamma_\omega. \tag{2}$$

The highest weight ω is associated with a *highest weight vector* which is both an eigenvector under the $\mathfrak{h}$ action and is annihilated by the nilpotent part of a maximal, solvable (Borel) subalgebra (which is the same for all Γ_ω in the decomposition).

Recall that irreducible representations of $SL_n(\mathbb{C})$ are associated with Young tableaux [3]: The irreducible of highest weight $\omega = (\omega_1, \ldots, \omega_{n-1})$ is associated with the tableau of row lengths $(\sum_{i=1}^{n-1} \omega_i, \sum_{i=2}^{n-1} \omega_i, \sum_{i=3}^{n-1} \omega_i, \ldots, \omega_{n-1})$.

We are going to define Hirota maps on tensor products of $\mathfrak{sl}_n(\mathbb{C})$ modules which are (almost) intertwining operators for the $\mathfrak{sl}_n(\mathbb{C})$ action. But to do this we need some explicit expressions for the e_{ij}. These would be rather complicated in general. However *all* modules appear in the decompositions of tensor products of modules of a relatively simple kind, which we denote Γ_N.

A basis of Γ_N is labeled by all n-tuples, $(k_1, k_2, \ldots, k_n)$, with positive integer entries satisfying

$$\sum_{i=1}^n k_i = N. \tag{3}$$

We use this label interchangeably with the basis element itself. The simplest such module is the invariant one, Γ_0. Γ_1 is n-dimensional and Γ_N itself the symmetric N-fold tensor product,

$$\Gamma_N = Sym^N \Gamma_1. \tag{4}$$

Then we can associate with Γ_N a homogenous polynomial of degree N in variables $x_1, x_2, \ldots, x_n$ with suitably normalized coefficients,

$$f^{(N)}(x_1, x_2, \ldots, x_n) = \sum_{k_1 + \cdots + k_n = N} \binom{N}{k_1 k_2 \ldots k_n} (k_1, k_2, \ldots, k_n) x_1^{k_1} x_2^{k_2} \ldots x_n^{k_n}, \tag{5}$$

linear in the basis elements. From the $SL_n(\mathbb{C})$ action on the variables x_i,

$$x_i \mapsto \sum_{j=1}^{n} g_{ij} x_j, \quad \det g = 1, \tag{6}$$

we obtain the $\mathfrak{sl}_n(\mathbb{C})$ operators

$$E_{ij} \equiv x_i \frac{\partial}{\partial x_j}, \quad i \neq j, \tag{7}$$

$$H_{ij} = [E_{ij}, E_{ji}] \equiv x_i \frac{\partial}{\partial x_i} - x_j \frac{\partial}{\partial x_i}, \quad i \neq j, \tag{8}$$

$$\tag{9}$$

and define the $\mathfrak{sl}_n(\mathbb{C})$ action, which we denote e_{ij}, on the basis elements by the requirement

$$E_{ij} f^{(N)}(x_1, \ldots, x_n) = e_{ij} f^{(N)}(x_1, \ldots, x_n), \tag{10}$$

$$H_{ij} f^{(N)}(x_1, \ldots, x_n) = h_{ij} f^{(N)}(x_1, \ldots, x_n). \tag{11}$$

For example, let $n = 3$ and $N = 2$. Then

$$f^{(2)}(x_1, x_2, x_3) = (2, 0, 0)x_1^2 + (0, 2, 0)x_2^2 + (0, 0, 2)x_3^2$$
$$+ 2(1, 1, 0)x_1 x_2 + 2(0, 1, 1)x_2 x_3 + 2(1, 0, 1)x_1 x_3$$

and

$$E_{12} f^{(2)}(x_1, x_2, x_3) = 2(0, 2, 0)x_1 x_2 + 2(1, 1, 0)x_1^2 + 2(0, 1, 1)x_1 x_3.$$

Comparison of coefficients of monomials yields actions

$$e_{12}(2, 0, 0) = 2(1, 1, 0),$$
$$e_{12}(1, 1, 0) = (0, 2, 0),$$
$$e_{12}(1, 0, 1) = (0, 1, 1),$$

the others vanishing.

This construction is slightly round the houses. After all we could simply extend the action on Γ_1 to the symmetric tensor product without introducing the artificial, auxiliary x_i variables. However, the present approach serves both to connect the construction with the classical approach to $\mathfrak{sl}_2(\mathbb{C})$ invariant theory represented in, say, Hilbert's classical lectures [4], where the explicit expressions are reminiscent of Hirota bi- and multilinear forms, and to connect the Hirota maps we shall define shortly with the Hirota derivative itself [5] in the limit that $N \to \infty$.

Quite generally for the $\mathfrak{sl}_n(\mathbb{C})$ module Γ_N, the e_{ij} act thus

$$e_{ij}(\ldots, k_i, \ldots, k_j, \ldots) = k_i(\ldots, k_i - 1, \ldots, k_j + 1, \ldots). \tag{12}$$

We can take a basis of $\mathfrak{h}$ to be the set of $n - 1$ elements

$$h_i = h_{ii+1}, \quad i = 1, \ldots, n - 1 \tag{13}$$

whose action on basis elements of Γ_N is

$$h_i(k_1, k_2, \ldots, k_n) = (k_{i+1} - k_i)(k_1, k_2, \ldots, k_n). \tag{14}$$

The *Hirota maps* are defined on g-fold tensor products of Γ_{N_i} but it is simplest to start with the case $g = 2$. For $i \neq j$,

$$\mathbb{D}_{ij}^{12} : \Gamma_{N_1} \otimes \Gamma_{N_2} \to \Gamma_{N_1+1} \otimes \Gamma_{N_2+1} \tag{15}$$
$$(\ldots k_i \ldots k_j \ldots) \otimes (\ldots l_i \ldots l_j \ldots) \mapsto (\ldots k_i + 1 \ldots k_j \ldots)$$
$$\otimes (\ldots l_i \ldots l_j + 1 \ldots) - (\ldots k_i \ldots k_j + 1 \ldots) \otimes (\ldots l_i + 1 \ldots l_j \ldots).$$

It is important to note that there are, up to linear dependence, $n - 1$ such $\mathbb{D}$ operators and that they alter the weights of the modules on which they act. Their crucial property is the following. They commute with the e_{ij} except when one of their indices coincides with the *first* of the indices on e_{ij}:

$$[\mathbb{D}_{ij}^{12}, e_{kl}] = \begin{cases} \mathbb{D}_{jl}^{12} & i = k, j \neq l \\ \mathbb{D}_{li}^{12} & j = k, i \neq l \\ 0 & \text{otherwise.} \end{cases} \tag{16}$$

and they commute with most of the Cartan subalgebra:

$$[\mathbb{D}_{ij}^{12}, h_l] = \begin{cases} \mathbb{D}_{ij}^{12} & i = l \neq j - 1 \\ -\mathbb{D}_{ij}^{12} & i - 1 = l \neq j \\ \mathbb{D}_{ij}^{12} & j = l \neq i - 1 \\ -\mathbb{D}_{ij}^{12} & j - 1 = l \neq i \\ 0 & \text{otherwise} \end{cases} \tag{17}$$

Consequently, if $v \in \Gamma_{N_1} \otimes \Gamma_{N_2}$ is a highest weight vector according to some choice of Borel subalgebra, $\mathfrak{B}$, there will be a subset, $\mathbb{D}_{\mathfrak{B}}$, of the Hirota operators defined above which commute with the nilpotent part of the $\mathfrak{B}$ action, so that $\mathbb{D}(v) \in \Gamma_{N_1+1} \otimes \Gamma_{N_2+1}$ is again highest weight, with respect to $\mathfrak{B}$, but with a different weight value (because of the nontrivial relations (17)). This will hold for any of the irreducible submodules in the tensor product.

Hirota maps $\mathbb{D}_{ij}^{IJ}$ on g-fold tensor products

$$\mathbb{D}_{ij}^{IJ} : \Gamma_{N_1} \otimes \cdots \otimes \Gamma_{N_I} \otimes \cdots \otimes \Gamma_{N_J} \otimes \cdots \otimes \Gamma_{N_g}$$
$$\rightarrow \Gamma_{N_1} \otimes \cdots \otimes \Gamma_{N_I+1} \otimes \cdots \otimes \Gamma_{N_J+1} \otimes \cdots \otimes \Gamma_{N_g} \qquad (18)$$

are defined by applying the rule (15) to the Ith and Jth terms in the tensor product. Their commutation rules are unchanged.

In passing it should be remarked that the definition of the $\mathbb{D}_{ij}^{IJ}$ given here is slightly different to that given in [6]. Firstly it applies to $sl_n(\mathbb{C})$ with n arbitrary whereas the former held only for $sl_2(\mathbb{C})$. Secondly in the present definition the tensor arguments are of arbitrary weight and only the weights of two of the arguments are altered. In the former case a tensor product of g copies of one module was mapped to a product of g copies of the module of higher weight.

To illustrate these ideas we give some examples from the $sl_2(\mathbb{C})$ and $sl_3(\mathbb{C})$ theories.

3   EXAMPLES FROM $sl_2(\mathbb{C})$

In the $sl_2(\mathbb{C})$ case life is quite simple. Irreducible modules, Γ_N, have bases $\{(N, 0), (N - 1, 1), (N - 2, 2), \ldots, (0, N)\}$ with the $sl_2(\mathbb{C})$ action,

$$e_{12}(i, j) = i(i - 1, j + 1),$$
$$e_{21}(i, j) = j(i + 1, j - 1),$$
$$h_1(i, j) = (j - i)(i, j),$$

and the single Hirota map (on 2-fold tensor products) $\mathbb{D}_{12}^{12}(i, j)$ abbreviated to $\mathbb{D}$,

$$\mathbb{D}\{(i, j) \otimes (l, m)\} = (i + 1, j) \otimes (i, j + 1) - (i, j + 1) \otimes (i + 1, j). \quad (19)$$

Consider $\Gamma_N \otimes \Gamma_M$ with $N \geq M$. Because, in this case ($n = 2$) only, $\mathbb{D}$ commutes with the full ($sl_2(\mathbb{C})$) action, it is a genuine intertwining operator and we can write,

$$\Gamma_N \otimes \Gamma_M \xleftarrow{\mathbb{D}} \Gamma_{N-1} \otimes \Gamma_{M-1} \xleftarrow{\mathbb{D}} \cdots \Gamma_{N-M+1} \otimes \Gamma_1 \xleftarrow{\mathbb{D}} \Gamma_{N-M} \otimes \Gamma_0. \quad (20)$$

For example,

$$\Gamma_N \otimes \Gamma_M \ni \begin{matrix} (1, N - 1) \otimes (0, M) \\ - (0, N) \otimes (1, M - 1) \end{matrix} \xleftarrow{\mathbb{D}} (0, M - 1) \otimes (0, N - 1)$$
$$\times \in \Gamma_{N-1} \otimes \Gamma_{M-1} \qquad (21)$$

and

$$(2, N - 2) \otimes (0, M)$$
$$\Gamma_N \otimes \Gamma_M \ni -2(1, N-1) \otimes (1, M-1) \overset{\mathbb{D}^2}{\leftarrow} (0, M-2) \otimes (0, N-2) \in \Gamma_{N-2} \otimes \Gamma_{M-2}$$
$$+ (0, N) \otimes (2, M - 2)$$

(22)

It is clear that

$$\Gamma_N \otimes \Gamma_M \cong \Gamma_{N+M} \oplus \mathbb{D}(\Gamma_{N-1} \otimes \Gamma_{M-1}). \tag{23}$$

The weight of Γ_N is N and its dimension $N + 1$. The modules $\Gamma_{N-p} \otimes \Gamma_{M-p}$ have highest weights $M + N - 2p$ and dimensions $M + N - 2p + 1$. The dimension of $\Gamma_N \otimes \Gamma_M$ is $(N + 1)(M + 1)$. The maps $\mathbb{D}^i$ give the plethysm

$$\Gamma_N \otimes \Gamma_M \cong \bigoplus_{i=0}^{M} \mathfrak{g}\mathbb{D}^i \{(0, N - i) \otimes (0, M - i)\} \tag{24}$$

where $\mathfrak{g}(\cdot)$ denotes the action of the lie algebra on a highest weight vector to generate an irreducible, and the identity

$$(N + 1)(M + 1) = \sum_{i=0}^{M} N - M + 1 + 2i \tag{25}$$

expresses this decomposition in terms of dimensions.

4   EXAMPLES FROM $\mathfrak{sl}_3(\mathbb{C})$

As with many issues in representation theory the general case is better represented by the $\mathfrak{sl}_3(\mathbb{C})$ theory. The irreducible modules, Γ_N, have bases $\{(i, j, k) | i + j + k = N\}$ with the $\mathfrak{sl}_2(\mathbb{C})$ action,

$$
\begin{aligned}
e_{12}(i, j, k) &= i(i - 1, j + 1, k), & e_{21}(i, j, k) &= j(i + 1, j - 1, k), \\
e_{13}(i, j, k) &= i(i - 1, j, k + 1), & e_{31}(i, j, k) &= k(i + 1, j, k - 1), \\
e_{23}(i, j, k) &= j(i, j - 1, k + 1), & e_{32}(i, j, k) &= k(i, j + 1, k - 1), \\
h_1(i, j, k) &= (j - i)(i, j, k), & h_2(i, j, k) &= (k - j)(i, j, k).
\end{aligned}
$$

There are three Hirota maps (on 2-fold tensor products)

$$
\begin{aligned}
\mathbb{D}_{12}^{12}(i, j, k) \otimes (l, m, n) &= (i + 1, j, k) \otimes (l, m + 1, n) \\
&\quad - (i, j + 1, k) \otimes (l + 1, m, n), \\
\mathbb{D}_{23}^{12}(i, j, k) \otimes (l, m, n) &= (i, j + 1, k) \otimes (l, m, n + 1) \\
&\quad - (i, j, k + 1) \otimes (l, m + 1, n), \\
\mathbb{D}_{31}^{12}(i, j, k) \otimes (l, m, n) &= (i, j, k + 1) \otimes (l + 1, m, n) \\
&\quad - (i + 1, j, k) \otimes (l, m, n + 1).
\end{aligned}
$$

We will choose the Borel subalgebra $\mathfrak{B} = \mathfrak{h} \oplus \{e_{12}, e_{23}, e_{13}\}$. The operator $\mathbb{D}_{23}^{12}$ commutes with the nilpotent part and satisfies the following relations on $\mathfrak{h}$:

$$[\mathbb{D}_{23}^{12}, h_1] = -\mathbb{D}_{23}^{12}, \qquad [\mathbb{D}_{23}^{12}, h_2] = 0. \tag{26}$$

Consequently if $v \in \Gamma_N \otimes \Gamma_M$ is a highest weight vector of weight $(p, q) \in \mathbb{Z}^2$, $\mathbb{D}_{23}^{12}(v)$ will be a highest weight vector in $\Gamma_{N+1} \otimes \Gamma_{M+1}$ of weight $(p + 1, q)$.

The module Γ_N has weight $(0, N)$ but not all sl_3 modules are of this type. $\Gamma_1 \otimes \Gamma_1 \equiv \Gamma_{(0,1)} \otimes \Gamma_{(0,1)}$ is a nine-dimensional module. The obvious highest weight vector (given our choice of $\mathfrak{B}$, is $(0, 0, 1) \otimes (0, 0, 1)$ which has weight $(0, 2)$,

$$h_1\{(0, 0, 1) \otimes (0, 0, 1)\} = 0, \tag{27}$$
$$h_2\{(0, 0, 1) \otimes (0, 0, 1)\} = 2(0, 0, 1) \otimes (0, 0, 1),$$

and application of sl_3 generates the six-dimensional module $\Gamma_{(0,2)} \equiv \Gamma_2$. $\Gamma_0 \otimes \Gamma_0$ has a single element $(0, 0, 0) \otimes (0, 0, 0)$ of weight $(0, 0)$ which is mapped into an element of weight $(0, 0) + (1, 0) = (1, 0)$ in $\Gamma_1 \otimes \Gamma_1$ by $\mathbb{D}_{23}^{12}$:

$$\mathbb{D}_{23}^{12}\{(0, 0, 0) \otimes (0, 0, 0)\} = (0, 1, 0) \otimes (0, 0, 1) - (0, 0, 1) \otimes (0, 1, 0). \tag{28}$$

This element generates a three-dimensional irreducible module. Thus,

$$\Gamma_{(0,1)} \otimes \Gamma_{(0,1)} \cong \Gamma_{(0,2)} \oplus \Gamma_{(1,0)}, \tag{29}$$

or

$$\tag{30}$$

The next obvious case is $\Gamma_2 \otimes \Gamma_2 \equiv \Gamma_{(0,2)} \otimes \Gamma_{(0,2)}$. The element $(0, 0, 2) \otimes (0, 0, 2)$ generates a $\Gamma_4 \equiv \Gamma_{(0,4)}$ of dimension fifteen. The two highest weight elements in $\Gamma_1 \otimes \Gamma_1$ are $(0, 0, 1) \otimes (0, 0, 1)$ mapping under $\mathbb{D}_{23}^{12}$ to $(0, 1, 1) \otimes (0, 0, 2) - (0, 0, 2) \otimes (0, 1, 1) \in \Gamma_2 \otimes \Gamma_2$ and $(0, 0, 1) \otimes (0, 1, 0) - (0, 1, 0) \otimes (0, 0, 1)$ mapping under $\mathbb{D}_{23}^{12}$ to $-(0, 0, 2) \otimes (0, 2, 0) + 2(0, 1, 1) \otimes (0, 1, 1) - (0, 2, 0) \otimes (0, 0, 2)$. These elements have weights $(1, 2)$ and $(2, 0)$ respectively and generate modules of dimensions fifteen and six:

$$\Gamma_{(0,2)} \otimes \Gamma_{(0,2)} \cong \Gamma_{(0,4)} \oplus \Gamma_{(1,2)} \oplus \Gamma_{(2,0)}, \tag{31}$$

or, diagrammatically,

$$\tag{32}$$

By similar arguments one obtains

$$\Gamma_{(0,1)} \otimes \Gamma_{(0,2)} \cong \Gamma_{(0,3)} \oplus \Gamma_{(1,1)}, \tag{33}$$

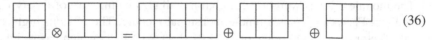 (34)

and

$$\Gamma_{(0,2)} \otimes \Gamma_{(0,3)} \cong \Gamma_{(0,5)} \oplus \Gamma_{(1,3)} \oplus \Gamma(2, 1), \tag{35}$$

(36)

and all other plethysms of the form $\Gamma_{(0,N)} \otimes \Gamma_{(0,M)}$.

It is clear that the action of $\mathbb{D}_{23}$ on Young tableaux is simple: it adds a box to the top row. In this it seems to be behaving as a very simple type of vertex operator [7].

5   EXAMPLES FROM $\mathfrak{sl}_4(\mathbb{C})$

Briefly, for $\mathfrak{sl}_4(\mathbb{C})$, $\mathfrak{h}$ is three-dimensional and the choice of Borel subalgebra spanned by $\mathfrak{h}$ and the e_{ij} with $i < j$ determines exactly one Hirota map, $\mathbb{D}_{34}$. This Hirota map augments the weight of a highest weight vector by $\delta = (0, 1, 0)$. Thus we obtain the decompositions:

$$\Gamma_{(0,0,1)} \otimes \Gamma_{(0,0,1)} = \Gamma_{(0,0,2)} \oplus \Gamma_{(0,1,0)}$$
$$\Gamma_{(0,0,2)} \otimes \Gamma_{(0,0,2)} = \Gamma_{(0,0,4)} \oplus \Gamma_{(0,1,2)} \oplus \Gamma_{(0,2,0)}$$
$$\Gamma_{(0,0,3)} \otimes \Gamma_{(0,0,3)} = \Gamma_{(0,0,6)} \oplus \Gamma_{(0,1,4)} \oplus \Gamma_{(0,2,2)} \oplus \Gamma_{(0,3,0)}$$

Diagrammatically,

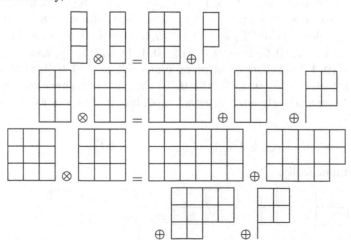

and for this choice of Hirota map, $\mathbb{D}_{34}$, the effect is seen to be to add a single box to each of the top two rows because the tableau associated with $\delta = (0, 1, 0)$ is

6 THE CLASSICAL HIROTA DERIVATIVE

We will consider only the cases of $\mathfrak{sl}_2(\mathbb{C})$ and $\mathfrak{sl}_3(\mathbb{C})$ in this section since these are the cases of direct relevance to the classical Hirota derivative and will confine ourselves to some remarks concerning the general case. We shall also give expressions for tensor products of order two with the understanding that everything can be extended to arbitrary tensor products in the manner described in earlier sections.

We start by defining a Hirota-like derivative on tensor products of homogeneous polynomials of the kind defined in (5). Because the Hirota map changes the weights of modules we are led to consider a sequence of such polynomials with degrees $N \in \mathbb{Z}^+$. The polynomial $f^{(N)}(x, y)$ has coefficients $\binom{i+j}{i}$ (i, j) where $i + j = N$. Define a D operator

$$(N + 1)(M + 1)D_{xy}^{12}\{f^{(N)}(x, y) \otimes g^{(M)}(x, y)\}$$
$$= \partial_x f^{(N+1)}(x, y) \otimes \partial_y g^{(M+1)}(x, y) - \partial_y f^{(N+1)}(x, y) \otimes \partial_x g^{(M+1)}(x, y).$$

It is easy to check that

$$D_{xy}^{12}\{f^{(N)}(x, y) \otimes g^{(M)}(x, y)\} = \mathbb{D}_{12}^{12}\{f^{(N)}(x, y) \otimes g^{(M)}(x, y)\} \quad (37)$$

where the D operator acts on the variables and the $\mathbb{D}$ operator acts on the coefficients, (i, j).

We now show that we recover exactly the classical Hirota derivative when we look at the projective variable $\zeta = Nx/y$ and allow $N \to \infty$. Then the polynomials in x and y are replaced by analytic functions in z and the $\mathfrak{sl}_2(\mathbb{C})$ action is replaced by an action of the Heisenberg algebra,

$$[\partial_z, z] = \natural \quad (38)$$

where $\natural$ is a "counting" operator.

To this end put

$$f^{(N)}(x, y) = y^N \phi^{(N)}(\zeta) \quad (39)$$

so that

$$\phi^{(N)}(\zeta) = \sum_{k_1+k_2=N} \frac{N!}{k_1!K_2!}(k_1, k_2)\frac{\zeta^{k_1}}{N^{k_1}} \tag{40}$$

$$\xrightarrow{N\to\infty} \sum_{k=0}^{\infty} \frac{1}{k!}(k)\zeta^k \equiv \phi(\zeta), \tag{41}$$

where we have abbreviated (k_1, k_2) to (k) for $k_1 = k$ and $k_2 = N - k$.

The $\mathfrak{sl}_2(\mathbb{C})$ action on $\phi^{(N)}$ is easily obtained by a change of variables:

$$e_{12} = \zeta - \frac{1}{N}\zeta^2\,\frac{\partial}{\partial\zeta}, \tag{42}$$

$$e_{21} = N\,\frac{\partial}{\partial\zeta}, \tag{43}$$

$$h_1 = -N + 2\zeta\,\frac{\partial}{\partial\zeta}; \tag{44}$$

and as $N \to \infty$ we define

$$\tilde{e}_{12} = e_{12} \to \zeta, \tag{45}$$

$$\tilde{e}_{21} = \frac{1}{N}e_{12} \to \frac{\partial}{\partial\zeta}, \tag{46}$$

$$\tilde{h}_1 = \frac{1}{N}H_1 \to \natural, \tag{47}$$

where $\natural$ is understood to be the "unit" derivation, e.g., $\natural(a \otimes b) = 1.a \otimes b + a \otimes 1.b = 2.(a \otimes b)$. It effectively counts the number of entries in the tensor product.

We also need to understand the Hirota map in this limit of infinite-dimensional representations. Using (37) one shows that

$$\mathbb{D}_{12}^{12}\{y^N\phi^{(N)} \otimes y^M\psi^{(M)}\} = y^N\partial_\zeta\phi^{(N)} \otimes y^M\psi^{(M)} - y^N\phi^{(N)} \otimes y^M\,\partial_\zeta\psi^{(M)}, \tag{48}$$

and assuming we may take factors of y through the tensor product *in the infinite limit* we recover the identity

$$\mathbb{D}\{\phi \otimes \psi\} = \partial_\zeta\phi \otimes \psi - \phi \otimes \partial_\zeta\psi \equiv D_\zeta\{\phi \otimes \psi\} \tag{49}$$

where $\mathbb{D}$ acts on the coefficients (k) in the analytic expansions of $\phi(\zeta)$ and $\psi(\zeta)$ and where we recognize D as the conventional Hirota derivative. More precisely, if we symmetrize over the tensor product,

$$\mathrm{Sym}\, D_\zeta(\phi \otimes \psi) = \phi_\zeta\psi - \phi\psi_\zeta$$
$$\mathrm{Sym}\, D_\zeta^2(\phi \otimes \psi) = \phi_{\zeta\zeta}\psi - 2\phi_\zeta\psi_\zeta + \phi\psi_{\zeta\zeta}$$

In this limit the Hirota derivative, D_ζ or $\mathbb{D}$, intertwines with the Heisenberg action:

$$[D_\zeta, \partial_\zeta] = 0, \qquad [D_\zeta, \zeta] = 0, \qquad [D_\zeta, \sharp] = 0. \tag{50}$$

This property of the classical Hirota derivative has been discussed and exploited in [8].

In verifying these relations directly one must take care that the Heisenberg acts by derivation over tensor products. Thus, for example,

$$
\begin{aligned}
[D_\zeta, \zeta](\phi \otimes \psi) &= D_\zeta(\zeta\phi \otimes \psi + \phi \otimes \zeta\psi) - \zeta(\phi_\zeta \otimes \psi - \phi \otimes \psi_\zeta) \\
&= (\zeta\phi)_\zeta \otimes \psi - \zeta\phi \otimes \psi_\zeta + \phi_\zeta \otimes \zeta\psi - \phi \otimes (\zeta\psi)_\zeta \\
&\quad - \zeta\phi_\zeta \otimes \psi - \phi_\zeta \otimes \zeta\psi + \zeta\phi \otimes \psi_\zeta + \phi \otimes \zeta\psi_\zeta \\
&= 0.
\end{aligned}
$$

As an aside, it seems logical to introduce a further "Hirota operator" related to ζ as D is related to ∂_ζ, namely

$$Z(\phi \otimes \psi) = \zeta\phi \otimes \psi - \phi \otimes \zeta\psi. \tag{51}$$

This operator is also intertwining for the Heisenberg action,

$$[Z, \partial_\zeta] = 0, \quad [Z, \zeta] = 0, \quad [Z, \sharp] = 0. \tag{52}$$

but is trivialized by symmetrization. Further D_ζ and Z satisfy the commutation relation

$$[D_\zeta, Z] = \sharp \tag{53}$$

Note that in [2] Hirota derivatives are taken to be differential operators defined by the relation (in conventional notation),

$$D_\zeta(e^{\epsilon\zeta}\tau(\zeta) \cdot e^{\epsilon\zeta}\sigma(\zeta)) = e^{2\epsilon\zeta} D_\zeta(\tau(\zeta) \cdot \sigma(\zeta)) \tag{54}$$

and since $e^{\epsilon\zeta}$ is the group element generated by the infinitesimal action of ζ we see that this is consistent with the definition given in this paper written in terms of the group action,

$$g(a \otimes b) = ga \otimes gb \tag{55}$$

so that

$$D_\zeta g = g D_\zeta \tag{56}$$

where g is any element of the full group $SL_2(\mathbb{C})$. Thus in addition to the condition (54) we should also require commutation with the translation operator $e^{\epsilon\partial_\zeta}$:

$$D(\tau(\zeta + \epsilon) \cdot \sigma(\zeta + \epsilon)) = D(\tau \cdot \sigma)(\zeta + \epsilon). \tag{57}$$

Moving on now to the case of $sl_3(\mathbb{C})$ the irreducible modules are labeled by pairs of positive integers, (M, N), so that there is a one-parameter family of limits labeled by the ratio $\lambda = \frac{M}{N}$ as both M and N tend to infinity. Actions on polynomials in three variables become actions on analytic functions of two variables, $\phi(\zeta, \eta)$. The rôle of the Heisenberg in $sl_2(\mathbb{C})$ is now played by the larger algebra $sl_2(\mathbb{C}) \ltimes H_2$ where H_2 is the two-dimensional Heisenberg. The $sl_2(\mathbb{C})$ action is given by

$$\zeta \partial_\eta, \qquad \eta \partial_\zeta, \qquad \zeta \partial_\zeta - \eta \partial_\eta; \tag{58}$$

and the H_2 by

$$(2 + \frac{1}{\lambda})\eta, \qquad \lambda \partial_\eta, \qquad -(2\lambda + 1)\sharp, \qquad (2 + \frac{1}{\lambda})\zeta, \qquad \lambda \partial_\zeta. \tag{59}$$

The *three* Hirota maps become Hirota derivatives:

$$D_\zeta(\phi \otimes \psi) = \phi_\zeta \otimes \psi - \phi \otimes \psi_\zeta, \tag{60}$$

$$D_\eta(\phi \otimes \psi) = \phi_\eta \otimes \psi - \phi \otimes \psi_\eta, \tag{61}$$

$$D_{\zeta\eta}(\phi \otimes \psi) = \phi_\zeta \otimes \psi_\eta - \phi_\eta \otimes \psi_\zeta. \tag{62}$$

They do not commute completely with the $sl_2(\mathbb{C}) \ltimes H_2$ action. In fact $D_{\zeta\eta}$ commutes with the $sl_2(\mathbb{C})$ part, but not with the H_2 part:

$$[D_{\zeta\eta}, \eta] = -D_\zeta, \tag{63}$$

$$[D_{\zeta\eta}, \zeta] = D_\eta; \tag{64}$$

while D_ζ and D_η commute with the H_2 but not the $sl_2(\mathbb{C})$ part:

$$[D_\zeta, \zeta \partial_\eta] = D_\eta \qquad [D_\zeta, \zeta \partial_\zeta - \eta \partial_\eta] = D_\zeta, \tag{65}$$

$$[D_\eta, \eta \partial_\zeta] = D_\zeta \qquad [D_\eta, \zeta \partial_\zeta - \eta \partial_\eta] = -D_\eta. \tag{66}$$

The Hirota operator $D_{\zeta\eta}$ is therefore not a Hirota derivative in the sense of [2]. Nevertheless it arises in the infinite-dimensional limit in the same way as the conventional Hirota derivatives D_ζ and D_η and plays a similar rôle in the representation theory of two-dimensional systems, as we shall see in the next section. For example,

$$Sym \, D_{\zeta\eta}^2(\phi \otimes \phi) = \begin{vmatrix} \phi_{\zeta\zeta} & \phi_{\zeta\eta} \\ \phi_{\eta\zeta} & \phi_{\eta\eta} \end{vmatrix} \tag{67}$$

$$= \phi_{\zeta\zeta}\phi_{\eta\eta} - \phi_{\eta\zeta}^2 \tag{68}$$

a common invariant appearing in, say, the Monge–Ampére equation.

In the case of infinite-dimensional limits of representations of $sl_n(\mathbb{C})$ we will obtain an $sl_{n-1}(\mathbb{C}) \ltimes H_{n-1}$ action with Hirota derivatives $D_1, D_2, \ldots, D_{n-1}$, commuting with the Heisenberg part, H_{n-1}, and a set of "mixed" Hirota maps, D_{ij}, commuting with the $sl_{n-1}(\mathbb{C})$ part.

7 HYPERELLIPTIC FUNCTIONS OF GENUS 2

The situation described in the previous section, the infinite-dimensional limit of $sl_3(\mathbb{C})$ modules, actually occurs in a very specific and important situation [9, 10].

The family of genus 2, hyperelliptic curves,

$$y^2 = g_6 x^6 + 6g_5 x^5 + 15g_4 x^4 + 20g_3 x^3 + 15g_2 x^2 + 6g_1 x + g_0 \qquad (69)$$

is permuted under transformations

$$x \mapsto \frac{\alpha x + \beta}{\gamma x + \delta} \qquad (70)$$

$$y \mapsto \frac{y}{(\gamma x + \delta)^3} \qquad (71)$$

in such a way that the coefficients g_i transform as a seven-dimensional, irreducible representation of $SL_2(\mathbb{C})$.

Variables u_1 and u_2 on the associated Jacobian variety are defined by the differential relations

$$du_1 = \frac{dx_1}{y_1} + \frac{dx_2}{y_2} \qquad (72)$$

$$du_2 = \frac{x_1 dx_1}{y_1} + \frac{x_2 dx_2}{y_2} \qquad (73)$$

where (x_1, y_1) and (x_2, y_2) are a pair of (regular) points on the curve.

Under the transformation described above the pair (u_1, u_2) transform as a two-dimensional, irreducible representation of $SL_2(\mathbb{C})$.

Three double-index objects, $\{\wp_{11}, \wp_{12}, \wp_{22}\}$, can be defined [11] as rational functions in the x_i and y_i and satisfying integrability conditions,

$$\frac{\partial}{\partial u_1} \wp_{22} = \frac{\partial}{\partial u_2} \wp_{12}, \qquad (74)$$

$$\frac{\partial}{\partial u_1} \wp_{12} = \frac{\partial}{\partial u_2} \wp_{11}. \qquad (75)$$

Consequently the $\wp_{ij}$ are second derivatives, with respect to the u_i, of some scalar ($SL_2(\mathbb{C})$ invariant) potential and themselves form a three-dimensional representation.

They satisfy the following set of five fourth-order partial differential equations,

$$-\frac{1}{3}\wp_{2222} + 2\wp_{22}^2 = g_2g_6 - 4g_3g_5 + 3g_4^2 + g_4\wp_{22}$$

$$-2g_5\wp_{12} + g_6\wp_{11} \tag{76}$$

$$-\frac{1}{3}\wp_{2221} + 2\wp_{22}\wp_{12} = \frac{1}{2}(g_1g_6 - 3g_2g_5 + 2g_3g_4) + g_3\wp_{22}$$

$$-2g_4\wp_{12} + g_5\wp_{11}$$

$$-\frac{1}{3}\wp_{2211} + \frac{2}{3}\wp_{22}\wp_{11} + \frac{4}{3}\wp_{12}^2 = \frac{1}{6}(g_0g_6 - 9g_1g_5 + 8g_3^2) + g_2\wp_{22}$$

$$-2g_3\wp_{12} + g_4\wp_{11}$$

$$-\frac{1}{3}\wp_{2111} + 2\wp_{21}\wp_{11} = \frac{1}{2}(g_0g_5 - 3g_1g_4 + 2g_2g_3) + g_1\wp_{22}$$

$$-2g_2\wp_{12} + g_3\wp_{11}$$

$$-\frac{1}{3}\wp_{1111} + 2\wp_{11}^2 = g_0g_4 - 4g_1g_3 + 3g_2^2 + g_0\wp_{22}$$

$$-2g_1\wp_{12} + g_2\wp_{11}$$

The terms in these equations can be associated with basis elements of irreducible representations of $SL_2(\mathbb{C})$. For example, the five four-index objects $\wp_{2222}$, $\wp_{2221}$ etc. are a five-dimensional representation, as are the quadratic terms $\wp_{22}^2$, $\wp_{22}\wp_{12}$, and so on. Likewise the terms quadratic in the g_i. Schematically we might represent the five equations (76) as

$$\mathbf{P}_5 \oplus (\mathbf{P}_3 \otimes \mathbf{P}_3)_5 = (\mathbf{G}_7 \otimes \mathbf{G}_7)_5 \oplus (\mathbf{G}_7 \otimes \mathbf{P}_3)_5, \tag{77}$$

where the subscripts denote dimensions or projections onto irreducibles.

It can be further shown that

$$\wp_{ij} = -\frac{\partial^2}{\partial u_i \, \partial u_j} \ln \sigma(u_1, u_2)$$

where σ is an entire function on the Jacobian, analogous to the Weierstraß σ-function. Consequently,

$$\wp_{ij} = -\frac{1}{2\sigma^2} D_i D_j \sigma \cdot \sigma \tag{78}$$

$$\wp_{ijkl} - 2\wp_{ij}\wp_{kl} - 2\wp_{ik}\wp_{jl} - 2\wp_{il}\wp_{jk} = -\frac{1}{2\sigma^2} D_i D_j D_k D_l \sigma \cdot \sigma \tag{79}$$

Then, and remarkably this formulation appears explicitly in Baker's 1907 book [11], σ satisfies the following *bilinear* equations:

$$\left(\frac{1}{6} D_{u_2}^4 + \frac{1}{2} g_4 D_{u_2}^2 - g_5 D_{u_1} D_{u_2} + \frac{1}{2} g_6 D_{u_1}^2 - g_2 g_6 + 4 g_3 g_5 - 3 g_4^2 \right) \sigma \cdot \sigma = 0$$

$$\left(\frac{1}{3} D_{u_2}^3 D_{u_1} + g_3 D_{u_2}^2 - 2 g_4 D_{u_1} D_{u_2} + g_5 D_{u_1}^2 - g_1 g_6 + 3 g_2 g_5 - 2 g_3 g_4 \right) \sigma \cdot \sigma = 0$$

$$\left(D_{u_2}^2 D_{u_1}^2 + 3 g_2 D_{u_2}^2 - 6 g_3 D_{u_1} D_{u_2} + 3 g_4 D_{u_1}^2 - g_0 g_6 + 9 g_1 g_5 - 8 g_3^2 \right) \sigma \cdot \sigma = 0$$

$$\left(\frac{1}{3} D_{u_2} D_{u_1}^3 + g_1 D_{u_2}^2 - 2 g_2 D_{u_1} D_{u_2} + g_3 D_{u_1}^2 - g_0 g_5 + 3 g_1 g_4 - 2 g_2 g_3 \right) \sigma \cdot \sigma = 0$$

$$\left(\frac{1}{6} D_{u_1}^4 + \frac{1}{2} g_0 D_{u_2}^2 - g_1 D_{u_1} D_{u_2} + \frac{1}{2} g_2 D_{u_1}^2 - g_0 g_4 + 4 g_1 g_3 - 3 g_2^2 \right) \sigma \cdot \sigma = 0$$

$$(80)$$

But now we have a pair of conventional Hirota derivatives, D_{u_1} and D_{u_2}, together with an $sl_2(\mathbb{C})$ action on the pair (u_1, u_2), that is, we have exactly a $sl_2(\mathbb{C}) \ltimes H_2$ action with which they commute as in the preceding section. There is therefore an accompanying Hirota derivative of the form $D_{u_1 u_2}$ which commutes with the $sl_2(\mathbb{C})$ action but not the H_2 action. A treatment of the theory which reflects this underlying group action is naturally formulated using Hirota derivatives: that is, all the identities satisfied by the σ-function are expressed via the Hirota derivatives. Not only this, but all identities can be classified according to the irreducible representations of $sl_2(\mathbb{C})$ and "highest weight" identities are naturally constructed using the intertwining properties of the Hirota derivatives.

It is also, of course, clear that there is an underlying finite-dimensional $sl_3(\mathbb{C})$ action, because the curve is the canonical form of a family of projective varieties, homogeneous of degree 6, in variables X, Y, and Z, say. But it is not immediately clear how this collapses to the above infinite-dimensional limit.

A specific instance, other than Eq. (76), is the quartic identity satisfied by the $\wp_{ij}$, the equation of the Kummer surface (the Jacobian variety factored by the involution, $y_i \longmapsto -y_i$. The leading term is $[\text{Sym } D_{u_1 u_2}^2 (\wp \otimes \wp)]^2$ and the remaining terms are invariants constructed in a more complex manner:

$$0 = \begin{vmatrix} g_6 & -3 g_5 & 3 g_4 + 2 \wp_{22} & -g_3 - 2 \wp_{12} \\ -3 g_5 & 9 g_4 - 4 \wp_{22} & -9 g_3 + 2 \wp_{12} & 3 g_2 + 2 \wp_{11} \\ 3 g_4 + 2 \wp_{22} & -9 g_3 + 2 \wp_{12} & 9 g_2 - 4 \wp_{11} & -3 g_1 \\ -g_3 - 2 \wp_{12} & 3 g_2 + 2 \wp_{11} & -3 g_1 & g_0 \end{vmatrix}$$

$$= 16 (\wp_{12}^2 - \wp_{11} \wp_{22})^2 + \cdots$$

$$(81)$$

Further either of the top or bottom equations of (80) are easily seen to be equivalent to a Boussinesq equation. The full system is then a restricted Boussinesq equation for which a Lax pair can be written down by applying the group action to that for the Boussinesq itself.

All these issues are discussed fully in the references [9, 10].

8   CONCLUSIONS AND FURTHER WORK

In this paper we have emphasised the relationship between the representation theory of $sl_n(\mathbb{C})$ and certain Hirota maps which are directly related to Hirota derivatives in the infinite-dimensional limit. We have made no attempt to explain *how* these observations contribute to an understanding of the solutions of bilinear equations. To do so it would appear to be necessary to think of such equations as the infinite-dimensional limits of relations between generic finite-dimensional representations, relations which are themselves open to geometric interpretation perhaps along Grassmanian lines [12]. On a related tack it is clearly of interest to see if the Hirota derivatives D_ζ, D_η, $D_{\zeta\eta}$, and so on provide useful constructions when the Heisenberg algebra is embedded in infinite-dimensional algebras such as the Virasoro or Kac–Moody algebras.

ACKNOWLEDGMENTS

I should like to record my thanks to Chris Eilbeck, Victor Enolskii, Jarmo Hietarinta, Jon Nimmo, and Jan Sanders for many useful conversations; to the Isaac Newton Institute, the University of Cambridge, and particularly the organizers and participants of the 2001 Integrable Systems programme where a large part of this work was born; and to NATO and the University of Glasgow for funding my visit to the Island of Elba.

REFERENCES

1. Kac, V. G. (1990) *Infinite Dimensional Lie Algebras*, CUP Cambridge, England.
2. Grammaticos, B., Ramani, A., and Hietarinta, J. (1994) Multilinear operators: the natural extension of Hirota's bilinear formalism, *Phys. Lett. A.* **190**, pp. 65–70.
3. Fulton, W. and Harris, J. (1991) Representation theory. A first course, *Graduate Texts in Mathematics*, Vol. 129, Readings in Mathematics, Springer-Verlag, New York.
4. Hilbert, D. (1993) *Theory of Algebraic Invariants*, CUP.
5. Hirota, R. (1982) Bilinearization of soliton equations, *J. Phys. Soc. Japan* **51**, pp. 323–331.

6. Athorne, C. (2001) Hirota derivatives and representation theory in Integrable systems: linear and nonlinear dynamics (Islay, 1999), *Glasg. Math. J.* **43A**, pp. 1–8.
7. Date, E., Jimbo, M., Kashiwara, M., and Miwa, T. (1982) Transformation groups for soliton equations. Euclidean Lie algebras and reduction of the KP hierarchy, *Publ. Res. Inst. Math. Sci.* **18**, pp. 1077–1110.
8. Olver, P. J. and Sanders, J. A. (2000) Transvectants, modular forms, and the Heisenberg algebra, *Adv. in Appl. Math.* **25**(3), pp. 252–283.
9. Athorne, C., Eilbeck, J. C., and Enolskii, V. Z. (xxxx) An $SL_2(\mathbb{C})$ covariant theort of genus 2 hyper-elliptic functions, *Math. Proc. Camb. Phil. Soc.*, (in press).
10. Athorne, C., Eilbeck, J. C., and Enolskii, V. Z. (2003) Identities for classical, genus 2 $\wp$ functions, *J. Geom. and Phys.*, **48**, pp. 354–368.
11. Baker, H. F. (1907) *Multiply Periodic Functions*, CUP London.
12. Pressley, A. and Segal, G. (1986) *Loop Groups*, Clarendon Press, Oxford.
13. Athorne, C. (1999) Algebraic invariants and generalized Hirota derivatives, *Phys. Lett. A.* **256**, pp. 20–24.

BOUNDARY STATES IN SUSY SINE-GORDON MODEL

Z. Bajnok, L. Palla*, and G. Takács

Abstract   After reviewing briefly the basic concepts and problems of boundary integrable theories we outline a boostrap solution leading to the spectrum of boundary states in SUSY sine-Gordon model with supersymmetric integrable boundary condition.

Keywords: Integrable field theory, field theory with boundary, bootstrap, supersymmetry, sine-Gordon model

1 INTRODUCTION

The aim of this investigation [1] is to determine the spectrum of boundary states and the associated reflection amplitudes in $N = 1$ SUSY sine-Gordon model with supersymmetry and integrability preserving boundary conditions. This boundary super sine-Gordon model (BSSG) is an integrable boundary theory, therefore to put our problem into proper context we review first the basic concepts and problems of boundary integrable theories.

The simplest way to obtain a boundary integrable theory is to take a bulk integrable one and restrict it to the $x \leq 0$ half line by imposing integrability preserving boundary conditions at the $x = 0$ end [2]. Technically it means that the action of the boundary theory is written as

$$S = \int\limits_{-\infty}^{\infty} dt \int\limits_{-\infty}^{0} dx \mathcal{L}[\Phi(x, t)] + \int\limits_{-\infty}^{\infty} dt V_B[\Phi(x = 0, t)], \qquad (1)$$

where $\mathcal{L}$ is the Lagrangian of the (integrable) bulk theory, and the role of the boundary potentail V_B is to impose the boundary conditions. The theory defined by this action is said to be integrable, if it admits conserved higher spin quantities. Since the Lagrangian is a local quantity the excitations (particles) of the original bulk theory are present also in the boundary theory, furthermore their local interactions (S matrices) are also not effected by the boundary.

---

* Conference speaker, e-mail: palla@ludens.elte.hu

L. Faddeev et al. (eds.),
Bilinear Integrable Systems: From Classical to Quantum, Continuous to Discrete, 35–42.
© 2006 *Springer. Printed in the Netherlands.*

Nevertheless, because the presence of the boundary, these bulk particles may now reflect on the boundary. The integrability of the model guarantees, that the number of particles is conserved in these reflection processes, thus they can be characterized by the reflection amplitudes $R_b^a(\theta)$; and a particle of type a with rapidity θ reflecting as particle of type b is described as

$$A^a(\theta)|B\rangle = R_b^a(\theta)A^b(-\theta)|B\rangle.$$

The reflection factor has to satisfy three rather restricting, complicated algebraic equations [2] namely the boundary versions of the Yang–Baxter equation

$$R^{c_2}{}_{a_2}(\theta_2)S_{a_1c_2}^{c_1d_2}(\theta_1+\theta_2)R^{d_1}{}_{c_1}(\theta_1)S_{d_2d_1}^{b_2b_1}(\theta_1-\theta_2)$$

$$= S_{a_1c_2}^{c_1c_2}(\theta_1-\theta_2)R^{d_1}{}_{c_1}(\theta_1)S_{c_2d_1}^{d_2b_1}(\theta_1+\theta_2)R^{b_2}{}_{d_2}(\theta_2),$$

unitarity

$$R^a{}_b(\theta)R^b{}_c(-\theta) = \delta^a{}_c,$$

and crossing

$$R^a{}_b\left(\frac{i\pi}{2}-\theta\right) = S_{a'b'}^{ab}(2\theta)R^{b'}{}_{a'}\left(\frac{i\pi}{2}+\theta\right),$$

that contain as an input the bulk S matrix. Now suppose a solution of this system of equations is found. If it has poles in the physical strip $0 \le \Im m\theta \le \frac{\pi}{2}$ that can not be explained by the boundary version of the Coleman–Thun mechanism, then they must be interpreted as signaling the presence of new, "excited" type of boundary states [2]. If the pole is at θ_0 then we can envisage that a particle of type a with rapidity θ_0, when reflecting on $|B\rangle$ becomes 'bound' to it, transforming it into a new state $|\tilde{B}\rangle$ with energy (mass) $E = m_a \cosh\theta_0$. Once the existence of this new state is established the problem of bulk particles reflecting on the new boundary emerges. The solution of this problem is given by the bootstrap procedure: exploiting the model's integrability one obtains:

$$|\tilde{B}\rangle R^d{}_c(\theta) = S_{ca}^{xy}(\theta-\theta_0)R^z{}_x(\theta)S_{yz}^{ad}(\theta+\theta_0).$$

If this new reflection factor has poles in the physical strip that must be interpreted as even "higher excited" boundary states, $|\tilde{B}\rangle$, then the bootstrap must be repeated once more to obtain the new reflection factors. This procedure ends only—i.e., the bootstrap becomes "closed", if in one of the new sets of reflection amplitudes there are no new, unexplained poles.

We carry out this procedure in the BSSG model, when the Lagrangian is that of the bulk SUSY sine-Gordon model:

$$\mathcal{L} = \frac{1}{2}\partial_\mu\Phi\partial^\mu\Phi + i\bar{\Psi}\gamma^\mu\partial_\mu\Psi + m\bar{\Psi}\Psi\cos\frac{\beta}{2}\Phi + \frac{m^2}{\beta^2}\cos\beta\Phi,$$

with Φ being a real scalar and $\Psi = \binom{\bar\psi}{\psi}$ a Majorana fermion fields, and when the boundary potential is [3]

$$V_B^{\pm} = (\pm\bar\psi\psi + ia\partial_t a - 2f^{\pm}(\Phi)a(\psi \mp \bar\psi) + \mathcal{B}(\Phi))|_{x=0}.$$

The functions f and $\mathcal{B}$ are fixed up to two free parameters by the requirement of integrability and supersymmetry. A surprising feature of this boundary potential is the appearance of the boundary fermionic degree of freedom $a(t)$, which is necessary to obtain the two two parameter sets of integrable and supersymmetric boundary conditions corresponding to the two choices of the signs.

The main idea is to look for the reflection amplitudes in this model in a form where there is no mixing between the supersymmetric and other internal quantum numbers. This means an Ansatz for the reflection amplitudes as a product of two terms one of which is the ordinary (bosonic) sine-Gordon reflection amplitude, while the other describes the scattering of the SUSY degrees of freedom. These ideas are motivated on the one hand by the successful description of the bulk scattering while on the other by the fact that the spectrum of boundary states together with the the reflection amplitudes in the ordinary sine-Gordon model are known [4].

2　BULK SUSY SINE-GORDON MODEL

The spectrum consists of the soliton/antisoliton multiplet, realizing supersymmetry in a nonlocal way, and possibly a few breather multiplets (that are bound states of a soliton with an antisoliton) upon which supersymmetry acts in the standard way. The supersymmetric solitons are described by RSOS kinks $K_{ab}^{\epsilon}(\theta)$ of mass M and rapidity θ, where the RSOS labels a, b take the values 0, $\frac{1}{2}$ and 1 with $|a - b| = 1/2$, and describe the supersymmetric structure, while $\epsilon = \pm$ corresponds to topological charge ± 1. The multikink states obey the nontrivial "adjacency" condition. The two-particle scattering process allowed by this condition

$$K_{ab}^{\epsilon_1}(\theta_1) + K_{bc}^{\epsilon_2}(\theta_2) \to K_{ad}^{\epsilon_2'}(\theta_2) + K_{dc}^{\epsilon_1'}(\theta_1)$$

has an amplitude of the form [5, 6]:

$$S_{\text{SUSY}}\begin{pmatrix} a & d \\ b & c \end{pmatrix} \theta_1 - \theta_2 \Bigg) \times S_{\text{SG}}(\theta_1 - \theta_2, \lambda)_{\epsilon_1\epsilon_2}^{\epsilon_1'\epsilon_2'},$$

where S_{SUSY} is identical to the S matrix of the tricritical Ising model perturbed by the primary field of dimension $\frac{3}{5}$ and S_{SG} coincides with the usual sine-Gordon S matrix (but the relation $\lambda = \frac{8\pi}{\beta^2} - \frac{1}{2}$ is different from the sine-Gordon case).

S_{SUSY} has no poles in the physical strip, thus the only poles that describe bound states (breathers) come from the sine-Gordon part.

The bulk theory has two supersymmetry charges of opposite chirality Q and $\bar{Q}$; their algebra contains a central charge $\{Q, \bar{Q}\} = 2MZ$, that can take the values 0 or ± 1. In the basis $\{K_{0\frac{1}{2}}, K_{1\frac{1}{2}}, K_{\frac{1}{2}0}, K_{\frac{1}{2}1}\}$ the central charge and the fermion number can be written as

$$
Z = \begin{pmatrix} 1 & 0 & 0 & 0 \\ 0 & 1 & 0 & 0 \\ 0 & 0 & -1 & 0 \\ 0 & 0 & 0 & -1 \end{pmatrix} \qquad \Gamma = (-1)^F = \begin{pmatrix} 0 & 1 & 0 & 0 \\ 1 & 0 & 0 & 0 \\ 0 & 0 & 0 & 1 \\ 0 & 0 & 1 & 0 \end{pmatrix}.
$$

This representation of SUSY describes BPS saturated objects. The SUSY action on breather states and the breather S matrices can be derived using the bootstrap. It turns out that the central charge Z (as well as the topological charge T) vanishes identically for the breathers. For further details we refer to [5, 6].

3   REFLECTIONS IN BOUNDARY SINE-GORDON MODEL

The most general reflection factor of the soliton antisoliton multiplet $|s, \bar{s}\rangle$ on the ground state boundary, satisfying the boundary versions of the Yang–Baxter, unitarity and crossing equations was found by Ghoshal and Zamolodchikov [2]

$$
R = \begin{pmatrix} P_0^+ & Q_0 \\ Q_0 & P_0^- \end{pmatrix} R_0(u) \frac{\sigma(\eta, u)\sigma(i\vartheta, u)}{\cos \eta \cosh \vartheta} = \begin{pmatrix} P^+ & Q \\ Q & P^- \end{pmatrix}
$$
$$
P_0^{\pm} = \cos \lambda u \cosh \vartheta \cos \eta \mp \sin \lambda u \sinh \vartheta \sin \eta
$$
$$
Q_0 = -\sin \lambda u \cos \lambda u
$$

where $u = -i\theta$, λ is the parameter in the bulk S matrix while η and ϑ are two real parameters that characterize the solution.

The spectrum of excited boundary states was determined in [7, 4]. It can be parametrized by a sequence of integers $|n_1, n_2, \ldots, n_k\rangle$, whenever the

$$
\frac{\pi}{2} \geq v_{n_1} > w_{n_2} > \cdots \geq 0 \tag{2}
$$

condition holds, where

$$
v_n = \frac{\eta}{\lambda} - \frac{\pi(2n+1)}{2\lambda} \quad \text{and} \quad w_k = \pi - \frac{\eta}{\lambda} - \frac{\pi(2k-1)}{2\lambda}.
$$

The mass of such a state is

$$
m_{|n_1, n_2, \ldots, n_k\rangle} = M \sum_{i \text{ odd}} \cos(v_{n_i}) + M \sum_{i \text{ even}} \cos(w_{n_i}).
$$

v_n and w_k give the poles of certain reflection amplitudes, thus the condition in (2) guarantees that these poles are in the physical strip and cannot be explained by the boundary Coleman–Thun mechanism. The soliton/antisoliton reflection amplitudes on excited boundaries $Q_{|n_1,n_2,...,n_k\rangle}(\eta, \vartheta, u)$, $P^{\pm}_{|n_1,n_2,...,n_k\rangle}(\eta, \vartheta, u)$ are obtained from the ground state ones by multiplying them with appropriate CDD factors [4]. The breather sector can be obtained again by bootstrap.

4 BOOTSTRAP IN THE BSSG MODEL

The last bit of information needed to start the bootstrap in the BSSG model is the form of the single boundary supercharge $\tilde{Q}_+$ or $\tilde{Q}_-$ corresponding to the choice of sign in the boundary potential. In [3] it is shown, that

$$\tilde{Q}_{\pm} = \int_{-\infty}^{0} (q(x, t) \pm \bar{q}(x, t))\, dx + Q_B(x = 0, t),$$

where q and $\bar{q}$ are the local densities of Q and $\bar{Q}$ and Q_B is the boundary contribution. One can argue [1], that the relation $\tilde{Q}_{\pm}^2 = 2(\tilde{H} \pm M\tilde{Z})$ holds between the boundary supercharge, boundary Hamiltonian and central charge. The action of the boundary supercharge on asymptotic states is expected to be given by

$$\tilde{Q}_{\pm} = \tilde{Q} \pm \bar{Q} + Q_B, \quad Q_B = \gamma\Gamma,$$

where Q, $\bar{Q}$ act on the particles as in the bulk theory and γ is some unknown function of the parameters appearing in the boundary potential. This choice is supported by the classical considerations in [3] and also guarantees that $\tilde{Q}_{\pm}$ commutes with the bulk S matrix.

Now according to our main idea we suppose that the reflection matrix factorizes as $R_{\text{SUSY}}(\theta) \times R_{\text{SG}}(\theta)$. In this special form the constraints as unitarity, boundary Yang-Baxter equation and crossing-unitarity relation [2] can be satisfied separately for the two factors. Since the sine-Gordon part already fulfills these requirements, we concentrate on the supersymmetric part.

From the RSOS nature of the bulk S-matrix it is clear that the boundary must also have RSOS labels and the adjacency conditions between the nearest kink and the boundary must also hold. Thus the following reflections are possible:

$$K_{ba}(\theta)|B_a\rangle = \sum_c R^b_{ac}(\theta)K_{bc}(-\theta)|B_c\rangle,$$

or in detail

$$K_{a\frac{1}{2}}(\theta)|B_{\frac{1}{2}}\rangle = R^a_{\frac{1}{2}\frac{1}{2}}(\theta)K_{a\frac{1}{2}}(-\theta)|B_{\frac{1}{2}}\rangle; \quad a = 0, 1,$$

and (for $b \neq a$ $a, b = 0, 1$)

$$K_{\frac{1}{2}a}(\theta)|B_a\rangle = R^{\frac{1}{2}}_{aa}(\theta)K_{\frac{1}{2}a}(-\theta)|B_a\rangle + R^{\frac{1}{2}}_{ab}(\theta)K_{\frac{1}{2}b}(-\theta)|B_b\rangle.$$

In the second process the label of the boundary state has changed, which shows that $|B_0\rangle$ and $|B_1\rangle$ form a doublet. All of the constraints mentioned above factorize in the sense that they give independent equations for the reflections on the boundary $|B_{1/2}\rangle$ and on the doublet $|B_{0,1}\rangle$. Since the ground state boundary is expected to be nondegenerate we assume it is a $|B_{1/2}\rangle$ state and first concentrate on reflection factors off the singlet boundary $|B_{1/2}\rangle$. For the boundary supercharge we need the action of Q, $\bar{Q}$, and Γ on the boundary ground state $|B_{1/2}\rangle$:

$$Q|B_{\frac{1}{2}}\rangle = 0, \quad \bar{Q}|B_{\frac{1}{2}}\rangle = 0, \quad \Gamma|B_{\frac{1}{2}}\rangle = |B_{\frac{1}{2}}\rangle.$$

Combining this with the square of $\tilde{Q}_\pm$ gives the interpretation of γ: $\gamma^2/2$ is nothing else but the ground state energy.

The solutions of the boundary Yang Baxter, crossing and unitarity conditions for $R^a_{\frac{1}{2}\frac{1}{2}}$ have been discussed in the literature [8, 9]; the new angle we add is that we insist on boundary SUSY from the onset - somewhat similarly as in [10]. The two choices $\tilde{Q}_\pm$ give different solutions. If the boundary supercharge $\tilde{Q}_+$ commutes with the reflections ($BSSG^+$ theory) then we obtain

$$R^0_{\frac{1}{2}\frac{1}{2}}(\theta) = R^1_{\frac{1}{2}\frac{1}{2}}(\theta) = 2^{-\theta/\pi i} P(\theta)$$

$$= 2^{-\theta/\pi i} \prod_{k=1}^{\infty} \left[\frac{\Gamma\left(k - \frac{\theta}{2\pi i}\right)\Gamma\left(k - \frac{\theta}{2\pi i}\right)}{\Gamma\left(k - \frac{1}{4} - \frac{\theta}{2\pi i}\right)\Gamma\left(k + \frac{1}{4} - \frac{\theta}{2\pi i}\right)} \middle/ \{\theta \leftrightarrow -\theta\} \right]$$

If, however, it is $\tilde{Q}_-$ that commutes with the reflections ($BSSG^-$) then the result is

$$R^0_{\frac{1}{2}\frac{1}{2}}(\theta) = \left(\cos\frac{\xi}{2} + i\sinh\frac{\theta}{2}\right) K(\theta - i\xi)K(i\pi - \theta - i\xi)2^{-\theta/\pi i} P(\theta)$$

$$R^1_{\frac{1}{2}\frac{1}{2}}(\theta) = \left(\cos\frac{\xi}{2} + i\sinh\frac{\theta}{2}\right) K(\theta - i\xi)K(i\pi - \theta - i\xi)2^{-\theta/\pi i} P(\theta),$$

where ξ is related to γ as $\gamma = -2\sqrt{M}\cos\frac{\xi}{2}$. Note that symmetry of the reflection under Γ requires $R^0_{\frac{1}{2}\frac{1}{2}}\theta = R^1_{\frac{1}{2}\frac{1}{2}}(\theta)$ thus in the first case (BSSG$^+$) the reflections also commute with the operator Γ, while in the other case (BSSG$^-$) they do not. Remarkably there are no poles in the physical strip in any of these reflection factors and they contain no free parameters.

We start the quest for boundary states with the analysis of the complete ground state reflection factors $R^a_{\frac{1}{2}\frac{1}{2}}(\theta) \times R(\theta)$, where the SUSY component has one of the two forms above. Since the only poles of these reflection factors are due to the sine-Gordon part (R) their explanation has to be similar to that in the bosonic theory. The boundary dependent poles of R, which describe boundary states, are located at $-i\theta = v_n$. At the position of these poles we associate boundary bound states to the reflection amplitudes $R^a_{\frac{1}{2}\frac{1}{2}}$, $a = 0, 1$

$$|a, 1/2|n\rangle = \frac{1}{g^{|1/2\rangle}_{|a,1/2|n\rangle}} K_{a\frac{1}{2}}(i\, v_n)\left|\frac{1}{2}\right\rangle, \quad \text{where} \quad \left|\frac{1}{2}\right\rangle \equiv \left|B_{\frac{1}{2}}\right\rangle,$$

where the g-factor is the SUSY part of the boundary coupling. The two states ($a = 0, 1$) for a given n form a doublet in two senses: on the one hand it is the $K_{\frac{1}{2}a}$ kinks that can scatter on it, while on the other they span a two dimensional representation space for the boundary supercharge.

The SUSY reflection factors of $K_{\frac{1}{2}a}$ off $|a, 1/2|n\rangle$ can be computed from the bootstrap principle with the result (in the case of the simpler $BSSG^+$ theory):

$$R^{\frac{1}{2}}_{ab}(\theta) = P(\theta)K(\theta + iv_n)K(\theta - iv_n)\frac{g_b^{\frac{1}{2}}}{g_a^{\frac{1}{2}}}\left(\delta_{ab}\cos\left(\frac{v_n}{2}\right) + \delta_{a,1-b}\sin\left(\frac{\theta}{2i}\right)\right),$$

having no poles in the physical strip. The full reflection factor on the $|a, 1/2|n\rangle$ excited boundary can be obtained by multiplying this result with the appropriate excited bosonic reflection factor: $R^{\frac{1}{2}}_{ab}(\theta) \times Q_{|n\rangle}(\eta, \vartheta, \theta)$ or $R^{\frac{1}{2}}_{ab}(\theta) \times P^{\pm}_{|n\rangle}(\eta, \vartheta, \theta)$. The poles of these expressions that describe boundary states come again from the sine-Gordon factor at precisely the same condition ($w_m < v_n$) as in the non SUSY theory.

Repeating the bootstrap procedure for the next level excited states [1] made it clear that the general boundary bound state has the structure

$$\left|a_k \dots \frac{1}{2}, a_1, \frac{1}{2}\middle| n_k \dots, m_1, n_1\right\rangle \quad \text{or} \quad \left|\frac{1}{2}, a_k \dots \frac{1}{2}, a_1, \frac{1}{2}\middle| m_k, n_k \dots, m_1, n_1\right\rangle,$$

i.e., is characterized by a sequence of integers and an RSOS sequence. This shows that in the supersymmetric case the excited boundary states have a non-trivial degeneracy in contrast to the bosonic theory, the degeneracy being labeled by the RSOS sequences. The energy of the boundary states depends only on the integers and is identical to the result obtained in the sine-Gordon model. The associated reflection factors can be computed from successive application of the bootstrap procedure [1]. Thus a two parameter solution of closing the bootsrap is found in the $BSSG^{\pm}$ models.

As open problems we mention the clarification of the relation between the η, ϑ, γ parameters appearing in bootstrap and the ones in the boundary potential or the pCFT description of BSSG$^{\pm}$.

ACKNOWLEDGMENT

L.P thanks the organizers for the kind invitation.

REFERENCES

1. Bajnok, Z., Palla, L., and Takács, G. Spectrum of boundary states in N = 1 SUSY sine-Gordon theory, hep-th/0207099 to appear in *Nucl. Phys*: **B**.
2. Ghoshal, S. and Zamolodchikov, A. B. (1994) *Int. J. Mod. Phys*. **A9**, pp. 3841–3886.
3. Nepomechie, R. I. (2001) *Phys. Lett*. **B509**, p. 183.
4. Bajnok, Z., Palla, L., Takács, G. and Tóth, G. Zs. (2002) *Nucl. Phys*. **B622**, pp. 548–564.
5. Ahn, C. (1991) *Nucl. Phys*. **B354**, pp. 57–84.
6. Hollowood, T. J. and Mavrikis, E. (1997) *Nucl. Phys*. **B484**, pp. 631–652, hep-th/9606116.
7. Mattsson, P. and Dorey, P. (2000) *J. Phys*. **A33**, pp. 9065–9094, hep-th/0008071.
8. Chim, L. (1996) *Int. J. Mod. Phys*. **A11**, pp. 4491–4512, hep-th/9510008.
9. Ahn, C. and Koo, W. M. (1996) *Nucl. Phys*. **B482**, p. 675, hep-th/9606003.
10. Nepomechie, R. I. Supersymmetry in the boundary tricritical Ising field theory, preprint UMTG-234, hep-th/0203123

GEOMETRY OF DISCRETE INTEGRABILITY. THE CONSISTENCY APPROACH

Alexander I. Bobenko*
Institut für Mathematik, Fakultät 2, Technische Universität Berlin,
Strasse des 17. Juni 136, 10623 Berlin, Germany

1 ORIGIN AND MOTIVATION: DIFFERENTIAL GEOMETRY

Long before the theory of solitons, geometers used integrable equations to describe various special curves, surfaces etc. At that time no relation to mathematical physics was known, and quite different geometries appeared in this context (we will call them integrable) were unified by their common geometric features:

- Integrable surfaces, curves etc. have *nice* geometric properties,
- Integrable geometries come with their *interesting* transformations (Bäcklund–Darboux transformations) acting within the class,
- These transformations are *permutable* (Bianchi permutability).

Since "nice" and "interesting" can hardly be treated as mathematically formulated features, let us address to the permutability property. We explain it for the classical example of surfaces with constant negative Gaussian curvature (K-surface) with their Bäcklund transformations.

Let $F : \mathbb{R}^2 \to \mathbb{R}^3$ be a K-surface and $F_{1,0}$ and $F_{0,1}$ its two Bäcklund transformed. The classical Bianchi permutability theorem claims that there exists a unique K-surface $F_{1,1}$ which is the Bäcklund transformed of $F_{1,0}$ and $F_{0,1}$. Proceeding further this way for a given point $F_{0,0}$ on the original K-surface one obtains a $\mathbb{Z}^2$ lattice $F_{k,\ell}$ of permutable Bäcklund transformations. From the geometric properties of the Bäcklund transformations it is easy to see [1] that $F_{k,\ell}$ defined this way is a discrete K-surface.

The discrete K-surfaces have the same properties and transformations as their smooth counterparts [2]. There exist deep reasons for that. The classical

---
* Partially supported by the SFB 288 "Differential geometry and quantum physics" and by the DFG research center "Mathematics for key technologies" (FZT 86) in Berlin

L. Faddeev et al. (eds.),
Bilinear Integrable Systems: From Classical to Quantum, Continuous to Discrete, 43–53.
© 2006 *Springer. Printed in the Netherlands.*

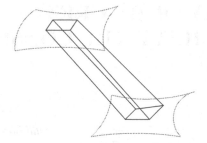

Figure 1. Surfaces and their transformations as a limit of multidimensional lattices

differential geometry of integrable surfaces may be obtained from a unifying multidimensional discrete theory by a refinement of the coordinate mesh-size in some of the directions.

Indeed, by refining of the coordinate mesh-size,

$$F : (\epsilon \mathbb{Z})^2 \to \mathbb{R}^3 \quad \longrightarrow \quad F : \mathbb{R}^2 \to \mathbb{R}^3,$$

discrete surface $\epsilon \to 0$ smooth surface

in the limit one obtains classical smooth K-surfaces from discrete K-surfaces. Starting with an n-dimensional net of permutable Bäcklund transformations

$$F : (\epsilon_1 \mathbb{Z}) \times \cdots \times (\epsilon_n \mathbb{Z}) \to \mathbb{R}^3$$

in the limit $\epsilon_1 \to 0, \epsilon_2 \to 0, \epsilon_3 = \cdots = \epsilon_n = 1$ one arrives to a smooth K-surface with its $n - 2$-dimensional discrete family of permutable Bäcklund transformations:

$$F : \mathbb{R}^2 \times \mathbb{Z}^{n-2} \to \mathbb{R}^3.$$

This simple idea is quite fruitful. In the discrete case all directions of the multidimensional lattices appear in quite symmetric way. It leads to:

- A *unification* of surfaces and their transformations. Discrete surfaces and their transformations are indistinguishable.
- A fundamental *consistency principle*. Due to the symmetry of the discrete setup the same equations hold on all elementary faces of the lattice. This leads us beyond the pure differential geometry to a new understanding of the integrability, classification of integrable equations and derivation of the zero curvature (Lax) representation from the first principles.
- Interesting *generalizations* to: $n > 2$-dimensional systems, quantum systems, discrete systems with the fields on various lattice elements (vertices, edges, faces etc.).

As it was mentioned above, all this suggests that it might be possible to develop the classical differential geometry, including both the theory of surfaces and of their transformations, as a mesh refining limit of the discrete constructions. On the other hand, the good quantitative properties of approximations delivered by the discrete differential geometry suggest that they might be put at the basis of the practical numerical algorithms for computations in the differential geometry. However until recently there were no rigorous mathematical statements supporting this observation.

The first step in closing this gap was made in the paper [3] where the convergence of the corresponding integrable geometric numerical scheme has been proven for nonlinear hyperbolic systems (including the K-surfaces and the sine–Gordon equation).

Thus, summarizing we arrive at the following philosophy of discrete differential geometry: surfaces and their transformations can be obtained as a special limit of a discrete master-theory. The latter treats the corresponding discrete surfaces and their transformation in absolutely symmetric way. This is possible because these are merged into multidimensional nets such that their all sublattices have the same geometric properties. The possibility of this multidimensional extension results to *consistency* of the corresponding difference equations characterizing the geometry. The latter is the main topic of this paper.

2   EQUATIONS ON QUAD-GRAPHS. INTEGRABILITY AS CONSISTENCY

Traditionally discrete integrable systems were considered with fields defined on the $\mathbb{Z}^2$ lattice. One can define integrable systems on arbitrary graphs as flat connections with the values in loop groups. However, one should not go that far with the generalization. As we have shown in [4], there is a special class of graphs, called *quad-graphs*, supporting the most fundamental properties of the integrability theory. This notion turns out to be a proper generalization of the $\mathbb{Z}^2$ lattice as far as the integrability theory is concerned.

Definition 1   *A cellular decomposition $\mathcal{G}$ of an oriented surface is called a* quad-graph, *if all its faces are quadrilateral.*

Note that if one considers an arbitrary cellular decomposition C jointly with its dual C^* one obtains a quad-graph $\mathcal{D}$ connecting by the edges the neighboring vertices of C and C^*. Let us stress that the edges of the quad-graph $\mathcal{D}$ differ from the edges of C and C^*.

For the integrable systems on quad-graphs we consider in this section the fields $z : V(\mathcal{D}) \mapsto \hat{\mathbb{C}}$ are attached to the vertices of the graph. They are subject to an equation $Q(z_1, z_2, z_3, z_4) = 0$, relating four fields sitting on the four vertices

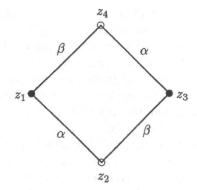

Figure 2. A face of the labelled quad-graph

of an arbitrary face from $F(\mathcal{D})$. The Hirota equation

$$\frac{z_4}{z_2} = \frac{\alpha z_3 - \beta z_1}{\beta z_3 - \alpha z_1} \tag{1}$$

is such an example. We observe that the equation carries parameters α, β which can be naturally associated to the edges, and the opposite edges of an elementary quadrilateral carry equal parameters (see Figure 2). At this point we specify the setup further. The example illustrated in Figure 2 can be naturally generalized. An integrable system on a quad-graph

$$Q(z_1, z_2, z_3, z_4; \alpha, \beta) = 0 \tag{2}$$

is parametrized by a function on the edges of the quad-graph which takes equal values on the opposite edges of any elementary quadrilateral. We call such a function a *labelling* of the quad-graph.

An elementary quadrilateral of a quad-graph can be viewed from various directions. This implies that the system (2) is well defined on a general quad-graph only if it possesses the rhombic symmetry, i.e., each of the equations

$$Q(z_1, z_4, z_3, z_2; \beta, \alpha) = 0, \quad Q(z_3, z_2, z_1, z_4; \beta, \alpha) = 0$$

is equivalent to (2).

2.1 3D-Consistency

Now we introduce a crucial property of discrete integrable systems which later on will be taken as a characteristic one.

Let us extend a quad–graph $\mathcal{D}$ into the third dimension. We take the second copy $\mathcal{D}'$ of $\mathcal{D}$ and add edges connecting the corresponding vertices of $\mathcal{D}$ and $\mathcal{D}'$. Elementary building blocks of so obtained "three-dimensional quad-graph" **D**

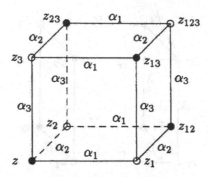

Figure 3. Elementary cube

are "cubes" as shown in Figure 3. The labelling of $\mathcal{D}$ can be extended to $\mathbf{D}$ so that the opposite edges of all elementary faces (including the "vertical" ones) carry equal parameters (see Figure 3).

Now, the fundamental property of discrete integrable system mentioned above is the *three-dimensional consistency*.

Definition 2 *Consider an elementary cube, as on Figure 3. Suppose that the values of the field z are given at the vertex z and at its three neighbors $z_1, z_2,$ and z_3. Then the Eq. (2) uniquely determines the values $z_{12}, z_{23},$ and z_{13}. After that the same Eq. (2) delivers three a priori different values for the value of the field z_{123} at the eighth vertex of the cube, coming from the faces $[z_1, z_{12}, z_{123}, z_{13}], [z_2, z_{12}, z_{123}, z_{23}],$ and $[z_3, z_{13}, z_{123}, z_{23}],$ respectively. The Eq.(2) is called* 3D-consistent *if these three values for z_{123} coincide for any choice of the initial data z, z_1, z_2, z_3.*

Proposition 3 *The Hirota equation*

$$\frac{z_{12}}{z} = \frac{\alpha_2 z_1 - \alpha_1 z_2}{\alpha_1 z_1 - \alpha_2 z_2}$$

is 3D-consistent.

This can be checked by a straightforward computation. For the field at the eighth vertex of the cube one obtains

$$z_{123} = \frac{(l_{21} - l_{12})z_1 z_2 + (l_{32} - l_{23})z_2 z_3 + (l_{13} - l_{31})z_1 z_3}{(l_{23} - l_{32})z_1 + (l_{31} - l_{13})z_2 + (l_{12} - l_{21})z_3}, \tag{3}$$

where $l_{ij} = \frac{\alpha_i}{\alpha_j}$.

In [4, 5] we suggested to treat the consistency property (in the sense of Definition 2) as the characteristic one for discrete integrable systems. Thus we come to the central.

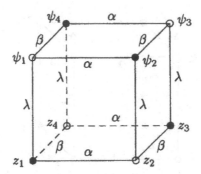

Figure 4. Zero curvature representation from the consistency

Definition 4   *A discrete equation is called* integrable *if it is* consistent.

Note that this definition of the integrability is *conceptually transparent* and *algorithmic*: for any equation it can be easily checked whether it is integrable or not.

2.2   Zero Curvature Representation from the 3D-Consistency

Our Definition 2 of discrete integrable systems is more fundamental then the traditional one as systems having a zero curvature representation in a loop group. Here we demonstrate how the corresponding flat connection in a loop group can be derived from the equation. Independently this was found in [6].

We get rid of our symmetric notations, consider the system

$$Q(z_1, z_2, z_3, z_4; \alpha, \beta) = 0 \tag{4}$$

on the base face of the cube and choose the vertical direction to carry an additional (spectral) parameter λ (see Figure 4).

Assume the left-hand-side of (4) is affine in each z_k. This gives z_4 as a fractional–linear (Möbius) transformation z_2 with the coefficients depending on z_1, z_3 and α, β. One can of course freely interchange $z_1, \ldots, z_4$ in this statement. Consider now the equations on the vertical faces of the cube in Figure 4. One gets ψ_2 as a Möbius transformation of ψ_1

$$\psi_2 = L(z_2, z_1; \alpha, \lambda)[\psi_1],$$

with the coefficients depending of the fields z_2, z_1, on the parameter α in the system (4) and on the additional parameter λ which is to be treated as the spectral parameter. The mapping $L(z_2, z_1; \alpha, \lambda)$ is associated to the oriented edge (z_1, z_2). Going from ψ_1 to ψ_3 in two different ways and using the arbitrariness of ψ_1 we get

$$L(z_3, z_2; \beta, \lambda)L(z_2, z_1; \alpha, \lambda) = L(z_3, z_4; \alpha, \lambda)L(z_4, z_1; \beta, \lambda). \tag{5}$$

Using the matrix representation of Möbius transformations

$$\frac{az+b}{cz+d} = L[z], \quad \text{where} \quad L = \begin{pmatrix} a & b \\ c & d \end{pmatrix},$$

and normalizing the matrices (for example by the condition det $L = 1$) we arrive at the zero curvature representation (5).

Let us apply this derivation method to the Hirota equation. Equation (1) can be written as $Q = 0$ with the affine

$$Q(z_1, z_2, z_3, z_4; \alpha, \beta) = \alpha(z_2 z_3 + z_1 z_4) - \beta(z_3 z_4 + z_1 z_2).$$

Performing the computations as above in this case we derive the well known zero curvature representation (5) with the matrices

$$L(z_2, z_1, \alpha, \lambda) = \begin{pmatrix} \alpha & -\lambda z_2 \\ \dfrac{\lambda}{z_1} & -\alpha \dfrac{z_2}{z_1} \end{pmatrix} \tag{6}$$

for the Hirota equation.

3 CLASSIFICATION

Here we classify all integrable (in the sense of Definition 2) one-field equations on quad-graphs satisfying some natural symmetry conditions.

We consider equations

$$Q(x, u, v, y; \alpha, \beta) = 0, \tag{7}$$

on quad-graphs. Equations are associated to elementary quadrilaterals, the fields $x, u, v, y \in \mathbb{C}$ are assigned to the four vertices of the quadrilateral, and the parameters $\alpha, \beta \in \mathbb{C}$ are assigned to its edges. We now list more precisely the assumptions under which we classify the equations.

1. **Consistency**. Equation (7) is integrable (in the sense it is 3D-consistent).
2. **Linearity**. The function $Q(x, u, v, y; \alpha, \beta)$ is linear in each argument (affine linear):

$$Q(x, u, v, y; \alpha, \beta) = a_1 xuvy + \cdots + a_{16}, \tag{8}$$

where coefficients a_i depend on α, β. This is equivalent to the condition that Eq. (7) can be uniquely solved for any one of its arguments $x, u, v, y \in \hat{\mathbb{C}}$.

3. **Symmetry**. The Eq. (7) is invariant under the group D_4 of the square symmetries, that is function Q satisfies the symmetry properties

$$Q(x, u, v, y; \alpha, \beta) = \varepsilon Q(x, v, u, y; \beta, \alpha) = \sigma Q(u, x, y, v; \alpha, \beta) \qquad (9)$$

with $\varepsilon, \sigma = \pm 1$.

4. **Tetrahedron property**. The function $z_{123} = f(z, z_1, z_2, z_3; \alpha_1, \alpha_2, \alpha_3)$, existing due to the three-dimensional consistency, actually does not depend on the variable z, that is, $f_x = 0$. This property holds (3) for the Hirota equation as well as for all other known integrable examples.

The proof of the classification theorem is rather involved and is given in [5].

Theorem 5 *Up to common Möbius transformations of the variables z and point transformations of the parameters α, the three-dimensionally consistent quad-graph equations (7) with the properties (2–4) (linearity, symmetry, tetrahedron property) are exhausted by the following three lists Q, H, A ($x = z, u = z_1, v = z_2, y = z_{12}, \alpha = \alpha_1, \beta = \alpha_2$).*
    List Q:

(Q1) $\alpha(x - v)(u - y) - \beta(x - u)(v - y) + \delta^2 \alpha \beta (\alpha - \beta) = 0$,
(Q2) $\alpha(x - v)(u - y) - \beta(x - u)(v - y) + \alpha\beta(\alpha - \beta)(x + u + v + y)$
    $- \alpha\beta(\alpha - \beta)(\alpha^2 - \alpha\beta + \beta^2) = 0$,
(Q3) $(\beta^2 - \alpha^2)(xy + uv) + \beta(\alpha^2 - 1)(xu + vy) - \alpha(\beta^2 - 1)(xv + uy)$
    $- \delta^2(\alpha^2 - \beta^2)(\alpha^2 - 1)(\beta^2 - 1)/(4\alpha\beta) = 0$,
(Q4) $a_0 xuvy + a_1(xuv + uvy + vyx + yxu) + a_2(xy + uv) + \bar{a}_2(xu + vy)$
    $+ \bar{\bar{a}}_2(xv + uy) + a_3(x + u + v + y) + a_4 = 0$,

where the coefficients a_i are expressed through (α, a) and (β, b) with $a^2 = r(\alpha), b^2 = r(\beta), r(x) = 4x^3 - g_2 x - g_3$, by the following formulae:

$$a_0 = a + b, \qquad a_1 = -\beta a - \alpha b, \qquad a_2 = \beta^2 a + \alpha^2 b,$$

$$\bar{a}_2 = \frac{ab(a + b)}{2(\alpha - \beta)} + \beta^2 a - \left(2\alpha^2 - \frac{g_2}{4}\right) b,$$

$$\bar{\bar{a}}_2 = \frac{ab(a + b)}{2(\beta - \alpha)} + \alpha^2 b - \left(2\beta^2 - \frac{g_2}{4}\right) a,$$

$$a_3 = \frac{g_3}{2} a_0 - \frac{g_2}{4} a_1, \qquad a_4 = \frac{g_2^2}{16} a_0 - g_3 a_1.$$

    List H:

(H1) $(x - y)(u - v) + \beta - \alpha = 0$,
(H2) $(x - y)(u - v) + (\beta - \alpha)(x + u + v + y) + \beta^2 - \alpha^2 = 0$,
(H3) $\alpha(xu + vy) - \beta(xv + uy) + \delta(\alpha^2 - \beta^2) = 0$.

List A:

(A1) $\alpha(x + v)(u + y) - \beta(x + u)(v + y) - \delta^2 \alpha \beta(\alpha - \beta) = 0$,
(A2) $(\beta^2 - \alpha^2)(xuvy + 1) + \beta(\alpha^2 - 1)(xv + uy) - \alpha(\beta^2 - 1)(xu + vy) = 0$.

Remarks

1. The list A can be dropped down by allowing an extended group of Möbius transformations, which act on the variables x, y differently than on u, v. So, really independent equations are given by the lists Q and H.
2. In both lists Q, H the last equations are the most general ones. This means that Eqs. (Q1)–(Q3) and (H1), (H2) may be obtained from (Q4) and (H3), respectively, by certain degenerations and/or limit procedures. This resembles the situation with the list of six Painlevé equations and the coalescences connecting them.
3. Note that the list contains the fundamental equations only. A discrete equation which is derived as a corollary of an equation with the consistency property usually loose this property.

4 GENERALIZATIONS: MULTIDIMENSIONAL AND NON-COMMUTATIVE (QUANTUM) CASES

4.1 Yang–Baxter Maps

It should be mentioned, however, that to assign fields to the vertices is not the only possibility. Another large class of two-dimensional systems on quad–graphs build those with the fields assigned to the *edges*.

In this situation each elementary quadrilateral carries a map $R : \mathcal{X}^2 \mapsto \mathcal{X}^2$, where $\mathcal{X}$ is the space where the fields take values. The question on the three–dimensional consistency of such maps is also legitimate and, moreover, began to be studied recently. The corresponding property can be encoded in the formula

$$R_{23} \circ R_{13} \circ R_{12} = R_{12} \circ R_{13} \circ R_{23}, \tag{10}$$

where each $R_{ij} : \mathcal{X}^3 \mapsto \mathcal{X}^3$ acts as the map R on the factors i,j of the cartesian product $\mathcal{X}^3$ and acts identically on the third factor. The maps with this property were introduced by Drinfeld [7] under the name of "set-theoretical solutions of the Yang-Baxter equations", an alternative name is "Yang-Baxter maps" used by Veselov in his recent study [8].

The problem of classification of Yang–Baxter maps, like the one achieved in the previous section, is under current investigation.

4.2 Four-Dimensional Consistency of Three-Dimensional Systems

The consistency principle can be obviously generalized to an arbitrary dimension. We say that

a d–dimensional discrete equation possesses the consistency property,
if it may be imposed in a consistent way on all d–dimensional sublattices
of a $(d + 1)$–dimensional lattice

In the three–dimensional context there are also *a priori* many kinds of systems, according to where the fields are defined: on the vertices, on the edges, or on the elementary squares of the cubic lattice. Consider three–dimensional systems with the fields sitting on the vertices. In this case each elementary cube carries just one equation

$$Q(z, z_1, z_2, z_3, z_{12}, z_{23}, z_{13}, z_{123}) = 0, \tag{11}$$

relating the fields in all its vertices. The four–dimensional consistency of such equations is defined in the same way as in Section 2.1 for the case of one dimension lower.

It is tempting to accept the four–dimensional consistency of equations of the type (11) as the constructive definition of their integrability. It is important to solve the correspondent classification problem.

We present here just one example of the equation appeared first in [9].

Proposition 6 *Equation*

$$\frac{(z_1 - z_3)(z_2 - z_{123})}{(z_3 - z_2)(z_{123} - z_1)} = \frac{(z - z_{13})(z_{12} - z_{23})}{(z_{13} - z_{12})(z_{23} - z)}. \tag{12}$$

is four–dimensionally consistent.

4.3 Noncommutative (Quantum) Cases

As we have shown in [10] the consistency approach works also in the noncommutative case, where the participating fields live in an arbitrary associative (not necessary commutative) algebra $\mathcal{A}$ (over the field $\mathcal{K}$).

In particular the noncommutative Hirota equation

$$yx^{-1} = \frac{1 - (\beta/\alpha)uv^{-1}}{(\beta/\alpha) - uv^{-1}}. \tag{13}$$

belongs to this class. Now $x, u, v, y \in \mathcal{A}$ are the fields assigned to the four vertices of the quadrilateral, and $\alpha, \beta \in \mathcal{K}$ are the parameters assigned to its edges. Note that Eq. (13) preserves the Weil commutation relations. This yields the quantum Hirota equation studied in [11].

Proposition 7 *The noncommutative Hirota equation is 3D-consistent.*

Similar to the commutative case the Lax representation can be derived from the equation and the consistency property. It turns out that finding the zero curvature representation does not hinge on the particular algebra $\mathcal{A}$ or on prescribing some particular commutation rules for fields in the neighboring vertices. The fact that some commutation relations are preserved by the evolution, is thus conceptually separated from the integrability.

REFERENCES

1. Bobenko, A. I. (1999) Discrete Integrable systems and geometry, in: *12th International Congress of Mathematical Physics ICMP '97*, (Brisbane, Australia, July 1997) eds. D. De Wit, A.J. Bracken, M.D. Gould, and P.A. Pearce, International Press, Boston, pp. 219–226.
2. Bobenko, A. I. and Pinkall, U. (1996) Discrete surfaces with constant negative Gaussian curvature and the Hirota equation, *J. Diff. Geom.* **43**, pp. 527–611.
3. Bobenko, A. I., Matthes, D., and Suris, Yu. B. (xxxx) Nonlinear hyperbolic equations in surface theory: integrable discretizations and approximation results, arXiv:math.NA/0208042.
4. Bobenko, A. I. and Suris, Yu. B. (2002) Integrable systems on quad-graphs, *Intern. Math. Res. Notices.* **11**, pp. 573–612.
5. Adler, V. E., Bobenko A. I., and Suris, Yu. B. (2003) Classification of integrable equations on quad-graphs. The consistency approach, *Comm. Math. Phys.* **233**, pp. 513–543.
6. Nijhoff, F. (2002) Lax pair for the Adler (lattice Krichever-Novikov system), *Phys. Lett. A* **297**, pp. 49–58.
7. Drinfeld, V. G. (1992) On some unsolved problems in quantum group theory, In: *Lecture Notes Math.* **1510**, pp. 1–8.
8. Veselov, A. P. (xxxx) Yang–Baxter maps and integrable dynamics, arXiv: math.QA/0205335.
9. Nimmo, J. J. C. and Schief, W. K. (1998) An integrable discretization of a $2 + 1$ dimensional sine-Gordon equation, *Stud. Appl. Math.* **100**, pp. 295–309.
10. Bobenko, A. I. and Suris, Yu. B. (2002) Integrable non-commutative equations on quad-graphs. The consistency approach, *Lett. Math. Phys.* **61**, pp. 241–254.
11. Faddeev, L. D. and Volkov, A. Yu. (1994) Hirota equation as an example of an integrable symplectic map, *Lett. Math. Phys.* **32**, pp. 125–135.
12. Hirota, R. (1977) Nonlinear partial difference equations. III. Discrete sine-Gordon equation, *J. Phys. Soc. Japan.* **43**, pp. 2079–2084.

HOMOCLINIC ORBITS AND DRESSING METHOD

E.V. Doktorov
B.I. Stepanov Institute of Physics, 220072 Minsk, Belarus

V.M. Rothos
*Department of Mathematics and Computer Sciences, Leicester University,
Leicester, LE1 7RH, U.K.*

Abstract   Chaos is frequently associated with orbits homoclinic to unstable modes
of deterministic nonlinear PDEs. Bilinear Hirota method was success-
fully employed to obtain homoclinic solutions for NLS with periodic
boundaries. We propose a new method to analytically generate homoclinic solutions for integrable nonlinear PDEs. This approach resembles
the dressing method known in the theory of solitons. The pole positions in the dressing factor are given by the complex double points of
the Floquet spectrum associated with unstable modes of the nonlinear
equation. As an example, we reproduce first the homoclinic orbit for
NLS, and then obtain the homoclinic solution for the modified nonlinear
Schrödinger equation solvable by the Wadati–Konno–Ichikawa spectral
problem.

1   INTRODUCTION

Alongside with solitons as stable solutions of nonlinear integrable PDEs with
important applications in physics and mathematics, these equations can admit
unstable waves such as homoclinic orbits. The existence of homoclinic solutions
serves as an indicator of chaotic behavior in a perturbed deterministic nonlinear
dynamical system.

The role of homoclinic solutions in the generation of chaos was shown for
the damped-driven sine-Gordon [1, 2] and perturbed NLS [3–6] equations with
periodic boundary conditions. Extended reviews of analytical and numerical
methods in this topic are given in [7, 8]. Different approaches have been proposed for derivation of homoclinic solutions for integrable PDEs: while the
bilinear Hirota method [9] was used in [3], the Bäcklund transformations were
employed in [1, 2, 5]. Up to now homoclinic structures were obtained for nonlinear equations associated with the Zakharov–Shabat spectral problem, which

L. Faddeev et al. (eds.),
Bilinear Integrable Systems: From Classical to Quantum, Continuous to Discrete, 55–64.
© 2006 *Springer. Printed in the Netherlands.*

is exemplified by the linear dependence on the spectral parameter, including the Manakov [10] and Davey–Stewartson [11] equations.

We propose here a new regular method to construct homoclinic solutions for nonlinear integrable wave equations with periodic boundaries which makes it possible to go beyond the Zakharov–Shabat spectral problem. It resembles closely the dressing method [12] developed for generating soliton solutions. In order to explain basic ideas, we first reproduce in Section 1 the known homoclinic solution of NLS by our method. Then in Section 2 we consider the modified NLS (MNLS) equation which is integrated by means of the Wadati–Konno–Ichikawa spectral problem [13] and has important applications in nonlinear optics [14] and plasma physics [15]. It should be stressed that MNLS provides the first example of treating the homoclinic orbits for the spectral problem with nonlinear dependence on the spectral parameter.

2 HOMOCLINIC SOLUTIONS FOR NLS

The NLS equation

$$iu_t = u_{xx} + 2(|u|^2 - \omega)u, \quad \omega \in Re \tag{1}$$

arises as a compatibility condition for the Lax pair equations $\psi_x = U\psi$ and $\psi_t = V\psi$ with the matrices U and V of the form

$$U = ik\sigma_3 + iQ, \qquad Q = \begin{pmatrix} 0 & u \\ \bar{u} & 0 \end{pmatrix},$$

$$V = i(2k^2 - Q^2 + \omega)\sigma_3 + 2ikQ + \sigma_3 Q_x,$$

k is a spectral parameter, ω is a real constant. We are interested in periodic solutions of NLS (1) with a spatial period L, $u(x + L, t) = u(x, t)$. Hence, the Floquet theory should be applied to the spectral equation $\psi_x = U\psi$. The fundamental matrix $M(x, k)$ is defined as a solution of the spectral equation with the boundary condition $M(0, k) = I$, I is the unit matrix. The Floquet discriminant is defined as $\Delta(k) = \text{tr}M(L, k)$ and the bounded eigenfunctions of the spectral problem correspond to $\Delta(k)$ satisfying the condition $-2 \leq \Delta(k) \leq 2$. The Floquet spectrum is characterized by the simple periodic points $\{k_j^s | \Delta(k_j^s) = \pm 2, (d\Delta/dk)_{k_j^s} \neq 0\}$ and the double points $\{k_j^d | \Delta(k_j^d) = \pm 2, (d\Delta/dk)_{k_j^s} = 0, (d^2\Delta/dk^2)_{k_j^d} \neq 0\}$. We will deal with the complex double points indicating linearized instability of solutions of a nonlinear wave equation because these points label the orbits homoclinic to unstable solutions.

We will consider the orbits homoclinic to the periodic plane wave solution u_0 of NLS (1) taken in the form

$$u_0 = c\exp[-2i(c^2 - \omega)t], \tag{2}$$

where c is a real amplitude. Simple calculation gives the fundamental matrix as

$$M(x, k) = \begin{pmatrix} \cos \mu x + i(k/\mu) \sin \mu x & i(c/\mu)e^{-2i(c^2-\omega)t} \sin \mu x \\ i(c/\mu)e^{2i(c^2-\omega)t} \sin \mu x & \cos \mu x - i(k/\mu) \sin \mu x \end{pmatrix},$$

$$\mu^2 = c^2 + k^2,$$

and hence $\Delta(k) = 2 \cos \mu L$. Thereby, the Floquet spectrum comprises the real axis of the k-plane (the main spectrum) and a part of the imaginary axis lying between the simple periodic points $\pm ic$. Besides, there exists an infinite sequence of real double points $k_n^d = [(n\pi/L)^2 - c^2]^{1/2}$, $c^2 \le (n\pi/L)^2$, n are integers, and a finite amount of complex double points k_j^d, j are integers, situated on the imaginary axis within the interval $(ic, -ic)$, $(j\pi/L)^2 < c^2$. In what follows we choose c and L in such a way that we will have a single pair of complex double points $k_1^d = \pm i[c^2 - (\pi/L)^2]^{1/2}$, that is a single unstable mode of the solution.

Diagonalizing the transfer matrix $M(L, k)$, $R^{-1}M(L, k)R = \text{diag}(e^{i\mu L}, e^{i\mu L})$, we define the Bloch solution $\tilde{\chi} = M(L, k)R$ of the spectral problem. Demanding the Bloch solution to satisfy both equations of the Lax pair, we obtain it explicitly as

$$\tilde{\chi} = e^{-i(c^2-\omega)t\sigma_3} \begin{pmatrix} 1 & -\dfrac{\mu - k}{c} \\ \dfrac{\mu - k}{c} & 1 \end{pmatrix} e^{i\mu(x+2kt)\sigma_3}.$$

In the following, it will be more convenient to work with a modified Bloch function $\chi = \tilde{\chi} \exp[-ikx\sigma_3 - i(2k^2 + \omega)t\sigma_3]$ which satisfies the equations

$$\chi_x = U\chi - ik\chi\sigma_3, \qquad \chi_t = V\chi - i(2k^2 + \omega)\chi\sigma_3 \qquad (3)$$

and admits the asymptotic expansion started with the unit matrix, $\chi = I + k^{-1}\chi^{(1)} + \mathcal{O}(k^{-2})$, while the potential Q is reconstructed via

$$Q = -[\sigma_3, \chi^{(1)}]. \qquad (4)$$

Suppose now that a solution of NLS homoclinic to the plane wave (2) can be obtained from Eq. (4) with the Bloch function χ being a result of dressing the Bloch function χ_0 which satisfies Eq. (3) with $u = u_0$:

$$\chi = D\chi_0. \qquad (5)$$

Here $D(k, x, t)$ is the dressing factor,

$$D = I - \frac{k_1 - \bar{k}_1}{k - \bar{k}_1} P, \qquad D^{-1} = I + \frac{k_1 - \bar{k}_1}{k - k_1} P, \qquad (6)$$

P is a projector of rank 1, i.e., $P^2 = P$ and $P = (|n\rangle\langle n|)/\langle n|n\rangle$, $\langle n| = |n\rangle\dagger$, $|n\rangle = (n_1, n_2)^t$ is a two-component vector. As regards the choice of the pole k_1 in (6), it is the point where we encounter a crucial difference from the standard applications of the dressing method. The positions of the poles in the dressing factors are usually taken quite arbitrary, without a reference to the seed solution u_0. Contrary, it is the seed solution u_0 which determines these poles in our case. Namely, we take the complex double points as the poles of the dressing factors. Therefore, $k_1 = k_1^d$.

Expanding Eq. (5) in the asymptotic series w.r.t. k^{-1} we obtain new solution of NLS in terms of the old one and the dressing factor:

$$Q = Q_0 - [\sigma_3, D^{(1)}], \tag{7}$$

where $D = I + k^{-1}D^{(1)} + \mathcal{O}(k^{-2})$. Hence, we need know the vector $|n\rangle$ to obtain new solution Q.

Differentiating (5) w.r.t. x gives

$$U(k) = -D(\partial_x - U_0(k))D^{-1}, \tag{8}$$

where $U_0 = U(u_0)$. Evidently, L.H.S. of (8) is regular in points k_1 and $\bar{k}_1$, while R.H.S. has simple poles in these points because of the dressing factors. From the condition of vanishing residue in the point k_1 we obtain $|n\rangle_x = U_0(k_1)|n\rangle$, and, similarly, $|n\rangle_t = V_0(k_1)|n\rangle$. These equations are easily integrated and we obtain

$$|n\rangle = e^{-i(c^2-\omega)t\sigma_3} \begin{pmatrix} A\exp(i\mu_1 x - 2k_0\mu_1 t) - \frac{\mu_1-ik_0}{c}\exp(-i\mu_1 x + 2k_0\mu_1 t) \\ A\frac{\mu_1-ik_0}{c}\exp(i\mu_1 x - 2k_0\mu_1 t) + \exp(-i\mu_1 x + 2k_0\mu_1 t) \end{pmatrix}. \tag{9}$$

Here $A = \text{const}$, $\mu_1 = \mu(k_1)$, $k_1 = ik_0$. Evidently, $D^{(1)} = -(k_1 - \bar{k}_1)P$ and hence $u = u_0 + 2(k_1 - \bar{k}_1)P_{12}$ with $P_{12} = (n_1\bar{n}_2)/(|n_1|^2 + |n_2|^2)$. Inserting here the vector $|n\rangle$ and introducing notations

$$A = \exp(\rho + i\beta), \qquad \tau = \sigma t - \rho, \qquad \phi = \beta - \pi/2,$$
$$\sigma = 4k_0\mu_1, \qquad \mu_1 + i\lambda_0 = ce^{ip},$$

we obtain the homoclinic solution in the form

$$u_h = \frac{\cos 2p - \sin p \, \text{sech}\tau \cos(2\mu_1 x + \phi) - i\sin 2p \, \tanh\tau}{1 + \sin p \, \text{sech}\tau \cos(2\mu_1 x + \phi)} ce^{-2i(c^2-\omega)t}, \tag{10}$$

which coincides with the solution obtained in [5] by means of the Bäcklund transformation.

It is easy to see that this solution is indeed homoclinic to the plane wave:

$$t \to \pm\infty: \quad u_h \to \exp(\pm 2ip)ce^{-2i(c^2-\omega)t}.$$

In the case of N unstable modes the above procedure can be iterated. However, more efficient way to deal with multiple double points will be described in the next section.

3 HOMOCLINIC SOLUTIONS FOR MNLS

3.1 Floquet Spectrum and Bloch Solutions

The MNLS equation

$$iu_t = u_{xx} + i\alpha(|u|^2 u)_x + 2(|u|^2 - \omega)u, \tag{11}$$

α and ω are real constants, admits the Lax representation with the matrices U and V of the form

$$U = i\Lambda\sigma_3 + ikQ, \qquad \Lambda(k) = \frac{1}{\alpha}(1 - k^2), \qquad Q = \begin{pmatrix} 0 & u \\ \bar{u} & 0 \end{pmatrix}, \tag{12}$$

$$V = i\Omega\sigma_3 + 2ik\Lambda Q - ik^2 Q^2\sigma_3 + k\sigma_3 Q_x + i\alpha k Q^3, \qquad \Omega(k) = 2\Lambda^2 + \omega.$$

As for the NLS equation, we take the plane wave solution of MNLS (11)

$$u_0 = c\exp[-2i(c^2 - \omega)t] \tag{13}$$

as a periodic solution with a spatial period L. The fundamental matrix $M(x, k)$ is obtained in the form

$$M(x, k) = \begin{pmatrix} \cos\mu x + i\dfrac{\Lambda}{\mu}\sin\mu x & i\dfrac{ck}{\mu}e^{-2i(c^2-\omega)t}\sin\mu x \\[2ex] i\dfrac{ck}{\mu}e^{2i(c^2-\omega)t}\sin\mu x & \cos\mu x - i\dfrac{\Lambda}{\mu}\sin\mu x \end{pmatrix},$$

where $\mu = (\Lambda^2 + c^2 k^2)^{1/2}$. Hence, $\Delta(k) = 2\cos\mu L$ and four complex double points $k_j = \pm(1 - (1/2)\alpha^2 c^2 \pm i\alpha c l_j)^{1/2}$, $l_j = [1 - (1/4)\alpha^2 c^2 - (j\pi/cL)^2]^{1/2}$, lying in four quadrants of the k-plane, correspond to each unstable mode. We choose c and L in such a way that only the single unstable mode exists, i.e., $l_1^2 > 0$ and $l_j^2 < 0$ for $j > 1$. The linearized stability analysis confirms that the above four complex double points $k_1 = (1 - (1/2)\alpha^2 c^2 - i\alpha c l_1)^{1/2}$, $k_2 = -k_1$, $k_3 = \bar{k}_1$, and $k_4 = -\bar{k}_1$ are associated with the exponential instability.

The Bloch function which solves both Lax equations takes a surprisingly simple form:

$$\tilde{\chi}_0(k, x, t) = \exp[-i(c^2 - \omega)t\sigma_3] \begin{pmatrix} 1 & -\dfrac{\mu - \Lambda}{ck} \\[2ex] \dfrac{\mu - \Lambda}{ck} & 1 \end{pmatrix}$$

$$\times \exp\left[i\mu\left(x + (2\Lambda + \alpha c^2)t\right)\sigma_3\right].$$

Now we define a modified Bloch function $\chi = \tilde{\chi} \exp(-i\Lambda x - i\Omega t)\sigma_3$ which satisfies the linear equations $\chi_x = U\chi - i\Lambda\chi\sigma_3$, $\chi_t = V\chi - i\Omega\chi\sigma_3$, and admits the asymptotic expansion $\chi = \chi^{(0)} + k^{-1}\chi^{(1)} + \mathcal{O}(k^{-2})$. Therefore, we obtain for the plane wave potential

$$\chi_0(k, x, t) = \exp[-i(c^2 - \omega)t\sigma_3] \begin{pmatrix} 1 & -\dfrac{\mu - \Lambda}{ck} \\ \dfrac{\mu - \Lambda}{ck} & 1 \end{pmatrix}$$

$$\times \exp[i(\mu - \Lambda)(x + 2\Lambda t)\sigma_3] \exp\left[i(\alpha\mu c^2 - \omega)t\sigma_3\right]$$

and the leading term of the asymptotic series $\chi_0^{(0)} = e^{-(i/2)\alpha c^2(x+(3/2)\alpha c^2 t)\sigma_3}$. It should be stressed that this leading term is not a unit matrix, as for the NLS equation. The reason is that MNLS does not a canonical equation for the Wadati–Konno–Ichikawa spectral problem and hence does not admit the standard normalization of the associated Riemann–Hilbert problem [16]. Therefore, we perform now one more transformation of the Bloch solution, $\phi = \chi^{(0)-1}\chi$, to have the unit matrix in the asymptotic expansion: $\phi = I + k^{-1}\phi^{(1)} + \mathcal{O}(k^{-2})$. ϕ satisfies the linear equations

$$\phi_x = U'\phi - i\Lambda\phi\sigma_3, \qquad \phi_t = V'\phi - i\Omega\phi\sigma_3, \qquad (14)$$

where

$$U' = i\Lambda\sigma_3 + ikQ' + \frac{i}{2}\alpha\sigma_3 Q'^2, \qquad (15)$$

$$V' = i\Omega\sigma_3 + 2ik\Lambda Q' - ik^2 Q'^2\sigma_3 + k\sigma_3 Q'_x - \frac{\alpha}{2}[Q', Q'_x] - \frac{i}{4}\alpha^2\sigma_3 Q'^4.$$

Here new potential Q' relates with the initial one Q as

$$Q' = \chi^{(0)-1}Q\chi^{(0)}. \qquad (16)$$

Evidently, $Q'^2 = Q^2$. Besides, $\phi^{(1)}$ is expressed via the potential as follows:

$$\phi^{(1)} = \frac{\alpha}{2}\sigma_3 Q'. \qquad (17)$$

3.2 Dressing

Suppose that a new solution of the linear equations (14) follows from the known one ϕ_0 by dressing $\phi = D\phi_0$. Here $D(k, x, t)$ is the dressing factor:

$$D(k) = \left(I - \frac{k_2 - \bar{k}_2}{k - \bar{k}_2}P_2\right)\left(I - \frac{k_1 - \bar{k}_1}{k - \bar{k}_1}P_1\right), \qquad (18)$$

$$D^{-1}(k) = \left(I + \frac{k_1 - \bar{k}_1}{k - k_1}P_1\right)\left(I + \frac{k_2 - \bar{k}_2}{k - k_2}P_2\right).$$

$P_j(j = 1, 2)$ is a projector of rank 1, i.e., $P_j^2 = P_j$, which is represented by means of a 2-vector $|p_j\rangle = (p_j^{(1)}, p_j^{(2)})^{\perp}$ as

$$P_j = \frac{|p_j\rangle \langle p_j|}{\langle p_j | p_j \rangle}, \quad \langle p_j | p_j \rangle = |p_j^{(1)}|^2 + |p_j^{(2)}|^2, \quad \langle p_j| = |p_j\rangle|^{\dagger}.$$

As before, we take the complex double points as the poles of the dressing factors. Therefore, four multipliers in the dressing factors (18) correspond to four complex double points ($k_j = k_j^d$). Expanding the relation $\phi = D\phi_0$ in the asymptotic series gives in accordance with (17) and (16) $Q = \chi^{(0)}(\chi_0^{(0)-1} Q_0 \chi_0^{(0)} + (2/\alpha)\sigma_3 D^{(1)})\chi^{(0)-1}$. Because $\phi(k = 0)$ and $\chi_0^{(0)-1}$ satisfy the same equation, we get $\chi^{(0)} = \chi_0^{(0)} D_0^{-1}$, $D_0 = D(k = 0)$ which results in the connection between new and old solutions of MNLS:

$$u = \frac{(D_0)_{22}}{(D_0)_{11}} \left[u_0 + \frac{2}{\alpha} \exp\left(-i\alpha c^2 x - \frac{3}{2}\alpha^2 c^4 t \right) D_{12}^{(1)} \right]. \qquad (19)$$

A successive application of both elementary multipliers in the dressing factor (18) to the solution ϕ_0 would not be an optimal strategy. Instead we decompose the dressing factor into simple fractions which results into

$$D(k) = I - \sum_{j,l=1}^{2} \frac{1}{k - \bar{k}_l} |j\rangle \left(S^{-1}\right)_{jl} \langle l|, \qquad S_{lj} = \frac{\langle l|j\rangle}{k_j - \bar{k}_l}, \qquad (20)$$

where for simplicity we right $|j\rangle$ instead of $|p_j\rangle$. Differentiating the relation $\phi = D\phi_0$ w.r.t. x gives $U'(k) = -D(\partial_x - U_0'(k))D^{-1}$. From the condition of vanishing residues in the points k_1 and k_2 we obtain the equations

$$|j\rangle_x = U_0'(k_j)|j\rangle, \qquad |j\rangle_t = V_0'(k_j)|j\rangle, \qquad j = 1, 2. \qquad (21)$$

Note that the vector $|2\rangle$ is expressed in terms of $|1\rangle$ because of the symmetry [16] $U_0'(k_2) = U_0'(-k_1) = \sigma_3 U_0'(k_1)\sigma_3 : |2\rangle = \sigma_3|1\rangle$. Hence,

$$S_{11} = \frac{\langle 1|1\rangle}{k_1 - \bar{k}_1} = -S_{22}, \qquad S_{21} = \frac{\langle 2|1\rangle}{k_1 - \bar{k}_2} = \frac{\langle 1|\sigma_3|1\rangle}{k_1 + \bar{k}_1} = -S_{12}.$$

As a result, we obtain D_0 and $D^{(1)}$ entering the formula (19) in the form

$$D_{01} = 1 + \frac{2}{\bar{k}_1} \frac{(|1\rangle\langle 1|)_{11}}{S_{11} - S_{21}}, \qquad D_{02} = 1 + \frac{2}{\bar{k}_1} \frac{(|1\rangle\langle 1|)_{22}}{S_{11} + S_{21}},$$

$$D^{(1)} = -2 \begin{pmatrix} 0 & (|1\rangle\langle 1|)_{12}(S_{11} - S_{21})^{-1} \\ (|1\rangle\langle 1|)_{21}(S_{11} + S_{21})^{-1} & 0 \end{pmatrix}.$$

Because S_{ij} are expressed in terms of the vector $|1\rangle$ (see (20)), we have to obtain it explicitly. In the next section we will account for the explicit

(x, t)-dependence of the vector $|1\rangle$ and justify the name "homoclinic" for the solution (19).

3.3 Homoclinic Solution

Integrating linear equations (21), we obtain

$$
\begin{aligned}
|1\rangle = \exp\left(\frac{i}{2}\alpha c^2 x \sigma_3\right) \exp\left[-i\left(c^2 - \omega - \frac{3}{4}\alpha^2 c^4\right) t \sigma_3\right] \exp\left[\frac{1}{2}(\gamma + i\beta)\right] \\
\times \left(\begin{array}{c} e^{i\xi-\tau} + e^{-i\xi+\tau} \\ [e^{-(\tau+\Phi)}e^{i(\xi-\lambda_-)} - e^{\tau+\Phi}e^{-i(\xi-\lambda_+)}]e^{i(\delta/2)} \end{array}\right).
\end{aligned}
$$

Here γ and β are real integration constants, $\mu_1 = \mu(k_1) = \pi/L$, $\xi = \mu_1(x + 2\alpha c^2 t) + (1/2)\beta$, $\tau = 2c\mu_1 l_1 t - (1/2)\gamma$,

$$
\Phi = \frac{1}{4}\log\frac{1+\alpha\mu_1}{1-\alpha\mu_1}, \qquad \tan\lambda_{\pm} = \frac{l_1}{\frac{\mu_1}{c} + \frac{1}{2}\alpha c}, \qquad \tan\delta = \frac{\alpha c l_1}{1 - \frac{1}{2}\alpha^2 c^2}.
$$

Hence, the matrix elements are written as

$$
S_{11} = 2e^{\gamma}(A + B)(k_1 - \bar{k}_1)^{-1}, \qquad S_{21} = 2e^{\gamma}(A - B)(k_1 + \bar{k}_1)^{-1},
$$

where

$$
\begin{aligned}
A(\xi, \tau) &= \cosh 2\tau + \cos 2\xi, \\
B(\xi, \tau) &= \cosh 2(\tau + \Phi) - \cos(2\xi - \lambda_+ - \lambda_-),
\end{aligned}
$$

and

$$
D_{01} = \frac{k_1}{\bar{k}_1}\frac{k_1 A + \bar{k}_1 B}{\bar{k}_1 A + k_1 B} = e^{i(\Theta-\delta)}, \qquad D_{02} = \frac{k_1}{\bar{k}_1}\frac{\bar{k}_1 A + k_1 B}{k_1 A + \bar{k}_1 B} = e^{-i(\Theta+\delta)},
$$

$$
\tan\frac{\Theta}{2} = \frac{1}{i}\frac{k_1 - \bar{k}_1}{k_1 + \bar{k}_1}\frac{A - B}{A + B},
$$

$$
\exp(-i\alpha c^2 x)\exp\left(-\frac{3}{2}i\alpha^2 c^4 t\right) D_{12}^{(1)}
$$

$$
= -\frac{i\alpha l_1 \exp(-i\delta/2)}{\bar{k}_1 A + k_1 B}\left[\left(e^{2\tau} + e^{2i\xi}\right)e^{\Phi-i\lambda_+} - \left(e^{-2\tau} + e^{-2i\xi}\right)e^{-\Phi+i\lambda_-}\right] u_0.
$$

Substituting the above formulas into (19), we obtain explicitly the homoclinic solution of MNLS equation:

$$
u = \left\{1 - \frac{2il_1 e^{-i(\delta/2)}}{\bar{k}_1 A + k_1 B}\left[\left(e^{2\tau} + e^{2i\xi}\right)e^{\Phi-i\lambda_+} - \left(e^{-2\tau} + e^{-2i\xi}\right)e^{-\Phi+i\lambda_-}\right]\right\} u_0 e^{-2i\Theta}.
$$

$$
\tag{22}
$$

The solution (22) is indeed homoclinic to the plane wave because

$$\tau \to \pm\infty: \quad u \to u_0 \exp[-2i(\Theta_\pm + \Phi_\pm)],$$

$$\Theta_\pm = \lim_{\tau \to \pm\infty} \Theta = \pm \arctan \frac{\alpha^2 c l_1 \mu_1}{2\left(1 - \frac{1}{2}\alpha^2 c^2\right)\left(1 - \frac{1}{4}\alpha^2 c^2\right) + (\alpha c l_1)^2},$$

$$\Phi_\pm = \pm\frac{1}{2} \arctan \frac{2 c l_1 \mu_1}{\mu_1^2 - c^2 l_1^2}.$$

In the $\alpha \to 0$ limit consisting in representing the spectral parameter k as $k = 1 - \frac{1}{2}\alpha \lambda_{NLS} + O(\alpha^2)$, where λ_{NLS} is that for the NLS equation, the solution (22) reproduces the NLS homoclinic orbit (10).

4 CONCLUSION

We have elaborated a new method to derive homoclinic solution in time for a soliton equation with periodic boundaries. Our method has much in common with the well-known dressing method for generating soliton solutions, with the important difference that the positions of poles in the dressing factor are not arbitrary but coincide with the positions of the complex double points of the Floquet spectrum for the seed solution of the nonlinear equation. We have demonstrated the method on the example of the modified NLS equation which is integrable by means of the Wadati–Konno–Ichikawa spectral problem with the quadratic dependence on the spectral parameter. We especially stress the distinctive role of the Bloch solution to the Lax representation. For the case of N unstable modes in the seed solution, the only computational difficulty consists in inverting some $2N \times 2N$ matrix. Though we restrict our consideration to soliton equations solvable via the quadratic bundle, it is evident that because of a wide applicability of the dressing method, our approach is feasible to more general class of nonlinear equations.

ACKNOWLEDGMENTS

It is pleasure to thank the organizers of the workshop "Bilinear Integrable Systems" for their kind invitation and warm hospitality. This work was partly supported by the London Mathematical Society under Grant No. 5707 and EPSRC Grant No. GR/R02702/01. E.D. thanks the Department of Mathematical Sciences, Loughborough University where part of this work was done.

REFERENCES

1. McLaughlin, D. W., Bishop, A. R., and Overman, E. A. (1988) Coherence and chaos in the driven damped sine-Gordon equation: measurement of the soliton spectrum, *Physica. D.* **19**, pp. 1–41.
2. McLaughlin, D. W., Bishop, A. R., Forest, M. G., and Overman, E. A. (1988) A quasiperiodic route to chaos in a near-integrable PDE, *Physica. D.* **23**, pp. 293–328.
3. Ablowitz, M. J. and Herbst, B. M. (1990) On homoclinic structure and numerically induced chaos for the nonlinear Schrödinger equation, *SIAM J. Appl. Math.* **50**, pp. 339–351.
4. Ablowitz, M. J., Schober, C. M., and Herbst, B. M. (1993) Numerical chaos, roundoff errors and homoclinic manifolds, *Phys. Rev. Lett.* **71**, pp. 2683–2686.
5. Li, Y. and McLaughlin, D. W. (1994) Morse and Melnikov functions for NLS PDE's, *Comm. Math. Phys.* **162**, pp. 175–214.
6. Haller, G., Menon, G., and Rothos, V. M. (1999) Shilnikov manifolds in coupled nonlinear Schrödinger equations, *Phys. Lett. A.* **263**, pp. 175–185.
7. McLaughlin, D. W. and Overman, E. A. II (1995) Whiskered tori for integrable PDE's: Chaotic behaviour in near integrable PDE's, *Surveys in Appl. Math.* **1**, pp. 83–200.
8. Ablowitz, M. J., Herbst, B. M., and Schober, C. M. (2001) Discretizations, integrable systems and computation, *J. Phys. A* **34**, pp. 10671–10693.
9. Hirota, R. (1980) Direct methods in soliton theory, in: Springer Topics in Current Physics, R. K. Bullough and P. J. Caudrey (eds.) *Solitons*, Vol. 17, Springer-Verlag, Heidelberg, pp. 157–174.
10. Wright, O. C. and Forest, M. G. (2000) On the Bäcklund–Gauge transformation and homoclinic orbits of a coupled nonlinear Schrödinger system, *Physica. D.* **141**, pp. 104–116.
11. Li, Y. (2000) Bäcklund–Darboux transformations and Melnikov analysis for Davey–Stewartson II equations, *J. Nonlin. Sci.* **10**, pp. 103–131.
12. Novikov, S. P., Manakov, S. V., Pitaevski, L. P., and Zakharov, V. E. (1984) *Theory of Solitons, the Inverse Scattering Method*, Consultant Bureau, New York.
13. Wadati, M., Konno, K., and Ichikawa, Y. H. (1979) A generalization of inverse scattering method, *J. Phys. Soc. Jpn* **46**, pp. 1965–1966.
14. Gerdjikov, V. S., Doktorov, E. V., and Yang, J. (2001) Adiabatic interaction of N ultrashort solitons: universality of the complex Toda chain model, *Phys. Rev. E.* **64**, 056617, 15 p.
15. Mio, K., Ogino, T., Minami, K., and Takeda, S. (1976) Modified nonlinear Schrödinger equation for Alfvén waves propagating along the magnetic field in cold plasma, *J. Phys. Soc. Jpn.* **41**, pp. 265–271.
16. Shchesnovich, V. S. and Doktorov, E. V. (1999) Perturbation theory for the modified nonlinear Schrödinger solitons, *Physica. D.* **115**, pp. 115–129.

RIEMANN-HILBERT PROBLEM AND ALGEBRAIC CURVES

V. Enolskii
Institute of Magnetism NASU, Vernadsky str. 36, O3142, Kiev, Ukraine

T. Grava
International School of Advanced Studies (SISSA), Via Beirut n. 2–4, 34014 Trieste, Italy

There are many problems in pure and applied mathematics that can be solved in terms of a Riemann-Hilbert (R-H problem). The list includes the remarkable class of nonlinear integrable equations, namely nonlinear equations that can be written as the compatibility conditions of linear equations. This class contains a large variety of equations: ODE's, PDE's difference equations, etc. Furthermore, the R-H problem formulation provides a powerful technique for obtaining asymptotic results for solutions of ODE's and PDE's of this class [1].

A remarkable application of this asymptotic technique is the derivation of asymptotic for orthogonal polynomials which is related to the universality conjecture in one-matrix models (see [2] and [3] and reference therein). In this case the associated rank two R-H problems are formulated on hyperelliptic Riemann surfaces. The integrable structure of multiorthogonal polynomials and in particular biorthogonal polynomials was pointed out in [4]. Asymptotic results for multiorthogonal polynomials necessarily involves nonhyperelliptic curves and higher rank R-H problems, which now attract much attention, because of their application to multimatrix models and approximation theory. Regarding two-matrix models, some asymptotic results have been obtained for the genus zero case (namely the analog of the one-cut case in one-matrix models) and the corresponding genus zero nonhyperelliptic Riemann surface has been derived in terms of the external potential [5]. Regarding approximation theory for multiple orthogonal polynomials some results have been obtained in [6], [7] and more recently, using Riemann-Hilbert techniques in [8].

The principal aim of our investigation is to give effective and explicit solutions to a class of higher rank R-H problems associated with nonhyperelliptic curves. This article is a review of the paper [9].

The Riemann-Hilbert problem in its classical formulation consists of deriving a linear differential equation of Fuchsian type with a given set $\mathcal{D}$ of singular

L. Faddeev et al. (eds.),
Bilinear Integrable Systems: From Classical to Quantum, Continuous to Discrete, 65–76.
© 2006 *Springer. Printed in the Netherlands.*

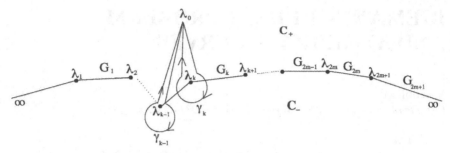

Figure 1. The contour $\mathcal{L}$

points and a given monodromy representation

$$\mathcal{M} : \pi_1(\mathbf{CP}^1 \backslash \mathcal{D}, \lambda_0) \to GL(N, \mathbf{C}), \quad N \geq 2. \tag{1}$$

An element γ of the group $\pi_1(\mathbf{CP}^1 \backslash \mathcal{D}, \lambda_0)$ is a loop contained in $\mathbf{CP}^1 \backslash \mathcal{D}$ with initial and end point λ_0, $\lambda_0 \notin D$ (see Figure 1). Not all the representation (1) can be realised as the monodromy representation of a Fuchsian system, [10, 11]. In dimension $N = 2$ the R-H problem is always solvable [12] for any number of singular points. For $N \geq 3$, every irreducible representation (1) can be realised as the monodromy representation of some Fuchsian system [13, 14]. In general, among the solvable cases, the solution of the matrix R-H problem cannot be computed analytically in terms of known special functions [15, 16]. Nevertheless, there are special cases when the R-H problem can be solved explicitly in terms of θ-functions [17–19]. Our article discusses one of these cases.

The method of solution proposed by Plemelj [20] consists of reducing the R-H problem to a homogeneous boundary value problem in the complex plane for a $N \times N$ matrix function $Y(\lambda)$. For this reason, boundary value problems in the complex plane are often referred to as R-H problems in the modern literature. In the case under study, the boundary can be chosen in the form of a polygon line $\mathcal{L}$. Assuming that the set of points $\mathcal{D} = \{\lambda_1, \ldots, \lambda_{2m+1}, \lambda_{2m+2} = \infty\}$ satisfy the relation

$$\mathrm{Re}\,\lambda_1 < \mathrm{Re}\,\lambda_3 < \mathrm{Re}\,\lambda_3 < \cdots < \mathrm{Re}\,\lambda_m < \mathrm{Re}\,\lambda_{2m+1}.$$

the oriented polygon line $\mathcal{L}$ is given by the union of segments

$$\mathcal{L} = [\infty, \lambda_1] \cup [\lambda_1, \lambda_2] \cup [\lambda_2, \lambda_3] \cup \cdots \cup [\lambda_{2m}, \lambda_{2m+1}] \cup [\lambda_{2m+1}, \infty].$$

The line $\mathcal{L}$ divides the complex plane into two domains, $\mathbf{C}_-$ and $\mathbf{C}_+$ (see Figure 1). Let $\gamma_1, \gamma_2, \ldots, \gamma_{2m+2}$ denote the set of generators of the fundamental group $\pi_1(\mathbf{CP}^1 \backslash \mathcal{D}, \lambda_0)$, i.e., the homotopy class γ_k corresponds to a small clock-wise loop around the point λ_k (see Figure 1). Then the matrices $\mathcal{M}(\gamma_k) = M_k \in SL(N, \mathbf{C}), k = 1, \ldots, 2m + 2$, form a set of generators of the monodromy group. Since the homotopy relation

$$\gamma_1 \circ \gamma_2 \circ \ldots \circ \gamma_{2m+2} \simeq \lambda_0,$$

the generators M_k satisfy the cyclic relation

$$M_\infty M_{2m+1} \ldots M_1 = 1_N,$$

where $M_{2m+2} = M_\infty$. Let us construct the matrices G_k defined by

$$G_k = M_k M_{k-1} \ldots M_1, \quad k = 1, \ldots, 2m + 2. \tag{2}$$

The homogeneous boundary value problem is formulated as follows [20] find the $N \times N$ matrix function $Y(\lambda)$ which satisfies

(i) $Y(\lambda)$ is analytic in $\mathbf{CP}^1 \backslash \mathcal{L}$;
(ii) The L_2-limits $Y_\pm(\lambda)$ as $\lambda \to \mathcal{L}_\pm$ satisfy the jump conditions

$$Y_-(\lambda) = Y_+(\lambda), \quad \text{for} \quad \lambda \in (\infty, \lambda_1)$$

and

$$Y_-(\lambda) = Y_+(\lambda)G_k \quad \text{for} \quad \lambda \in (\lambda_k, \lambda_{k+1}), \quad k = 1, \ldots, 2m + 1;$$

(iii) for $0 \le \epsilon < 1$;

$$Y(\lambda)\left(\frac{1}{\lambda}\right)^\epsilon \to 0 \quad \text{as } \lambda \to \infty,$$

$$Y_\pm(\lambda)(\lambda - \lambda_k)^\epsilon \to 0 \quad \text{as } \lambda \to \lambda_k, \quad k = 1, \ldots, 2m + 2,$$

over $\mathbf{C}_+$ or $\mathbf{C}_-$ respectively;
(iv) $Y(\lambda_0) = 1_N, \quad \lambda_0 \in \mathbf{C}_+ \backslash \mathcal{D}$.

There is always a solution of (i)–(iv) such that $\det Y(\lambda) \neq 0$ for $\lambda \neq \mathcal{D}$. The analytic continuation of the solution $Y(\lambda)$ along a small loop γ_k around λ_k is determined by the matrix $M_k = G_k G_{k-1}^{-1}$, namely

$$Y(\gamma_k(\lambda)) = Y(\lambda)M_k, \quad \lambda \in \mathbf{C} + \backslash \mathcal{D}, \quad k = 1, \ldots, 2m + 2.$$

The solution $Y(\lambda)$ of the R-H problem (i)–(iv) satisfies a Fuchsian equation

$$\frac{dY(\lambda)}{d\lambda} = \sum_{k=1}^{2m+1} \frac{A_k}{\lambda - \lambda_k} Y(\lambda), \tag{3}$$

if one of the monodromy matrices is diagonalisable [10] and [11]. Without this condition, Plemelj original argument does not go through.

The method of [20] was used by Deift, Its, Kapaev, and Zhou [17] to solve the 2×2 matrix R-H problem when all the matrices G_{2k} are diagonal and all the matrices $G_{2k-1}, k = 1, \ldots, m + 1$, are off-diagonal. The idea of the construction in [17] is to consider a hyperelliptic covering $\mathcal{C}$ over $\mathbf{CP}^1$ which is ramified in D and uses the natural monodromy of the hyperelliptic curve. The application of methods of finite-gap integration [21, 22] allows to obtain a θ-functional solution for the problem depending on $2m$ parameters. Similar results were obtained by Kitaev and Korotkin [18] by another method.

The extension of the 2×2 matrix R-H problem to higher dimensional matrices leads naturally to non-hyperelliptic curves. This fact was pointed out by Zverovich [23], who considered the $N \times N$ problem (i)–(iv) when all the matrices G_{2k} are diagonal, and the nonzero entries of the matrices $G_{2k-1}, k = 1, \ldots, m + 1$ are

$$(G_{2k-1})_{i,i-1} \neq 0, \quad i = 2, \ldots, N,$$
$$(G_{2k-1})_{1,N} \neq 0.$$

The solubility of the corresponding $N \times N$ matrix R-H problem is proved by lifting it to a scalar problem on the N-sheeted Riemann surface $\mathcal{C}_{N,m}$

$$\mathcal{C}_{N,m} := \left\{ (\lambda, y), \quad y^N = q^{N-1}(\lambda) p(\lambda) \right\}, \tag{4}$$

$$q(\lambda) = \prod_{j=1}^{m} (\lambda - \lambda_{2k}), \quad p(\lambda) = \prod_{j=0}^{m} (\lambda - \lambda_{2k+1}).$$

Such surface can be identified with N copies (sheets) of the complex λ-plane cut along the segments $\mathcal{L}_0 = \cup_{k=1}^{m+1} [\lambda_{2k-2}, \lambda_{2k}]$ and glued together according to the permutation rule $\begin{pmatrix} 1 & 2 & \ldots & N-1 & N \\ 2 & 3 & \ldots & N & 1 \end{pmatrix}$, that is the first sheet is pasted to the second, the second to the third and so on. On this surface, the projection map (which we still denote by λ) onto $\mathbf{CP}^1$ is a function of degree N and we think of y as a function of the point $P \in \mathcal{C}_{N,m}$ in the sense that y sends $P \in \mathcal{C}_{N,m}$ into $y(\lambda(P))$. Let $P_0^{(s)}, s = 1, \ldots, N$, be the points on the s-th sheet of $\mathcal{C}_{N,m}$ having the same λ_0 projection onto $\mathbf{CP}^1$. We define $y_0 := y(\lambda(P_0^{(1)}))$. Then $y(\lambda(P^{(s)})) = e^{2\pi i \frac{s-1}{N}} y_0$.

The algebraic-geometrical approach to the R-H-problem was developed further by Korotkin [19]. He showed that for quasi-permutation monodromy matrices (in which each row and each column have only one non zero element), the

R-H problem can be solved in terms of the Szegö kernel of a Riemann surface. The procedure to obtain the Riemann surface from the monodromy matrices relies on Riemann existence theorem. First, all the nonzero entries of the monodromy matrices M_k are set equal to one, so that all the monodromies become permutation matrices. Then it is proved that there is a one-to-one correspondence between permutation representations of $\pi_1(\mathbf{CP}^1 \backslash \mathcal{D}, \lambda_0)$ and N-sheeted compact Riemann surfaces realised as a ramified covering of $\mathbf{CP}^1$ with projection of branch points over $\mathbf{CP}^1$ equal to $\mathcal{D}$ (see e.g. [24]. In our case the permutation representation obtained by setting equal to one all the nonzero entries of the matrix G_k, $k = 1, \ldots, 2m + 2$, is

$$\gamma_{2k-1} \rightarrow \begin{pmatrix} 1 & 2 & \ldots & N-1 & N \\ 2 & 3 & \ldots & N & 1 \end{pmatrix} \quad k = 1, \ldots, m+1, \tag{5}$$

$$\gamma_{2k} \rightarrow \begin{pmatrix} 1 & 2 & \ldots & N-1 & N \\ N & 1 & \ldots & N-2 & N-1 \end{pmatrix} \quad k = 1, \ldots, m+1. \tag{6}$$

In this case it is straightforward to verify that the above permutation representation corresponds to the curves $\mathcal{C}_{N,m}$. However, the Riemann existence theorem is just an *existence* theorem, that is, it does not produce explicitly algebraic equations for the coverings. Tools for the computation of families of coverings from the permutation representation, are given in [25].

Our derivation of the solution of the R-H problem (i)–(iv), incorporates both the method of [17], implemented for hyperelliptic curves and ideas of [19]. First we solve the so-called canonical R-H problem, namely the problem (i)–(iv) when all the matrices G_{2k} are set equal to the identity and all the matrices G_{2k+1} are set equal to the quasi-permutation $\mathcal{P}_N$, where

$$\mathcal{P}_N = \begin{pmatrix} 0 & 0 & \ldots & 0 & 0 & (-1)^{N-1} \\ 1 & 0 & 0 & \ldots & 0 & 0 \\ 0 & 1 & 0 & \ldots & 0 & 0 \\ \ldots & & \ldots & \ldots & 0 & 0 \\ 0 & 0 & \ldots & 1 & 0 & 0 \\ 0 & 0 & \ldots & 0 & 1 & 0 \end{pmatrix}. \tag{7}$$

More precisely the canonical R-H problem consists of finding a matrix valued function $X(\lambda)$ analytic in the complex plane off the segment $\mathcal{L}_0 = \cup_{k=1}^{m+1} [\lambda_{2k-1}, \lambda_{2k}]$ such that

$$X_-(\lambda) = X_+(\lambda)\mathcal{P}_N, \quad \lambda \in \mathcal{L}_0, \tag{8}$$

$$X(\lambda_0) = 1_N, \quad \lambda_0 \in \mathbf{C}_+. \tag{9}$$

The entries of the matrix $X(\lambda)$ can be expressed in terms of the Szegö kernel with zero characteristic, $S[0](P, Q)$, defined on $\mathcal{C}_{N,m}$. We show that

$$S[0](P, Q) = \frac{1}{N} \frac{\sqrt{d\lambda(Q)d\lambda(P)}}{\lambda(P) - \lambda(Q)} \sum_{k=0}^{N-1} \left(\frac{q(\lambda(P))p(\lambda(Q))}{p(\lambda(Q))q(\lambda(P))} \right)^{-\frac{k}{N} + \frac{N-1}{2N}}, \qquad (10)$$

where $P, Q \in \mathcal{C}_{N,m}$. Then the $N \times N$ matrix $X(\lambda)$ with entries

$$X_{rs}(\lambda) = S[0](P^{(s)}, P_0^{(r)}) \frac{\lambda(P) - \lambda(Q)}{\sqrt{d\lambda(Q)d\lambda(P)}} \qquad (11)$$

$$= \frac{1}{N} \sum_{k=0}^{N-1} \left(e^{2\pi i \frac{(s-r)}{N}} \sqrt[N]{\frac{q(\lambda)}{p(\lambda)} \frac{p(\lambda_0)}{q(\lambda_0)}} \right)^{-k + \frac{N-1}{2}}, \quad \lambda_0 \notin \mathcal{D},$$

where $P^{(s)} = (\lambda, e^{2\pi i \frac{s-1}{N}} y)$ and $P_0^{(r)} = (\lambda_0, e^{2\pi i \frac{r-1}{N}} y_0)$, $r, s = 1, \ldots, N$, denote the points on the s-th and r-th sheet of $\mathcal{C}_{N,m}$ respectively, solves the R-H problem (8). When $N = 2$ and $\sqrt[4]{\frac{q(\lambda_0)}{p(\lambda_0)}} = 1$, such formula coincides with the canonical solution obtained in [17].

The quasi permutation matrices G_k, $k = 1, \ldots, 2m + 2$, of our problem are parametrised by a set of $2(N - 1)m$ nonzero complex constants $c_1, \ldots, c_{(N-1)m}$ and $d_1, \ldots, d_{(N-1)m}$ as follows as

$$G_{2k-1} = \begin{pmatrix} 0 & 0 & \cdots & 0 & 0 & (-1)^{N-1}c_k \\ \frac{c_{k+m}}{c_k} & 0 & 0 & \cdots & 0 & 0 \\ 0 & \frac{c_{k+2m}}{c_{k+m}} & 0 & \cdots & 0 & 0 \\ \cdots & & \cdots & \cdots & 0 & 0 \\ 0 & 0 & \cdots & \frac{c_{k+(N-2)m}}{c_{k+(N-3)m}} & 0 & 0 \\ 0 & 0 & \cdots & 0 & \frac{1}{c_{k+(N-2)m}} & 0 \end{pmatrix},$$

for $k = 1, \ldots, m$ and $G_{2m+1} = \mathcal{P}_N$, where $\mathcal{P}_N$ has been defined in (7); the diagonal matrix G_{2k} reads

$$G_{2k} = \mathrm{dig}\left(d_k, d_{k+m}, \ldots, d_{k+(N-2)m}, \prod_{j=0}^{N-2} \frac{1}{d_{k+jm}} \right),$$

for $k = 1, \ldots, m$ and $G_{2m+2} = 1_N$.

The solution $Y(\lambda)$, of the full R-H problem (i)–(iv), where the constant matrices G_k, $k = 1, \ldots, 2m + 1$, are parametrised by $2(N - 1)m$ arbitrary complex constants, is obtained, following [19], using the Szegö kernel with nonzero characteristics. From the relation (10), we are able to write the global solution

$Y(\lambda)$ of the R-H problem (i)–(iv) in the form

$$Y_{rs}(\lambda) = X_{rs}(\lambda) \frac{\theta \begin{bmatrix} \delta \\ \epsilon \end{bmatrix} \left(\int_{P_0^{(r)}}^{P^{(s)}} d\mathbf{v}; \Pi \right)}{\theta \left(\int_{P_0^{(r)}}^{P^{(s)}} d\mathbf{v}; \Pi \right)} \frac{\theta(\mathbf{0}; \Pi)}{\theta \begin{bmatrix} \delta \\ \epsilon \end{bmatrix} (\mathbf{0}; \Pi)}, \quad r, s = 1, \ldots, N,$$

(12)

where $d\mathbf{v}$ is the basis of normalised holomorphic differentials on $\mathcal{C}_{N,m}$, Π is the period matrix of $d\mathbf{v}$, $\theta \begin{bmatrix} \delta \\ \epsilon \end{bmatrix}$ is the canonical θ-function with characteristics ϵ and δ defined by the relations

$$\epsilon_{k+sm} = \frac{1}{2\pi_1} \log \frac{c_{k+sm}}{c_{k+1+sm}}, \quad s = 0, \ldots, N-2, \ k = 1, \ldots, m-1,$$

$$\epsilon_{sm} = \frac{1}{2\pi_1} \log c_{sm}, \quad s = 1, \ldots, N-1,$$

$$\delta_k = \frac{1}{2\pi_1} \log d_k, \quad k = 1, \ldots, (N-1)m.$$

We remark that the genus of the curve $\mathcal{C}_{N,m}$ is equal to $g = (N-1)m$ and we have introduced a set of $2m(N-1)$ parameters in the monodromy matrices. In this way, there is a one to one correspondence between the $2(N-1)m$ θ-function characteristics and $2(N-1)m$ monodromy parameters.

It can be proved that $\det Y(\lambda) \equiv 1$ thus showing that the solution (12) is non singular. The solution (12) exists if

$$\theta \begin{bmatrix} \delta \\ \epsilon \end{bmatrix} (\mathbf{0}; \Pi) \neq 0,$$

that is if $\Pi\delta + \epsilon \notin (\Theta)$, where (Θ) is the θ-divisor in the Jacobian of the Riemann surface. The monodromy matrices $M_k = G_k G_{k-1}^{-1}$ can be written in the form

$$M_k = C_k^{-1} e^{2\pi 1 \sigma_N} C_k, \quad k = 1, \ldots, 2m+1, \quad C_k \in GL(N, \mathbf{C}), \quad (13)$$

where the matrix σ_N reads

$$\sigma_N = \text{diag} \left(\frac{-N+1}{2N}, \frac{-N+3}{2N}, \ldots, \frac{N-3}{2N}, \frac{N-1}{2N} \right). \quad (14)$$

The matrix $Y(\lambda)$ has regular singularities of the following form near the points λ_k

$$Y(\lambda) = \hat{Y}_k(\lambda)(\lambda - \lambda_k)^{\sigma_N} C_k^{\pm}, \quad \lambda \in \mathbf{C}_{\pm}, \ k = 1, \ldots, 2m+1,$$

where the matrices $\hat{Y}_k(\lambda)$ are holomorphic and invertible at $\lambda = \lambda_k$, $C_k^+ = C_k$ and $C_k^- = C_k G_{k-1}$, $k = 1, \ldots, 2m + 1$.

It follows from the above expansion that $\frac{dY(\lambda)}{d\lambda} Y^{-1}(\lambda)$ is meromorphic in $\mathbf{CP}^1$ with simple poles at $\lambda_1, \lambda_2, \ldots, \lambda_{2m+2}$. Therefore $Y(\lambda)$ satisfies the Fuchsian equation

$$\frac{dY(\lambda)}{d\lambda} = \sum_{k=1}^{2m+1} \frac{A_k}{\lambda - \lambda_k} Y(\lambda), \tag{15}$$

where

$$A_k = A_k(\lambda_1, \ldots, \lambda_{2m+1} | M_1, \ldots M_{2m+1}) = \mathrm{Res}_{[\lambda = \lambda_k]} \left[\frac{dY(\lambda)}{d\lambda} Y(\lambda) \right]$$

$$= \hat{Y}_k(\lambda_k) \sigma_N \hat{Y}_k^{-1}(\lambda_k), \quad k = 1, \ldots, 2m + 1.$$

If none of the monodromy matrices depend on the position of the singular points λ_k, $k = 1, \ldots, 2m + 1$, the function $Y(\lambda; \lambda_1, \ldots, \lambda_{2m+1})$ in addition to (15) satisfies the following equations

$$\frac{\partial}{\partial \lambda_k} Y(\lambda) = \left(\frac{A_k}{\lambda_0 - \lambda_k} - \frac{A_k}{\lambda - \lambda_k} \right) Y(\lambda), \quad k = 1, \ldots, 2m + 1. \tag{16}$$

Compatibility condition of (15) and (16) is described by the system of Schlesinger equations

$$\frac{\partial}{\partial \lambda_j} A_k = \frac{[A_k, A_j]}{\lambda_k - \lambda_j} - \frac{[A_k, A_j]}{\lambda_0 - \lambda_j}, \quad j \neq k, \tag{17}$$

$$\frac{\partial}{\partial \lambda_k} A_k = - \sum_{j \neq k, j=1}^{2m+1} \left(\frac{[A_k, A_j]}{\lambda_k - \lambda_j} - \frac{[A_k, A_j]}{\lambda_0 - \lambda_j} \right).$$

Thus the solution of the R-H problem (i)–(iv) leads immediately to the particular solution (16) of the Schlesinger system (17).

The τ-function corresponding to the particular solution (16) of the Schlesinger system, has the form

$$\tau(\lambda_1, \ldots, \lambda_{2m+1}) = \frac{\theta \begin{bmatrix} \delta \\ \epsilon \end{bmatrix} (0; \Pi)}{\theta(0; \Pi)}$$

$$\times \frac{\displaystyle\prod_{k<i,i,k=0}^{m} (\lambda_{2k+1} - \lambda_{2i+1})^{\frac{N^2-1}{6N}} \prod_{k<i,k,i=1}^{m} (\lambda_{2k} - \lambda_{2i})^{\frac{N^2-1}{6N}}}{\displaystyle\prod_{i<j,i,j=1}^{2m+1} (\lambda_i - \lambda_j)^{\frac{N^2-1}{12N}}}.$$

$$\tag{18}$$

We remark that the factor not containing $\theta\begin{bmatrix} \delta \\ \epsilon \end{bmatrix}(0; \Pi)$ in the above expression, is related to the Bergmann projective connection of the surface. For $N = 2$ the above formula coincides with the one obtained in [18].

Finally, we investigate in detail the case $N = 3$ and $m = 1$. The monodromy matrices read

$$M_1 = \begin{pmatrix} 0 & 0 & c_1 \\ \frac{c_2}{c_1} & 0 & 0 \\ 0 & (c_2)^{-1} & 0 \end{pmatrix}, \quad M_2 = \begin{pmatrix} 0 & \frac{c_1 d_1}{c_2} & 0 \\ 0 & 0 & c_2 d_2 \\ \frac{1}{c_1 d_1 d_2} & 0 & 0 \end{pmatrix},$$

$$M_3 = \begin{pmatrix} 0 & 0 & d_1 d_2 \\ \frac{1}{d_1} & 0 & 0 \\ 0 & \frac{1}{d_2} & 0 \end{pmatrix}, \quad M_\infty = \begin{pmatrix} 0 & 1 & 0 \\ 0 & 0 & 1 \\ 1 & 0 & 0 \end{pmatrix},$$

where c_1, c_2, d_1, d_2 are nonzero constants. Then the solution of the R-H problem (i)–(iv) is defined in terms of the Szegö kernel of the genus two Riemann surface

$$C_{3,1} : y^3 = (\lambda - \lambda_1)(\lambda - \lambda_2)^2(\lambda - \lambda_3).$$

The automorphism group of $C_{3,1}$ is isomorphic to the dihedral group D_3. For this reason $C_{3,1}$ is a two-sheeted cover of two elliptic curves that are 3-misogynous. As a result the solution of the R-H problem and of the Schlesinger equations can be expressed explicitly in terms of Jacobi' ϑ-functions. The τ-function of the Schlesinger system reads

$$\tau(\lambda_1 \lambda_2, \lambda_3) = \left(\frac{\lambda_1 - \lambda_3}{(\lambda_1 - \lambda_2)(\lambda_2 - \lambda_3)} \right)^{\frac{2}{9}} e^{2\pi i [T(\delta_1^2 + \delta_1 \delta_2 + \delta_2^2) + \epsilon_1 \delta_1 + \epsilon_2 \delta_2]}$$

$$\times \frac{\sum_{k=2}^{3} \vartheta_k(\epsilon_1 + \epsilon_2 + 3T(\delta_1 + \delta_2); 6T)\vartheta_k(\epsilon_1 - \epsilon_2 + T(\delta_1 - \delta_2); 2T)}{\vartheta_3(0; 6T)\vartheta_3(0; 2T) + \vartheta_2(0; 6T)\vartheta_2(0; 2T)},$$

where ϑ_i, $i = 2, 3$ are the Jacobi's ϑ-functions,

$$\epsilon_i = \frac{1}{2\pi_1} \log c_i, \quad \delta_i = \frac{1}{2\pi_1} \log d_i, \quad i = 1, 2$$

and

$$T = \frac{i\sqrt{3}}{3} \frac{F\left(\frac{1}{3}, \frac{1}{3}, 1; 1 - t\right)}{F\left(\frac{1}{3}, \frac{2}{3}, 1; t\right)}, \quad t = \frac{\lambda_3 - \lambda_2}{\lambda_3 - \lambda_1}$$

Here $F\left(\frac{1}{3}, \frac{1}{3}, 1; 1 - t\right)$ and $F\left(\frac{1}{3}, \frac{2}{3}, 1; t\right)$ are two independent solutions of the hypergeometric equation

$$t(1 - t)F'' + (1 - 2t)F' - \frac{2}{9}F = 0.$$

The function $T = T(t)$ is in general not single-valued. For T belonging to the Siegel upper half space modulo the modular group $\Gamma_0(3)$ the function $t = t(T)$ is single-valued and reads

$$t = 27\vartheta_3^4(0; 3T)\frac{(\vartheta_3^4(0; 3T) - \vartheta_3^4(0; T))^2}{(3\vartheta_3^4(0; 3T) + \vartheta_3^4(0; T))^3}.$$

Clearly, the above expression is automorphic under the action of the group $\Gamma_0(3)$. From the classical theory of the hypergeometric equation it follows that the function $t = t(T)$ satisfies the Schwartz equation (see for example [26])

$$\{t, T\} + \frac{\dot{t}^2}{2}V(t) = 0,$$

where $\dot{t} = \frac{dt}{dT}$, $\{,\}$ is the Schwartzian derivative,

$$\{t, T\} = \frac{d}{dT}\frac{\ddot{t}}{\dot{t}} - \frac{1}{2}\left(\frac{\ddot{t}}{\dot{t}}\right)^2$$

and the potential $V(t)$ is given by

$$V(t) = \frac{1 - \beta^2}{t^2} + \frac{1 - \gamma^2}{(t - 1)^2} + \frac{\beta^2 + \gamma^2 - \alpha^2 - 1}{t(t - 1)}, \quad \alpha = \frac{1}{3}, \quad \beta = \gamma = 0.$$

From the function $t = t(T)$ it is possible to derive an expression for the solution of the corresponding general Halphen system equivalent to the one derived in [27] in terms of Dedekind η-function [26]. Indeed the functions

$$\omega_1 = -\frac{1}{2}\frac{d}{dT}\ln\frac{\dot{t}}{t(t - 1)}, \quad \omega_2 = -\frac{1}{2}\frac{d}{dT}\ln\frac{\dot{t}}{t - 1}, \quad \omega_3 = -\frac{1}{2}\frac{d}{dT}\ln\frac{\dot{t}}{t},$$

solve the general Halphen system

$$\dot{\omega}_1 = \omega_2\omega_3 - \omega_1(\omega_2 + \omega_3) + R,$$
$$\dot{\omega}_2 = \omega_1\omega_3 - \omega_2(\omega_1 + \omega_3) + R,$$
$$\dot{\omega}_3 = \omega_1\omega_2 - \omega_1(\omega_1 + \omega_2) + R,$$

where

$$R = \frac{1}{9}(\omega_1 - \omega_2)(\omega_3 - \omega_1).$$

ACKNOWLEDGMENTS

The authors are grateful to B. Dubrovin, J. Harnad, A. Its, D. Korotkin, J. McKay, and F. Nijhoff for discussions of the results. We are also grateful to Yu. Brezhnev for drawing our attention to the paper [28]. During the preparation of the paper the authors used the Maple software by B. Deconnink and M. van Hoeij [29] to study algebraic curves.

REFERENCES

1. Deift, P. and Zhou, X. (1993) A steepest descent method for oscillatory Riemann-Hilbert problems. Asymptotics for the MKdV equation. *Ann. of Math.* **137**, pp. 295–368.
2. Deift, P., Its, A., and Zhou, X. (1997) A Riemann-Hilbert approach to asymptotic problems arising in the theory of random matrix models, and also in the theory of integrable statistical mechanics. *Ann. of Math.* **146** (1), pp. 149–235.
3. Deift, P., Kriecherbauer, T., McLaughlin, K.T.-R., Venakides, S., and Zhou, X. (1999) Uniform asymptotics for polynomials orthogonal with respect to varying exponential weights and applications to universality questions in random matrix theory. *Comm. Pure Appl. Math.* **52** (11), pp. 1335–1425.
4. Adler, M. and van Moerbeke, P. (1999) The spectrum of coupled random matrices, *Ann. of Math.* **149**, pp. 921–976.
5. Bertola, M. (2003) Free energy of the two-matrix model/dToda tau-function, *Nuclear Phys. B* **669** (3), pp. 435–461.
6. Aptekarev, A. I. (1999) Strong asymptotic of polynomials of simultaneous orthogonality for Nikishin systems (Russian), *Mat. Sb.* **190** (5), pp. 3–44; translation in *Sb. Math.* **190** (5–6), pp. 631–669.
7. Nikishin, E. M. (1986) Asymptotic behavior of linear forms for simultaneous Padé approximants. (Russian) *Izv. Vyssh. Uchebn. Zaved. Mat.* (2), pp. 33–41.
8. Kuijlaars, A., Stahl, H., Van Assche, W., and Wielonsky, F. (2003) Asymptotic behavior of quadratic Hermite-Padé approximants of the exponential function and Riemann-Hilbert problems *C. R. Math. Acad. Sci. Paris* **336** (11), pp. 893–896.
9. Enolskii, V. and Grava, T. *Singular Z_N curves, Riemann-Hilbert problem and modular solutions of the Schlesinger system, Intern. Math. Res.* Notices to appear.
10. Treibich Kohn, A. (1983) Un result de Plemelj, *Progr. Math.* **37**, pp. 307–312.
11. Arnol'd, V. I. and Il'yashenko, Yu. S. (1985) Ordinary differential equations, in: *Encyclopaedia of Mathematical Sciences*, eds. V. I. Arnold and D. V. Anosov, Springer Verlag, Berlin, Title of the Russian edition: *Itogi nauki i tekhniki, Sovremennye problemy matematiki, Fundamrntal'nye napravlenia*, Vol. 1, Dinamicheskie sistemy I, pp. 7–149.
12. Dekkers, W. (1979) *The matrix of a connection having regular singularities on a vector bundle of rank 2 on CP^1*, Lecture Notes in Mathematics 712, pp. 33–43.
13. Anosov, D. V. and Bolibruch, A. A. (1994) The Riemann-Hilbert problem, in: *Aspects of Mathematics*, Vol. **E22**, Vieweg and Sohn, Braunschweig.
14. Kostov, V. P. (1992) Fuchsian system on $\mathbf{CP}^1$ and the Riemann-Hilbert problem, *C.R. Acad. Sci. Paris Ser. I Math* **315**, pp. 207–238.
15. Okamoto, K. (1987) Studies on the Painlevé equations. I. Sixth Painlevé equation P_{VI}, *Annali Mat. Pura Appl.* **146**, pp. 337–381.

16. Umemura, H. (1990) *Second proof of the irreducibility of the first differential equation of Painlevé*, Nagoya Math. J. **117**, pp. 125–171.

17. Deift, P., Its, A., Kapaev, A., and Zhou, X. (1999) On the algebro-geometric integration of the Schlesinger equations, *Commun. Math. Phys.* **203**, pp. 613–633.

18. Kitaev, A. and Korotkin, D. (1998) On solutions of Schlesinger equations in terms of theta-functions, *Intern. Math. Res. Notices* **17**, pp. 877–905.

19. Korotkin, D. *Matrix Riemann-Hilbert problems related to branched coverings of* **CP**^{1}, in: *Operator Theory: Advances and Applications, Proceedings of the Summer School on Factorization and Integrable Systems, Algarve, September 6–9, 2000*, N. Manojlovic I. Gohberg, A. F. dos Santos, eds. Boston, 2002. Birkhäuser. xxx.lanl.gov/math-ph/0106009.
Solution of matrix Riemann-Hilbert problems with quasi-permutation monodromy matrices, xxx.lanl.gov/math-ph/0306061.

20. Plemelj, J. (1964) *Problems in the sense of Riemann and Klein*, Interscience, New York, London, Sydney.

21. Dubrovin, B. A., Krichever, I. M., and Novikov, S. P. (2001) Integrable systems. I Dynamical systems, IV, 177–332, in: *Encyclopaedia Math. Sci.* 4, Springer, Berlin.

22. Belokolos, E. D., Bobenko, A. I., Enolskii V. Z., Its, A. R., and Matveev, V. B. (1994) *Algebro Geometrical Approach to Nonlinear Integrable Equations*, Springer, Berlin.

23. Zverovich, E. I. (1971) *Boundary problems of the theory of analytic functions*, Uspekhi. Matem. Nauk **31** (5), pp. 113–181.

24. Miranda, R. (1995) *Algebraic Curves and Riemann Surfaces*, in: *Graduate Studies in Mathematics*. Amer. Math. Soc. 5 Providence, R.I.

25. Couveignes, J.-M. (1999) Tools for computing families of coverings, in: *Aspects of Galois theory* (Gainesville, FL, 1996), 38–65, *London Math. Soc. Lecture Note Ser.* 256, Cambridge University Press, Cambridge.

26. Ford, L. (1929) *Automorphic Functions*, McGraw–Hill, New York.

27. Harnad, J. and McKay, J. (2000) Modular solutions to equations of generalized Halphen type, *R. Soc. Lond. Proc. Ser. A. Math. Phys.* **456**, pp. 261–294.

28. Hutchinson, J. I. (1902) On a class of automorphic functions, *Trans. Amer. Math. Soc.* **3**, pp. 1–11.

29. Deconinck, B. and van Hoeij, M. (2001) *Computing Riemann matrices of algebraic curves*, Physica D **152–153**, pp. 28–46.

30. Schlesinger, L. (1912) Über eine Klasse von Differentialsystemen beliebiger Ordnung mit festen kritischen Punkten, *J. reine angew. Math.* **141**, pp. 96–145.

ANALYTIC AND ALGEBRAIC ASPECTS OF TODA FIELD THEORIES AND THEIR REAL HAMILTONIAN FORMS

V.S. Gerdjikov
Institute for Nuclear Research and Nuclear Energy
72 Tzarigradsko chaussee, 1784 Sofia, Bulgaria

Abstract   One of the paradigms which stimulated the development of powerful methods in mathematical physics is the Toda field theory in $1 + 1$ dimensions. It enhanced the development of graded and Kac-Moody algebras and the method of reductions in the inverse scattering method. Here we outline the basic ideas and some of their latest developments which allowed to relate to each TFT a family of its real Hamiltonian forms.

Keywords: Solitons, Toda field theory, Hamiltonian systems, Tau function.

1   INTRODUCTION

The famous paper by Hirota [1] started a new trend in soliton theory. It stimulated both the construction of soliton solutions of new nonlinear evolution equations (NLEE) and the development of the algebraic approach to soliton theory via the method of the tau-functions, see [1, 2]. Special role here play the conformal and the affine Toda field theories in $0 + 1$ and $1 + 1$ space-time dimensions [3, 4, 5].

In the present paper we analyze some of the analytic and algebraic aspects of the affine Toda field theories (ATFT) in $1 + 1$ dimensions. Next we show that the well known ATFT can be viewed just as a member in the family of real Hamiltonian forms (RHF) of these theories. A nontrivial example of RHF for the ATFT is given explicitly.

2   PRELIMINARIES

To each simple Lie algebra $\mathfrak{g}$ one can relate both conformal and affine versions of a TFT in $1 + 1$ dimensions. It allows Lax representation:

$$[L, M] = 0 \tag{1}$$

77

L. Faddeev et al. (eds.),
Bilinear Integrable Systems: From Classical to Quantum, Continuous to Discrete, 77–83.
© 2006 *Springer. Printed in the Netherlands.*

where L and M are first order ordinary differential operators whose potentials take values in $\mathfrak{g}$:

$$L\psi \equiv \left(i\frac{d}{dx} - iq_x(x, t) - \lambda J_0 \right) \psi(x, t, \lambda) = 0, \tag{2}$$

$$M\psi \equiv \left(i\frac{d}{dt} - \frac{1}{\lambda}I(x, t) \right) \psi(x, t, \lambda) = 0. \tag{3}$$

Here $q(x, t) \in \mathfrak{h}$—the Cartan subalgebra of $\mathfrak{g}$, $\vec{q}(x, t) = (q_1, \ldots, q_r)$ is its dual r-component vector, $r = \mathrm{rank}\, \mathfrak{g}$, and

$$J_0 = \sum_{\alpha \in \pi} E_\alpha, \qquad I(x, t) = \sum_{\alpha \in \pi} e^{-(\alpha, \vec{q})} E_{-\alpha}. \tag{4}$$

If π is the set of simple roots $\pi = \{\alpha_1, \ldots, \alpha_r\}$ of $\mathfrak{g}$ then we get the conformal TFT; if π is the set of admissible roots, i.e., $\pi = \{\alpha_0, \alpha_1, \ldots, \alpha_r\}$ where α_0 is the minimal root of $\mathfrak{g}$ then the corresponding TFT is the affine one. The equations of motion in the latter case is of the form:

$$\frac{\partial^2 \vec{q}}{\partial x \partial t} = \sum_{j=0}^{r} n_j \alpha_j e^{-(\alpha_j, \vec{q})}, \tag{5}$$

where n_j are the minimal positive integers for which $\sum_{j=0}^{r} n_j \alpha_j = 0$.

The operators L and M are invariant with respect to the reduction group $\mathcal{G}_{\mathbb{R}} \simeq \mathbb{D}_h$ [3] where h is the Coxeter number of $\mathfrak{g}$. This reduction group is generated by two elements satisfying $g_1^h = g_2^2 = (g_1 g_2)^2 = 1$ which allow realizations both as elements in $\mathrm{Aut}\, \mathfrak{g}$ and in $\mathrm{Conf}\, \mathbb{C}$. The invariance condition has the form:

$$C_k(U(x, t, \kappa_k(\lambda))) = U(x, t, \lambda), \quad C_k(V(x, t, \kappa_k(\lambda))) = V(x, t, \lambda) \tag{6}$$

where $U(x, t, \lambda) = -iq_x(x, t) - \lambda J_0$ and $V(x, t, \lambda) = -\frac{1}{\lambda}I(x, t)$. Here C_k are automorphisms of finite order of $\mathfrak{g}$, i.e., $C_1^h = C_2^2 = (C_1 C_2)^2 = 1$ while $\kappa_k(\lambda)$ are conformal mappings of the complex λ-plane. The algebraic constraint (6) are automatically compatible with the evolution.

Lemma 1 *Let $\mathfrak{g}$ be a simple Lie algebra from one of the classical series A_r, B_r, C_r or D_r and let h be its Coxeter number and N_0—the dimension of the typical representation. Then the characteristic equation of J_0 taken in the typical representation has the form:*

$$\zeta^{r_0}(\zeta^h - 1) = 0, \quad r_0 = N_0 - h. \tag{7}$$

Remark 2 The constant $r_0 = 0$ for $\mathfrak{g} \simeq A_r, C_r$; $r_0 = 1$ for $\mathfrak{g} \simeq B_r$ and $r_0 = 2$ for $\mathfrak{g} \simeq D_r$. Solving the inverse scattering problem in the last two cases requires special treatment of the subspaces related to $\zeta = 0$.

3 THE SPECTRAL PROPERTIES OF L

The reduction conditions (6) lead to rather special properties of the operator L. Along with L we will use also the equivalent system:

$$\tilde{L}m(x,t,\lambda) \equiv i\frac{dm}{dx} + iq_x m(x,t,\lambda) - \lambda[J_0, m(x,t,\lambda)] = 0, \tag{8}$$

where $m(x,t,\lambda) = \psi(x,t,\lambda)e^{iJ_0 x\lambda}$. Combining the ideas of [6] with the symmetries of the potential (6) we can construct a set of $2h$ fundamental analytic solutions (FAS) $m_\nu(x,t,\lambda)$ of (1.8) and prove that:

1. the continuous spectrum Σ of L fills up $2h$ rays l_ν passing through the origin: $\lambda \in l_\nu$: $\arg\lambda = (\nu-1)\pi/h$;
2. $m_\nu(x,t,\lambda)$ is a FAS of (1.8) analytic with respect to λ in the sector Ω_ν: $(\nu-1)\pi/h \le \arg\lambda \le \nu\pi/h$ satisfying $\lim_{\lambda\to\infty} m_\nu(x,t,\lambda) = 1$;
3. to each l_ν one relates a subalgebra $\mathfrak{g}_\nu \subset \mathfrak{g}$ such that $\mathfrak{g}_\nu \cap \mathfrak{g}_\mu = \emptyset$ for $\nu \ne \mu$ mod (h) and $\cup_{\nu=1}^h \mathfrak{g}_\nu = \mathfrak{g}$. The symmetry ensure that each of the subalgebras $\mathfrak{g}_\nu$ is a direct sum of $sl(2)$-subalgebras;
4. on Σ the FAS $m_\nu(x,t,\lambda)$ satisfy

$$m_\nu(x,t,\lambda) = m_{\nu-1}(x,t,\lambda)G_\nu(x,t,\lambda), \quad \lambda \in l_\nu, \tag{9}$$
$$G_\nu(x,t,\lambda) = e^{-i(\lambda J_0 x + f(\lambda))t}G_{0,\nu}(\lambda)\,e^{i(\lambda J_0 x + f(\lambda))t} \in \mathcal{G}_\nu, \tag{10}$$

where $\mathcal{G}_\nu$ is the subgroup with Lie algebra $\mathfrak{g}_\nu$ and $f(\lambda)$ is determined by the dispersion law of the NLEE: $f(\lambda) = \sum_{k=0}^r E_{-\alpha_k}/\lambda$;
5. the FAS of (8) satisfy:

$$\tilde{C}_1(m_\nu(x,t,\omega\lambda)) = m_{\nu-2}(x,t,\lambda), \quad \lambda \in l_{\nu-2}, \tag{11}$$

where $\tilde{C}_1$ is equivalent to the Coxeter automorphism:

$$\tilde{C}_1(X) \equiv C_1^{-1}XC_1, \quad C_1 = e^{\frac{2\pi i}{h}H_\rho}, \quad \rho = \frac{1}{2}\sum_{\alpha>0}\alpha; \tag{12}$$

obviously $C_1^h = 1$ and $\tilde{C}_1(J_0) = \omega^{-1}J_0$;
6. the FAS $m_\nu(x,t,\lambda)$ satisfy one of the following two involutions:

$$\tilde{C}_2(m_\nu(x,t,\lambda^*))^\dagger = C_2(m_{2h-\nu+2}^{-1}(x,t,\lambda)), \tag{13}$$

where C_2, $C_2^2 = 1$ is conveniently chosen Weyl group element, or

$$(m_\nu(x,t,-\lambda^*))^* = m_{h-\nu+2}(x,t,\lambda). \tag{14}$$

These relations lead to the following constraints for the sewing functions $G_{0,\nu}(\lambda)$ and the minimal set of scattering data:

$$\tilde{C}_1(G_{0,\nu}(\omega\lambda)) = G_{0,\nu-2}(\lambda), \tag{15}$$

$$\tilde{C}_2(G_{0,\nu}^\dagger(\lambda^*)) = G_{0,2h-\nu+2}^{-1}(\lambda), \tag{16}$$

$$G_{0,\nu}^*(-\lambda^*) = G_{0,h-\nu+2}(\lambda). \tag{17}$$

If L has no discrete eigenvalues the minimal set of scattering data is provided by the coefficients of $G_{0,1}(\lambda)$, $\lambda \in l_1$ and $G_{0,2}(\lambda)$, $\lambda \in l_2$. All other sewing functions $G_{0,\nu}(\lambda)$ can be determined from them by applying (15), (16) or (15), (17).

4 THE REAL HAMILTONIAN FORMS OF ATFT

The Lax representations of the ATFT models widely discussed in the literature (see e.g., [3–5, 7] and the references therein) are related mostly to the normal real form of the Lie algebra $\mathfrak{g}$. Our aim here is to:

1. generalize the ATFT to complex-valued fields $\vec{q}^C = \vec{q}_0 + i\vec{q}_1$, and
2. describe the family of RHF of these ATFT models.

We also provide a tool to construct new inequivalent RHF's of the ATFT. This tool is a natural generalization of the one in [8] to $1 + 1$-dimensional systems. Indeed, the ATFT can be written down as an infinite-dimensional Hamiltonian system as follows:

$$\frac{dq_k}{dt} = \{q_k, H\}, \quad \frac{dp_k}{dt} = \{p_k, H\}, \tag{18}$$

$$H_{\mathrm{ATFT}} = \int_{-\infty}^{\infty} dx \left(\frac{1}{2}(\vec{p}(x,t), \vec{p}(x,t)) + \sum_{k=0}^{r} n_k e^{-(\vec{q}(x,t),\alpha_k)} \right),$$

where $\vec{p} = d\vec{q}/dx$ and $\vec{q}$ are the canonical momenta and coordinates satisfying canonical Poisson brackets:

$$\{q_k(x,t), p_j(y,t)\} = \delta_{jk}\delta(x - y). \tag{19}$$

Next we introduce an involution $\mathcal{C}$ acting on the phase space $\mathcal{M} \equiv \{q_k(x), p_k(x)\}_{k=1}^n$ and satisfying:

1) $\mathcal{C}(F(p_k, q_k)) = F(\mathcal{C}(p_k), \mathcal{C}(q_k))$,
2) $\mathcal{C}(\{F(p_k, q_k), G(p_k, q_k)\}) = \{\mathcal{C}(F), \mathcal{C}(G)\}$,
3) $\mathcal{C}(H(p_k, q_k)) = H(p_k, q_k)$.

It is important also that the Hamiltonian $H(p_k, q_k)$ is an analytic functional of the fields $q_k(x, t)$ and $p_k(x, t)$.

The complexification of the ATFT is rather straightforward. The resulting complex ATFT (CATFT) can be written down as standard Hamiltonian system with twice as many fields $\vec{q}_a(x, t)$, $\vec{p}_a(x, t)$, $a = 0, 1$:

$$\vec{p}^{\mathbb{C}}(x, t) = \vec{p}^0(x, t) + i\vec{p}^1(x, t), \quad \vec{q}^{\mathbb{C}}(x, t) = \vec{q}^0(x, t) + i\vec{q}^1(x, t), \quad (20)$$

$$\{q_k^0(x, t), p_j^0(y, t)\} = -\{q_k^1(x, t), p_j^1(y, t)\} = \delta_{kj}\delta(x - y). \quad (21)$$

The densities of the corresponding Hamiltonian and symplectic form equal

$$\mathcal{H}^{\mathbb{C}}_{\text{ATFT}} \equiv \text{Re}\,\mathcal{H}_{\text{ATFT}}(\vec{p}^0 + i\vec{p}^1, \vec{q}^0 + i\vec{q}^1) \quad (22)$$

$$= \frac{1}{2}(\vec{p}^0, \vec{p}^0) - \frac{1}{2}(\vec{p}^1, \vec{p}^1) + \sum_{k=0}^{r} n_k e^{-(\vec{q}^0, \alpha_k)} \cos((\vec{q}^1, \alpha_k)),$$

$$\omega^{\mathbb{C}} = (d\vec{p}^0 \wedge id\vec{q}^0 - d\vec{p}^1 \wedge d\vec{q}^1). \quad (23)$$

The family of RHF then are obtained from the CATFT by imposing an invariance condition with respect to the involution $\tilde{\mathcal{C}} \equiv \mathcal{C} \circ *$ where by $*$ we denote the complex conjugation. The involution $\tilde{\mathcal{C}}$ splits the phase space $\mathcal{M}^{\mathbb{C}}$ into a direct sum $\mathcal{M}^{\mathbb{C}} \equiv \mathcal{M}^{\mathbb{C}}_+ \oplus \mathcal{M}^{\mathbb{C}}_-$ where

$$\mathcal{M}^{\mathbb{C}}_+ = \mathcal{M}_0 \oplus i\mathcal{M}_1, \quad \mathcal{M}^{\mathbb{C}}_- = i\mathcal{M}_0 \oplus \mathcal{M}_1, \quad (24)$$

The phase space of the RHF is $\mathcal{M}_{\mathbb{R}} \equiv \mathcal{M}^{\mathbb{C}}_+$. By $\mathcal{M}_0$ and $\mathcal{M}_1$ we denote the eigensubspaces of $\mathcal{C}$, i.e., $\mathcal{C}(u_a) = (-1)^a u_a$ for any $u_a \in \mathcal{M}_a$.

Thus to each involution $\mathcal{C}$ satisfying 1)–3) one can relate a RHF of the ATFT. Due to the condition 3) $\mathcal{C}$ must preserve the system of admissible roots of $\mathfrak{g}$; such involutions can be constructed from the $\mathbb{Z}_2$-symmetries of the extended Dynkin diagrams of $\mathfrak{g}$ studied in [7].

Example 1 We choose $\mathfrak{g} \simeq A_{2r+1}$ and fix up the involution $\mathcal{C}$ by:

$$\mathcal{C}(q_k) = -q_{2r+2-k}, \quad \mathcal{C}(p_k) = -p_{2r+2-k}, \quad k = 1, \ldots, r,$$

$$\mathcal{C}(q_{r+1}) = -q_{r+1}, \quad \mathcal{C}(p_{r+1}) = -p_{r+1}. \quad (25)$$

The coordinates in $\mathcal{M}_\pm$ are given by:

$$q_k^\pm = \frac{1}{\sqrt{2}}(q_k \mp q_{2r+2-k}), \quad p_k^\pm = \frac{1}{\sqrt{2}}(p_k \mp p_{2r+2-k}),$$

$$q_{r+1}^- = q_{r+1}, \quad p_{r+1}^- = p_{r+1}, \quad (26)$$

where $k = 1, \ldots, r$, i.e., dim $\mathcal{M}_+ = 2r$ and dim $\mathcal{M}_- = 2r + 2$. Then the densities $\mathcal{H}^{\mathbb{R}}_{\text{AFEF}}$, $\omega^{\mathbb{R}}_{\text{ATFT}}$ for the RHF of AFTF equal:

$$
\mathcal{H}^{\mathbb{R}}_{\text{ATFT}} = \frac{1}{2} \sum_{k=1}^{r} p_k^{+2} - \frac{1}{2} \sum_{k=1}^{r+1} p_k^{-2} + 2e^{-q_{r-1}^+/\sqrt{2}} \cos\left(\frac{q_{r-1}^-}{\sqrt{2}} - q_{r+1}^-\right)
$$

$$
+ \sum_{k=1}^{r-1} 2e^{(q_{k+1}^+ - q_k^+)/\sqrt{2}} \cos\left(\frac{q_{k+1}^- - q_k^-}{\sqrt{2}}\right) + 2e^{q_1^+/\sqrt{2}} \cos\left(\frac{q_1^-}{\sqrt{2}} - q_{r+1}^-\right).
$$

$$
\omega^{\mathbb{R}}_{\text{ATFT}} = \sum_{k=1}^{r} dp_k^+ \wedge dq_k^+ - \sum_{k=1}^{r+1} dp_k^- \wedge dq_k^- \tag{27}
$$

where $\vec{p}_k^{\pm} = d\vec{q}_k^{\pm}/dx$. If we put $q_j^- = 0$ and $p_j^- = 0$ we get the reduced ATFT related to the Kac-Moody algebra $D_{r+1}^{(2)}$ [7].

The automorphism $\mathcal{C}$ is dual to an automorphism $\mathcal{C}^{\#}$ of the corresponding Lax pair and the Lie algebra $\mathfrak{g}$. In fact $\tilde{C}^{\#} = -C^{\#}(X^{\dagger})$ is a Cartan involution of $\mathfrak{g}$ and therefore the Lax pair of the RHF is related to a real form $\mathfrak{g}_{\mathbb{R}}$ of $\mathfrak{g}$. The reduction condition (13) (or (14)) picks up the real form of the related Kac-Moody algebra.

5   DISCUSSION

Though some examples of RHF of ATFT have been known before [9], the method proposed in [8] provides a tool for the systematic construction and classification of the RHF for any Hamiltonian system, not necessarily integrable. It can be proved that the RHF of an integrable system is again integrable [8]. The solutions depending analytically on the initial parameters go into solutions of the RHF. Such are, e.g., the soliton solutions derived by Hirota's method in [7]. Imposing the reduction conditions on these parameters one can obtain the soliton solutions of the RHF of the ATFT. Their properties (stability, asymptotic dynamics etc.) however, will be different and deserve separate treatment. The situation here is similar to the one for the Toda chain. The real Toda chain allows only asymptotically free particles while the complex Toda chain and its RHF contain also asymptotic bound states, see [10].

The consequences of these fact for the corresponding tau-functions has to be investigated. It is important to find out all types of soliton solutions that the RHF of ATFT possesses. The $\mathbb{Z}_2$- symmetries of the extended Dynkin diagrams and the relevant reduction of the ATFT [7] can be used to classify all RHF of ATFT.

Other integrable models, as e.g., the $\mathbb{Z}_n$-nonlinear Schrödinger equation [11] and their RHF can be investigated along the same lines.

ACKNOWLEDGMENTS

It is my pleasure to thank Prof. F. Lambert and Prof. P. van Moerbeke for the possibility to take part in the conference. I am grateful to Prof. G. Marmo and Prof. G. Vilasi for numerous useful discussions and for their hospitality at the universities of Napoli and Salerno where part of these results were obtained. Financial support in part from Gruppo collegato di INFN at Salerno, Italy and project PRIN 2000 (contract 323/2002) is also acknowledged.

REFERENCES

1. Hirota, R. (1974) *Prog. Theor. Phys.* **52**, 1498.
2. Adler, M., Shiota, T., and van Moerbeke, P. (2002) *Math. Annalen.* **322**, pp. 423–476. Adler, M. and van Moerbeke, P. (1999) *Commun. Math. Phys.* **203**, pp. 185–210.
3. Mikhailov, A. V. (1981) *Physica D* **3**, pp. 73–117.
4. Mikhailov, A, V., Olshanetzky, M. A., and Perelomov, A. M. (1981) *Commun. Math. Phys.* **79**, pp. 473–490.
5. Olive, D., Turok, N., and Underwood, J. W. R. (1993) *Nucl. Phys.* **B401**, 663; **B409**, 509.
6. Gerdjikov, V. S. and Yanovski, A. B. J. (1994) *Math. Phys.* **35**, pp. 3687–3725.
7. Khastgir, S. P. and Ryu Sasaki. (1996) *Progr. Theor. Phys.* **95**, pp. 485–501; pp. 503–511.
8. Gerdjikov, V. S., Kyuldjiev, A. V., Marmo, G., and Vilasi, G. (2002) *Eur. Phys. J* **B29**, 177.
9. Evans, J. (1993) *Nucl. Phys.* **B390**, p. 225; Evans, J. and Madsen, J. (1996) *Phys. Lett.* **B384**, p. 131.
10. Gerdjikov, V. S., Evstatiev, E. G., and Ivanov, R. I. (1998) *J. Phys. A: Math & Gen.* **31**, pp. 8221–8232; **33**, pp. 975–1006 (2000).
11. Gerdjikov, V. S. (1981) In Nonlinear evolution equations: integrability and spectral methods, ed. A. P. Fordy, A. Degasperis, M. Lakshmanan, Manchester University Press, pp. 367–379.

BILINEAR AVATARS OF THE DISCRETE PAINLEVÉ II EQUATION

B. Grammaticos
GMPIB, Université Paris VII, case 7021, 75251 Paris, France

A. Ramani and A.S. Carstea*
CPT, Ecole Polytechnique, CNRS, UMR 7644, 91128 Palaiseau, France

Abstract   We present the list of all known to date difference and q-discrete forms of the Painlevé II equations. We show that, with the exception of two of them, all these equations are symmetric reductions of equations with more than one parameter and give the bilinearization of these systems. Finally we point out the existence of another one-parameter discrete Painlevé equation: the one-parameter d-P_{III}.

1   INTRODUCTION

The relation of (continuous) Painlevé equations to bilinearization goes a long way back, to before the bilinear formalism was introduced. Having the Painlevé property, the solutions of the Painlevé equations are meromorphic functions [1]. This means that they can be expressed as ratios of entire functions. But, precisely, the ansatz expressing the nonlinear variable as a ratio of entire, τ, functions lies at the heart of the bilinear approach. We can illustrate this in the case of the continuous P_{II}. We start with the equation

$$w'' = 2w^3 + tw + \alpha \tag{1}$$

and introduce the ansatz $w = (log(F/G))'$. Substituting into (1) we find that it is possible to split it into a system of two equations:

$$F''G + FG'' = 2F'G' \tag{2a}$$
$$F'''G - FG''' + 3(F'G'' - F''G') = t(F'G - FG') + \alpha FG \tag{2b}$$

These two equations constitute the bilinearization of P_{II}. The bilinear approach has been applied with success to all Painlevé equations, resulting into the

---

* *Permanent address: Department of Theoretical Physics, Institute of Physics and Nuclear Engineering, Magurele, Bucharest, Romania*

L. Faddeev et al. (eds.),
Bilinear Integrable Systems: From Classical to Quantum, Continuous to Discrete, 85–96.
© 2006 *Springer. Printed in the Netherlands.*

bilinear forms which can be of the utmost usefulness for constructing auto-Bäcklund transformations and special solutions.

In the case of discrete Painlevé equations (d-$\mathbb{P}$) we cannot rely on analyticity arguments. However another property of d-$\mathbb{P}$'s comes to the rescue of the bilinear approach. As a matter of fact, discrete equations, integrable through spectral methods (as is the case for the d-$\mathbb{P}$'s), possess the singularity confinement property [2]. This means that any spontaneously appearing singularity disappears after some (a few) iteration steps. From a practical point of view this means that if we try to represent the nonlinear dependent variable as a ratio of functions which have only zeros (which would be the discrete analogue of entire functions) only a finite number of such terms is needed. The bilinear approach for discrete $\mathbb{P}$'s was systematically introduced in [3] and has been extensively used since. The bilinearization of what we have dubbed the standard d-$\mathbb{P}$ family, was given in [4]. An interesting result of feedback was given in that paper, where using the bilinear form of discrete P_{VI} we have obtained the bilinearization of its continuous analogue (a result that had eluded all previous studies). Finally the bilinear formalism is the key ingredient in the geometrical description and classification of d-$\mathbb{P}$'s in terms of affine Weyl groups [5].

One remarkable property of discrete $\mathbb{P}$'s is that there are so many of them. (In [6] we have argued that an infinite number of d-$\mathbb{P}$'s may exist). Since the names of d-$\mathbb{P}$'s are based on their continuous limits and the latter are limited by the existence of just six $\mathbb{P}$'s it is natural to have more than one discrete analogue for each continuous Painlevé equation. In a previous study [7] we have presented a list of over 15 different forms of d-P_I (and still the list is certainly incomplete). In this paper we shall present a similar study for d-P_{II}. We shall give the various forms of equations already identified as the discrete analogues of P_{II}. We shall show that for the majority of them the P_{II} continuous limit results from an artificial restriction of the degrees of freedom of the equation to just one. In the present analysis the full freedom will be restored (which allows us to identify the genuine difference and q form of P_{II}). The bilinearization of all these equations will be equally obtained in full generality.

2   THE VARIOUS DISCRETE FORMS OF PAINLEVÉ II EQUATION

As we have explained in the introduction, some equations are written as one-parameter P_{II} but they can, in fact, be extended to equations with a higher number of parameters. The best example, in order to illustrate this point, is what we call the "standard" d-P_{II}. In full generality we have

$$x_{n+1} + x_{n-1} = \frac{(\alpha n + \beta)x_n + \gamma + \delta(-1)^n}{1 - x_n^2} \tag{3}$$

The presence of the $(-1)^n$ term suggests that we separate even and odd terms. Putting $X_m = x_{2m}$ and $Y_m = x_{2m+1}$ we have

$$X_{m+1} + X_m = \frac{Z_{m+1/4}Y_m + A}{1 - Y_m^2} \tag{4a}$$

$$Y_m + Y_{m-1} = \frac{Z_{m-1/4}X_m + B}{1 - X_m^2} \tag{4b}$$

where $Z_m = (2\alpha m + \beta - \alpha/2)$, $A = \gamma - \delta$, and $B = \gamma + \delta$, Thus the equation which is a d-P$_{II}$ when $\delta = 0$ requires an extra parameter for $\delta \neq 0$ and was shown to go over to P$_{III}$ at the continuous limit [8].

Equations (3) and (4) are what, in the QRT [9] terminology, are called *symmetric* and *asymmetric* forms. The equations we shall list below possess for the most part degrees of freedom which would suggest an asymmetric form. Still for better legibility we shall restrict ourselves to symmetric forms. However in every case we shall explicitly give the number of parameters of the equation (and the way they enter) as well as the affine Weyl group under which the equation can be classified. The way the full freedom of the equation will be obtained is through the application of the singularity confinement criterion. The procedure is perfectly legitimate (as we have amply explained in previous publications [10]). In every case we start from a mapping which is integrable in its autonomous form and use the singularity confinement criterion just for its deautonomization. In this setting singularity confinement is a sufficient integrability criterion.

1. The first example we shall give is that of the standard d-P$_{II}$. This equation was first obtained through an orthogonal polynomial method by Periwal and Shevitz (albeit in a zero-parameter form) in a field-theoretical model. Simultaneously Nijhoff and Papageorgiou derived the same equation from the similarity reduction of the discrete modified KdV equation (in a nice parallel to what happens in the continuous case). The full form was obtained in [11], and as we just said above (3) is

$$x_{n+1} + x_{n-1} = \frac{z_n x_n + a_n}{1 - x_n^2} \tag{5}$$

where $z_n = \alpha n + \beta$ and $a_n = \gamma + \delta(-1)^n$. As we explained above this equation has P$_{III}$ as continuous limit. Its geometrical description was given in [12] in terms of the $A_3^{(1)}$ affine Weyl group.

2. The "alternate" d-P$_{II}$

$$\frac{z_n}{x_{n+1}x_n + 1} + \frac{z_n - 1}{x_n x_{n-1} + 1} = -x_n + \frac{1}{x_n} + z_n + \mu \tag{6}$$

where $z_n = \alpha n + \beta$ and μ is a constant, with no further degree of freedom. This equation was first obtained in [13] as a contiguity of the continuous P_{III}. It was studied in detail in [14] where it was shown that it possess the property of self-duality. This means that the discrete equation governing the evolution along the parameter μ (evolution obtained through the Schlesinger transformations) is precisely the "alternate" d-P_{II} itself. This observation has made possible the geometrical description of not only (6) (which was given in [15] in terms of the affine Weyl group $2A_1^{(1)}$) but of all the d-$\mathbb{P}$'s.

3. The q-discrete equation

$$x_{n+1}x_{n-1} = \frac{a_n(x_n - b_n)}{x_n(x_n - 1)} \tag{7}$$

where $\log a_n = \alpha n + \beta + (-1)^n \gamma$, $\log b_n = \alpha n + \delta - (-1)^n \gamma$ (so the quantity $a_n b_n$ does not exhibit any even–odd dependence) was obtained in symmetrical form in [15] and in full freedom in [16]. This equation is one of the few examples not possessing the property of self-duality. Its geometrical description is given in terms of the affine Weyl group $A_2^{(1)} + A_1^{(1)}$. Its detailed study is due to Kajiwara, Noumi, and Yamada [17].

4. Another q-discrete equation, of the same family exists:

$$x_{n+1}x_{n-1} = \frac{a_n(x_n - b_n)}{x_n - 1} \tag{8}$$

where $\log a_n = \alpha n + \beta + (-1)^n \gamma + j^n \delta + j^{2n} \zeta$, ($j$ being a cubic root of unity), $\log b_n = 2\alpha n + \theta - j^n \delta - j^{2n} \zeta$ (so the quantity $a_n b_n$ does not exhibit any three-fold dependence). This equation has five degrees of freedom and its description can be given in terms of the affine Weyl group $D_5^{(1)}$. The full form of (8) was obtained in [16] where we have argued that this mapping is a Miura transformation of the asymmetric q-P_{III}, discrete P_{VI}, equation of Jimbo and Sakai [18].

5. A difference equation of the P_{IV} family also exists as a d-P_{II} when restricted to symmetric form:

$$(x_{n+1} + x_n)(x_n + x_{n-1}) = \frac{(x_n + z_n + k_n)(x_n^2 - b^2)}{x_n - 2z_n} \tag{9}$$

The full freedom this equation is $z_n = \alpha n + \beta + j^n \gamma + j^{2n} \delta$, $k_n = \zeta + (-1)^n \theta - 3j^n \gamma - 3j^{2n} \delta$, so k is a constant (as implied in [15]) only in the "symmetric" case $\gamma = \delta = \theta = 0$. When all five parameters taken into account, the description of the equation is in terms of the affine Weyl group $E_6^{(1)}$.

Three different equations from the q-P_V family also exist as discrete P_{II}'s.
6. The first is

$$(x_{n+1}x_n - 1)(x_n x_{n-1} - 1) = \frac{a_n x_n}{x_n - b_n} \tag{10}$$

where $a_n = a_0 \lambda^{2n}$ and $b_n = b_0 \lambda^n$. The quantities a, b are related through $b_{n+1}/b_{n-1} = a_n/a_{n-1}$. This would suggest an even/odd degree of freedom for b. However it is straightforward to absorb this extra parameter into a trivial gauge and thus Eq. (10) has just one genuine parameter: it is a q-discrete form of P_{II} (something that was never before pointed out although this equation is known for quite a few years). The geometrical description of (10) can be given in terms of the affine Weyl group $A_1^{(1)} + A_1^{(1)}$. This is also one of the few cases which are not self-dual.
7. The second equation is

$$(x_{n+1}x_n - 1)(x_n x_{n-1} - 1) = a_n(x_n - b_n) \tag{11}$$

(as a matter of fact this equation does not appear in the classification of d-$\mathbb{P}$'s presented in [15]: this is an ommission). The full freedom of (11) is obtained through $\log a_n = 3\alpha n + \beta$, $\log b_n = -\alpha n + \gamma + j^n \delta + j^{2n} \zeta$. Its geometrical description can be given in the framework of the affine Weyl group $A_4^{(1)}$. In the symmetric case ($\delta = \zeta = 0$) the continuous limit of (11) is obtained through $x = (1 + \epsilon w)/\sqrt{3}$, $\alpha = \epsilon^3$, $e^\beta = -8/(3\sqrt{3})$, $e^\gamma = (1 - c\epsilon^3/6)\sqrt{3}/2$ and at the limit $\epsilon \to 0$ we find (with $t = n\epsilon$), $d^2w/dt^2 = 2w^3 + 12tw + c$.
8. Finally one last discrete P_{II} was identified in [15] in the form

$$(x_{n+1}x_n - 1)(x_n x_{n-1} - 1) = (1 - a_n x_n)(1 - x_n b_n) \tag{12}$$

The singularity analysis of this equation results into the following expressions for a, b compatible with the confinement property: $\log a_n = \alpha n + \beta + (-1)^n(n\gamma + \delta) + i^n \zeta + (-i)^n \eta$, $\log b_n = \alpha n + \theta + (-1)^n(n\gamma + \kappa) - i^n \zeta - (-i)^n \eta$. The interesting thing in this case is the presence of the term $e^{(-1)^n(n\gamma+\delta)}$ (or $\delta \to \kappa$). One of the $e^{(-1)^n \delta}$ term can be gauged out (only the quantity $\delta - \kappa$ is fixed) so the total number of parameters is 5, and the geometry is that of the affine Weyl group $E_6^{(1)}$. However, the remaining $e^{(-1)^n n\gamma}$ cannot be gauged out and is indeed corresponds to one of the genuine parameters, resulting into a term $\rho^{\pm n}$ (the sign depending on the parity of n). This is really a unique feature: we do not know of any other example where such a term appears.

To summarize, we have eight various forms of discrete P_{II}, where two are genuine discrete analogues of P_{II}, (2) and (6), and the remaining six are symmetric reductions of equations with more degrees of freedom. Still, as they stand all eight constitute discretizations of P_{II}.

3 BILINEARIZING THE DISCRETE P_{II} EQUATIONS

The bilinearizing ansatz, i.e., expressing the nonlinear dependent variable in terms of τ-functions, can be guided by the singularity structure of the equation. As we have explained in [3] one can guess the minimal number of τ-functions necessary for the bilinearization on the basis of the various singularity patterns. (However, it often turns out that a complete bilinearization necessitates also the introduction of auxiliary τ-functions).

Let us present in detail the bilinearization of the standard d-P_{II} which will serve as a guide for the remaining cases. In the case of d-P_{II} we have two singularity patterns, and so we expect two τ-functions to appear in the expression of x. We start with the pattern $\{-1, \infty, +1\}$. The diverging x may be, related to a vanishing τ-function, say F, in the denominator. In order to ensure that x_{n-1} and x_{n+1} are respectively -1 and $+1$, we choose x_n in the form $x_n = -1 + \frac{F_{n+1}}{F_n} p = 1 + \frac{F_{n-1}}{F_n} q$, where p, q must be expressed in terms of the second τ-function G. We turn now to the second pattern $\{+1, \infty, -1\}$ related to the vanishing of the τ-function G. We find, in this case, $x_n = 1 + \frac{G_{n+1}}{G_n} r = -1 + \frac{G_{n-1}}{G_n} s$, where r, s are expressed in terms of F. Combining the two expressions in terms of F and G we find, with the appropriate choice of gauge, the following simple expression for x:

$$x_n = -1 + \frac{F_{n+1}G_{n-1}}{F_n G_n} = 1 - \frac{F_{n-1}G_{n+1}}{F_n G_n} \tag{13}$$

which satisfies both singularity patterns. Thanks to this particular choice of gauge the relative sign is such that the continuous limit of (13), obtained through $x = \epsilon w$, is $w = \partial_t z \log \frac{F}{G}$, i.e., precisely the transformation in the case of P_{II} we encountered in the introduction. Since, two τ-functions are present here, we expect d-P_{II} to be given as a system of two bilinear relations. Equation (13) does indeed provide the first equation of the system. By eliminating the denominator $F_n G_n$ we obtain

$$F_{n+1}G_{n-1} + F_{n-1}G_{n+1} - 2F_n G_n = 0 \tag{14}$$

In order to obtain the second equation we rewrite d-P_{II} as $(x_{n+1} + x_{n-1})(1 - x_n)(1 + x_n) = z_n x_n + a_n$. We use the two possible definitions of x_n in terms of F, G in order to simplify the expressions $1 - x_n$ and $1 + x_n$. Next, we obtain two equations by using these two definitions for x_{n+1} combined with the alternate definition for x_{n-1}. We obtain thus

$$F_{n+2}F_{n-1}G_{n-1} - F_{n-2}F_{n+1}G_{n+1} = F_n^2 G_n(z_n x_n + a_n) \tag{15a}$$

$$G_{n-2}G_{n+1}F_{n+1} - G_{n+2}G_{n-1}F_{n-1} = G_n^2 F_n(z_n x_n + a_n) \tag{15b}$$

Finally, we add Eq. (15a) multiplied by G_{n+2} and (15b) multiplied by F_{n+2}. Up to the use of the upshift of (14), a factor $F_{n+1}G_{n+1}$ appears in both sides of the resulting expression. After simplification, the remaining equation is indeed bilinear:

$$F_{n+2}G_{n-2} - F_{n-2}G_{n+2} = z_n(F_{n+1}G_{n-1} - F_{n-1}G_{n+1}) + 2a_n F_n G_n \quad (16)$$

where a symmetric expression was used for x in the R.H.S., obtained as the arithmetic mean of the two R.H.S. of (13). Equations (14) and (16), constitute the bilinearization of d-P_{II}.

The bilinearization of the "alternate" d-P_{II} was presented in [14]. Two singular patterns exist $\{0, \infty\}$ and $\{\infty, 0\}$. This suggests the introduction of two τ-functions and the substitution:

$$x_n = \frac{F_n G_{n-1}}{F_{n-1}G_n} \quad (17)$$

From the forms that appear in the denominators of the L.H.S. of (6) it is clear that the only hope for a simplification is when there exists some relation between numerator and denominator. This leads to the introduction of a first bilinear condition:

$$F_{n+1}G_{n-1} + F_{n-1}G_{n+1} = z_n F_n G_n \quad (18)$$

However, even with the use of (18), the equation we obtain is quadrilinear. In order to simplify it further we introduce a third, (auxiliary), τ-function E:

$$F_{n+1}F_{n-1} = F_n^2 + G_n E_n \quad (19)$$

Using (18) and (19) we can bilinearize (6). We obtain finally

$$G_n E_{n-1} - G_{n-1}E_n = \mu F_n F_{n-1} \quad (20)$$

Equations (18), (19), and (20) are the bilinear expression of alternate d-P_{II}.

We turn now to the q-P_{II} equations, cases (3) and (4) above. Equation (7) has the singularity patterns $\{1, \infty, 0, b\}$ and $\{b, 0, \infty, 1\}$. This suggests the following ansatz:

$$x_n = \frac{F_n G_{n+1}}{F_{n+1}G_n} = 1 + a_n \frac{F_{n+2}G_{n-1}}{F_{n+1}G_n} = b_n + \frac{F_{n-1}G_{n+2}}{F_{n+1}G_n} \quad (21)$$

These three different definitions of x yield two bilinear equations:

$$F_n G_{n+1} = F_{n+1}G_n + a_n F_{n+2}G_{n-1} = b_n F_{n+1}G_n + F_{n-1}G_{n+2} \quad (22)$$

Similarly for equation (8) we have the singularity patterns $\{1, \infty, \underline{a}, 0, b\}$ and $\{b, 0, \overline{a}, \infty, 1\}$. By under- and overlining we mean that the values of a are not the relevant local ones but rather the precise values taken one step backward and forward respectively. The bilinear ansatz is

$$x_n = \frac{F_{n-1}G_{n+1}}{F_{n+1}G_{n-1}} = 1 + a_n \frac{F_{n+2}G_{n-2}}{F_{n+1}G_{n-1}} = b_n + \frac{F_{n-2}G_{n+2}}{F_{n+1}G_{n-1}} \quad (23)$$

leading to

$$F_{n-1}G_{n+1} = F_{n+1}G_{n-1} + a_n F_{n+2}G_{n-2} = b_n F_{n+1}G_{n-1} + F_{n-2}G_{n+2} \qquad (24)$$

and Eq. (8) is identically satisfied.

For the bilinearization of (9), case (5), we start from the singularity patterns $\{-z - k, \underline{z + k}, \infty, 2b\}$, $\{2z, \infty, \overline{z + k}, -z - k\}$, $\{b, -b\}$, and $\{-b, b\}$, where under- and overlining have to be intepreted as above. We introduce the ansätze:

$$x_n = 2z_n - \frac{F_{n+1}G_{n-1}}{F_n G_n} = -z_n - k_n + \frac{F_{n-2}G_{n+2}}{F_n G_n} = b + \frac{M_n N_n}{F_n G_n}$$

$$= -b + \frac{M_{n-1}N_{n+1}}{F_n G_n} \qquad (25)$$

and introduce two auxiliary τ-functions through

$$x_n = z_{n-1} + k_{n-1} + \frac{H_{n-1}G_{n+1}}{F_n G_n} = z_{n+1} + k_{n+1} + \frac{F_{n-1}K_{k+1}}{F_n G_n} \qquad (26)$$

We have thus five equations for the six τ-functions. The remaining equation can be obtained by substituting into (9). We find

$$G_{n+2}(F_{n-2}F_{n+1}G_{n+1} + H_n F_n G_n)G_{n+1}((F_{n-3}F_n G_n + H_{n-1}F_{n-1}G_{n-1})$$
$$= (F_{n-1}G_{n+1}F_{n-2}G_{n+2} + M_n M_{n-1}N_n N_{n+1}) \qquad (27)$$

we remark readily that both sides are products of the form $\Omega_n \Omega_{n-1}$. This suggests a separation which leads finally to the trilinear equation:

$$F_{n-2}F_{n+1}G_{n+1} + H_n F_n G_n + M_n N_{n+1}F_{n-1} = 0 \qquad (28)$$

It turns out that it is possible, introducing the appropriate combination and using some of the five equations mentioned above, to reduce the trilinear equation to a bilinear one:

$$G_{n-2}F_{n+2} + 2(z_n + k_n)F_n G_n = G_{n+1}H_{n-1} \qquad (29)$$

As a matter of fact, we could have obtained this equation directly if we had introduced the ansatz:

$$x_n = 2z_n + 2k_n - z_{n-1} - k_{n-1} + \frac{F_{n+2}G_{n-2}}{F_n G_n} \qquad (30)$$

(but there is no way one could have guessed that this was the proper ansatz before obtaining (29) the hard way).

For Eq. (10), case (6), we have the singularity patterns $\{0, \infty, b\}$ and $\{b, \infty, 0\}$. We thus introduce the τ-functions through

$$x_n = \frac{F_{n+1}G_{n-1}}{F_n G_n} = b_n + \frac{F_{n-1}G_{n+1}}{F_n G_n} \qquad (31)$$

yielding a first bilinear equation

$$F_{n+1}G_{n-1} = b_n F_n G_n + F_{n-1}G_{n+1} \tag{32}$$

Substituting into the mapping (10) we find

$$(F_{n+2}G_{n-1} - F_n G_{n+1})(F_{n+1}G_{n-2} - F_{n-1}G_n) = a_n F_{n+1}G_n F_n G_{n-1} \tag{33}$$

Putting $a_n = \rho_n \rho_{n-1}$ we can separate this equation and obtain

$$F_{n+2}G_{n-1} - F_n G_{n+1} = \rho_n F_{n+1}G_n \tag{34}$$

Note that, with this definition of ρ_n, we have $b_{n+1}/b_{n-1} = a_n/a_{n-1} = \rho_n/\rho_{n-2}$ so $b_{n+1} = \rho_n$ is a constant if an appropriate gauge is used to eliminate the spurious even–odd degree of freedom.

For case (7), Eq. (11) we have the singularity patterns $\{b, 1/b, \infty, 0\}$ and $\{0, \infty, \overline{1/b}, b\}$. This leads to the ansatz

$$x_n = \frac{F_{n-1}G_{n+1}}{F_n G_n} = b + \frac{F_{n+2}G_{n-2}}{F_n G_n} \tag{35}$$

and a first equation

$$F_{n-1}G_{n+1} = bF_n G_n + F_{n+2}G_{n-1} \tag{36}$$

The second equation is obtained exactly as in the case (6) above, introducing $a_n = \rho_n \rho_{n-1}$ and splitting the equation obtained by a direct substitution of the ansatz for x into the mapping. We find thus

$$F_{n-1}G_{n+2} - F_{n+1}G_n = \rho_n F_{n+2}G_{n-2} \tag{37}$$

The final case we shall examine is (8), i.e., Eq. (12). The singularity patterns in this case are $\{1/a, \underline{a}, \infty, \overline{b}, 1/b\}$ and $\{1/b, \underline{b}, \infty, \overline{a}, 1/a\}$. In this case auxiliary τ-functions, related to the central patterns $\{\underline{a}, \infty, \overline{b}\}$ and $\{\underline{b}, \infty, \overline{a}\}$ must be introduced. We have in all

$$x_n = \frac{1}{a_n}\left(1 + \frac{F_{n-2}G_{n+2}}{F_n G_n}\right) = \frac{1}{b_n}\left(1 + \frac{F_{n+2}G_{n-2}}{F_n G_n}\right) \tag{38}$$

$$= a_{n+1}\left(1 + \frac{F_{n-1}H_{n+1}}{F_n G_n}\right) = b_{n-1}\left(1 + \frac{F_{n+1}K_{n-1}}{F_n G_n}\right)$$

$$= b_{n+1}\left(1 + \frac{G_{n-1}M_{n+1}}{F_n G_n}\right) = a_{n-1}\left(1 + \frac{G_{n+1}N_{n-1}}{F_n G_n}\right)$$

We have thus five equations for the six τ-functions. The sixth equation can be obtained by direct substitution into the mapping (12). Working with the auxiliary τ-function K we find that, as in the previous cases, the equation can be written as a product of the form $\Omega_n\Omega_{n-1}$. We introduce $a_n b_n = \rho_n \rho_{n-1}$ and separate obtaining the trilinear equation:

$$K_n F_n G_n + G_{n-2} F_{n+1} G_{n+1} + F_{n+2} G_{n-2} K_n = \rho_n G_{n+2} F_{n-1} G_{n-1} \quad (39)$$

Similarly, working with any of the H,M,N we could have obtained equations involving these τ-functions. On the other hand one could have dispensed altogether with them, work with just F, G, and the auxiliary K and obtain just three equations which fully describe (12) in terms of τ-functions.

4 CONCLUSIONS

In the preceding sections we have investigated the various forms of discrete P_{II} equations. Eight different equations were analyzed. We have shown that the equations

$$\frac{z_n}{x_{n+1}x_n + 1} + \frac{z_{n-1}}{x_n x_{n-1} + 1} = -x_n + \frac{1}{x_n} + z_n + \mu \quad (40)$$

and

$$(x_{x+1}x_n - 1)(x_n x_{n-1} - 1) = \frac{a_n x_n}{x_n - b_n} \quad (41)$$

are indeed discrete (difference and q respectively) forms of P_{II}. All the cases correspond to symmetric reductions of equations with a higher number of parameters. In two cases, (9) and (12), the discrete Painlevé equations involve five parameters and their geometry is described by the affine Weyl group $E_6^{(1)}$. At this point we must stress that more discrete forms of P_{II} certainly exist, in particular forms related to the affine Weyl groups $E_7^{(1)}$ and $E_8^{(1)}$ which have not been studied in detail yet.

A final remark is in order here. As in the continuous case d-P_{II} is not the only one-parameter discrete Painlevé equation. As we have explained in [19] the one-parameter P_{III} is a Painlevé equation in its own right (its solution introduces a Painlevé transcendent different from the other six ones). Discrete form of one-parameter P_{III} are known. No systematic search for equations that have the one-parameter P_{III} as their continuous limit in the symmetric case (i.e., when all, or maybe only some, periodic degrees of freedom are not taken into account). But, quite expectedly there exist difference and q-forms of genuine one parameter P_{III}. The first is

$$\left(1 + \frac{a_n}{x_n + x_{n-1}}\right)\left(1 + \frac{a_{n+1}}{x_{n+1} + x_n}\right) = \frac{1}{x_n^2 - b^2} \quad (42)$$

with $a_n = \alpha n + \beta$ and b is the single constant parameter. The second one cannot be written at all as a single component equation but only as a system:

$$x_n x_{n+1} = \frac{1 + a_n y_n}{y_n(by_n + 1)} \tag{43a}$$

$$y_n y_{n-1} = \frac{1 + x_n}{b x_n^2} \tag{43b}$$

with $\log a_n = \alpha n + \beta$.

As this study clearly shows the richness of the dP's surpasses even the most optimistic predictions. Many more studies are certainly needed in order to complete the exploration of this domain which was terra incognita a mere 10 years ago.

REFERENCES

1. Painlevé, P. (1902) Sur les équations différentielles du second ordre et d'ordre supérieur dont l'intégrale générale est uniforme, *Acta Math.* **25**, pp. 1–85.
2. Grammaticos, B., Ramani, A., and Papageorgiou, V. (1991) Do integrable mappings have the Painlevé, property?, *Phys. Rev. Lett.* **67**, pp. 1825–1828.
3. Ramani, A., Grammaticos, B., and Satsuma, J. (1995) Bilinear discrete Painlevé equations, *J. Phys. A.* **28**, pp. 4655–4665.
4. Ohta, Y., Ramani, A., Grammaticos, B., and Tamizhmani, K. M. (1996) From discrete to continuous Painlevé equations: a bilinear approach, *Phys. Lett. A.* **216**, pp. 255–261.
5. Sakai, H. (2001) Rational surfaces associated with affine root systems and geometry of the Painlevé equations, *Comm. Math. Phys.* **220**, pp. 1165–229.
6. Ohta, Y., Ramani, A., and Grammaticos, B. (2002) Elliptic discrete Painlevé equations, *J. Phys. A.* **35**, pp. L653–L659.
7. Grammaticos, B., Tamizhmani, T., Ramani, A., Carstea, A. S., and Tamizhmani, K. M. (2002) A bilinear approach to the discrete Painlevé I equations, *J. Phys. Soc. Japan* **71**, pp. 443–447.
8. Grammaticos, B., Nijhoff, F. W., Papageorgiou, V., Ramani, A., and Satsuma, J. (1994) Linearization and solution of the discrete Painlevé-III equation, *Phys. Lett. A.* **185**, pp. 446–452.
9. Quispel, G. R. W, Roberts, J. A. G., and Thompson, C. J. (1988) Integrable mappings and soliton equations, *Phys. Lett. A.* **126**, pp. 419–421.
10. Ohta, Y., Tamizhmani, K. M., Grammaticos, B., and Ramani, A. (1999) Singularity confinement and algebraic entropy: the case of the discrete Painlevé equations, *Phys. Lett. A.* **262**, pp. 152–156.
11. Ramani, A., Grammaticos, B., and Hietarinta, J. (1991) Discrete versions of the Painlevé equations, *Phys. Rev. Lett.* **67**, pp. 1829–1832.
12. Ramani, A., Ohta, Y., Satsuma, J., and Grammaticos, B. (1998) Self-duality and Schlesinger chains for the, asymmetric d-PII and q-PIII equations, *Comm. Math. Phys.* **192**, pp. 67–76.

13. Grammaticos, B. and Ramani, A. (1993) From continuous to discrete Painlevé equations, A. Fokas, *J. Math. Anal. and Appl.* **180**, pp. 342–360.
14. Nijhoff, F. W., Satsuma, J., Kajiwara, K., Grammaticos, B., and Ramani, A. (1996) A study of the alternate discrete Painlevé-II equation, *Inv. Probl.* **12**, pp. 697–716.
15. Ramani, A. and Grammaticos, B. (1996) Discrete Painlevé equations: coalescences, limits and degeneracies, *Physica. A.* **228**, pp. 160–171.
16. Kruskal, M. D., Tamizhmani, K. M., Grammaticos, B., and Ramani, A. (2000) Asymmetric discrete Painlevé equations, *Reg. Chaot. Dyn.* **5**, pp. 1–8.
17. Kajiwara, K., Noumi, M., and Yamada, Y. (2001) A study on the fourth q-Painlevé equation, *J. Phys. A.* **34**, pp. 8563–8581.
18. Jimbo, M. and Sakai, H. (1996) A q-analog of the sixth Painlevé equation, *Lett. Math. Phys.* **38**, pp. 145–154.
19. Ramani, A., Grammaticos, B., Tamizhmani, T., and Tamizhmani, K. M. (2000) On a transcendental equation related to Painlevé III equation and its discrete forms, *J. Phys. A.* **33**, pp. 579–590.
20. Ramani, A. and Grammaticos, B. (1996) The grand scheme for discrete Painlevé equations, in: *Lecture at the Toda 96 symposium.*

ORTHOGONAL POLYNOMIALS SATISFYING Q-DIFFERENCE EQUATIONS

Luc Haine

Department of Mathematics, Université catholique de Louvain,
Chemin du Cyclotron 2, 1348 Louvain-la-Neuve Belgium

Abstract   The Askey–Wilson polynomials are the most general orthogonal polynomials, which are eigenfunctions of a second order q-difference operator.
I survey recent results aiming at constructing all orthogonal polynomials
which are eigenfunctions of a q-difference operator of an arbitrary order,
by means of the lattice Darboux transformation.

1   INTRODUCTION

The method of the Darboux transformation that was systematized by Matveev
and Salle (see [1] and references therein), is a basic tool in the theory of integrable systems which generates the soliton solutions from the "vacuum". It is
intimately related with Hirota's bilinear method, via the theory of vertex operators, see [2, 3]. In this note, I want to explain that the method turns out to be
very useful to handle some problems in the theory of orthogonal polynomials,
going back to S. Bochner and H.L. Krall. The role of the "vacuum" will be
played by the classical orthogonal polynomials. A purely continuous version
of these problems was first considered and completely solved by Duistermaat
and Grünbaum [4]. Since then, the name "bispectral problem" is often attached
to this general area of research. In recent years, bispectral problems have been
shown to be related with various areas of mathematics, like representation
theory [5], noncommutative algebraic geometry [6] and combinatorics [7].

In 1938, H. L. Krall [8] posed the problem to find all families of orthogonal
polynomials $\{p_n(x)\}_{n=0}^{\infty}$, which are also eigenfunctions of a differential operator
$\sum_{i=0}^{m} a_i(x)(d/dx)^i$ of an arbitrary order. In particular, he showed that the order
m must be even. When $m = 2$, the problem was already solved by Bochner
[9] in 1929. The only solutions are provided by the Hermite, the Laguerre, the
Jacobi and the (lesser known) Bessel polynomials. The next case in line, $m = 4$,
was completely solved by H. L. Krall [10] in 1940. He discovered that, besides
the previous solutions, there are three new families of orthogonal polynomials,

L. Faddeev et al. (eds.),
Bilinear Integrable Systems: From Classical to Quantum, Continuous to Discrete, 97–112.

which are eigenfunctions of a differential operator of order 4. The easiest way to describe them is via their weight functions, which are obtained by adding one or two mass points at the boundary of the interval of orthogonality of some instances of the classical orthogonal polynomials. Namely:

$$e^{-x}dx + r\delta(x) \text{ on } [0, \infty[, \text{ Krall-Laguerre weight,}$$
$$(1+x)^{\beta}dx + r\delta(x-1) \text{ on } [-1, 1], \quad \beta > -1, \text{ Krall-Jacobi weight,}$$
$$dx + r\{\delta(x+1) + \delta(x-1) \text{ on } [-1, 1], \text{ Krall-Legendre weight,}$$

with $r > 0$, *a free parameter*, and $\delta(x)$ Dirac's delta function.

In 1996, jointly with F.A. Grünbaum [11], I was able to fit this result within the apparatus of the lattice Darboux transform. This paper was the source of a renewed interest in Krall's problem, see [12–17]. In the sequel, I denote by $\mathbb{N}$ the set of positive integers, and by $\mathbb{Z}$ the set of non-negative integers. For the purpose of this note, I shall only mention the following generalization of Krall's result:

Theorem 1 ([14, 15]) *The orthogonal polynomials with weight distributions given by*

$$x^{\alpha}e^{-x}dx + \sum_{\text{finitely many } k\in\mathbb{Z}_+} r_k\delta^{(k)}(x) \text{ on } [0, \infty[, \quad \alpha \in \mathbb{Z}_+,$$

Krall-Laguerre type weight distribution,

$$(1-x)^{\alpha}(1+x)^{\beta}dx + \sum_{\text{finitely many } k\in\mathbb{Z}_+} r_k\delta^{(k)}(x-1) \text{ on } [-1, 1],$$

$$\alpha \in \mathbb{Z}_+, \quad \beta > -1, \text{ or}$$

$$(1-x)^{\alpha}(1+x)^{\beta}dx + \sum_{\text{finitely many } k\in\mathbb{Z}_+} r_k\delta^{(k)}(x-1)$$

$$+ \sum_{\text{finitely many } k\in\mathbb{Z}_+} s_k\delta^{(k)}(x+1) \text{ on } [-1, 1], \quad \alpha, \beta \in \mathbb{Z}_+,$$

Krall–Jacobi type weight distribution,

where $r_k, s_k > 0$ *denote arbitrary free parameters, and* $\delta^{(k)}$ *denotes the k-th. derivative of Dirac's delta function, solve Krall's problem, that is they are eigenfunctions of a differential operator.*

Some of the solutions described in Theorem 1 had already been found by followers of H.L. Krall, see [18, 19–22] and references therein. However, situations involving derivatives of Dirac's delta function were not considered in these works. The reason might be that in these cases, there is no orthogonality *measure*. My main message in this note is that by replacing in Krall's original problem the differential operator by a q-difference operator, *the added mass points split up* and we do get an orthogonality measure. It is only in the limit $q \to 1$ that all the mass points accumulate at the boundary of the interval of orthogonality. The role of the Laguerre and the Jacobi polynomials will now be taken up by the celebrated Askey–Wilson polynomials.

2   CHRISTOFFEL AND GERONIMUS TRANSFORMS

In the literature on orthogonal polynomials, Darboux transformations are better known under the names of Christoffel and Geronimus transformations. In this section, I review some basic facts about these transforms following [11, 13, 14, 17]. Consider a sequence $\{p_n(x)\}_{n=0}^{\infty}$ of monic polynomials defined by a three-term recurrence relation

$$p_{n+1}(x) = (x - b_n)p_n(x) - c_n p_{n-1}(x), \quad p_{-1}(x) = 0, \quad p_0(x) = 1, \quad (1)$$

with $b_n, c_n \in \mathbb{C}$, and $c_n \neq 0, \forall n \geq 1$. A classical theorem of Favard, see [23], asserts that, up to a non-zero constant factor, there is a unique quasi-definite moment functional $\mathcal{L}$ for which the sequence $p_n(x), n \geq 0$, is an orthogonal polynomial sequence (in short OPS), that is

$$\mathcal{L}[p_m(x)p_n(x)] = 0 \quad \text{for } m \neq n \quad \text{and} \quad \mathcal{L}[p_n^2(x)] \neq 0.$$

We remind the reader that given a sequence $\{\mu_n\}_{n=0}^{\infty}$ of complex numbers, a moment functional $\mathcal{L}$ is a complex valued linear functional defined on the vector space of all polynomials by $\mathcal{L}[x^n] = \mu_n, n = 0, 1, 2, \ldots$, and extended by linearity. It is called quasi-definite if and only if all the Hankel determinants $\Delta_n = \det(\mu_{i+j})_{i,j=0}^{n} \neq 0$.

We denote by J the tridiagonal Jacobi matrix

$$J = \begin{pmatrix} b_0 & 1 & 0 & 0 & \cdots \\ c_1 & b_1 & 1 & 0 & \cdots \\ 0 & c_2 & b_2 & 1 & \cdots \\ \vdots & \vdots & \vdots & & \ddots \end{pmatrix}.$$

It will be convenient to use the notations of difference operators. Denoting by E and E^{-1}, respectively, the customary backward and forward shift operators, acting on a function h_n by

$$Eh_n = h_{n+1}, \quad n \geq 0, \quad E^{-1}h_0 = 0, \quad E^{-1}h_n = h_{n-1}, \quad n \geq 1,$$

we can write

$$J \equiv J(n, E) = c_n E^{-1} + b_n \text{Id} + E, \quad (2)$$

with Id the identity operator. With the same conventions, we introduce the lower and upper semi-infinite matrices L and U (with two diagonals) by

$$L \equiv L(n, E) = \alpha_n E^{-1} + \text{Id}, \quad U \equiv U(n, E) = \beta_n \text{Id} + E. \quad (3)$$

Given $\lambda \in \mathbb{C}$, the so-called *kernel polynomials*

$$p_n^*(\lambda; x) = (x - \lambda)^{-1} \left[p_{n+1}(x) - \frac{p_{n+1}(\lambda)}{p_n(\lambda)} p_n(x) \right], \qquad (4)$$

are defined as long as $p_n(\lambda) \neq 0, \forall n$. These polynomials form an OPS with respect to the quasi-definite moment functional

$$\mathcal{L}^* = (x - \lambda)\mathcal{L}, \qquad (5)$$

where $(x - \lambda)\mathcal{L}$ is defined on an arbitrary polynomial $\pi(x)$ by $(x - \lambda)$ $\mathcal{L}[\pi(x)] = \mathcal{L}[(x - \lambda)\pi(x)]$. They can be obtained in terms of what is called in [13] a Darboux transformation without free parameter and, according to [17], should be called a Christoffel transform.

Proposition 2 ([13], [17]). *Assume that $p_n(\lambda) \neq 0, \forall n$. Then, the matrix $J - \lambda \operatorname{Id}$ can be uniquely factorized as a product of a lower and an upper matrix as defined in* (3). *The Jacobi matrix J^* defined by the transform*

$$J = \lambda \operatorname{Id} + LU \to J^* = \lambda \operatorname{Id} + UL, \qquad (6)$$

gives the coefficients of the three-term recurrence relation satisfied by the kernel polynomials (4).

In [11], we observed that if one performs an upper-lower factorization

$$J - \lambda \operatorname{Id} = UL,$$

instead of a lower-upper one, there is a free parameter in the factorization. Indeed, solving inductively for the entries of L and U one finds

$$\beta_n = \frac{c_n}{b_{n-1} - \beta_{n-1} - \lambda}, \quad \alpha_n = b_{n-1} - \beta_{n-1} - \lambda, \quad n \geq 1,$$

with β_0 *a free parameter*. In [11], we called the transform

$$J = \lambda \operatorname{Id} + UL \to \tilde{J} = \lambda \operatorname{Id} + LU, \qquad (7)$$

a Darboux transform with free parameter, since this version of the Darboux process can be made to fit within the framework proposed by Matveev and Salle [1] in the context of a doubly infinite tridiagonal matrix. According to [17], this transform was first considered in 1940 by Ya. L. Geronimus and should be called a Geronimus transform.

Proposition 3 ([14]). *If $\tilde{J}$ satisfies the hypotheses of Favard's theorem, then the unique (up to a non-zero constant) moment functional $\tilde{\mathcal{L}}$ making the* *polynomials $\tilde{p}_n(x)$ defined by $x\tilde{p} = \tilde{J}\tilde{p}$ a OPS, is given by*

$$\tilde{\mathcal{L}}[1] = \frac{\mu_0}{\beta_0}, \tag{8}$$

$$\tilde{\mathcal{L}}[x^n(x - \lambda)] = \mu_n, \; n \geq 0, \tag{9}$$

with $\mu_n = \mathcal{L}[x^n]$ and $\mathcal{L}$ the moment functional of the OPS formed by the $p_n(x), n \geq 0$.

Proof The monic polynomials defined by the recurrence relation $\tilde{J}\tilde{p} = x\tilde{p}$ are given by $\tilde{p} = Lp$. Hence $xp = Jp = \lambda p + ULp = \lambda p + U\tilde{p}$, that is

$$p_n(x) = (x - \lambda)^{-1}[\tilde{p}_{n+1}(x) + \beta_n\tilde{p}_n(x)], \quad \forall n \geq 0. \tag{10}$$

This shows that, up to a non-zero constant factor,

$$\mathcal{L} = (x - \lambda)\tilde{\mathcal{L}}, \tag{11}$$

which establishes (9). From (10) with $n = 0$, we get $x - \lambda = \tilde{p}_1(x) + \beta_0$, which gives

$$\tilde{\mathcal{L}}[x - \lambda] = \beta_0\tilde{\mathcal{L}}[1]. \tag{12}$$

From (11) we have $\mathcal{L}[1] = \tilde{\mathcal{L}}[x - \lambda]$, which combined with (12) gives (8), completing the proof. $\qquad\square$

Defining for any polynomial $\pi(x)$

$$\left(\frac{1}{x - \lambda}\mathcal{L}\right)[\pi(x)] = \mathcal{L}\left[\frac{\pi(x) - \pi(\lambda)}{x - \lambda}\right], \quad \delta(x - \lambda)[\pi(x)] = \pi(\lambda),$$

the functional $\tilde{\mathcal{L}}$ defined by (8) and (9) can be written as follows

$$\tilde{\mathcal{L}} = \frac{1}{x - \lambda}\mathcal{L} + \frac{\mu_0}{\beta_0}\delta(x - \lambda). \tag{13}$$

3 THE ASKEY–WILSON POLYNOMIALS

I use the standard notations for basic hypergeometric series, following the book of Gasper and Rahman [24]. In particular, I write

$$(a_1, a_2, \ldots, a_r; q)_k = \prod_{i=1}^{r}(a_i; q)_k,$$

with

$$(a; q)_k = \frac{(a; q)_\infty}{(aq^k; q)_\infty} \quad \text{and} \quad (a; q)_\infty = \prod_{i=0}^{\infty}(1 - aq^i),$$

for products of q-shifted factorials, where $0 < q < 1$. The series expansion

$$_r\phi_s \begin{bmatrix} a_1, a_2, \ldots, a_r \\ b_1, b_2, \ldots, b_s \end{bmatrix}; q, z \end{bmatrix} = \sum_{k=0}^{\infty} \frac{(a_1, a_2, \ldots, a_r; q)_k}{(q, b_1, \ldots, b_s; q)_k} \left[(-1)^k q^{\binom{k}{2}} \right]^{1+s-r} z^k$$

defines the $_r\phi_s$ basic hypergeometric series.

The celebrated Askey–Wilson polynomials [23] are defined as follows in terms of basic hypergeometric series

$$P_n(z; a, b, c, d) = {}_4\phi_3 \begin{bmatrix} abcdq^{n-1}, q^{-n}, az, az^{-1} \\ ab, ac, ad \end{bmatrix}; q, q \end{bmatrix}, \quad n \geq 0. \quad (14)$$

They satisfy the three-term recurrence relation

$$A_{n;a,b,c,d} P_{n+1}(z; a, b, c, d) + B_{n;a,b,c,d} P_n(z; a, b, c, d)$$
$$+ C_{n;a,b,c,d} P_{n-1}(z; a, b, c, d) = (z + z^{-1}) P_n(z; a, b, c, d), \quad (15)$$

with

$$A_{n;a,b,c,d} = \frac{(1 - abq^n)(1 - acq^n)(1 - adq^n)(1 - abcdq^{n-1})}{a(1 - abcdq^{2n-1})(1 - abcdq^{2n})}, \quad (16)$$

$$C_{n;a,b,c,d} = \frac{a(1 - q^n)(1 - bcq^{n-1})(1 - bdq^{n-1})(1 - cdq^{n-1})}{(1 - abcdq^{2n-2})(1 - abcdq^{2n-1})}, \quad (17)$$

$$B_{n;a,b,c,d} = a + a^{-1} - (A_{n;a,b,c,d} + C_{n;a,b,c,d}). \quad (18)$$

The Askey–Wilson polynomials are also eigenfunctions of a *second order q-difference operator* (they form, in fact, the most general class of orthogonal polynomials with this property, see [25, 26]). This can be seen from a *duality relation* that these polynomials satisfy. Putting

$$a' = \sqrt{q^{-1}abcd}, \quad b' = \frac{ab}{a'}, \quad c' = \frac{ac}{a'}, \quad d' = \frac{ad}{a'} \quad \text{and} \quad x = a'q^n,$$

we have

$$P_n(z; a, b, c, d) = {}_4\phi_3 \begin{bmatrix} az, az^{-1}, a'x, a'x^{-1} \\ a'b', a'c', a'd' \end{bmatrix}; q, q \end{bmatrix}$$
$$= P_m(x; a', b', c', d'), \quad \text{if } z = aq^m, m \in \mathbb{Z}_+.$$

Hence,

$$P_n(aq^m, a, b, c, d) = P_m(a'q^n; a', b', c', d'), \quad \forall n, m \in \mathbb{Z}_+. \quad (19)$$

From the duality relation (19), one deduces immediately the next result.

Proposition 4   ([25]) *The Askey–Wilson polynomials satisfy*

$$A(z) P_n(qz; a, b, c, d) - [A(z) + A(z^{-1})] P_n(z; a, b, c, d)$$
$$+ A(z^{-1}) P_n(q^{-1}z; a, b, c, d) = \Lambda_{n;a,b,c,d} P_n(z; a, b, c, d), \quad (20)$$

with

$$A(z) = \frac{(1-az)(1-bz)(1-cz)(1-dz)}{(1-z^2)(1-qz^2)}, \tag{21}$$

$$\Lambda_{n;a,b,c,d} = q^{-n}(1-q^n)(1-abcdq^{n-1}). \tag{22}$$

Notice from (15) that the Askey–Wilson polynomials are polynomials in $z + z^{-1}$. Putting $2x = z + z^{-1}$, one checks easily that, in order for the polynomials to be monic in x, one must put

$$p_n(x;a,b,c,d) = \frac{(ab,ac,ad;q)_n}{2^n a^n (abcdq^{n-1};q)_n} P_n(z;a,b,c,d). \tag{23}$$

The coefficients of the recurrence relation (1) satisfied by $p_n(x;a,b,c,d)$ are given by

$$b_n = \frac{a + a^{-1} - (A_{n;a,b,c,d} + C_{n;a,b,c,d})}{2}, \quad c_n = \frac{(A_{n-1;a,b,c,d} C_{n;a,b,c,d})}{4} \tag{24}$$

We denote the associated Jacobi matrix (2) by $J_{a,b,c,d}$. It is immediate to check from (24) that

$$J_{a,b,c,d} = \frac{a + a^{-1}}{2} \mathrm{Id} + L_{a,b,c,d} U_{a,b,c,d}, \tag{25}$$

with

$$L_{a,b,c,d}(n, E) = \left(-\frac{C_{n;a,b,c,d}}{2} E^{-1} + \mathrm{Id} \right),$$

$$U_{a,b,c,d}(n, E) = \left(-\frac{A_{n;a,b,c,d}}{2} \mathrm{Id} + E \right). \tag{26}$$

In the rest of the paper, I shall assume that a,b,c,d are real and that $\max(|a|, |b|, |c|, |d|) < 1$. Then, the Askey-Wilson polynomials are orthogonal on the interval $[-1, 1]$, with respect to the weight function

$$w(x;a,b,c,d) = \frac{h(x,1)h(x,-1)h(x,q^{\frac{1}{2}})h(x,-q^{\frac{1}{2}})}{h(x,a)h(x,b)h(x,c)h(x,d)} \frac{dx}{\sqrt{1-x^2}}, \tag{27}$$

with $h(x,\alpha) = (\alpha z, \alpha z^{-1}; q)_\infty$. One checks easily that

$$w(x, aq, b, c, d) = -2a\left(x - \frac{a + a^{-1}}{2} \right) w(x;a,b,c,d). \tag{28}$$

From (5), it follows that $p_n(x;aq, b, c, d)$ are the kernel polynomials defined in (4) associated with the OPS $p_n(x;a,b,c,d)$, for $\lambda = \frac{a+a^{-1}}{2}$. From (25), (26) and (6) in Proposition 2, we deduce then that $J_{aq,b,c,d}$ is a Christoffel transform of $J_{a,b,c,d}$. Precisely:

Proposition 5

$$J_{aq,b,c,d} = \frac{a + a^{-1}}{2}\mathrm{Id} + U_{a,b,c,d}L_{a,b,c,d}. \tag{29}$$

In the sequel, it will be crucial to use the standard normalization for the Askey–Wilson polynomials. I shall write the difference operator going with the three-term recurrence relation (15) as

$$R_{a,b,c,d} \equiv R_{a,b,c,d}(n, E) = A_{n;a,b,c,d}E + B_{n;a,b,c,d}\mathrm{Id} + C_{n;a,b,c,d}E^{-1}. \tag{30}$$

From (23), one computes easily that the effect of the Christoffel transform, described in Proposition 5, on $R_{a,b,c,d}$ reads

$$R_{a,b,c,d} = (a + a^{-1})\mathrm{Id} + S_{a,b,c,d}T_{a,b,c,d} \rightarrow$$
$$R_{aq,b,c,d} = (a + a^{-1})\mathrm{Id} + T_{a,b,c,d}S_{a,b,c,d}, \tag{31}$$

with $S_{a,b,c,d}$ and $T_{a,b,c,d}$ the difference operators

$$S_{a,b,c,d}(n, E) = (A_{n;a,b,c,d}\mathrm{Id} - C_{n;a,b,c,d}E^{-1})\frac{1}{\varphi_{n;a,b,c,d}}, \tag{32}$$

$$T_{a,b,c,d}(n, E) = \varphi_{n;a,b,c,d}(E - \mathrm{Id}), \tag{33}$$

with

$$\varphi_{n;a,b,c,d} = \frac{1}{q^{-n} - abcdq^n}.$$

4   RATIONAL DARBOUX TRANSFORMATIONS FROM THE ASKEY–WILSON POLYNOMIALS

In this section, I sketch the main steps for constructing orthogonal polynomials satisfying q-difference equations from the Askey–Wilson polyomials. The crucial new idea is the concept of a *rational* Darboux transformation.

We start from the pair of equations satisfied by the Askey–Wilson polynomials, normalized as in (14)

$$R_{a,b,c,d}(n, E)P_n(z; a, b, c, d) = (z + z^{-1})P_n(z; a, b, c, d), \tag{34}$$
$$B_{a,b,c,d}(z, D_z)P_n(z; a, b, c, d) = \Lambda_{n;a,b,c,d}P_n(z; a, b, c, d), \tag{35}$$

with $R_{a,b,c,d}$ as in (30), $\Lambda_{n;a,b,c,d}$ as in (22) and $B_{a,b,c,d}(z, D_z)$ the second order Askey–Wilson q-difference operator defined in (20)

$$B_{a,b,c,d}(z, D_z) = A(z)D_z - [A(z) + A(z^{-1})]\mathrm{Id} + A(z^{-1})D_z^{-1}, \tag{36}$$

with $A(z)$ as in (21), and D_z and D_z^{-1}, respectively the forward and backward q-shift operators acting on a function $h(z)$ by $D_z h(z) = h(qz)$ and $D_z^{-1} h(z) = h(q^{-1}z)$.

Let us denote by

$$\mathcal{B} = \langle R_{a,b,c,d}, \Lambda_{a,b,c,d} \rangle, \tag{37}$$

the algebra of difference operators generated by $R_{a,b,c,d}$ and $\Lambda_{a,b,c,d}$, with $\Lambda_{a,b,c,d}$ the diagonal operator $\Lambda_{a,b,c,d}(n, E) = \Lambda_{n;a,b,c,d}$ Id. Similarly

$$\mathcal{B}' = \langle z + z^{-1}, B_{a,b,c,d} \rangle, \tag{38}$$

will denote the algebra of q-difference operators generated by the operator of multiplication by $z + z^{-1}$ and the operator $B_{a,b,c,d}$ defined in (36). Formulas (34) and (35) serve to define an anti-isomorphism $\flat : \mathcal{B} \to \mathcal{B}'$ between these two algebras, given on the generators by

$$\flat(R_{a,b,c,d}) = z + z^{-1} \quad \text{and} \quad \flat(\Lambda_{a,b,c,d}) = B_{a,b,c,d}, \tag{39}$$

i.e., $X P_n(z; a, b, c, d) = \flat(X) P_n(z; a, b, c, d), \forall X \in \mathcal{B}$. We also need the commutative subalgebras (the algebras of "functions") of $\mathcal{B}$ and $\mathcal{B}'$

$$\mathcal{K} = \langle \Lambda_{a,b,c,d} \rangle \quad \text{and} \quad \mathcal{K}' = \langle z + z^{-1} \rangle. \tag{40}$$

The next theorem summarizes the technology of *rational Darboux transformations*. It was initiated in [27] and [28], in the context of a study of bispectral rings of commutative differential operators (see also [29]), and was adapted and applied to the case of difference operators in [13, 14].

Theorem 6 ([30], see also [13, 14) *Let $\mathcal{L}$ be a constant coefficient polynomial in $R_{a,b,c,d}$, which factorizes rationally as*

$$\mathcal{L} = \mathcal{P}\mathcal{Q}, \tag{41}$$

in such a way that

$$\mathcal{P} = W\Gamma^{-1}, \quad \mathcal{Q} = \Theta^{-1}V, \tag{42}$$

with $V, W \in \mathcal{B}$, and $\Theta, \Gamma \in \mathcal{K}$. Let

$$\tilde{\mathcal{L}} = \mathcal{Q}\mathcal{P}, \tag{43}$$

be the Darboux transform of $\mathcal{L}$. Then, defining $\mu(z) = \flat(\mathcal{L}) \in \mathcal{K}'$ and $\tilde{P} = \mathcal{Q}P$, with $P = (P_0(z; a, b, c, d), P_1(z; a, b, c, d), \ldots)^T$ satisfying (34) and (35), we have

$$\tilde{\mathcal{L}}\tilde{P} = \mu(z)\tilde{P}, \tag{44}$$

$$\tilde{\mathcal{B}}\tilde{P} = \Theta\Gamma\tilde{P}, \tag{45}$$

with

$$\tilde{B} = b(V)b(W)\mu(z)^{-1}. \tag{46}$$

Despite the apparent simplicity of Theorem 6, it is a priori very complicated to recognize that an operator admits a *rational* Darboux factorization as defined by (41), (42). An important new idea was introduced in [15] in the context of the Jacobi polynomials, in order to make Theorem 6 more effective. In [30], this idea is used to deal with the case of the Askey–Wilson polynomials. The crucial observation is that a Laurent polynomial $p(q^n) = \sum_{k=-m}^{n} c_k q^{kn}$ in q^n, is a polynomial in $\Lambda_{n;a,b,c,d}$, i.e. it belongs to $\mathcal{K}$ as defined in (40), if and only if it is invariant under the involution

$$I(q^n) = \frac{q^{1-n}}{abcd}. \tag{47}$$

This is an easy exercise using the relation $q^n + (abcd)^{-1}q^{1-n} = q(abcd)^{-1} \times (\Lambda_{n;a,b,c,d} + 1 + abcd \, q^{-1})$, which follows from the definition (22).

Let us denote by $\mathcal{R}$ the algebra of difference operators of the form $T(n, E) = \sum_{j=m_1}^{m_2} h_j(q^n)E^j$, with coefficients $h_j(q^n)$ rational functions in q^n. We extend the involution I to $\mathcal{R}$ by

$$(Ih_j)(q^n) = h_j(I(q^n)) \quad \text{and} \quad I(E^j) = E^{-j}. \tag{48}$$

A straightforward computation from (17) and (18) gives $I(A_{n;a,b,c,d}) = C_{n;a,b,c,d}$. Using (19), this shows that the operator $R_{a,b,c,d}$ in (30) is I-invariant, that is $I(R_{a,b,c,d}) = R_{a,b,c,d}$. The next theorem characterizes the rational Darboux transformations as those for which the two factors $\mathcal{P}$ and $\mathcal{Q}$ in (41) are I-invariant operators.

Theorem 7   ([27], see also [14]) *The following conditions on an operator $T \in \mathcal{R}$ are equivalent:*

  (i) *The operator T is I-invariant, i.e. $I(T) = T$;*
 (ii) *T has the form $\Theta^{-1}V$, for some $V \in \mathcal{B}$ and some $\Theta \in \mathcal{K}$;*
(iii) *T has the form $W\Gamma^{-1}$, for some $W \in \mathcal{B}$ and some $\Gamma \in \mathcal{K}$.*

Theorems 6 and 7 are the basic tools which allow us to construct orthogonal polynomials satisfying q-difference equations, by adding mass points to some instances of the weight function $w(x; a, b, c, d)$ (27) of the Askey–Wilson polynomials.

Theorem 8 ([27]) *Let $\alpha, \beta \in \mathbb{Z}_+, m, n \in \mathbb{N}$ and $0 < q < 1$. The orthogonal polynomials with weight functions on $[-1, 1]$ given by*

$$w(x; q^{\frac{m}{2}+\alpha}, b, c, q^{\frac{m}{2}}) + \sum_{\text{finitely many } k\in\mathbb{Z}_+} r_k\delta\left(x - \frac{q^{\frac{m}{2}+\alpha+k} + q^{-\frac{m}{2}-\alpha-k}}{2}\right),$$

with b, c, arbitrary, $\max(|b|, |c|) < 1$, (49)

$$w(x; q^{\frac{m}{2}+\alpha}, -q^{\frac{n}{2}+\beta}, -q^{\frac{n}{2}}, q^{\frac{m}{2}})$$

$$+ \sum_{\text{finitely many } k\in\mathbb{Z}_+} r_k\delta\left(x - \frac{q^{\frac{m}{2}+\alpha+k} + q^{-\frac{m}{2}-\alpha-k}}{2}\right)$$

$$+ \sum_{\text{finitely many } k\in\mathbb{Z}_+} s_k\delta\left(x + \frac{q^{\frac{n}{2}+\beta+k} + q^{-\frac{n}{2}-\beta-k}}{2}\right),$$ (50)

where $r_k, s_k > 0$ denote arbitrary free parameters, solve the q-version of Krall's problem, that is they are eigenfunctions of a q-difference operator.

One can show that, putting $m = 1, b = -q^{\frac{1}{2}+\beta}(\beta > -1), c = -q^{\frac{1}{2}}$ in (49), or $m = n = 1$ in (50), the corresponding orthogonal polynomials tend, when $q \to 1$, to the Krall–Jacobi type polynomials that we mentioned in Theorem 1. It is only when $q \to 1$ that all the mass points accumulate at the boundary ± 1 of the interval of orthogonality $[-1, 1]$, giving rise to weight distributions involving the derivatives of Dirac's delta function.

For a more complete account of the results above as well as detailed proofs, which involve the use of the Ismail–Rahman functions [31], I refer the reader to my joint works with F.A. Grünbaum [26] and P. Iliev [30].

5 THE SIMPLEST EXAMPLE

In this section I shall illustrate Theorem 8 on the simplest case, constructing explicitly the orthogonal polynomials with weight function

$$w(x; q^{\frac{1}{2}}, b, c, q^{\frac{1}{2}}) + r\delta\left(x - \frac{q^{\frac{1}{2}} + q^{-\frac{1}{2}}}{2}\right),$$ (51)

with $r > 0$ an arbitrary free parameter, b,c arbitrary, $\max(|b|, |c|) < 1$. This amounts to pick $\alpha = 0, m = 1, k = 0$ in (49).

Formulas (13) and (28) show that these polynomials can be obtained as the Geronimus transform (7) of $J_{q^{3/2}, b, c, q^{1/2}}$, with $\lambda = (q^{1/2} + q^{-1/2})/2$. Thus, using the normalization $R_{q^{3/2}, b, c, q^{1/2}}$ (30), the matrix $\tilde{R}$ defining the three-term recurrence relation satisfied by the orthogonal polynomials with weight function (51), is given (up to conjugation by a diagonal matrix) by

$$R_{q^{3/2},b,c,q^{1/2}} = (q^{1/2} + q^{-1/2})\text{Id} + UL \to \tilde{R} = (q^{1/2} + q^{-1/2})\text{Id} + LU.$$
$$(52)$$

Theorem 7 is not immediately applicable to this situation, since the two factors U and L (being respectively upper and lower matrices) cannot be I-invariant. As explained in (31), $R_{q^{3/2},b,c,q^{1/2}}$ can be obtained as a Christoffel transform from $R_{q^{1/2},b,c,q^{1/2}}$

$$R_{q^{1/2},b,c,q^{1/2}} = (q^{1/2} + q^{-1/2})\text{Id} + S_{q^{1/2},b,c,q^{1/2}} T_{q^{1/2},b,c,q^{1/2}} \to$$
$$R_{q^{3/2},b,c,q^{1/2}} = (q^{1/2} + q^{-1/2})\text{Id} + T_{q^{1/2},b,c,q^{1/2}} S_{q^{1/2},b,c,q^{1/2}}, \qquad (53)$$

with $S_{q^{1/2},b,c,q^{1/2}}, T_{q^{1/2},b,c,q^{1/2}}$ as in (32) and (33)

$$S_{q^{1/2},b,c,q^{1/2}}(n, E) = (A_{n;q^{1/2},b,c,q^{1/2}}\text{Id} - C_{n;q^{1/2},b,c,q^{1/2}}E^{-1}) \frac{1}{\varphi_{n;q^{1/2},b,c,q^{1/2}}},$$

$$T_{q^{1/2},b,c,q^{1/2}}(n, E) = \varphi_{n;q^{1/2},b,c,q^{1/2}}(E - \text{Id}),$$

$$\varphi_{n;q^{1/2},b,c,q^{1/2}} = \frac{1}{q^{-n} - bcq^{n+1}}.$$

From (52) and (53), we deduce the Darboux transformation (41), (43)

$$\mathcal{L} \equiv \left(R_{q^{1/2},b,c,q^{1/2}} - (q^{1/2} + q^{-1/2})\text{Id}\right)^2$$
$$= \mathcal{P}\mathcal{Q} \to \tilde{\mathcal{L}} \equiv \mathcal{Q}\mathcal{P} = (\tilde{R} - (q^{1/2} + q^{-1/2})\text{Id})^2, \qquad (54)$$

with

$$\mathcal{P} = S_{q^{1/2},b,c,q^{1/2}}U, \quad \mathcal{Q} = LT_{q^{1/2},b,c,q^{1/2}}. \qquad (55)$$

We are going to show that the Darboux transformation (54) fits within the framework of Theorems 6 and 7, that is the two factors $\mathcal{P}$ and $\mathcal{Q}$ in (5.5) are invariant for the involution (47), (48), with $a = d = q^{1/2}$:

$$I(q^n) = \frac{1}{bcq^n}, \quad I(E) = E^{-1}. \qquad (56)$$

The tricky part is to compute the Geronimus transformation (52), which involves a free parameter μ, linearly equivalent to r in (51). The result, which can be easily checked, is the following

$$U(n, E) = \left(A_{n;q^{3/2},b,c,q^{1/2}}E - C_{n;q^{3/2},b,c,q^{1/2}}\frac{\psi_{n-1}}{\psi_n}\text{Id}\right) \frac{1}{\psi_{n-1}\gamma_n}, \qquad (57)$$

$$L(n, E) = \psi_{n-1}\psi_n \left(\text{Id} - \frac{\psi_n}{\psi_{n-1}}E^{-1}\right), \qquad (58)$$

with

$$\psi_n = \frac{q^n}{(1 - bq^{n+\frac{1}{2}})(1 - cq^{n+\frac{1}{2}})} \left(\mu + \frac{q^n}{(1 - q^{n+1})(1 - bcq^n)} \right),$$

μ a free parameter, and $\gamma_n = q^{-n} - bcq^n$. The purpose of the factor $\psi_{n-1}\gamma_n$ in (58) and its inverse in (57) is to make the resulting operator $\tilde{R}$ in (52) I-invariant, with I as in (56). To alleviate the notations, I shall write $A_n = A_{n;q^{1/2}b,c,q^{1/2}}$, $C_n = C_{n;q^{1/2}b,c,q^{1/2}}$, $\varphi_n = \varphi_{n;q^{1/2}b,c,q^{1/2}}$. With these conventions, (53) is equivalent to

$$A_{n;q^{3/2}b,c,q^{1/2}} = \frac{\varphi_n}{\varphi_{n+1}} A_{n+1}, \quad C_{n;q^{3/2}b,c,q^{1/2}} = \frac{\varphi_n}{\varphi_{n-1}} C_n. \tag{59}$$

The I-invariance of $\mathcal{P}$ and $\mathcal{Q}$ in (55) is then easily checked by using

$$I(\gamma_n) = -\gamma_n, \, I(\varphi_n) = -\varphi_{n-1}, \, I(\psi_n) = \psi_{n-1}, \, I(A_n) = C_n,$$

and (59). Writing

$$R = R_{q^{1/2}b,c,q^{1/2}}, \quad \Lambda = \Lambda_{q^{1/2}b,c,q^{1/2}},$$

as predicted by Theorem 7, $\mathcal{Q}$ in (55) can be written as $\mathcal{Q} = \Theta^{-1}V$, $V \in \mathcal{B}$, $\Theta \in \mathcal{K}$, $\mathcal{B} =< R, \Lambda >$, $\mathcal{K} =< \Lambda >$ as in (37), (40), with

$$V = ((bc - 1)^2 \mathrm{Id} + 2(bc + 1)\Lambda + \Lambda^2)(x_0 \mathrm{Id} + x_1 \Lambda + x_2 R + x_3 \Lambda R + R\Lambda),$$
$$\Theta = y_1 \Lambda + y_2 \Lambda^2 + y_3 \Lambda^3 + y_4 \Lambda^4, \tag{60}$$

and

$$x_0 = \frac{(q - 1)}{q}((b + c)(q - 1) + 2q^{1/2}(bc - 1)) + \frac{(q + 1)^2}{\mu q^{3/2}},$$

$$x_1 = 2(q^{1/2} + q^{-1/2}),$$

$$x_2 = -\frac{(q - 1)(bcq - 1)}{q} - \frac{q + 1}{\mu q},$$

$$x_3 = -\frac{1 + q + q^2}{q}, \tag{61}$$

$$y_1 = \frac{(q^2 - 1)(q^{1/2} - b)(q^{1/2} - c)(q - bc)(bq^{1/2} - 1)(cq^{1/2} - 1)}{\mu q^{3/2}},$$

$$y_2 = -\frac{(q + 1)}{\mu q} \left[(b + c)(q + 1)(q^2 - 2bcq - 2q + bc) \right.$$
$$\left. - q^{1/2}((bc + 2)q^2 - (3b^2c^2 + c^2 + 6bc + b^2 + 3)q + bc(2bc + 1)) \right],$$

$$y_3 = \frac{(q + 1)(q^2 + (b + c)q^{3/2} - 3bcq - 3q + (b + c)q^{1/2} + bc)}{\mu q^{1/2}},$$

$$y_4 = -\frac{q^{1/2}(q + 1)}{\mu}. \tag{62}$$

Similarly, for $\mathcal{P}$ in (55), one finds $\mathcal{P} = W\Gamma^{-1}$, $W \in \mathcal{B}$, $\Gamma \in \mathcal{K}$, with

$$W = (x_0\mathrm{Id} + x_1\Lambda + x_2R + x_3R\Lambda + \Lambda R)\Theta,$$

$$\Gamma = (z_0\mathrm{Id} + z_1\Lambda + z_2\Lambda^2)((bc - 1)^2\mathrm{Id} + 2(bc + 1)\Lambda + \Lambda^2), \qquad (63)$$

x_i defined as in (61), Θ as in (60), (62) and

$$z_0 = \frac{(q + 1)^2(\mu(bc - 1)(q - 1) + 1)}{\mu^2 q^2},$$

$$z_1 = -\frac{(q + 1)^2(\mu q^2 - (\mu bc + \mu + 1)q + \mu bc - 1)}{\mu q^2}, \qquad z_2 = \frac{(q + 1)^2}{q}.$$

Notice that there are quite a few cancellations in the Darboux factorization (54) of $\mathcal{L}$. In fact, from (60) and (63), we finally obtain that

$$\mathcal{L} = (x_0\mathrm{Id} + x_1\Lambda + x_2R + x_3R\Lambda + \Lambda R)(z_0\mathrm{Id} + z_1\Lambda + z_2\Lambda^2)^{-1}$$
$$\times (x_0\mathrm{Id} + x_1\Lambda + x_2R + x_3\Lambda R + R\Lambda).$$

From Theorem 5, using the anti-isomorphism $\flat : \mathcal{B} \to \mathcal{B}'$ defined in (39), with $\mathcal{B}' = \langle z + z^{-1}, B = B_{q^{1/2},b,c,q^{1/2}} \rangle$ as in (38), it follows that the orthogonal polynomials with weight function (51) are eigenfunctions of the fourth order q-difference operator $\tilde{B}$ defined as in (46)

$$\tilde{B} = (x_0 + x_1B + x_2(z + z^{-1}) + x_3(z + z^{-1})B + B(z + z^{-1}))$$
$$\times (x_0 + x_1B + x_2(z + z^{-1}) + x_3B(z + z^{-1})$$
$$+ (z + z^{-1})B)(z + z^{-1} - (q^{1/2} + q^{-1/2}))^{-2},$$

with eigenvalues $\tilde{\Lambda}_n = z_0 + z_1\Lambda_{n;q^{1/2},b,c,q^{1/2}} + z_2\Lambda^2_{n;q^{1/2},b,c,q^{1/2}}$ as in (45).

REFERENCES

1. Matveev, V. B. and Salle, M. A. (1991) *Darboux Transformations and Solitons*, Springer Series in Nonlinear Dynamics, Springer-Verlag, New York.
2. Adler, M. and van Moerbeke, P. (1998) Toda-Darboux maps and vertex operators, *Int. Math. Res. Not.* **10**, pp. 489–511.
3. Miwa, T., Jimbo, M., and Date, E. (2000) Solitons, Differential equations, symmetries and infinite dimensional algebras, in: *Cambridge Tracts in Mathematics*, **Vol. 135**, Cambridge University Press, Cambridge.
4. Duistermaat, J. J. and Grünbaum, F. A. (1986) Differential equations in the spectral parameter, *Commun. Math. Phys.* **103**, pp. 177–240.
5. Bakalov, B., Horozov, E., and Yakimov, M. (1998) Highest weight modules over the $W_{1+\infty}$ algebra and the bispectral problem, *Duke Math. J.* **93**, pp. 41–72.
6. Berest, Y. and Wilson, G. (2000) Automorphisms and ideals of the Weyl algebra, *Math. Ann.* **318**, pp. 127–147.

7. Terwilliger, P. (2003) Introduction to Leonard pairs, *J. Comput. Appl. Math.* **153**, pp. 463–475.
8. Krall, H. L. (1938) Certain differential equations for Tchebycheff polynomials, *Duke Math. J.* **4**, pp. 705–718.
9. Bochner, S. (1929) Über Sturm-Liouvillesche Polynomsysteme, *Math. Z.* **29**, pp. 730–736.
10. Krall, H. L. (1940) On orthogonal polynomials satisfying a certain fourth order differential equation, *The Pennsylvania Sate College Studies*, **6**.
11. Grünbaum, F. A. and Haine, L. (1996) Orthogonal polynomials satisfying differential equations: The role of the Darboux transformation, in: *Symmetries and Integrability of Difference Equations, CRM Proceedings Lecture Notes*, **Vol. 9**, Amer. Math. Soc., D. Levi, L. Vinet and P. Winternitz eds. Providence, pp. 143–154.
12. Chen, Y. and Griffin, J. (2002) Krall-type polynomials via the Heine formula, *J. Phys. A: Math. Gen.* **35**, pp. 637–656.
13. Grünbaum, F. A. and Haine, L. (1997) Bispectral Darboux transformations: An extension of the Krall polynomials, *Int. Math. Res. Not.* **8**, pp. 359–392.
14. Grünbaum, F. A., Haine, L., and Horozov, E. (1999) Some functions that generalize the Krall-Laguerre polynomials, *J. Comput. Appl. Math.* **106**, pp. 271–297.
15. Grünbaum, F. A. and Yakimov, M. (2002) Discrete bispectral Darboux transformations from Jacobi operators, *Pacific J. Math.* **204**, pp. 395–431.
16. Haine, L. and Iliev, P. (2001) A rational analogue of the Krall polynomials, *J. Phys. A: Math. Gen.* **34**, pp. 2445–2457.
17. Spiridonov, V., Vinet, L., and Zhedanov, A. (1998) Bispectrality and Darboux transformations in the theory of orthogonal polynomials, in: *The Bispectral Problem, CRM Proc. Lecture Notes*, Amer. Math. Soc., 14, eds. J. Harnad and A. Kasman, Providence, pp. 111–122.
18. Everitt, W. N., Kwon, K. H., Littlejohn, L. L., and Wellman, R. (2001) Orthogonal polynomial solutions of linear ordinary differential equations, *J. Comput. Appl. Math.* **133**, pp. 85–109.
19. Koekoek, J. and Koekoek, R. (2000) Differential equations for generalized Jacobi polynomials, *J. Comput. Appl. Math.* **126**, pp. 1–31.
20. Koornwinder, T. H. (1984) Orthogonal polynomials with weight function $(1 - x)^\alpha (1 + x)^\beta + M\delta(x + 1) + N\delta(x - 1)$, *Canad. Math. Bull.* **27** (2), pp. 205–214.
21. Krall, A. M. (1980) Chebyshev sets of polynomials which satisfy an ordinary differential equation, *SIAM Rev.* **22**(4), pp. 436–441.
22. Littlejohn, L. L. (1982) The Krall polynomials: A new class of orthogonal polynomials, *Quaestiones Mathematicae* **5**, pp. 255–265.
23. Chihara, T. S. (1978) *An Introduction to Orthogonal Polynomials*, Gordon and Breach, New York.
24. Gasper, G. and Rahman, M. (1990) Basic hypergeometric series, in: *Encyclopedia of Mathematics and Its Applications*, Vol. 35, Cambridge University Press, Cambridge.
25. Askey, R. and Wilson, J. (1985) Some basic hypergeometric orthogonal polynomials that generalize Jacobi polynomials, *Mem. Amer. Math. Soc.* **319**.

26. Grünbaum, F. A. and Haine, L. (1997) Some functions that generalize the Askey-Wilson polynomials, *Commun. Math. Phys.* **184**, pp. 173–202.
27. Bakalov, B., Horozov, E., and Yakimov, M. (1996) General methods for constructing bispectral operators, *Phys. Lett. A* **222**, pp. 59–66.
28. Kasman, A. and Rothstein, M. (1997) Bispectral Darboux transformations: the generalized Airy case, *Physica D* **102**, pp. 159–176.
29. Wilson, G. (1993) Bispectral commutative ordinary differential operators, *J. reine angew. Math.* **442**, pp. 177–204.
30. Haine, L. and Iliev, P. (in press) Askey-Wilson type functions, with bound states, *The Ramanujan Journal*.
31. Ismail, M. E. H. and Rahman, M. (1991) The associated Askey-Wilson polynomials, *Trans. Amer. Math. Soc.*, **328**(1), pp. 201–237.

DISCRETIZATION OF COUPLED SOLITON EQUATIONS

Ryogo Hirota

Professor Emeritus Waseda University, Shinjyuku, Tokyo 160, Japan,
NATO ARW 15–19 September 2002

Abstract   We describe integrable discretization of coupled forms of the well-known soliton equations such as KdV equation, modified KdV equation, sine-Gordon equation and nonlinear Schrödinger equation.

1   INTRODUCTION

Discretization of integrable systems has been the focus of an intense activities [1–5]. Discretization of integrable nonlinear ordinary differential equations has been studied in [6–8]. Recent progress in integrable discrete systems has uncovered remarkable relationships in otherwise unrelated areas of research such as numerical algorithms [9], discrete geometry [10], cellular automaton [11], and quantum integrable systems [12].

In a series of papers [1], we have developed a method of discretizing integrable equations. The method is based on the bilinear formalism and follows three steps. Firstly a given differential equation is transformed into the bilinear equation by a dependent variable transformation. Secondly the bilinear equation is discretized with the help of gauge-invariance of the bilinear equation and the integrability of the discretized bilinear equation is determined by checking soliton solutions. Thirdly the discrete bilinear equation is transformed into a discrete nonlinear equation in the ordinary form by an associated dependent variable transformation.

Coupled forms of the well-known soliton equations have been proposed by several authors [14–18]:

1. Coupled KdV eq. [14]:

$$\frac{\partial u_i}{\partial t} + 6\left(\sum_{k=1}^{M} u_k\right)\frac{\partial u_i}{\partial x} + \frac{\partial^3 u_i}{\partial x^3} = 0, \quad \text{for } i = 1, 2, \ldots, M. \tag{1}$$

113

L. Faddeev et al. (eds.),
Bilinear Integrable Systems: From Classical to Quantum, Continuous to Discrete, 113–122.

2. Coupled modified KdV equation:
   Type (i) [15]

$$\frac{\partial v_i}{\partial t} + 6 \left(\sum_{k=1}^{M} v_k^2 \right) \frac{\partial v_i}{\partial x} + \frac{\partial^3 v_i}{\partial x^3} = 0, \quad \text{for } i = 1, 2, \ldots, M, \quad (2)$$

which is extended to the following form
Type (ii) [16]

$$\frac{\partial v_i}{\partial t} + 3 \left(\sum_{1 \le j < k \le M} c_{j,k} v_j v_k \right) \frac{\partial v_i}{\partial x} + \frac{\partial^3 v_i}{\partial x^3} = 0, \quad \text{for } i = 1, 2, \ldots, M,$$

$$(3)$$

where $c_{j,k}$ are arbitrary constants, which is extended furhter to the following form

$$\frac{\partial v_i}{\partial t} + 3 \left(\sum_{j,k=1}^{M} c_{j,k} v_j v_k \right) \frac{\partial v_i}{\partial x} + \frac{\partial^3 v_i}{\partial x^3} = 0, \quad \text{for } i = 1, 2, \ldots, M, \quad (4)$$

where $c_{j,k}$ are arbitrary constants.
3. Coupled nonlinear Klein-Gordon (Sine-Gordon) equation [15]:

$$\begin{cases} \frac{\partial^2}{\partial x \partial y} w_i - w_i + 2w_i \frac{\partial \Gamma}{\partial y} = 0, & \text{for } i = 1, 2, \ldots, M, \\ \frac{\partial \Gamma}{\partial x} = \sum_{j=1}^{M} w_j^2, \end{cases} \quad (5)$$

which is extended by introducing arbitrary constants, $c_{j,k}$ to

$$\begin{cases} \frac{\partial^2}{\partial x \partial y} w_i - w_i + 2w_i \frac{\partial \Gamma}{\partial y} = 0, & \text{for } i = 1, 2, \ldots, M, \\ \frac{\partial \Gamma}{\partial x} = \sum_{j,k=1}^{M} c_{j,k} w_j w_k. \end{cases} \quad (6)$$

4. Vector nonlinear Schrödinger equation [17]:

$$i q_t = q_{xx} + 2\|q\|^2 q, \quad (7)$$

which is extended to a generalized form [18] among which we choose the following form,

$$i \frac{\partial}{\partial t} \psi_i + \frac{\partial^2}{\partial x^2} \psi_i + \left(\sum_{j,k=1}^{M} c_{j,k} \psi_j \psi_k^* \right) \psi_i = 0, \quad (8)$$

with arbitrary constants $c_{j,k}$ satisfying the reality conditions $c_{j,k}^* = c_{k,j}$.

We shall discretize these coupled equations in the following sections. N-soliton solution to these M-coupled discrete equations are obtained, which will be published elsewhere. Here we present soliton solutions for the simplest case $N = 2$ and $M = 2$.

2 COUPLED KdV EQUATIONS

First we transform the coupled KdV eq. (1) into the bilinear form. Let $u_i = G_i/F$. Then we have

$$\frac{D_t G_i \cdot F}{F^2} + 6 \left(\sum_{k=1}^{M} \frac{G_k}{F} \right) \frac{D_x G_i \cdot F}{F^2} + \frac{D_x^3 G_i \cdot F}{F^2} - 3 \frac{D_x G_i \cdot F}{F^2} \frac{D_x^2 F \cdot F}{F^2} = 0,$$

for $i = 1, 2, \ldots, M$,

which is decoupled to the bilinear forms

$$\begin{cases} (D_t + D_x^3) G_j \cdot F = 0, & \text{for } j = 1, 2, \ldots, M, \\ D_x^2 F \cdot F = 2 \left(\sum_{j=1}^{M} G_j \right) F. \end{cases} \tag{9}$$

A coupled differential-difference KdV Equations are obtained by discretizing the spacial part of the bilinear eq. (9),

$$D_x^3 G \cdot F \rightarrow \frac{1}{\epsilon} (G_{n+1} F_{n-1} - G_{n-1} F_{n+1}),$$

$$D_x^2 F \cdot F \rightarrow \frac{2}{\epsilon^2} (F_{n+1} F_{n-1} - F_n^2).$$

We obtain

$$\begin{cases} D_t G_{j,n} \cdot F_n + \frac{1}{\epsilon} (G_{j,n+1} F_{n-1} - G_{j,n-1} F_{n+1}) = 0, \\ \qquad\qquad\qquad\qquad \text{for } j = 1, 2, \ldots, M, \\ F_{n+1} F_{n-1} - F_n^2 = \epsilon^2 \left(\sum_{j=1}^{M} G_{j,n} \right) F_n, \end{cases} \tag{10}$$

which is transformed into the ordinary form

$$\frac{\partial u_j}{\partial t} + \frac{1}{\epsilon} \left(1 + \epsilon^2 \sum_{j=1}^{M} u_j \right) (u_{j,n+1} - u_{j,n-1}) = 0, \tag{11}$$

for $j = 1, 2, \ldots, M$, through the transformation $G_{j,n} = u_{j,n} F_n$.
Time-discretization of the coupled KdV equations is obtained by replacing the differential operator $D_t G \cdot F$ by the corresponding difference operator:

$$D_t G \cdot F \rightarrow (G^{m+1} F^m - G^m F^{m+1})/\delta,$$

where $t = m\delta$, m being integers and δ a time-interval.

We postulate that discretized bilinear forms are invariant under the gauge transfomation:

$$\begin{cases} F_n^m \to F_n^m \exp(q_0 m + p_0 n), \\ G_n^m \to G_n^m \exp(q_0 m + p_0 n). \end{cases}$$

We find a gauge invariant discrete bilinear KdV equation

$$\begin{cases} G_{j,n}^{m+1} F_n^m - G_{j,n}^m F_n^{m+1} + \dfrac{\delta}{\epsilon}(G_{j,n+1}^{m+1} F_{n-1}^m - G_{j,n-1}^m F_{n+1}^{m+1}) = 0, \\ \qquad\qquad\qquad\qquad\qquad\qquad\qquad \text{for } j = 1, 2 \ldots, M, \\ F_{n+1}^m F_{n-1}^m - (F_n^m)^2 = \epsilon^2 (\sum_{j=1}^M G_{j,n}^m) F_n^m. \end{cases} \qquad (12)$$

We introduce an auxiliary variable Γ_n^m defined by

$$\Gamma_n^m = \frac{F_{n+1}^{m+1} F_{n-1}^m}{F_n^{m+1} F_n^m}. \qquad (13)$$

We note that Γ_n^m satisfies an identity

$$\Gamma_n^m = \frac{F_{n+1}^{m+1} F_{n-1}^{m+1}}{(F_n^{m+1})^2} \frac{(F_{n-1}^m)^2}{F_n^m F_{n-2}^m} \Gamma_{n-1}^m. \qquad (14)$$

Let $G_{j,n}^m = u_{j,n}^m F_n^m$. Then eq. (12) is transformed into the ordinary form with the help of the identity (14)

$$\begin{cases} u_{j,n}^{m+1} - u_{j,n}^m + \dfrac{\delta}{\epsilon}\Gamma_n^m(u_{j,n+1}^{m+1} - u_{j,n-1}^m) = 0, \quad \text{for } j = 1, 2, \ldots, M, \\ \Gamma_n^m = \dfrac{1 + \sum_{j=1}^M u_{j,n}^{m+1}}{1 + \sum_{j=1}^M u_{j,n-1}^m} \Gamma_{n-1}^m. \end{cases} \qquad (15)$$

Soliton solutions to the coupled discrete KdV eq. (15) are given by $u_{j,n}^m = G_{j,n}^m / F_n^m$ where F_n^m and $G_{j,n}^m$ are solutions to the bilinear form (12).

Let f_n^m satisfy a bilinear form of the discrete KdV equation [5],

$$f_{n+1}^{m+1} f_n^m - f_n^{m+1} f_{n+1}^m + (\delta/\epsilon)(f_{n+2}^{m+1} f_{n-1}^m - f_{n+1}^{m+1} f_n^m) = 0 \qquad (16)$$

and a bilinear form of the Lotka-Volterra equation [5],

$$D_s f_{n+1}^m \cdot f_n^m = f_{n+2}^m f_{n-1}^m - f_{n+1}^m f_n^m, \quad (D_s = \sum_{j=1}^M D_j), \qquad (17)$$

where we have introduced new variables s_j, s and the bilinear operators, $D_j f \cdot g = \frac{\partial f}{\partial s_j} g - f \frac{\partial g}{\partial s_j}$ and $D_s = \sum_{j=1}^{M} D_j$, respectively.
Then we find easily that F_n^m and $G_{j,n}^m$ defined by

$$\begin{cases} F_n^m = f_{n+1}^m f_n^m, \\[2mm] G_{j,n}^m = D_j f_{n+1}^m \cdot f_n^m, \end{cases} \tag{18}$$

satisfy the coupled discrete KdV equation (12).
   We have 2-soliton solutions to the equations (16) and (17), for example,

$$f_n^m = 1 + \exp \eta_1 + \exp \eta_2 + a_{1,2} \exp(\eta_1 + \eta_2), \tag{19}$$

where

$$\exp \eta_j = b_j p_j^n q_j^m \exp(\omega_j(s_j + s)), \tag{20}$$

$$q_j = (1 + (\delta/\epsilon)/p_j)/(1 + (\delta/\epsilon)p_j), \tag{21}$$

$$\omega_j = p_1 - 1/p_j, \quad \text{for } j = 1, 2, \ldots, M, \tag{22}$$

$$a_{j,k} = (p_j - p_k)^2/(p_j p_k - 1)^2, \quad \text{for } j, k = 1, 2, \ldots, M, \tag{23}$$

where a_j and p_j are arbitrary parameters.

3   COUPLED MODIFIED KdV EQUATIONS

Discretization of the coupled modified KdV equations is discussed in [4]. We present the results. Equation (4) is transformed into the bilinear form by the rational transformation $v_j = G_j/F$

$$\begin{cases} (D_t + D_x^3)G_j \cdot F = 0, \quad \text{for } j = 1, 2, \ldots, M, \\[3mm] D_x^2 F \cdot F = \sum_{j,k=1}^{M} c_{j,k} G_j G_k, \end{cases} \tag{24}$$

where $c_{j,k}$ are arbitrary coupling constants.
Following the procedure described in the previous section we discretize eq.(24) and obtain a discrete bilinear form:

$$\begin{cases} G_{j,n}^{m+1} F_n^m - G_{j,n}^m F_n^{m+1} + \dfrac{\delta}{\epsilon}(G_{j,n+1}^{m+1} F_{n-1}^m - G_{j,n-1}^m F_{n+1}^{m+1}) = 0, \\[3mm] \qquad\qquad \text{for } j = 1, 2, \ldots, M, \\[3mm] F_{n+1}^m F_{n-1}^m - (F_n^m)^2 = \epsilon^2 \sum_{j,k=1}^{M} c_{j,k} G_{j,n}^m G_{k,n}^m. \end{cases}$$

The bilinear equation is transformed, by using the identity (14), into a coupled discrete MKdV equation

$$
\begin{cases}
v_{j,n}^{m+1} - v_{j,n}^m + \dfrac{\delta}{\epsilon}\Gamma_n^m(v_{j,n+1}^{m+1} - v_{j,n-1}^m) = 0, & \text{for } j = 1, 2, \ldots, M, \\[2ex]
\Gamma_n^m = \dfrac{1 + \sum_{j,k=1}^M v_{j,n}^{m+1} v_{k,n}^{m+1}}{1 + \sum_{j,k=1}^M v_{j,n-1}^m v_{k,n-1}^m}\Gamma_{n-1}^m,
\end{cases}
$$

through the transformation $G_{j,n}^m = v_{j,n}^m F_n^m$.

4   COUPLED NONLINEAR KLEIN-GORDON (SINE-GORDON) EQUATIONS

The coupled nonlinear Klein-Gordon (Sine-Gordon) equation (6) is transformed into the following bilinear form through the rational transformation $w_j = \frac{G_j}{F}$,

$$
\begin{cases}
(D_x D_y - 1)G_j \cdot F = 0, & \text{for } j = 1, 2, \ldots, M, \\[2ex]
D_x^2 F \cdot F = 2\displaystyle\sum_{j,k=1}^M c_{j,k} G_j G_k.
\end{cases}
\tag{25}
$$

Discretizing $y(= m\delta)$ first and then $x(= n\delta)$ we obtain

$$
\begin{aligned}
D_x D_y G \cdot F \\
\to (1/\delta)D_x(G^{m+1} \cdot F^m - G^m \cdot F^{m+1}), \\
\to (1/\delta^2)(G_{n+1}^{m+1} F_n^m - G_n^{m+1} F_{n+1}^m - G_{n+1}^m F_n^{m+1} + G_n^m F_{n+1}^{m+1}).
\end{aligned}
\tag{26}
$$
$$
\tag{27}
$$

Taking the gauge invariance into account we tansform GF as

$$
GF \to (1/4)(G_{n+1}^{m+1} F_n^m + G_n^{m+1} F_{n+1}^m + G_{n+1}^m F_n^{m+1} + G_n^m F_{n+1}^{m+1}).
\tag{28}
$$

Then equation(25) is discretized as

$$
\begin{cases}
G_{j,n+1}^{m+1} F_n^m + G_{j,n}^m F_{n+1}^{m+1} - G_{j,n}^{m+1} F_{n+1}^m - G_{j,n+1}^m F_n^{m+1} \\
\qquad = \dfrac{\delta^2}{4}(G_{j,n+1}^{m+1} F_n^m + G_{j,n}^m F_{n+1}^{m+1} + G_{j,n}^{m+1} F_{n+1}^m + G_{j,n+1}^m F_n^{m+1}), \\
\qquad\qquad\qquad \text{for } j = 1, 2, \ldots, M, \\[2ex]
F_{n+1}^m F_{n-1}^m - (F_n^m)^2 = \delta^2 \displaystyle\sum_{j,k=1}^M c_{j,k} G_{j,n}^m G_{k,n}^m,
\end{cases}
\tag{29}
$$

which is transformed, by using an identity similar to the identity (14), into a coupled form of nonlinear discrete Klein-Gordon equations,

$$
\begin{cases}
w_{j,n+1}^{m+1} + w_{j,n}^m - (w_{j,n}^{m+1} + w_{j,n+1}^m)\Gamma_n^m \\[2mm]
= \dfrac{\delta^2}{4}(w_{j,n+1}^{m+1} + w_{j,n}^m + (w_{j,n}^{m+1} + w_{j,n+1}^m)\Gamma_n^m), \\[3mm]
\qquad\quad \text{for} \quad j = 1, 2, \ldots, M, \\[3mm]
\Gamma_n^m = \dfrac{1 + \delta^2 \sum_{j,k=1}^{M} c_{j,k} w_{j,n}^m w_{k,n}^m}{1 + \delta^2 \sum_{j,k=1}^{M} c_{j,k} w_{j,n}^{m+1} w_{k,n}^{m+1}} \Gamma_{n-1}^m,
\end{cases}
\tag{30}
$$

through the transformation $G_{j,n}^m = w_{j,n}^m F_n^m$.

Soliton solutions to the coupled discrete nonlinear Klein-Gordon equation (30) are given by $w_{j,n}^m = G_{j,n}^m / F_n^m$ where F_n^m and $G_{j,n}^m$ are solutions to the bilinear form (29). N-soliton solutions to the M-coupled equations are expressed by pfaffians, which will be published elsewhere. Here we give soliton solutions for the simplest case, $N = 2$ and $M = 2$,

$$
\begin{aligned}
F_n^m &= 1 + a_{20}\exp(2\eta_1) + a_{11}\exp(\eta_1 + \eta_2) \\
&\quad + a_{02}\exp(2\eta_2) + a_{22}\exp(2\eta_1 + 2\eta_2), \\
G_{1,n}^m &= \exp\eta_1(1 + a_{12}\exp(2\eta_2)), \\
G_{2,n}^m &= \exp\eta_2(1 + a_{21}\exp(2\eta_1)),
\end{aligned}
$$

where

$$
\begin{aligned}
&\exp\eta_j = a_j p_j^n q_j^m, \\
&q_j = (1 - p_j - (\delta/2)^2(1 + p_j))/(1 - p_j + (\delta/2)^2(1 + p_j)), \\
&a_{20} = \delta^2 c_{11} p_1^2/(p_1^2 - 1)^2, \qquad a_{02} = \delta^2 c_{22} p_2^2/(p_2^2 - 1)^2, \\
&a_{11} = \frac{\delta^2(c_{12} + c_{21})p_1 p_2}{2(p_1 p_2 - 1)^2}, \\
&a_{12} = a_{02} b_{12}, \qquad a_{21} = a_{20} b_{12}, \qquad a_{22} = a_{02} a_{20} b_{12}^2, \\
&b_{12} = (p_1 - p_2)^2/(p_1 p_2 - 1)^2,
\end{aligned}
$$

a_j, p_j for $j = 1, 2$ being arbitrary parameters.

5 COUPLED NONLINEAR SCHRÖDINGER EQUATION

We consider the following form of the coupled nonlinear Schrödinger equation

$$
i\frac{\partial}{\partial t}\psi_i + \frac{\partial^2}{\partial x^2}\psi_i + \left(\sum_{j,k=1}^{M} c_{j,k}\psi_j\psi_k^*\right)\psi_i = 0,
\tag{31}
$$

with arbitrary constants $c_{j,k}$ satisfying the reality conditions $c_{j,k}^* = c_{k,j}$ where $*$ denotes complex conjugate. Equation (31) is transformed into the bilinear form

$$\begin{cases} (iD_t + D_x^2)G_j \cdot F = 0, \quad \text{for} \quad j = 1, 2, \ldots, M, \\ D_x^2 F \cdot F = 2 \sum_{j,k=1}^{M} c_{j,k} G_j G_k^* \end{cases} \tag{32}$$

through the rational transformation $\Psi_i = G_i/F$, for real F.

We obtain a system of semi-discrete nonlinear Schrödinger equations replacing the bilinear operator, $D_x^2 G \cdot F \to (G_{n+1}F_{n-1} - 2G_n F_n + G_{n-1}F_{n+1})/\epsilon^2$ in equation (32)

$$\begin{cases} iD_t G_{j,n} \cdot F_n + \dfrac{1}{\epsilon^2}(G_{j,n+1}F_{n-1} + G_{j,n-1}F_{n+1} - 2G_{j,n}F_n) = 0, \\ \qquad\qquad \text{for} \quad j = 1, 2, \ldots, M, \\ F_{n+1}F_{n-1} - F_n^2 = \epsilon^2 \sum_{j,k=1}^{M} c_{j,k} G_{j,n} G_{k,n}^*. \end{cases} \tag{33}$$

Discretizing time and taking the gauge invariance of the bilinear equation into account, we obtain a system of discrete nonlinear Schrödinger equations

$$\begin{cases} i\left(G_{j,n}^{m+1} F_n^m - G_{j,n}^m F_n^{m+1}\right) \\ + \dfrac{\delta}{\epsilon^2}\left(G_{j,n+1}^{m+1} F_{n-1}^m + G_{j,n-1}^m F_{n+1}^{m+1} - G_{j,n}^m F_n^{m+1} - G_{j,n}^{m+1} F_n^m\right) = 0, \\ \qquad\qquad \text{for} \quad j = 1, 2, \ldots, M, \\ F_{n+1}^m F_{n-1}^m - (F_n^m)^2 = \epsilon^2 \sum_{j,k=1}^{M} c_{j,k} G_{j,n}^m G_{k,n}^{m*}. \end{cases} \tag{34}$$

Equations (33) and (34) are transformed, by using the identity (14), into the ordinary forms

$$i\frac{\partial \Psi_j}{\partial t} + \frac{1}{\epsilon^2}(\Psi_{j,n+1} + \Psi_{j,n-1} - 2\Psi_{j,n})$$

$$+ (\Psi_{j,n+1} + \Psi_{j,n-1})\left(\sum_{j,k=1}^{M} c_{j,k}\Psi_{j,n}\Psi_{j,k}^*\right) = 0, \quad \text{for} \quad j = 1, 2, \ldots, M, \tag{35}$$

and

$$\begin{cases} i(\Psi_{j,n}^{m+1} - \Psi_{j,n}^m) + \dfrac{\delta}{\epsilon^2}(\Gamma_n^m(\Psi_{j,n+1}^{m+1} + \Psi_{j,n-1}^m) - \Psi_{j,n}^m - \Psi_{j,n}^{m+1}) = 0, \\ \qquad\qquad \text{for} \quad j = 1, 2, \cdots, M, \\ \Gamma_n^m = \dfrac{1 + \displaystyle\sum_{j,k=1}^{M} c_{j,k}\Psi_{j,n}^{m+1}\Psi_{k,n}^{*m+1}}{1 + \displaystyle\sum_{j,k=1}^{M} c_{j,k}\Psi_{j,n-1}^m \Psi_{k,n-1}^{*m}}\Gamma_{n-1}^m, \end{cases} \tag{36}$$

through the transformations $\Psi_{j,n} = G_{j,n}F_n$ and $\Psi_{j,n}^m = G_{j,n}^m F_n^m$, respectively. Soliton solutions to the system of discrete nonlinear Schrödinger equations (36) are given by $\Psi_{j,n}^m = G_{j,n}^m / F_n^m$ where F_n^m and $G_{j,n}^m$ are solutions to the bilinear form (34). N-soliton solutions to the M-coupled equations are expressed by pfaffians, which will be published elsewhere. Here we give soliton solutions for the simplest case, $N = 2$ and $M = 2$,

$$F_n^m = 1 + a_{11}\exp(\eta_1 + \eta_1^*) + a_{12}\exp(\eta_1 + \eta_2^*) + a_{21}\exp(\eta_2 + \eta_1^*)$$
$$\qquad + a_{22}\exp(\eta_2 + \eta_2^*) + a_{1212}\exp(\eta_1 + \eta_2 + \eta_1^* + \eta_2^*),$$
$$G_{1,n}^m = \exp(\eta_1)(1 + b_{121}\exp(\eta_2 + \eta_1^*) + b_{122}\exp(\eta_2 + \eta_2^*)),$$
$$G_{2,n}^m = \exp(\eta_2)\left(1 + \hat{b}_{121}\exp(\eta_1 + \eta_1^*) + \hat{b}_{122}\exp(\eta_1 + \eta_2^*)\right),$$

where

$$\exp(\eta_j) = c_j\omega_j^m p_j^n,$$
$$\omega_j = (1 + i\delta(p_j - 1))/(1 - i\delta(1/p_j - 1)),$$
$$b_{121} = a_{21}b_{11}\beta_{12}, \qquad b_{122} = a_{22}b_{12}\beta_{12},$$
$$\hat{b}_{121} = a_{11}b_{21}\beta_{21}, \qquad \hat{b}_{122} = a_{12}b_{22}\beta_{21},$$
$$a_{1212} = [p_1 p_2 p_1^* p_2^*[c_{1,1}a_{22}(b_{12}\beta_{12} + b_{21}\beta_{12}^*) + c_{1,2}a_{21}(b_{11}\beta_{12} + b_{22}\beta_{21}^*)$$
$$\qquad + c_{2,1}a_{12}(b_{11}\beta_{12}^* + b_{22}\beta_{21}) + c_{2,2}a_{11}(b_{12}\beta_{12}^* + b_{21}\beta_{21})]$$
$$\qquad - a_{11}a_{22}(p_1 P_1^* - p_2 p_2^*)^2 - a_{12}a_{21}(p_1 p_2^* - p_2 p_1^*)^2]/(p_1 p_2 p_1^* p_2^* - 1)^2,$$
$$a_{jk} = c_{j,k}p_j p_k^*/(p_j p_k^* - 1)^2,$$
$$b_{jk} = [p_j + p_k^* + i\delta(p_j - p_k^*)]/(p_j p_k^* - 1),$$
$$\beta_{jk} = (p_j - p_k)/[p_j p_k + 1 + i\delta(p_j p_k - 1)].$$

6 CONCLUSION

Integrable discretization of coupled forms of the well-known soliton equations such as KdV equation, modified KdV equation, sine-Gordon equation and nonlinear Schrödinger equation is described.

REFERENCES

1. Hirota, R. (1978) Nonlinear partial difference equations. IV. Bäcklund transformation for the discrete-time toda equation, *J. Phys. Soc. Jpn.* **45**, pp. 321–332.
2. Hirota, R. (1981) Discrete analogue of a generalized toda equation, *J. Phys. Soc. Jpn.* **50**, pp. 3785–3791.
3. Hirota, R., Tsujimoto, S., and Imai, T. (1993) Difference scheme of soliton equations, in: *Future Directions of Nonlinear Dynamics in Physics and Biological Systems, eds. P.L. Christiansen et al., Plenum Press, New York,* pp. 7–15.
4. Hirota, R. (2000) Discretization of coupled modified KdV equations, *Chaos, Solitons and Fractals,* **11**, pp. 77–84.
5. Hirota, R. and Iwao, M. (2000) Time-discretization of soliton equations, *CRM Proceedings and Lecture Notes,* **25**, pp. 217–229.
6. Suris, Y. B. (1989) Integrable mappings of the standard type, *Funk. Anal. Prilozhen.,* **23**, pp. 84–85.
7. Hirota, R. and Kimura, K. (2000) Discretization of the Euler top, *J. Phys. Soc. Jpn.* **69**, pp. 627–630.
8. Kimura, K. and Hirota, R. (2000) Discretization of the Lagrange top, *J. Phys. Soc. Jpn.* **69**, pp. 3193–3199.
9. Papageorgiou, V., Gramaticos, B., and Ramani, A. (1993), Integrable lattices and convergence acceleration algorithms, *Phys. Lett. A,* **179**, p. 111–115.
10. Bobenko, A. I. and Pinkall, U. (1996) Discrete surfaces with constant negative gaussian curvature and the Hirota equation, *J. Diff. Geom.* **43**, pp. 527–611.
11. Tokihiro, T., Takahashi, D., Matukidaira, J., and Satsuma, J. (1996) From soliton equations to integrable cellular automata through a limiting procedure, *Phys. Rev. Lett.* **76**, pp. 3247–3250.
12. Zabrodin, A. (1997) Discrete Hirota's equation in quantum integrable models, *Int. J. Mod. Phys.* **B11**, pp. 3125–3158.
13. Quispel, G. R. W., Robert, J. A. G., and Thompson, C. J. (1989) Integrable mappings and Soliton equations II *Physica.* **D 34**, pp. 183–192.
14. Yoneyama, T. (1984) Interacting Korteweg-de Vries equations and attractive Soliton interaction, *Progr. Theor. Phys.* **72**, pp. 1081–1088.
15. Hirota, R. and Ohta, Y. (1991) Hierarchies of coupled Soliton equations. I, *J. Phys. Soc. Jpn.* **60**, pp. 798–809.
16. Iwao, M. and Hirota, R. (1997) Soliton solutions of a coupled modified KdV equations, *J. Phys. Soc. Jpn.,* **66**, pp. 577–588.
17. Manakov, S. V. (1974) On the theory of two-dimensional stationary self-focusing of electromagnetic waves, *Soviet Phy. JETP.* **38**(2), pp. 248–253.
18. Svinolupov, S. I. (1992) Generalized Schrödinger equations and Jordan Pairs, *Commun. Math. Phys.* **143**, pp. 559–575.

AN ADELIC W-ALGEBRA AND RANK ONE BISPECTRAL OPERATORS

E. Horozov

Institute of Mathematics and Informatics, Bulg. Acad. of Sci., Acad. G. Bonchev Str., Block 8, 1113 Sofia, Bulgaria

Abstract   We introduce a Lie algebra which we call adelic W-algebra. It is a central extension of the Lie algebra of the differential operators on the complex line with rational coefficients. We construct its natural bosonic representation similar to highest weight representation. Then we show that the rank one algebras of bispectral operators are in 1:1 correspondence with the tau-functions in this representation.

1 INTRODUCTION

In this note we give representation-theoretic description of rank one maximal commutative algebras of bispectral ordinary differential operators. This object has several quite different realizations, e.g., rational solutions of KP-hierarchy [1, 2]; the completed phase space of Calogero–Moser particle systems [3]; isomorphism classes of right ideals of the Weyl algebra [4, 5], etc. to mention few of them. Wilson has described it as a subset of Sato's Grassmannian and has named it "adelic Grassmannian" [6]. Some of the bispectral operators (not only rank one) were characterized in terms of representation theory of $W_{1+\infty}$-algebra [7]. Here we obtain a similar result for the entire set of rank one bispectral operators or in other words we characterize the rank one algebras of bispectral differential operators in terms of representations of a suitable infinite-dimensional Lie algebra, which will be called an adelic W-algebra. Before explaining in more details the main results we recall some of the notions that have been mentioned above.

Bispectral operators have been introduced by F.A. Grünbaum in his work on medical imaging [8] (see also [9]). An ordinary differential operator $L(x, \partial_x)$ is called bispectral if there exists an infinite-dimensional family of eigenfunctions $\psi(x, z)$, which are also eigenfunctions of another differential operator $\Lambda(z, \partial_z)$ in the spectral parameter z, i.e., for which the following identities hold:

$$L(x, \partial_x)\psi(x, z) = f(z)\psi(x, z), \tag{1}$$

$$\Lambda(z, \partial_z)\psi(x, z) = \theta(x)\psi(x, z), \tag{2}$$

123

L. Faddeev et al. (eds.),
Bilinear Integrable Systems: From Classical to Quantum, Continuous to Discrete, 123–136.
© 2006 *Springer. Printed in the Netherlands.*

with some nonconstant functions $f(z)$ and $\theta(x)$. G. Wilson [6] has classified all bispectral operators of rank one (see the next section for more details). Using slightly different terminology than in [6], they are all operators with rational coefficients that are Darboux transformations of operators with constant coefficients. Sato's theory associates with each operator (or rather with the maximal algebra of operators that commute with it) a plane in Sato's Grassmannian. The set of all planes corresponding to the rank one bispectral algebras of operators has been called by G. Wilson *an adelic Grassmannian* and denoted by Gr^{ad}. Originally G. Wilson has characterized the rank one bispectral algebras A as those whose spectral curve $SpecA$ is rational and its singularities are only cusps. In a different development [7] we have characterized those of bispectral algebras whose spectral curve has only one cusp in terms of representations of $W_{1+\infty}$-algebra. More precisely we have built certain bosonic highest weight modules of $W_{1+\infty}$. Denote the module corresponding to the rank one case by $\mathcal{M}_0$. Then the tau-functions of the bispectral operators (with the above restriction) lie in $\mathcal{M}_0$ and vice versa—all the tau functions in the module are tau-functions of bispectral operators.

A natural question (see [7]) is if a similar result holds for the entire set of rank one bispectral operators. The present paper gives an affirmative answer to this question. Obviously one has first to point out a suitable generalization of the $W_{1+\infty}$-algebra. The most natural candidate does the job—the algebra we look for is a central extension of the algebra of differential operators with rational coefficients. We call this new algebra *an adelic W-algebra*. Then we proceed as in [7, 10]. We construct a bosonic representation $\mathcal{M}^{ad}$ which is similar to a highest weight representation. Our main result is the following:

Theorem 1 *If an element $\tau \in \mathcal{M}^{ad}$ is a tau-function then the corresponding plane belongs to Gr^{ad}. Conversely, if $W \in Gr^{ad}$ then $\tau_W \in \mathcal{M}^{ad}$.*

Returning to the other realizations of Gr^{ad} we obtain other interesting connections. For example, using the fact that Gr^{ad} bijectively maps onto the set $\mathcal{R}$ of isomorphism classes of right ideals of the Weyl algebra A_1 (cf. [4, 5]) we obtain

Corollary 2 *The isomorphism classes of the right ideals of the Weyl algebra are in 1:1 correspondence with the tau-functions in $\mathcal{M}^{ad}$.*

(Recall that the Weyl algebra A_1 is the algebra of differential operators in one variable with polynomial coefficients. Two right ideals $I, J \in A_1$ are isomorphic iff they are isomorphic as right A_1-modules.)

For other interpretations see [5, 11]. Many of the constructions in the present paper are similar to those of [7]. The organization of the paper is the following. Section 2 contains preliminaries on Sato's Grassmannian, Darboux transformations, bispectral operators, $W_{1+\infty}$-algebra. In Section 3 we introduce the adelic

W-algebra together with a bosonic representation $\mathcal{M}^{ad}$. In Section 4 we give a sketch of Theorem 1. The detailed proofs will be presented elsewhere.

This paper is dedicated to Prof. R. Hirota.

2 PRELIMINARIES

Here we have collected some facts and notation needed throughout the paper. In particular we recall Sato's theory, Darboux transforms, and the bispectral problem, $W_{1+\infty}$-algebra.

2.1 Sato's Theory of KP-Hierarchy

In this subsection we recall some facts and notation from Sato's theory of KP-hierarchy [2–14] needed in the paper. We use the approach of V. Kac and D. Peterson based on infinite wedge products (see e.g., [15]) and the survey paper by P. van Moerbeke [16].

Consider the infinite-dimensional vector space of formal series

$$\mathbb{V} = \Big\{ \sum_{k \in \mathbb{Z}} a_k v_k \,\Big|\, a_k = 0 \quad \text{for } k \gg 0 \Big\}.$$

Sato's Grassmannian *Gr* (more precisely—its big cell) [12, 13] consists of all subspaces ("planes") $W \subset \mathbb{V}$ which have an admissible basis

$$w_k = v_k + \sum_{i < k} w_{ik} v_i, \quad k = 0, 1, 2, \ldots$$

Then define the *fermionic Fock space* $F^{(0)}$ consisting of formal infinite sums of semi-infinite wedge monomials

$$v_{i_0} \wedge v_{i_1} \wedge \cdots$$

such that $i_0 < i_1 < \cdots$ and $i_k = k$ for $k \gg 0$. The wedge monomial

$$\psi_0 = v_0 \wedge v_1 \wedge \cdots$$

plays a special role and is called the *vaccum*. The plane that corresponds to it will be denoted by W_0. There exists a well-known linear isomorphism, called a *boson-fermion correspondence*:

$$\sigma \colon F^{(0)} \to B, \tag{3}$$

(see [15]), where $B = \mathbb{C}[[t_1, t_2, \ldots]]$ is the bosonic Fock space.

To any plane $W \in Gr$ one naturally associates a state $|W\rangle \in F^{(0)}$ as follows

$$|W\rangle = w_0 \wedge w_1 \wedge w_2 \wedge \cdots,$$

where $w_0, w_1, \ldots$, form an admissible basis. One of the main objects of Sato's theory is the *tau-function* of W defined as the image of $|W\rangle$ under the boson-fermion correspondence (3)

$$\tau_W(t) = \sigma(|W\rangle) = \sigma(w_0 \wedge w_{-1} \wedge w_{-2} \wedge \cdots). \tag{4}$$

It is a formal power series in the variables $t_1, t_2, \ldots$, i.e., an element of $B :=$ $\mathbb{C}[[t_1, t_2, \ldots]]$. In particular the tau-function corresponding to the vacuum ψ_0 is $\tau_0 \equiv 1$. Using the tau-function one can define the other important function connected to W—the *Baker* or *wave function*

$$\Psi_W(t, z) = e^{\sum_{k=1}^{\infty} t_k z^k} \frac{\tau_W(t - [z^{-1}])}{\tau_W(t)}, \tag{5}$$

where $[z^{-1}]$ is the vector $(z^{-1}, z^{-2}/2, \ldots)$. Introducing the vertex operator

$$X(t, z) = \exp\left(\sum_{k=1}^{\infty} t_k z^k\right) \exp\left(-\sum_{k=1}^{\infty} \frac{1}{kz^k} \frac{\partial}{\partial t_k}\right) \tag{6}$$

the above formula (5) can be written as

$$\Psi_W(t, z) = \frac{X(t, z)\tau(t)}{\tau(t)}. \tag{7}$$

We often use the formal series $\Psi_W(x, z) = \Psi_W(t, z)|_{t_1=x, \, t_2=t_3=\cdots=0}$, which we call again wave function. The wave function, corresponding to the vacuum is

$$\Psi_0(x, z) = e^{xz}.$$

The wave function $\Psi_W(x, z)$ contains the whole information about W and hence about τ_W, as the vectors $w_{-k} = \partial_x^K \Psi_W(x, z)|_{x=0}$ form an admissible basis of W (if we take $v_k = z^k$ as a basis of $\mathbb{V}$).

2.2 Darboux Transforms and Bispectral Operators

We shall recall a version of Darboux transform from [17].

Definition 3 *We say that a plane W (or the corresponding wave function $\Psi_W(x, z)$, the tau-function τ_W) is a Darboux transformation of the vacuum (respectively—of the wave function $\Psi_0(x, z)$, the tau-function τ_0) if there exist polynomials $f(z)$, $g(z)$, and differential operators $P(x, \partial_x)$, $Q(x, \partial_x)$ such that*

$$\Psi_W(x, z) = \frac{1}{g(z)} P(x, \partial_x)\Psi_0(x, z), \tag{8}$$

$$\Psi_0(x, z) = \frac{1}{f(z)} Q(x, \partial_x)\Psi_W(x, z). \tag{9}$$

The Darboux transformation is called *polynomial* if the operators $P(x, \partial_x)$ and $Q(x, \partial_x)$ have rational coefficients.

Obviously

$$Q(x, \partial_x)P(x, \partial_x)\Psi_0 = g(z)f(z)\Psi_0, \qquad (10)$$

Denoting the polynomial $g(z)f(z)$ by $h(z)$ and recalling that $\Psi_0 = e^{xz}$ we see that

$$Q(x, \partial_x)P(x, \partial_x) = h(\partial_x).$$

On the other hand the wave function Ψ_W is an eigen-function of the differential operator

$$L(x, \partial_x) = P(x, \partial_x)Q(x, \partial_x).$$

Notice that the operator L is a traditional Darboux transform of the operator $h(\partial_x)$, which justifies the terminology of the definition. We will also say that the operator L is *a polynomial Darboux transform* of the operator ∂_x.

We shall need a second definition of the polynomial Darboux transformation. In the above notation let the polynomial $h(\partial_x)$ factorize as

$$h(\partial_x) = \prod_{j=1}^{m}(\partial_x - \lambda_j)^{d_j},$$

where λ_j are the different roots with multiplicities d_j. Then the kernel of $h(\partial_x)$ is given by

$$ker\,h(\partial_x) = \oplus_{j=1}^{m} W_j,$$

where

$$W_j = \{e^{\lambda_j x}, xe^{\lambda_j x}, \ldots, x^{d_j-1}e^{\lambda_j x}\}.$$

Definition 4 *The Darboux transform is polynomial iff the kernel of P has the form*

$$\ker P = \oplus_{j=1}^{m} K_j,$$

where K_j is a linear subspace of W_j.

The equivalence of the two definitions can be found in [18]. Each nonzero element $f \in K_j$ will be called (after Wilson) *condition supported at λ_j*. The Darboux transform will be called *monomial* iff all the conditions are supported at one point. Finally we recall the bispectral involution b, which in this case maps the operators with polynomial coefficients in the x-variable into operators with polynomial coefficients in the z-variable by the formulas

$$b(\partial_x) = z, \qquad b(x) = \partial_z,$$

i.e., in this case b is the formal Fourier transform. It will be used when the differential operators are applied to Ψ_0 as follows:

$$\partial_x \Psi_0 = z \Psi_0, \qquad x \Psi_0 = \partial_z \Psi_0$$

We end this subsection with the following important result of G. Wilson [6]:

Theorem 5 *Any polynomial Darboux transform of ∂_x is a rank one bispectral operator and vice versa.*

This theorem is formulated by G. Wilson in a different terminology. See [19] for an exposition using Darboux transforms.

Following G. Wilson we will call the set of all planes $W \subset Gr$ that are polynomial Darboux transforms of W_0 *the adelic Grassmannian* and denote it by Gr^{ad}.

2.3 $W_{1+\infty}$-Algebra

In this subsection we recall the definition of $W_{1+\infty}$, and some of its bosonic representations introduced in [10]. For more details see [20].

The algebra w_∞ of the additional symmetries of the KP-hierarchy is isomorphic to the Lie algebra of regular polynomial differential operators on the circle

$$w_\infty \equiv \mathcal{D} = \text{span}\{z^\alpha \partial_z^\beta | \alpha, \beta \in \mathbb{Z}, \beta \geq 0\}.$$

It was introduced in [21, 22] and was extensively studied by many authors (see, e.g., [23, 24], etc.). Its unique central extension is denoted by $W_{1+\infty}$.

Denote by c the central element of $W_{1+\infty}$ and by $W(A)$ the image of $A \in \mathcal{D}$ under the natural embedding $\mathcal{D} \hookrightarrow W_{1+\infty}$ (as vector spaces). The algebra $W_{1+\infty}$ has a basis

$$c, J_k^l = W(-z^{l+k}\partial_z^l), \quad l, k \in \mathbb{Z}, l \geq 0.$$

In [10] we constructed a family of highest weight modules of $W_{1+\infty}$. Here we need the most elementary one of them, for which the next theorem is an easy exercise.

Theorem 6 *The function τ_0 satisfies the constraints*

$$J_k^l \tau_0 = 0, \quad k \geq 0, l \geq 0, \tag{11}$$

$$W\left(z^{-k} P_k(D_z) D_z^l\right) \tau_0 = 0, \quad k \geq 0, l \geq 0, \tag{12}$$

where $P_k(D_z) = \prod_{j=0}^{k-1}(D_z - j), D_z = z\partial_z$.

The first constraint means that τ_0 is a highest weight vector with highest weight $\lambda(J_0^l) = 0$ of a representation of $W_{1+\infty}$ in the module

$$\mathcal{M}_0 = \text{span}\left\{ J_{k_1}^{l_1} \dots J_{k_p}^{l_p} \tau_0 \big| k_1 \leq \dots \leq k_p < 0 \right\}. \tag{13}$$

One easily checks that the central charge $c = 1$. The second constraint yields that the module $\mathcal{M}_0$ is quasifinite, i.e., it is finite-dimensional in each level.

3 AN ADELIC W-ALGEBRA

The adelic W-algebra is a Lie algebra that we intend to introduce in analogy with $W_{1+\infty}$. Most of the definitions and constructions are similar to those of $W_{1+\infty}$.

Instead of the Lie algebra w_∞ of regular operators on the circle we start with the Lie algebra $\mathcal{RD}$ of differential operators with rational coefficients on the complex line. We are going to use the following basis of $\mathcal{RD}$:

1. $z^{n+l}\partial_z^l$, $n \in \mathbb{Z}$, $l \geq 0$;   (14)
2. $(z - a)^{-n+l}\partial_z^l$, $-n + l < 0$, $l \geq 0$, $a \in \mathbb{C} - \{0\}$;   (15)

Usually we shall consider the elements from $\mathcal{RD}$ as differential operators with coefficients that are Laurent series in z^{-1} by expanding $(z - a)^{-n+l}$ around infinity. We would like to construct natural representations of $\mathcal{RD}$. We shall work with the space $\mathbb{V}$ where $v_k = z^k$. Obviously $\mathcal{RD}$ acts naturally on $\mathbb{V}$. Then we can associate with each operator $A \in \mathcal{RD}$ an infinite matrix having only finite number of diagonals below the principal one but having eventually infinite number above it. In other words the matrix $(a_{i,j})$, associated with A, has the property that $a_{i,j} = 0$ for $i - j \gg 0$. The Lie algebra of such matrices will be denoted by a'_∞. It can be considered as a completion of the algebra a_∞ of matrices having only finite number of diagonals (see [15]). Now we explain how to construct representations in the fermionic Fock space $F^{(0)}$. We recall that in the case considered here $F^{(0)}$ consists of formal series of semi-infinite wedge monomials:

$$z^{i_0} \wedge z^{i_1} \wedge z^{i_2} \wedge \cdots, \tag{16}$$

with $i_0 < i_1 < \cdots$ and $i_k = k$ for $k \gg 0$. We can define the action of $A \in a'_\infty$ by the standard definition (see [15]). First for matrices with only finite number of entries define

$$r(A)(z^{i_0} \wedge z^{i_1} \wedge \cdots) = Az^{i_0} \wedge z^{i_1} \wedge \cdots$$
$$+ z^{i_0} \wedge Az^{i_1} \wedge \cdots$$
$$\cdots$$

It is easy to check that if $A \in a'_\infty$ has no entries on the main diagonal $r(A)$ still makes sense, the image being infinite formal series. For matrices with infinite number of entries on the main diagonal the above definition is no longer me aningful. For that reason we need to modify it as follows. We put

$$\hat{r}(E_{i,j}) = r(E_{i,j}) \quad \text{for} \quad i \neq j \quad \text{or} \quad i = j > 0; \tag{17}$$

$$\hat{r}(E_{i,i}) = r(E_{i,i}) - Id \quad \text{for} \quad i \leq 0. \tag{18}$$

See [15] for more details.

This defines a representation of the central extension $a'_\infty \oplus \mathbb{C}c$. The corresponding central extension of the subalgebra $\mathcal{RD}$ of a'_∞ will be called *adelic W-algebra*. We will use the notation W^{ad}. The terminology and the notation are chosen to be similar to those of the adelic Grassmannian Gr^{ad}. The main result of the present paper naturally connects the two objects.

We shall describe in some more details W^{ad}. By $W(A)$ we shall denote the image of the element $A \in \mathcal{RD}$ under the natural embedding $\mathcal{RD} \subset W^{ad}$ (as vector spaces). Then for $a \in \mathbb{C}, l \geq 0, \quad n \in \mathbb{Z}$ put

$$J_n^l(a) = W(-(z-a)^{n+l} \partial_z^l) \tag{19}$$

For $a = 0$ we also shall use the notation $J_n^l = J_n^l(0)$. When a is fixed the above operators (19) together with the central charge c form a copy of $W_{1+\infty}$, which we shall denote by $W_{1+\infty}(a)$. Recall that $W_{1+\infty}(a)$ has a grading: the elements $J_n^l(a)$ have weight n. The elements with nonnegative grading are common for all a. In fact the common part is much larger: for all $n + l \geq 0$ the elements $J_n^l(a)$ are common. Thus we have the following basis for W^{ad}:

1. $J_n^l(0), \quad l \geq 0, \quad n + l \geq 0;$ $\qquad\qquad\qquad\qquad\qquad$ (20)

2. $J_n^l(a), \quad n + l < 0, \quad a \neq 0;$ $\qquad\qquad\qquad\qquad\qquad$ (21)

3. $c.$ $\qquad\qquad\qquad\qquad\qquad\qquad\qquad\qquad\qquad\qquad\qquad\qquad$ (22)

In complete analogy to the case of $W_{1+\infty}$ we can construct a representation of W^{ad} in the Fock spaces using the vacuum. We formulate the needed properties in the following theorem.

Theorem 7 *The tau-function τ_0 satisfies the following constrains*

1) $J_n^l \tau_0 = 0, \quad l \geq 0; \quad n \geq 0$ $\qquad\qquad\qquad\qquad\qquad\qquad$ (23)

2) $W((z-a)^{-k} P_k((z-a)\partial_z))((z-a)\partial_z)^l \tau_0 = 0,$ $\qquad\qquad$ (24)

where $P_k(u) = u(u-1)\ldots(u-k+1).$

We set

$$W_-^{ad} = span\{J_n^l, a \in \mathbb{C}, n < 0\}. \tag{25}$$

Then define the W_-^{ad}-module $\mathcal{M}^{ad}$ by

$$\mathcal{M}^{ad} = span\{J_{n_1}^{l_1}(a_1)\ldots J_{n_m}^{l_m}(a_m)\tau_0\}, \tag{26}$$

where $n_j + l_j < 0$ for $a_j \neq 0$ and $n_j < 0$ for $a_j = 0$.

Corollary 8 *The vector space $\mathcal{M}^{ad}$ is a space of representation of the Lie algebra W^{ad}.*

4 PROOF OF THE MAIN RESULT

In this section we give a sketch of the prove of Theorem 1. It will be split into two parts.

1. Assume that τ is a tau-function in the module $\mathcal{M}^{ad}$. We are going to show that it is polynomial Darboux transform of τ_0.

Let $\tau_W = u\tau_0$ where u is an element of the universal enveloping algebra. One can express the wave function $\Psi_W(t, z)$ in terms of u:

$$\Psi_W(t, z) = \frac{X(t, z)u\tau_0}{u\tau_0}|_{t_1=x,\, t_2=\cdots=0}. \tag{27}$$

Commuting u and $X(t, z)$ we obtain

$$\Psi_W = \frac{U(t, z)X(t, z)\tau_0}{u\tau_0}|_{t_1=x,\, t_2=\cdots=0}, \tag{28}$$

where

$$U(t, z) = \sum b_{k,l,a}\left(J_{-k_1}^{l_1}(a_1) + l_1 J_{-k_1+1,-1}^{l_1-1} - (z - a_1)^{-k_1+l_1}\partial_z^{l_1}\right)$$

$$\ldots\left(J_{-k_p}^{l_p}(a_p) + l_p J_{-k_p+1,-1}^{l_p-1} - (z - a_p)^{-k_p+l_p}\partial_z^{l_p}\right)$$

$$\left(J_{-k_{p+1}}^{l_{p+1}} + l_{p+1} J_{-k_{p+1}}^{l_p-1} + \delta_{k_{p+1},0}\delta_{l_{p+1},0} - z^{-k_{p+1}+l_{p+1}}\partial_z^{l_{p+1}}\right)$$

$$\ldots\left(J_{-k_{p+r}}^{l_{p+r}} + l_{p+r} J_{-k_{p+r}}^{l_r-1} + \delta_{k_{p+r},0}\delta_{l_{p+r},0} - z^{-k_{p+r}+l_{p+r}}\partial_z^{l_{p+r}}\right).$$

One can prove that the element $U(t, z)$ is equivalent to another element where $l_j < n_j$. Now we use that

$$J_{-k}^l|_{t_1=x,\, t_2=t_3=\cdots=0} = x^k\delta_{l+1,k} \quad if \ l < k. \tag{29}$$

The point is that there are no differentiations but only multiplications by powers of x. From the above formula we can derive for $J_{-k}^l(a)$ and for $J_{-k+1,-1}^l(a)$ the following ones:

$$J^l_{-k}(a)|_{t_1=x,\ t_2=t_3=\cdots=0} = x^k \delta_{l+1,k},\tag{30}$$

$$J^l_{-k+1,-1}(a)|_{t_1=x,\ t_2=t_3=0} = 0.\tag{31}$$

Both formulas follow from the expansion of the L.H.S. as infinite series and the fact that $l < k$. From (27) we obtain (using also the bispectral involution)

$$\Psi_W = \frac{P_1(x, \partial_x)\Psi_0}{g(z)},\tag{32}$$

where P_1 is an operator with rational coefficients.

We need also to express $\Psi_0(x, z)$ in terms of $\Psi_W(x, z)$. The adjoint involution a implies

$$\Psi_0(x, z) = \frac{P_2^*(x, \partial_x)\Psi_W}{g_2(z)},\tag{33}$$

which together with (32) gives the proof of the first part of the theorem.

2. Next we assume that τ_W is a bispectral tau-function.

To fix the notation let the Darboux transform of the corresponding wave function Ψ_W be given by

$$\Psi_W(x, z) = \frac{P(x, \partial_x)\Psi_0(x, z)}{g(z)}\tag{34}$$

$$\Psi_0(x, z) = \frac{Q(x, \partial_x)\Psi_W(x, z)}{f(z)},\tag{35}$$

where $Q \circ P = h(\partial_x)$ with some polynomial h. Let $\lambda_1, \ldots, \lambda_m$ be the different points, where the conditions are supported. Then we can suppose that the polynomial h is

$$h(z) = \prod_{j=0}^m (z - \lambda_j)^{d_j}.$$

Denote the degree of h by d. Let the number of the conditions supported at the point λ_j be r_j. Then

$$g(z) = \prod_{j=0}^m (z - \lambda_j)^{r_j}.$$

Put also $deg g(z) = r = r_1 + \cdots r_m$. Let $\{\Phi_i\}_{|i=1,\ldots,d}$ be the standard basis of $ker h(z)$, i.e.

$$\{\Phi_i\} = \bigcup_{j=1}^m \{e^{\lambda_j x}, \ldots, x^{d_j-1}e^{\lambda_j x}\}$$

Denote by $f_1, \ldots, f_r$ the functions forming the kernel of the operator P, i.e, defining the Darboux transform (34)–(35). Then

$$f_l(x) = \sum_{i=1}^{d} a_{l,i} \Phi_i(x), \quad l = 1, \ldots, r \qquad (36)$$

Denote by A the matrix formed by the above coefficients, i.e.

$$A = (a_{l,i}), \quad l = 1, \ldots, r, \quad i = 1, \ldots, d.$$

For any r-element subset $I\{i_1, \ldots, i_r\} \subset \{1, \ldots, d\}$ denote by A^I the following minor of A:

$$A^I = (a_{l,i_k})_{|l, k=1,\ldots,r}.$$

Put $\Phi = \{\Phi_{i_1,\ldots,i_r}\}$ and

$$\Psi_I(x, z) = \frac{Wr(\Phi_I, \Psi_0)}{g(z)Wr(\Phi_I)}. \qquad (37)$$

We need the following formula from [16]:

$$\Psi_W(x, z) = \frac{\sum_I \det A^I Wr(\Phi_I)\Psi_I(x, z)}{\sum_I \det^I Wr(\Phi_I)}. \qquad (38)$$

Notice that for any I we have

$$Wr(\Phi_I, \Psi_0) = e^{x \sum \lambda_j r_j} P_I(x, \partial_x) = e^{x \sum \lambda_j r_j} \sum_{j=0}^{r} p_{I,j}(x)\partial_x^j e^{xz},$$

where $p_{I,j}(x)$ are polynomials. Also we have

$$Wr(\Phi_I) = e^{x \sum \lambda_j r_j} q_I(x),$$

where the polynomial $q_I(x) = p_{I,r}(x)$. Notice that the exponential factor is the same everywhere. Then we have

$$\Psi_W = \frac{\sum_I \det A^I P_I(x, \partial_x)e^{xz}}{g(z) \sum_I \det^I q_I(x)}. \qquad (39)$$

Among the subsets I there is one that corresponds to the set of following functions from the kernel of $h(\partial_x)$:

$$\tilde{f}_1(x) = e^{\lambda_1 x}, \ldots, \tilde{f}_{r_1}(x) = x^{r_1-1}e^{\lambda_1 x}$$

$$\cdots \quad \cdots \quad \cdots \quad \cdots \quad \cdots \quad \cdots$$

$$\tilde{f}_{r-r_m+1} = e^{\lambda_m x}, \ldots, \tilde{f}_r = x^{r_m-1}e^{\lambda_r x}$$

Denote this subset by I_0. Notice that $P_{I_0} = \sum_{j=0}^{r} \beta_j \partial_x^j$, where $b_j \in \mathbb{C}$, i.e., P_{I_0} is an operator with constant coefficients. It is easy to check that $P_{I_0} \equiv g(\partial_x)$.

Introduce the matrix A_0 as follows. Let $I_0 = (i_1^0 < i_2^0 < \cdots < i_r^0)$. Then let $A_0 = (a_{j,i}\delta_{j,i_k^0})$. Now consider A as a deformation of A_0:

$$A(\epsilon) = \epsilon A + (1 - \epsilon)A_0.$$

Obviously $a_{j,i}(\epsilon) = a_{j,i}$ if $i \in I_0$ and $a_{j,i} = \epsilon a_{j,i}$ otherwise.

Generically $det A_0^I \neq 0$. We can assume that $det A_0^I = 1$. Using the bispectral involution and expanding the denominator in a series in ε one obtains

$$\Psi_{W(\epsilon)} = (1 + \sum_{j=1}^{\infty} \epsilon^j P_j(z, \partial_z))\Psi_0. \qquad (40)$$

The important fact here is that all the operators $P_j(z, \partial_z) \in W_-^{ad}$. The standard basis of W_0 is given by $w_k = \partial_x^k \Psi_0 = z^k, k = 0, 1, \ldots$. We need to find expression for the basis of $W(\epsilon)$. We have

$$\partial_x^k \Psi_{W(\epsilon)} = (1 + \sum_{j=1}^{\infty} \epsilon^j P_j(z, \partial_z))w_k.$$

Using the boson-fermion correspondence σ we get the tau-function $\tau_{W(\epsilon)}$:

$$\tau_{W(\epsilon)} = \sigma\Big((1 + \sum_{j=1}^{\infty} \epsilon^j P_j(z, \partial_z))w_0 \wedge (1 + \sum_{j=1}^{\infty} \epsilon^j P_j(z, \partial_z))w_1 \wedge \cdots\Big)$$

$$= \tau_0 + \epsilon r(P_1)\tau_0 + \epsilon^2 \left(r(P_2) + \frac{1}{2}r(P_1)^2 - \frac{1}{2}r(P_1^2)\right)\tau_0 + \cdots$$

Notice that the coefficients at the powers of ϵ are polynomials in $r(P_k^j)$, applied to τ_0. Hence all of them belong to the W^{ad}-module $\mathcal{M}^{ad}$. We shall show that the entire series belong to it. Once again we need a formula from [16]—this time for the tau-function. We have

$$\tau_{W(\epsilon)} = \frac{1 + \sum_{I \neq I_0} det A^I(\epsilon)\Delta_I \tau_I}{1 + \sum_{I \neq I_0} det A^I(\epsilon)\Delta_I},$$

where $\Delta_I = Wr(\Phi_I)(0)$. Multiplying $\tau_{W(\epsilon)}$ by the denominator, which is a polynomial in ϵ we get a polynomial in ϵ (the numerator), which, having a finite number of terms that belong to $\mathcal{M}^{ad}$ by the above argument, itself belongs to $\mathcal{M}^{ad}$ for all ϵ. In particular for $\epsilon = 1$ we get that $\tau_W \in \mathcal{M}^{ad}$.

The case when $det A_0^I = 0$ easily follows from the above.

ACKNOWLEDGMENTS

I am grateful to the organizers of the NATO ARW "Bilinear integrable systems: from classical to quantum, continuous to discrete" for the invitation to participate in it. This work was partially supported by Grant MM-1003/2000 of NFSR of Bulgarian Ministry of Education.

REFERENCES

1. Adler, M. and Moser, J. (1978) On a class of polynomials connected with the Korteweg–de Vries equation, *Commun. Math. Phys.* **61**, pp. 1–30.
2. Krichever, I. M. (1978) On rational solutions of Kadomtsev-Petviashvily equation and integrable systems of N particles on the line, *Funct. Anal. Appl.* **12**(1), pp. 76–78 (Russian), 59–61 (English).
3. Wilson, G. (1998) Collisions of Calogero–Moser particles and an adelic Grassmannian (with an appendix by I.G. Macdonald), *Invent. Math.* **133**, pp. 1–41.
4. Cannings, R. C. and Holland, M. P. (1994) Right ideals in rings of differential operators, *J. Algebra*, **167**, pp. 116–141.
5. Berest, Y. and Wilson, G. (2000) Automorphisms and ideals of the Weyl algebra, *Math. Ann.* **318**(1), pp. 127–147.
6. Wilson, G. (1993) Bispectral commutative ordinary differential operators, *J. Reine Angew. Math.* **442**, pp. 177–204.
7. Bakalov, B., Horozov, E., and Yakimov, M. (1998) Highest weight modules over $W_{1+\infty}$, and the bispectral problem, *Duke Math. J.* **93**, pp. 41–72.
8. Günbaum, F. A. (1982) The limited angle reconstruction problem in computer tomography, in: Processings of Symposium on Applied Mathematics, Vol. 27, AMS, ed. L. Shepp, pp. 43–61.
9. Duistermaat, J. J. and Grünbaum, F. A. (1986) Differential equations in the spectral parameter, *Commun. Math. Phys.* **103**, pp. 177–240.
10. Bakalov, B., Horozov, E., and Yakimov, M. (1996) Tau-functions as highest weight vectors for $W_{1+\infty}$ algebra, *J. Phys. A: Math. Gen.* **29**, pp. 5565–5573, hep-th/9510211.
11. Berest, Y. and Wilson, G. (2001) Ideal classes of the Weyl algebra and noncommutative projective geometry, arXiv.math.AG/0104240.
12. Sato, M. (1981) Soliton equations as dynamical systems on infinite dimensional Grassmann manifolds, *RIMS Kokyuroku* **439**, pp. 30–40.
13. Date, E., Jimbo, M., Kashiwara, M., and Miwa, T. (1983) Transformation groups for solution equations, in: *Proceedings of Research Institute for Mathematical Sciences Symposium on Nonlinear Integrable systems—Classical and Quantum Theory* (Kyoto 1981), World Scientific, Singapore, eds. M. Jimbo and T. Miwa, pp. 39–111.
14. Segal, G. and Wilson, G. (1985) Loop groups and equations of KdV type, *Publ. Math. IHES* **61**, pp. 5–65.
15. Kac, V. G. and Raina, A. (1987) Bombay lectures on highest weight representations of infinite dimensional Lie algebras, *Advanced Series in Mathematical Physics* **2**, World Scientific, Singapore.
16. van Moerbeke, P. (1994) Integrable foundations of string theory (CIMPA–Summer school at Sophia–Antipolis 1991), in: *Lectures on Integrable Systems*, eds. O. Babelon et al., World Scientific, Singapore.

17. Bakalov, B., Horozov, E., and Yakimov, M. (1996) Bäcklund–Darboux transformations in Sato's Grassmannian, *Serdica Math. J.* **4**, q-alg/9602010.
18. Bakalov, B., Horozov, E., and Yakimov, M. (1997) Bispectral algebras of commuting ordinary differential operators, *Comm. Mat. Phys.* **190**, pp. 331–373, q-alg/9602011.
19. Horozov, E. (2002) Dual algebras of differential operators, in: *Kowalevski property* (Montréal), CRM Proceedings Lecture Notes 32, Surveys from Kowalevski Workshop on Mathematical methods of Regular Dynamics, Leeds, April 2000, American Mathematical Society, Providence, pp. 121–148.
20. Kac, V. G. and Radul, A. (1993) Quasifinite highest weight modules over the Lie algebra of differential operators on the circle, *Commun. Math. Phys.* **157**, pp. 429–457, hep-th/9308153.
21. Fuchsteiner, B. (1983) Master-symmetries, higher order time-dependent symmetries and conserved densities of nonlinear evolution equations, *Progr. Theor. Phys.* **70**(6), pp. 1508–1522.
22. Orlov, A. Y. and Schulman, E. I. (1989) Additional symmetries for integrable and conformal algebra representation, *Lett. Math. Phys.* **12**, pp. 171–179.
23. Adler, M., Shiota, T., and van Moerbeke, P. (1995) A Lax representation for the vertex operator and the central extension, *Commun. Math. Phys.* **171**, pp. 547–588.
24. Dickey, L. (1991) *Soliton Equations and Integrable Systems*. World Scientific, Singapore.
25. Kac, V. G. and Peterson, D. H. (1981) Spin and wedge representations of infinite-dimensional Lie algebras and groups, *Proc. Natl. Acad. Sci. USA.* **78**, pp. 3308–3312.
26. Kasman, A. (1995) Bispectral KP solutions and linearization of Calogero–Moser particle systems, *Commun. Math. Phys.* **172**, pp. 427–448.

TOROIDAL LIE ALGEBRA AND BILINEAR IDENTITY OF THE SELF-DUAL YANG-MILLS HIERARCHY

Saburo Kakei
Department of Mathematics, Rikkyo University, Nishi-ikebukuro, Tokyo 171-8501, Japan

Abstract   Bilinear identity associated with the self-dual Yang-Mills hierarchy is discussed by using a fermionic representation of the toroidal Lie algebra $\mathfrak{sl}_2^{\mathrm{tor}}$.

1   INTRODUCTION

There have been many studies on soliton equations in higher dimensional space-time. Among other things, the self-duality equation of the Yang-Mills model in four-dimension plays prominent role, since it produces many integrable equations by dimensional reduction (see [1] and the references therein). In this sense, the self-dual Yang-Mills (SDYM) equation may be regarded as a "master equation" for soliton-type equations.

The SDYM equation can also be treated also by Hirota's bilinear method. Toward this aim, we shall take so-called "Yang's R-gauge" or the "J-formulation" of the SDYM [2]:

$$\partial_{\bar{y}}\left(J^{-1}\partial_y J\right) + \partial_{\bar{z}}\left(J^{-1}\partial_z J\right) = 0. \tag{1}$$

Following the work [3], Sasa, Ohta and Matsukidaira [4] considered the gauge field J of the form,

$$J = \frac{1}{f}\begin{bmatrix} 1 & -g \\ e & f^2 - eg \end{bmatrix}, \quad e = \frac{\tau_1}{\tau_5}, \quad f = \frac{\tau_2}{\tau_5}, \quad g = \frac{\tau_3}{\tau_5}. \tag{2}$$

The gauge field J of (2) solves (1) if the τ-functions satisfy the following seven Hirota-type equations [4],

$$\tau_5^2 + \tau_2\tau_8 - \tau_4\tau_6 = 0, \tag{3}$$

$$D_y\tau_1 \cdot \tau_5 = D_{\bar{z}}\tau_4 \cdot \tau_2, \tag{4}$$

$$D_y\tau_2 \cdot \tau_6 = D_{\bar{z}}\tau_5 \cdot \tau_3, \tag{5}$$

L. Faddeev et al. (eds.),
Bilinear Integrable Systems: From Classical to Quantum, Continuous to Discrete, 137–146.

$$D_y \tau_4 \cdot \tau_8 = D_{\bar{z}} \tau_5 \cdot \tau_7, \tag{6}$$

$$D_z \tau_1 \cdot \tau_5 = D_{\bar{y}} \tau_2 \cdot \tau_4, \tag{7}$$

$$D_z \tau_2 \cdot \tau_6 = D_{\bar{y}} \tau_3 \cdot \tau_5, \tag{8}$$

$$D_z \tau_4 \cdot \tau_8 = D_{\bar{y}} \tau_7 \cdot \tau_5, \tag{9}$$

where we have used Hirota derivatives,

$$D_x^n f(x) \cdot g(x) = (\partial_x - \partial_{x'})^n f(x) g(x')|_{x'=x},$$

and introduced auxiliary dependent variables $\tau_4, \tau_6, \tau_7, \tau_8$.

Due to the integrable structure of the SDYM equation, one can generate a hierarchy of higher order integrable equations associated with the SDYM equation [1, 5, 6]. In the case of the KP hierarchy, there is a method to introduce τ-function from the hierarchy structure [7]. However no general method has been found for defining τ-functions directly from the hierarchy structure of the SDYM. So a natural question may arise: What is the meaning of the τ-functions in (3)–(9)?

The aim of this article is to give an answer to the question. As shown below, we can reproduce (3)–(9) from representation theory of the toroidal Lie algebra $\mathfrak{sl}_2^{\text{tor}}$ [8]. We note that the relation between integrable hierarchies and toroidal algebras has been discussed several authors [9–13] by using vertex operator representations.

2 TOROIDAL LIE ALGEBRAS

We start with the definitions of $(M + 1)$-toroidal Lie algebras, which is the universal central extension of $(M + 1)$-fold loop algebras [14, 15]. Let $\mathfrak{g}$ be a finite-dimensional simple Lie algebra over $\mathbb{C}$. Let R be the ring of Laurent polynomials of $(M + 1)$ variables $\mathbb{C}[s^{\pm 1}, t_1^{\pm 1}, \ldots, t_M^{\pm 1}]$. Also assume $M \geq 0$. The module of Kähler differentials Ω_R of R is defined with the canonical derivation $d : R \to \Omega_R$. As an R-module, Ω_R is freely generated by $ds, dt_1, \ldots, dt_M$. Let $\bar{} : \Omega_R \to \Omega_R / dR$ be the canonical projection. Let κ denote Ω_R / dR. Let $(\cdot|\cdot)$ be the normalized Killing form on $\mathfrak{g}$. We define the Lie algebra structure on $\mathfrak{g}^{\text{tor}} \overset{\text{def}}{=} \mathfrak{g} \otimes R \oplus \mathcal{K}$ by

$$[X \otimes f, Y \otimes g] = [X, Y] \otimes fg + (X|Y)\overline{(df)g}, \quad [\mathcal{K}, \mathfrak{g}^{\text{tor}}] = 0. \tag{10}$$

This bracket defines a universal central extension of $\mathfrak{g} \otimes R$ [14, 15].

We have, for $u = s, t_1, \ldots, t_M$, the Lie subalgebras

$$\widehat{\mathfrak{g}_u} \overset{\text{def}}{=} \mathfrak{g} \otimes \mathbb{C}[u^{\pm 1}] \oplus \mathbb{C}\overline{d \log u}, \tag{11}$$

with the brackets given by

$$[X \otimes u^m, Y \otimes u^n] = [X, Y] \otimes u^{m+n} + m\delta_{m+n,0}(X|Y)K_u, \tag{12}$$

which are isomorphic to the affine Lie algebra $\widehat{\mathfrak{g}}$ with the canonical central element $K_u \overset{\text{def}}{=} \overline{d \log u}$. In terms of the generating series,

$$X(z) \overset{\text{def}}{=} \sum_{n \in \mathbb{Z}} X \otimes u^n \cdot z^{-n-1}, \tag{13}$$

the relation (12) is equivalent to the following *operator product expansion* (OPE, in short. See, for example, [16]):

$$X(z)Y(w) \sim \frac{1}{z-w}[X, Y](w) + \frac{1}{(z-w)^2}(X|Y)K_u. \tag{14}$$

We prepare the generating series of $\mathfrak{g}^{\text{tor}}$ as follows:

$$X_{\underline{m}}(z) \overset{\text{def}}{=} \sum_{n \in \mathbb{Z}} X \otimes s^n \underline{t}^{\underline{m}} \cdot z^{-n-1}, \tag{15}$$

$$K_{\underline{m}}^s(z) \overset{\text{def}}{=} \sum_{n \in \mathbb{Z}} \overline{s^n \underline{t}^{\underline{m}} d \log s} \cdot z^{-n}, \tag{16}$$

$$K_{\underline{m}}^{t_k}(z) \overset{\text{def}}{=} \sum_{n \in \mathbb{Z}} \overline{s^n \underline{t}^{\underline{m}} d \log t_k} \cdot z^{-n-1}, \tag{17}$$

where $X \in \mathfrak{g}, \underline{m} = (m_1, \ldots, m_M) \in \mathbb{Z}^M, \underline{t}^{\underline{m}} = t_1^{m_1} \ldots t_M^{m_M}$, and $k = 1, \ldots, M$. The relation $\overline{d(s^n \underline{t}^{\underline{m}})} = 0$ can be neatly expressed by these generating series as

$$\frac{\partial}{\partial z} K_{\underline{m}}^s(z) = \sum_{k=1}^M m_k K_{\underline{m}}^{t_k}(z), \tag{18}$$

and the bracket (10) as

$$X_{\underline{m}}(z)Y_{\underline{n}}(w) \sim \frac{1}{z-w}[X, Y]_{\underline{m}+\underline{n}}(w) + \frac{1}{(z-w)^2}(X|Y)K_{\underline{m}+\underline{n}}^s(w)$$

$$+ \sum_{k=1}^M \frac{m_k}{z-w}(X|Y)K_{\underline{m}+\underline{n}}^{t_k}(w). \tag{19}$$

Hereafter we consider only the sl_2^{tor}-case to treat the $SU(2)$-SDYM hierarchy. The generators of sl_2 is denoted by E, F and H as usual:

$$[E, F] = H, \quad [H, E] = 2E, \quad [H, F] = -2F. \tag{20}$$

We prepare the language of the 2-*component free fermions* [7, 17] to construct a fermionic representation of the affine Lie algebra $\widehat{sl}_2$. Let $\mathcal{A}$ be the associative

$\mathbb{C}$-algebra generated by $\psi_j^{(\alpha)}$, $\psi_j^{(\alpha)*}$ ($j \in \mathbb{Z}$, $\alpha = 1, 2$) with the relations,

$$\left[\psi_i^{(\alpha)}, \psi_j^{(\beta)*}\right]_+ = \delta_{ij}\delta_{\alpha\beta}, \quad \left[\psi_i^{(\alpha)}, \psi_j^{(\beta)}\right]_+ = \left[\psi_i^{(\alpha)*}, \psi_j^{(\beta)*}\right]_+ = 0. \quad (21)$$

In terms of the generating series defined as

$$\psi^{(\alpha)}(\lambda) = \sum_{n\in\mathbb{Z}} \psi_n^{(\alpha)}\lambda^n, \quad \psi^{(\alpha)*}(\lambda) = \sum_{n\in\mathbb{Z}} \psi_n^{(\alpha)*}\lambda^{-n}, \quad (\alpha = 1, 2), \quad (22)$$

the relation (21) are rewritten as

$$\left[\psi^{(\alpha)}(\lambda), \psi^{(\beta)*}(\mu)\right]_+ = \delta_{\alpha\beta}\delta(\lambda/\mu),$$
$$\left[\psi^{(\alpha)}(\lambda), \psi^{(\beta)}(\mu)\right]_+ = \left[\psi^{(\alpha)*}(\lambda), \psi^{(\beta)*}(\mu)\right]_+ = 0, \quad (23)$$

where $\delta(\lambda) \overset{\text{def}}{=} \sum_{n\in\mathbb{Z}} \lambda^n$ is the formal delta-function.

Consider a left $\mathcal{A}$-module with a cyclic vector $|\text{vac}\rangle$ satisfying

$$\psi_j^{(\alpha)}|\text{vac}\rangle = 0 \quad (j < 0), \quad \psi_j^{(\alpha)*}|\text{vac}\rangle = 0, \quad (j \geq 0). \quad (24)$$

This $\mathcal{A}$-module $\mathcal{A}|\text{vac}\rangle$ is called the fermionic Fock space, which we denote by $\mathcal{F}$. We also consider a right $\mathcal{A}$-module (the dual Fock space $\mathcal{F}^*$) with a cyclic vector $\langle\text{vac}|$ satisfying

$$\langle\text{vac}|\psi_j^{(\alpha)} = 0 \quad (j \geq 0), \quad \langle\text{vac}|\psi_j^{(\alpha)*} = 0, \quad (j < 0). \quad (25)$$

We further define the *generalized vacuum vectors* as

$$|s_2, s_1\rangle \overset{\text{def}}{=} \Psi_{s_2}^{(2)}\Psi_{s_1}^{(1)}|\text{vac}\rangle, \quad \langle s_1, s_2| \overset{\text{def}}{=} \langle\text{vac}|\Psi_{s_1}^{(1)*}\Psi_{s_2}^{(2)*}, \quad (26)$$

$$\Psi_s^{(\alpha)} \overset{\text{def}}{=} \begin{cases} \psi_s^{(\alpha)*} \cdots \psi_{-1}^{(\alpha)*} & (s < 0), \\ 1 & (s = 0), \\ \psi_{s-1}^{(\alpha)} \cdots \psi_0^{(\alpha)} & (s > 0), \end{cases} \quad \Psi_s^{(\alpha)*} \overset{\text{def}}{=} \begin{cases} \psi_{-1}^{(\alpha)} \cdots \psi_s^{(\alpha)} & (s < 0), \\ 1 & (s = 0), \\ \psi_0^{(\alpha)*} \cdots \psi_{s-1}^{(\alpha)*} & (s > 0). \end{cases}$$

There exists a unique linear map (the *vacuum expectation value*),

$$\mathcal{F}^* \otimes_A \mathcal{F} \to \mathbb{C} \quad (27)$$

such that $\langle\text{vac}| \otimes |\text{vac}\rangle \mapsto 1$. For a $a \in \mathcal{A}$ we denote by $\langle\text{vac}|a|\text{vac}\rangle$ the vacuum expectation value of the vector $\langle\text{vac}|a \otimes |\text{vac}\rangle (= \langle\text{vac}| \otimes a|\text{vac}\rangle)$ in $\mathcal{F}^* \otimes_A \mathcal{F}$. Using the expectation value, we prepare another important notion of the *normal ordering*: $:\psi_i^{(\alpha)}\psi_j^{(\beta)*}: \overset{\text{def}}{=} \psi_i^{(\alpha)}\psi_j^{(\beta)*} - \langle\text{vac}|\psi_i^{(\alpha)}\psi_j^{(\beta)*}|\text{vac}\rangle$.

Lemma 1 ([7, 17]) *The operators*

$$\begin{cases} E(z) = z^{-1}\psi^{(1)}(z)\psi^{(2)*}(z), \\ F(z) = z^{-1}\psi^{(2)}(z)\psi^{(1)*}(z), \\ H(z) = z^{-1}\{:\psi^{(1)}(z)\psi^{(1)*}(z): - :\psi^{(2)}(z)\psi^{(2)*}(z):\}, \end{cases} \quad (28)$$

satisfy the OPE (14) with $c = 1$, i.e., give a representation of $\widehat{\mathfrak{sl}}_2$ on the fermionic Fock space $\mathcal{F}$.

To construct a representation of $\mathfrak{sl}_2^{\text{tor}}$, we consider the space of polynomials,

$$F_y \stackrel{\text{def}}{=} \bigotimes_{k=1}^{M} \left(\mathbb{C}[y_j^{(k)}, j \in \mathbb{N}] \otimes \mathbb{C}[e^{\pm y_0^{(k)}}] \right). \tag{29}$$

We define the generating series

$$\varphi^{(k)}(z) \stackrel{\text{def}}{=} \sum_{n \in \mathbb{N}} n y_n^{(k)} z^{n-1}, \quad V_{\underline{m}}(\underline{y}; z) \stackrel{\text{def}}{=} \prod_{k=1}^{M} \exp \left[m_k \sum_{n \in \mathbb{N}} y_n^{(k)} z^n \right], \tag{30}$$

for each $k = 1, \ldots, M$, $\underline{m} \in \mathbb{Z}^M$. Using the representation (28), We can obtain a representation of $\mathfrak{sl}_2^{\text{tor}}$ on $\mathcal{F}_y^{\text{tor}} \stackrel{\text{def}}{=} \mathcal{F} \otimes F_y$.

Proposition 1 *The following operators satisfy the OPE* (19):

$$\begin{cases} X_{\underline{m}}(z) = X(z) \otimes V_{\underline{m}}(z) \quad (X = E, F, H), \\ K_{\underline{m}}^s(z) = 1 \otimes V_{\underline{m}}(z), \\ K_{\underline{m}}^{t_k}(z) = 1 \otimes \varphi^{(k)}(z) V_{\underline{m}}(z). \end{cases} \tag{31}$$

Proof By the OPE (14) and the property $V_{\underline{m}}(z) V_{\underline{n}}(z) = V_{\underline{m}+\underline{n}}(z)$, we obtain

$$(X(z) \otimes V_{\underline{m}}(z))(Y(w) \otimes V_{\underline{n}}(w))$$

$$\sim \left\{ \frac{1}{z-w}[X, Y](w) + \frac{c}{(z-w)^2}(X|Y) \right\}$$

$$\otimes \left\{ V_{\underline{m}}(w) + \frac{\partial V_{\underline{m}}(w)}{\partial w}(z-w) \right\} V_{\underline{n}}(w)$$

$$\sim \frac{1}{z-w}[X, Y](w) \otimes V_{\underline{m}+\underline{n}}(w) + \frac{c}{(z-w)^2}(X|Y) \otimes V_{\underline{m}+\underline{n}}(w)$$

$$+ \sum_{k=1}^{M} \frac{m_k c}{z-w}(X|Y)\varphi^{(k)}(w) V_{\underline{m}+\underline{n}}(w). \tag{32}$$

Comparing the last line to (19), we have the desirous result.   ∎

We will use this representation in what follows to derive the bilinear identity, which is a generating function of Hirota-type equations.

3 DERIVATION OF THE BILINEAR IDENTITY

In this section, we set $M = 1$ for simplicity. We first introduce the following operator acting on $\mathcal{F}_y^{\text{tor}} \otimes \mathcal{F}_{y'}^{\text{tor}}$:

$$\Omega^{\text{tor}} \stackrel{\text{def}}{=} \sum_{m \in \mathbb{Z}} \sum_{\alpha=1,2} \oint \frac{d\lambda}{2\pi i \lambda} \psi^{(\alpha)}(\lambda) V_m(\underline{y}; \lambda) \otimes \psi^{(\alpha)*}(\lambda) V_m(\underline{y}'; \lambda). \tag{33}$$

Lemma 2 *The operator Ω^{tor} enjoys the following properties:*

(i) $[\Omega^{\text{tor}}, \mathfrak{sl}_2^{\text{tor}} \otimes 1 + 1 \otimes \mathfrak{sl}_2^{\text{tor}}] = 0,$ (34)

(ii) $\Omega^{\text{tor}}\{(|s_2, s_1\rangle \otimes 1) \otimes (|s_2 + 1, s_1 + 1\rangle \otimes 1)\}$

$$= \sum_{m \in \mathbb{Z}} \left\{ (|s_2 + 1, s_1\rangle \otimes e^{my_0}) \otimes \left(|s_2, s_1 + 1\rangle \otimes e^{-my_0'}\right) \right. \tag{35}$$

$$\left. - (|s_2, s_1 + 1\rangle \otimes e^{my_0}) \otimes \left(|s_2 + 1, s_1\rangle \otimes e^{-my_0'}\right) \right\}$$

Proof Since the representation of $\mathfrak{sl}_2^{\text{tor}}$ under consideration is constructed from Lemma 1, it is enough to show

$$\left[\Omega^{\text{tor}}, \, \psi^{(\alpha)}(p)\psi^{(\beta)*}(p)V_n(\underline{y}; p) \otimes 1 + 1 \otimes \psi^{(\alpha)}(p)\psi^{(\beta)*}(p)V_n(\underline{y}'; p) \right] = 0, \tag{36}$$

for $\alpha, \beta = 1, 2$ and $n \in \mathbb{Z}$. From (23), we have

$$\left[\psi^{(\alpha)}(p)\psi^{(\beta)*}(q), \, \psi^{(\gamma)}(\lambda)\right] = \delta_{\beta\gamma}\delta(q/\lambda)\psi^{(\alpha)}(p),$$

$$\left[\psi^{(\alpha)}(p)\psi^{(\beta)*}(q), \, \psi^{(\gamma)*}(\lambda)\right] = -\delta_{\alpha\gamma}\delta(p/\lambda)\psi^{(\beta)}(q). \tag{37}$$

These equations and the relation $V_m(\underline{y}; \lambda)V_n(\underline{y}; \lambda) = V_{m+n}(\underline{y}; \lambda)$ give the commutativity above.   ∎

If we translate Lemma 2 into bosonic language, then it comes out a hierarchy of Hirota bilinear equations. To do this, we present a summary of the *boson-fermion correspondence* in the two-component case. Define the operators $H_n^{(\alpha)}$ as $H_n^{(\alpha)} \stackrel{\text{def}}{=} \sum_{j \in \mathbb{Z}} \psi_j^{(\alpha)} \psi_{j+n}^{(\alpha)*}$ for $n = 1, 2, \ldots, \alpha = 1, 2$, which obey the canonical commutation relation $[H_m^{(\alpha)}, H_n^{(\beta)}] = m\delta_{m+n,0}\delta_{\alpha\beta} \cdot 1$. The operators $H_n^{(\alpha)}$ generate the Heisenberg subalgebra (*free bosons*) of $\mathcal{A}$, which is isomorphic to the algebra with the basis $\{nx_n^{(\alpha)}, \partial/\partial x_n^{(\alpha)} (\alpha = 1, 2, n = 1, 2, \ldots)\}$.

Lemma 3 ([7, 17])   *For any $|v\rangle \in \mathcal{F}$ and $s_1, s_2 \in \mathbb{Z}$, we have the following formulas,*

$$\langle s_1, s_2 | e^{H(\underline{x}^{(1)}, \underline{x}^{(2)})} \psi^{(1)}(\lambda)|v\rangle$$
$$= (-)^{s_2}\lambda^{s_1-1}e^{\xi(\underline{x}^{(1)}, \lambda)}\langle s_1 - 1, s_2 | e^{H(\underline{x}^{(1)}-[\lambda^{-1}], \underline{x}^{(2)})}|v\rangle, \tag{38}$$

$$\langle s_1, s_2 | e^{H(\underline{x}^{(1)}, \underline{x}^{(2)})} \psi^{(1)*}(\lambda)|v\rangle$$
$$= (-)^{s_2}\lambda^{-s_1}e^{-\xi(\underline{x}^{(1)}, \lambda)}\langle s_1 + 1, s_2 | e^{H(\underline{x}^{(1)}+[\lambda^{-1}], \underline{x}^{(2)})}|v\rangle, \tag{39}$$

$$\langle s_1, s_2 | e^{H(\underline{x}^{(1)}, \underline{x}^{(2)})} \psi^{(2)}(\lambda)|v\rangle$$
$$= \lambda^{s_2-1}e^{\xi(\underline{x}^{(2)}, \lambda)}\langle s_1, s_2 - 1 | e^{H(\underline{x}^{(1)}, \underline{x}^{(2)}-[\lambda^{-1}])}|v\rangle, \tag{40}$$

$$\langle s_1, s_2 | e^{H(\underline{x}^{(1)}, \underline{x}^{(2)})} \psi^{(2)*}(\lambda)|v\rangle$$
$$= \lambda^{-s_2}e^{-\xi(\underline{x}^{(2)}, \lambda)}\langle s_1, s_2 + 1 | e^{H(\underline{x}^{(1)}, \underline{x}^{(2)}+[\lambda^{-1}])}|v\rangle, \tag{41}$$

where the "Hamiltonian" $H(\underline{x}^{(1)}, \underline{x}^{(2)})$ is defined as

$$H(x^{(1)}, x^{(2)}) \overset{\text{def}}{=} \sum_{\alpha=1,2} \sum_{n=1}^{\infty} x_n^{(\alpha)} H_n^{(\alpha)}, \tag{42}$$

and

$$\xi(\underline{x}; \lambda) \overset{\text{def}}{=} \sum_{n=1}^{\infty} x_n \lambda^n. \tag{43}$$

We prepare one more lemma due to Billig [9].

Lemma 4 ([9], Proposition 3. See also [11]) *Let $P(m) = \sum_{j \geq 0} m^j P_j$, where P_j are differential operators that may not depend on z. If*

$$\sum_{m \in \mathbb{Z}} z^m P(m) f(z) = 0$$

for some function $f(z)$, then

$$P(\epsilon - z \partial_z) f(z)|_{z=1} = 0$$

as a polynomial in ϵ.

Now we are in position to state the bilinear identity for the $SU(2)$-SDYM hierarchy. Let $\mathrm{SL}_2^{\text{tor}}$ denote a group of invertible linear transformations on $\mathcal{F}_y^{\text{tor}}$ generated by the exponential action of the elements in $\mathfrak{sl}_2 \otimes R$ acting locally nilpotently. Define the τ-function associated with $g \in \mathrm{SL}_2^{\text{tor}}$ as

$$\tau_{s_2,s_1}^{s_1',s_2'}(\underline{x}^{(1)}, \underline{x}^{(2)}, \underline{y}) \overset{\text{def}}{=} {}^{\text{tor}}\langle s_1', s_2'| e^{H(\underline{x}^{(1)}, \underline{x}^{(2)})} g(\underline{y}) |s_2, s_1 \rangle^{\text{tor}}, \tag{44}$$

where $|s_2, s_1\rangle^{\text{tor}} \overset{\text{def}}{=} |s_2, s_1\rangle \otimes 1$ and ${}^{\text{tor}}\langle s_1', s_2'| \overset{\text{def}}{=} \langle s_1', s_2'| \otimes 1$. Hereafter we shall omit the superscripts "tor" if it does not cause confusion. Since $g \in \mathrm{SL}_2^{\text{tor}}$, the τ-function (44) have the following properties [17]:

$$\tau_{s_2+1,s_1+1}^{s_1+\ell+1,s_2-\ell+1} = (-1)^\ell \tau_{s_2,s_1}^{s_1+\ell,s_2-\ell}, \tag{45}$$

$$\left(\frac{\partial}{\partial x_j^{(1)}} + \frac{\partial}{\partial x_j^{(2)}} \right) \tau_{s_2,s_1}^{s_1',s_2'} = 0, \tag{46}$$

i.e., the τ-function depends only on $\{x_j \overset{\text{def}}{=} x_j^{(1)} - x_j^{(2)}\}$ and $\{y_j\}$. Since the first one y_0 of $\{y_j\}$ plays a special role, we will use the notation $\check{y} = (y_1, y_2 \ldots)$.

Proposition 2 *The τ-functions (44) associated with $\mathbf{g} \in SL_2^{\mathrm{tor}}$ satisfy*

$$(-1)^{s_2'+s_2''} \oint \frac{d\lambda}{2\pi i} \lambda^{s_1'-s_1''-2} e^{\xi((\underline{x}-\underline{x}')/2,\lambda)}$$

$$\times \tau_{s_2,s_1}^{s_1'-1,s_2'}(\underline{x}-[\lambda^{-1}], \underline{y}-\underline{b}_\lambda)\tau_{s_2+1,s_1+1}^{s_1''+1,s_2''}(\underline{x}'+[\lambda^{-1}], \underline{y}+\underline{b}_\lambda)$$

$$+ \oint \frac{d\lambda}{2\pi i} \lambda^{s_2'-s_2''-2} e^{\xi((\underline{x}'-\underline{x})/2,\lambda)}$$

$$\times \tau_{s_2,s_1}^{s_1',s_2'-1}(\underline{x}+[\lambda^{-1}], \underline{y}-\underline{b}_\lambda)\tau_{s_2+1,s_1+1}^{s_1'',s_2''+1}(\underline{x}'-[\lambda^{-1}], \underline{y}+\underline{b}_\lambda)$$

$$= \tau_{s_2+1,s_1}^{s_1',s_2'}(\underline{x}, y_0, \check{y}-\check{b})\tau_{s_2,s_1+1}^{s_1'',s_2''}(\underline{x}', y_0, \check{y}+\check{b})$$

$$- \tau_{s_2,s_1+1}^{s_1',s_2'}(\underline{x}, y_0, \check{y}-\check{b})\tau_{s_2+1,s_1}^{s_1'',s_2''}(\underline{x}', y_0, \check{y}+\check{b}), \tag{47}$$

where $\underline{b}_\lambda$ denotes $(b_0, b_1, b_2, \ldots)$ with the constraint $b_0 = -\xi(\check{b}, \lambda)$.

Proof Applying

$$\langle s_1', s_2'|e^{H(\underline{x}^{(1)},\underline{x}^{(2)})}\mathbf{g}(\underline{y}) \otimes \langle s_1'', s_2''|e^{H(\underline{x}^{(1)'},\underline{x}^{(2)'})}\mathbf{g}(\underline{y}')$$

to (35) from the left and using Lemma 3, we have

$$(-1)^{s_2'+s_2''} \oint \frac{d\lambda}{2\pi i} \lambda^{s_1'-s_1''-2} e^{\xi((\underline{x}-\underline{x}')/2,\lambda)}$$

$$\times \sum_{m\in\mathbb{Z}} \left\{ \tau_{s_2,s_1}^{s_1'-1,s_2'}(\underline{x}-[\lambda^{-1}], \underline{y}-\underline{b})\tau_{s_2+1,s_1+1}^{s_1''+1,s_2''}(\underline{x}'+[\lambda^{-1}], \underline{y}+\underline{b}) \right\} V_m(2\underline{b};\lambda)$$

$$+ \oint \frac{d\lambda}{2\pi i} \lambda^{s_2'-s_2''-2} e^{\xi((\underline{x}'-\underline{x})/2,\lambda)}$$

$$\times \sum_{m\in\mathbb{Z}} \left\{ \tau_{s_2,s_1}^{s_1',s_2'-1}(\underline{x}+[\lambda^{-1}], \underline{y}-\underline{b})\tau_{s_2+1,s_1+1}^{s_1'',s_2''+1}(\underline{x}'-[\lambda^{-1}], \underline{y}+\underline{b}) \right\} V_m(2\underline{b};\lambda)$$

$$= \sum_{m\in\mathbb{Z}} \left\{ \tau_{s_2+1,s_1}^{s_1',s_2'}(\underline{x}, \underline{y}-\underline{b})\tau_{s_2,s_1+1}^{s_1'',s_2''}(\underline{x}', \underline{y}+\underline{b}) \right.$$

$$\left. - \tau_{s_2,s_1+1}^{s_1',s_2'}(\underline{x}, \underline{y}-\underline{b})\tau_{s_2+1,s_1}^{s_1'',s_2''}(\underline{x}', \underline{y}, \underline{b}) \right\} V_m(2\underline{b};\lambda). \tag{48}$$

where we have replaced $\underline{y} \mapsto \underline{y}-\underline{b}$, $\underline{y}' \mapsto \underline{y}+\underline{b}$. If we use Lemma 4 with $z = e^{2b_0}$, we obtain the desired result. ∎

Expanding (47) and applying (45), we can obtain the following Hirota-type equations,

$$(\tau_{s_2,s_1}^{s_1,s_2})^2 + \tau_{s_2+1,s_1}^{s_1+1,s_2}\tau_{s_2,s_1+1}^{s_1,s_2+1} - \tau_{s_2,s_1+1}^{s_1+1,s_2}\tau_{s_2+1,s_1}^{s_1,s_2+1} = 0, \tag{49}$$

$$D_{y_0}\tau_{s_2,s_1}^{s_1+1,s_2-1} \cdot \tau_{s_2,s_1}^{s_1,s_2} = D_{y_1}\tau_{s_2,s_1+1}^{s_1+1,s_2} \cdot \tau_{s_2+1,s_1}^{s_1,s_2}, \tag{50}$$

$$D_{y_0}\tau_{s_2,s_1}^{s_1-1,s_2+1} \cdot \tau_{s_2,s_1}^{s_1,s_2} = D_{y_1}\tau_{s_2,s_1+1}^{s_1,s_2+1} \cdot \tau_{s_2+1,s_1}^{s_1,s_2+1}, \tag{51}$$

which agree with (3)–(6) if we set

$$\bar{y} = y_0, \quad z = y_1,$$

$$\tau_1 = \tau_{1,-1}^{0,0}, \quad \tau_2 = i\tau_{1,0}^{0,1}, \quad \tau_3 = \tau_{0,0}^{-1,1}, \quad \tau_4 = i\tau_{1,0}^{1,0}, \tag{52}$$

$$\tau_5 = \tau_{0,0}^{0,0}, \quad \tau_6 = i\tau_{0,1}^{0,1}, \quad \tau_7 = \tau_{0,0}^{1,-1}, \quad \tau_8 = i\tau_{0,1}^{1,0}.$$

If we introduce another set of variables $\{z_j (j = 0, 1, \ldots)\}$ that play the same role as $\{y_j\}$ and set $\bar{z} = z_0, y = -z_1$, the corresponding τ-functions solve (3)–(9) simultaneously. We remark that the introduction of the variables $\{z_j\}$ corresponds to the symmetry of the 3-toroidal Lie algebra as mentioned in Section 2.

To consider the reality condition for the $SU(2)$-gauge fields, we introduce an anti automorphism κ as

$$\kappa(\psi_n^{(\alpha)}) = \psi_n^{(\alpha)*}, \quad \kappa(\psi_n^{(\alpha)*}) = \psi_n^{(\alpha)}, \quad (n \in \mathbb{Z}, \alpha = 1, 2), \tag{53}$$

which have the following properties:

- $\kappa^2 = \mathrm{id}$,
- $\langle \mathrm{vac}|\kappa(g)|\mathrm{vac}\rangle = \langle \mathrm{vac}|g|\mathrm{vac}\rangle, \quad {}^\forall g \in \mathrm{SL}_2^{\mathrm{tor}}$.

Using κ, we impose the following condition on $g = g(\underline{y}, \underline{z})$:

$$\kappa(g(\underline{y}, \underline{z})) = \overline{g(\underline{y}, \underline{z})}. \tag{54}$$

Then we find that the τ-function (44) with $\underline{x}^{(1)} = \underline{x}^{(2)} = 0$ obeys

$$\overline{\langle s_1', s_2'|g(\underline{y}, \underline{z})|s_2, s_1\rangle} = \langle s_1, s_2|g(\underline{y}, \underline{z})|s_2', s_1'\rangle, \tag{55}$$

and that e, f, and g of (2) satisfies

$$\overline{f} = -f, \quad \overline{e} = g. \tag{56}$$

If we define $\tilde{J}$ as

$$\tilde{J} = \begin{bmatrix} \omega & 0 \\ 0 & \omega^{-1} \end{bmatrix} J \begin{bmatrix} \omega & 0 \\ 0 & \omega^{-1} \end{bmatrix}, \quad \omega = \frac{1+i}{\sqrt{2}}, \tag{57}$$

then $\tilde{J}$ satisfies (1) and the reality condition $\bar{\tilde{J}} = {}^t\tilde{J}$.

4 CONCLUDING REMARKS

We have described the hierarchy structure associated with the $SU(2)$-SDYM equation based on the representation theory of the toroidal Lie algebra sl_2^{tor}.

The hierarchy considered in the present article includes several interesting equations such as a (2+1)-dimensional generalized nonlinear Schrödinger equation. This topic is discussed in [8].

We have restricted ourselves to the $SU(2)$-case, and it may be straight-forward to generalize the results if we start with multicomponent fermions. We will discuss this subject elsewhere.

REFERENCES

1. Ablowitz, M. and Clarkson, P. A. (1991) *Solitons, nonlinear evolution equations and inverse scattering*, Cambridge University Press, Cambridge.
2. Yang, C. N. (1977) *Phys. Rev. Lett.* **38**, pp. 1377–1379.
3. Corrigan, E. F., Fairlie, D. B., Yates, R. G., and Goddard, P. (1978) *Commun. Math. Phys.* **58**, pp. 223–240.
4. Sasa, N., Ohta, Y., and Matsukidaira, J. (1998) *J. Phys. Soc. Jpn.* **67**, pp. 83–86.
5. Nakamura, Y. (1991) *J. Math. Phys.* **32**, pp. 382–385.
6. Takasaki, K. (1990) *Commun. Math. Phys.* **127**, pp. 225–238.
7. Date, E., Jimbo, M., Kashiwara, M., and Miwa, T. (1983) in: *Proceedings of RIMS symposium*, World Scientific, Singapore, eds. M. Jimbo and T. Miwa pp. 39–120.
8. Kakei, S., Ikeda, T., and Takasaki, K. (2002) *Annales Henri Poincaré* **3**, pp. 817–845.
9. Billig, Y. (1999) *J. Algebra* **217**, pp. 40–64.
10. Iohara, K., Saito, Y., and Wakimoto, M. (1999) *Phys. Lett.* **A254**, pp. 37–46.
11. Iohara, K., Saito, Y., and Wakimoto, M. (1999) *Progr. Theor. Phys. Suppl.* **135**, pp. 166–181.
12. Ikeda, T. and Takasaki, K. (2001) *Intl. Math. Res. Notices*, (7), pp. 329–369.
13. Kakei, S. and Ohta, Y. (2001) *J. Phys.* **A34**, pp. 10585–10592.
14. Kassel, C. (1985) *J. Pure Appl. Algebra* **34**, pp. 265–275.
15. Moody, R. V., Eswara Rao, S., and Yokonuma, T. (1990) *Geom. Ded.* **35**, pp. 283–307.
16. Kac, V. G. (1998) *Vertex algebras for beginners*, 2nd ed., AMS, Providence.
17. Jimbo, M. and Miwa, T. (1983) *Publ. RIMS, Kyoto Univ.* **19**, pp. 943–1001.

FROM SOLITON EQUATIONS TO THEIR ZERO CURVATURE FORMULATION

F. Lambert
Dienst Theoretische Natuurkunde, Vrije Universiteit Brussel
Pleinlaan 2, B-1050 Brussels, Belgium

J. Springael
Vakgroep M.T.T., Faculteit T.E.W., Universiteit Antwerpen
Prinsstraat 13, B-2000 Antwerpen, Belgium

1 INTRODUCTION

A key property of classical (continuous) soliton systems is the fact that they correspond to nonlinear partial differential equations (NLPDE's) which happen to be expressible as the integrability condition for a system of linear equations. Linear eigenvalue problems and associated t-evolutions have produced classes of soliton equations [1, 2] and have led to the disclosure of major integrability features (such as the existence of multisoliton solutions and infinite sequences of conserved quantities).

A still open "inverse" problem is to get access to the linear system, associated with a given soliton system, starting from the NLPDE itself. Its solution requires a skillful decomposition of an appropriate "Bäcklund condition" into a set of linear constraints on some new dimensionless dependent variables. Insight into the nature of these "primary" variables can sometimes be obtained by means of Painlevé techniques [3], or by looking for a bilinear Bäcklund transformation (BT) acting at the level of a Hirota representation of the original NLPDE [4, 5]. Yet, from the practical point of view, these manipulations are not as "direct" as one should wish.

Here, we present a reformulation of Hirota's bilinear BT method capable of producing systematically (if not algorithmically) the zero curvature formulation of sech squared soliton systems that may be derived from a single equation in quadratic Hirota form. The method is also shown to work for less elementary soliton systems (such as the Lax-KdV$_5$ equation), as it allows an elementary and systematic search for fundamental members of multidimensional integrable hierarchies leading (through reduction) to $1 + 1$ dimensional sech squared soliton systems.

L. Faddeev et al. (eds.),
Bilinear Integrable Systems: From Classical to Quantum, Continuous to Discrete, 147–159.
© *2006 Springer. Printed in the Netherlands.*

We start our presentation by reformulating the Hirota procedure (and the basic Hirota-Bäcklund ansatz) in terms of exponential polynomials (Section 2 and 3). We then apply the resulting method on 3 examples (KdV, KP, and BKP) leading to a unified treatment of several well known $1 + 1$ dimensional soliton systems (KdV, Boussinesq, Sawada–Kotera, Ramani, Lax–KdV$_5$). The calculations are presented in detail in order to exhibit the elementary nature of the procedure.

2   BILINEAR BÄCKLUND TRANSFORMATIONS

The simplest soliton equations, from "bilinear" point of view, are NLPDE's for a dependent variable $u(x, t)$ which, through the bilinearizing transformation $u = 2\partial_x^2 \ln f$, can be derived from a single quadratic Hirota equation of the form (m_j and n_j are integer or zero, c_j are constants):

$$\mathcal{F}(f, f) \equiv \sum_j c_j D_x^{m_j} D_t^{n_j} f \cdot f = 0, \quad m_j + n_j = \text{even}, \tag{1}$$

in which the D-operators are defined as follows:

$$D_x^m D_t^n f \cdot g \equiv (\partial_x - \partial_{x'})^m (\partial_t - \partial_{t'})^n f(x, t) g(x', t')|_{x'=x, t'=t} \tag{2}$$

Equations of the form (1) are known [6] to admit solitary wave solutions:

$$f_{\text{sol}} = 1 + e^\theta, \quad \theta = kx + \omega t + \tau, \text{with} \sum_j c_j k^{m_j} \omega^{n_j} = 0, \tag{3}$$

as well as "two soliton solutions":

$$f_2 = 1 + e^{\theta_1} + e^{\theta_2} + A_{12} e^{\theta_1 + \theta_2}, \quad \theta_i = k_i x + \omega_i t + \tau_i, \tag{4}$$

with

$$\sum_j c_j k_i^{m_j} \omega_i^{n_j} = 0, \quad A_{12} = -\frac{\sum_j c_j (k_1 - k_2)^{m_j} (\omega_1 - \omega_2)^{n_j}}{\sum_j c_j (k_1 + k_2)^{m_j} (\omega_1 + \omega_2)^{n_j}}. \tag{5}$$

These Hirota equations can also produce a Bäcklund transformation (BT) for the original NLPDE if there exists a pair of bilinear equations ($p_i, q_i, r_i, s_i = $ integer or zero, $c_{ij} = $ constant):

$$\mathcal{F}_1(f', f) \equiv \sum_i c_{i1} D_x^{p_i} D_t^{q_i} f' \cdot f = 0 \tag{6}$$

$$\mathcal{F}_2(f', f) \equiv \sum_i c_{i2} D_x^{r_i} D_t^{s_i} f' \cdot f = 0 \tag{7}$$

which are compatible if f satisfies Eq. (1), and which imply the relation

$$f^{-2} \mathcal{F}(f, f) - f'^{-2} \mathcal{F}(f', f') = 0. \tag{8}$$

A current, but tricky procedure to find out whether a given NLPDE admits such a bilinear BT is to start from the Hirota representation (1) (if one can find it) and to try to decompose the condition (8)—we call it the Hirota–Bäcklund (HB) ansatz—into a pair of eqs. (6), (7) with the help of appropriate "exchange formulas" [4]. One must then verify that the compatibility of the obtained bilinear system is subject to a condition on f which is satisfied as a result of Eq. (1).

Here we present a simpler method, based on the use of two classes of exponential polynomials (generalizations of the Bell [7] or Faá di Bruno [8] polynomials), which will be seen to lead (in a systematic way) to the zero curvature formulation of the given NLPDE.

3 EXPONENTIAL POLYNOMIALS AND ZERO CURVATURE FORMULATIONS

We start our discussion by noticing that a Hirota equation of the form (1) can be mapped by the transformation $f = \exp(\frac{Q}{2})$ onto a corresponding *primary* NLPDE for Q:

$$E(Q) \equiv \sum_j c_j P_{m_j x, n_j t}(Q) = 0, \tag{9}$$

with

$$P_{mx,nt}(Q) \equiv e^{-Q(x,t)} \partial_x^m \partial_t^n e^{Q(x,t)} |_{Q_{rx,st}=0} \text{ if } r+s=\text{odd}. \tag{10}$$

The link between Eqs. (1) and (9) is the identity [9]:

$$f^{-2} D_x^m D_t^n f \cdot f \equiv P_{mx,nt}(Q = 2\ln f), \quad m+n = \text{even}. \tag{11}$$

P-polynomials and primary NLPDE's of type (9) are easy to recognize on account of their simple partitional balance: linear (even order) terms $Q_{px,qt}$, with $p+q > 2$, are accompanied by nonlinear terms which correspond precisely to the even part partitions of (p, q):

$$P_{2x}(Q) = Q_{2x}, \quad P_{x,t} = Q_{xt}, \quad P_{2x,2t}(Q) = Q_{2x,2t} + Q_{2x}Q_{2t} + 2Q_{xt}^2,$$
$$P_{3x,t}(Q) = Q_{3x,t} + 3Q_{2x}Q_{xt}, \quad P_{4x}(Q) = Q_{4x} + 3Q_{2x}^2, \tag{12}$$
$$P_{6x}(Q) = Q_{6x} + 15Q_{2x}Q_{4x} + 15Q_{2x}^3, \ldots$$

Each polynomial carries a particular weight, determined by its order and the dimensions of the independent variables (the dependent variable being dimensionless).

We now observe that bilinear equations of the form (6) or (7) can be mapped onto equations which are linear with respect to "binary" exponential polynomials $\mathcal{Y}_{px,qt}(v, w)$ defined in terms of two "mixing" variables $v = \ln(f'/f)$ and $w = \ln(f'f)$:

$$\mathcal{Y}_{px,qt}(v, w) \equiv e^{-y(x,t)} \partial_x^p \partial_t^q e^{y(x,t)} \Big|_{y_{rx,st}=} \begin{cases} v_{rx,st} \text{ if } r + s = \text{odd} \\ w_{rx,st} \text{ if } r + s = \text{even} \end{cases} \tag{13}$$

Thus, making use of the two field generalization [9] of the identity (11)

$$(f'f)^{-1} D_x^p D_t^q f' \cdot f \equiv \mathcal{Y}_{px,qt}(v = \ln f'/f, \ w = \ln f'f), \tag{14}$$

we find that the bilinear system (6, 7) corresponds to the following $\mathcal{Y}$-system:

$$E_1(v, w) \equiv \sum_i c_{i1} \mathcal{Y}_{p_i x, q_i t}(v, w) = 0 \tag{15}$$

$$E_2(v, w) \equiv \sum_i c_{i2} \mathcal{Y}_{r_i x, s_i t}(v, w) = 0 \tag{16}$$

The map:

$$Q = 2 \ln f = w - v, \quad Q' = 2 \ln f' = w + v \tag{17}$$

is also seen to transform the HB ansatz (8) into the simpler condition:

$$E(Q' = w + v) - E(Q = w - v) = 0. \tag{18}$$

The problem of obtaining a bilinear BT from a given $\mathcal{F}(f, f)$ can therefore be reformulated as the problem of obtaining a decomposition of the condition (18) into a pair of $\mathcal{Y}$-constraints (15, 16) which are compatible for every Q that satisfies the primary NLPDE (9).

We shall see that this problem can be tackled systematically (if not algorithmically) by expressing:

$$E(w + v) - E(w - v) \tag{19}$$

in terms of $\mathcal{Y}$-polynomials and their derivatives.

An important point is the logarithmic linearizability of the above $\mathcal{Y}$-systems. This follows from the property [9]:

$$\mathcal{Y}_{px,qt}(v = \ln \psi, \ w = Q + \ln \psi) = \psi^{-1} L_{p,q}(Q) \psi, \tag{20}$$

with

$$L_{p,q}(Q) = \sum_{r=0}^{p} \sum_{s=0}^{q} \binom{p}{r} \binom{q}{s} \mathcal{Y}_{rx,st}(0, Q) \partial_x^{p-r} \partial_t^{q-s}, \tag{21}$$

on account of which every system (15, 16) is mapped by the transformation

$$w = v + Q, \quad v = \ln \psi \tag{22}$$

onto a linear system for ψ:

$$\sum_i c_{i1} L_{p_i, q_i}(Q) \psi = 0 \tag{23}$$

$$\sum_i c_{i2} L_{r_i, s_i}(Q) \psi = 0. \tag{24}$$

Thus, having found a decomposition of the HB ansatz (18) into a pair of $\mathcal{Y}$-constraints, it is a straightforward matter to check whether their compatibility is subject to a condition on Q which is satisfied as a result of Eq. (9). This enables us to undertake a systematic search for linear equations which may provide a zero curvature formulation of the original NLPDE.

4 KP, BKP AND THEIR REDUCTIONS

The simplest sech squared soliton system is the KdV equation:

$$\text{KdV}(u) \equiv u_t - u_{3x} - 6uu_x = 0, \tag{25}$$

which, by setting $u = Q_{2x}$ (Q is the simplest dimensionless alternative to u) is seen to correspond to a primary NLPDE of weight 4 (we choose the dimension of x to be equal to 1 and define the weight of $P_{px,qt}(Q)$ as $p + q$ dim t):

$$E_{\text{KdV}}(Q) \equiv Q_{xt} - (Q_{4x} + 3Q_{2x}^2) \equiv P_{x,t}(Q) - P_{4x}(Q) = 0, \tag{26}$$

or to the equivalent Hirota equation ($Q = 2 \ln f$):

$$(D_x D_t - D_x^4) f \cdot f = 0. \tag{27}$$

The disclosure of a Bäcklund transformation for Eq. (25) is an easy matter which has been dealt with in [10]: it suffices to decouple the condition

$$\begin{aligned} C_{\text{KdV}}(v, w) &\equiv E_{\text{KdV}}(w + v) - E_{\text{KdV}}(w - v) \\ &\equiv 2v_{xt} - 2v_{4x} - 12v_{2x} w_{2x} = 0 \end{aligned} \tag{28}$$

into a pair of $\mathcal{Y}$-constraints, and to verify that their compatibility is subject to the condition: $\partial_x E_{\text{KdV}}(Q) = 0$. This is easily done by expressing $C_{\text{KdV}}(v, w)$ in terms of $\mathcal{Y}$-polynomials and their first order derivatives by means of the formulas[1]:

$$\begin{aligned} v_{2x} &= \partial_x \mathcal{Y}_x, \quad v_{xt} = \partial_x \mathcal{Y}_t, \quad w_{2x} = \mathcal{Y}_{2x} - \mathcal{Y}_x^2, \\ v_{3x} &= \mathcal{Y}_{3x} - 3\mathcal{Y}_x \mathcal{Y}_{2x} + 2\mathcal{Y}_x^3. \end{aligned} \tag{29}$$

---

[1] They follow straight away from def.(13). We write $\mathcal{Y}_{px,qt}$ instead of $\mathcal{Y}_{px,qt}(v, w)$.

Thus, it is seen that

$$C_{KdV}(v, w) = 2\partial_x[\mathcal{Y}_t - \mathcal{Y}_{3x}] + 6W_x[\mathcal{Y}_x, \mathcal{Y}_{2x}], \tag{30}$$

with $W_x[f, g] \equiv f(\partial_x g) - g(\partial_x f)$, indicating that the condition (28) is satisfied if v and w obey the system (λ = constant of weight 2):

$$\mathcal{Y}_{2x}(v, w) = \lambda \tag{31}$$
$$\mathcal{Y}_t(v) = \mathcal{Y}_{3x}(v, w) + 3\lambda \mathcal{Y}_x(v). \tag{32}$$

Its compatibility is subject to that of the equivalent system for $\psi = e^v$ (setting $w = v + Q$):

$$\psi_{2x} + (Q_{2x} - \lambda)\psi = 0 \tag{33}$$
$$\psi_t - \psi_{3x} - 3(Q_{2x} + \lambda)\psi_x = 0 \tag{34}$$

i.e., to the condition

$$(Q_{xt} - Q_{4x} - 3Q_{2x}^2)x = 0. \tag{35}$$

Let us now consider the more challenging example of a homogeneous 1 + 2-dimensional primary NLPDE which involves a linear combination of all P-polynomials of weight 4 that may be defined with respect to a set of independent variables t_p of weight $p = 1, 2, 3, \ldots$ (with $t_1 = x$):

$$E_4(Q) \equiv P_{4x}(Q) + \alpha P_{x,t_3}(Q) + \beta P_{2t_2}(Q) = 0. \tag{36}$$

The corresponding expression:

$$C_4(v, w) = E_4(w + v) - E_4(w - v) = 2v_{4x} + 12v_{2x}w_{2x} + 2\alpha v_{xt_3} + 2\beta v_{2t_2} \tag{37}$$

can again be expressed in terms of $\mathcal{Y}$-polynomials and their first order derivatives:

$$C_4(v, w) = 2\partial_x[\alpha \mathcal{Y}_{t_3} + \mathcal{Y}_{3x}] + 2\beta \partial_{t_2}[\mathcal{Y}_{t_2}] + 6W_x[\mathcal{Y}_{2x}, \mathcal{Y}_x]. \tag{38}$$

In order to find out whether $C_4(v, w)$ can be reduced to the x-derivative of a linear combination including all $\mathcal{Y}$-polynomials of weight 3, say (a = undetermined constant):

$$2\partial_x[\alpha \mathcal{Y}_{t_3} + \mathcal{Y}_{3x} + a\mathcal{Y}_{x,t_2}] \tag{39}$$

by means of another homogeneous "auxiliary" $\mathcal{Y}$-constraint, it suffices to consider the only possible candidate which is a constraint of weight 2 (b = undetermined constant):

$$\mathcal{Y}_{t_2} + b\mathcal{Y}_{2x} = 0. \tag{40}$$

It suggests that $C_4(v, w)$ should be rewritten in the form:

$$C_4(v, w) = 2\partial_x[\alpha \mathcal{Y}_{t_3} + \mathcal{Y}_{3x} + a\mathcal{Y}_{x,t_2}] + 2\beta \partial_{t_2}[\mathcal{Y}_{t_2} + b\mathcal{Y}_{2x}] + R_4(v, w), \tag{41}$$

with

$$R_4(v, w) = 6W_x[\mathcal{Y}_{2x}, \mathcal{Y}_x] - 2a\partial_x\mathcal{Y}_{x,t_2} - 2b\beta\partial_{t_2}\mathcal{Y}_{2x} \tag{42}$$

or

$$R_4(v, w) = 2W_x[3\mathcal{Y}_{2x} + b\beta\mathcal{Y}_{t_2}, \mathcal{Y}_x] - 2(a + b\beta)\partial_x\mathcal{Y}_{x,t_2} \tag{43}$$

on account of the identity:

$$\partial_{t_2}\mathcal{Y}_{2x} \equiv \partial_x\mathcal{Y}_{x,t_2} + W_x[\mathcal{Y}_x, \mathcal{Y}_{t_2}]. \tag{44}$$

It follows that the desired reduction of $C_4(v, w)$ can be obtained by means of Eq. (40) with $b = \pm\sqrt{\frac{3}{\beta}}$ and $a = \mp\sqrt{3\beta}$.

We conclude that the HB-ansatz (18) associated with Eq. (36) is satisfied if v and w obey the system:

$$\mathcal{Y}_{t_2}(v) \pm \sqrt{\frac{3}{\beta}}\mathcal{Y}_{2x}(v, w) = 0 \tag{45}$$

$$\alpha\mathcal{Y}_{t_3}(v) + \mathcal{Y}_{3x}(v, w) \mp \sqrt{3\beta}\mathcal{Y}_{x,t_2}(v, w) = 0. \tag{46}$$

Its compatibility is subject to that of the equivalent system for $\psi = e^v$ (setting $w = v + Q$):

$$\psi_{t_2} = \mp\sqrt{\frac{3}{\beta}}(\psi_{2x} + Q_{2x}\psi) \tag{47}$$

$$\alpha\psi_{t_3} = -4\psi_{3x} - 6Q_{2x}\psi_x - (3Q_{3x} \mp \sqrt{3\beta}Q_{xt_2})\psi, \tag{48}$$

i.e., to a condition on Q which is easily seen to coincide with $\partial_x E_4(Q)$. Setting $\beta = 3$ and $\alpha = -4$ we see that the system:

$$\psi_{t_2} = L_2(Q)\psi, \quad L_2(Q) = \partial_x^2 + Q_{2x} \tag{49}$$

$$\psi_{t_3} = L_3(Q)\psi, \quad L_3(Q) = \partial_x^3 + \frac{3}{2}Q_{2x}\partial_x + \frac{3}{4}(Q_{3x} + Q_{xt_2}) \tag{50}$$

provides us with a zero curvature formulation of a distinguished member of the family (36) known as the KP equation [11]

$$E_{KP}(Q) \equiv P_{4x}(Q) + 3P_{2t_2}(Q) - 4P_{x,t_3}(Q) = 0. \tag{51}$$

The operators L_2 and L_3 can be shown [12] to define a pair of Darboux covariant t-evolutions.

As a t_2-reduction of Eq. (51) we recover, as expected, the primary KdV Eq. (26) with $t = \frac{1}{4}t_3$. The actual "Lax pair" for KdV is obtained from Eqs. (49), (50) by setting $Q_{t_2} = 0$ and $\psi_{t_2} = \lambda\psi$:

$$L_2(Q)\psi = \lambda\psi \tag{52}$$

$$\psi_{t_3} = B(Q)\psi, \quad B(Q) = \partial_x^3 + \frac{3}{2}Q_{2x}\partial_x + \frac{3}{4}Q_{3x} \tag{53}$$

As a t_3-reduction of Eq. (51) we obtain:

$$E_{Bq}(Q) \equiv P_{4x}(Q) + 3P_{2t_2}(Q) = 0 \tag{54}$$

The Lax pair of the corresponding Boussinesq equation ($u = Q_{2x}$):

$$3u_{2t_2} + u_{4x} + 3(u^2)_{xx} = 0 \tag{55}$$

is obtained from Eqs. (49, 50) by setting $\psi_{t_3} = \mu\psi$ (μ being a parameter of weight 3):

$$L_3(Q)\psi = \mu\psi \tag{56}$$

$$\psi_{t_2} = L_2(Q)\psi \tag{57}$$

As a second $1 + 2$ dimensional example we consider a primary NLPDE which involves a linear combination of all P-polynomials of weight 6 that may be defined with respect to the subset of odd-dimensional variables $t_1 = x$, t_3, t_5, ...

$$E_6(Q) \equiv P_{6x} + \alpha P_{x,t_5}(Q) + \beta P_{3x,t_3}(Q) + \gamma P_{2t_3}(Q) = 0 \tag{58}$$

The corresponding HB ansatz

$$\begin{aligned}
C_6(v, w) &\equiv 2v_{6x} + 30(v_{2x}w_{4x} + v_{4x}w_{2x}) + 30v_{2x}^3 + 90v_{2x}w_{2x}^2 \\
&\quad + 2\alpha v_{xt_5} + 2\beta v_{3x,t_3} + 6\beta(v_{2x}w_{xt_3} + v_{xt_3}w_{2x}) \\
&\quad + 2\gamma v_{2t_3} = 0
\end{aligned} \tag{59}$$

can again be expressed in terms of $\mathcal{Y}$-polynomials and their derivatives by means of standard substitutions, and subsequently reduced to (see Appendix):

$$\begin{aligned}
C_6(v, w) &\equiv \partial_x[2\alpha\mathcal{Y}_{t_5} - 3\mathcal{Y}_{5x} + 3\beta\mathcal{Y}_{2x,t_3}] + 2\partial_{t_3}[\gamma\mathcal{Y}_{t_3} - \beta\mathcal{Y}_{3x}] \\
&\quad + \partial_x^3[5\mathcal{Y}_{3x} + \beta\mathcal{Y}_{t_3}] - 3[5\mathcal{Y}_{3x} + \beta\mathcal{Y}_{t_3}](\partial_x\mathcal{Y}_{2x}) \\
&\quad + (9\mathcal{Y}_{2x} - 6\mathcal{Y}_x^2)\partial_x[5\mathcal{Y}_{3x} + \beta\mathcal{Y}_{t_3}] = 0
\end{aligned} \tag{60}$$

If $\gamma = -\frac{\beta^2}{5}$ it is possible to satisfy this condition by means of two homogeneous $\mathcal{Y}$-constraints on v and w. These constraints are:

$$\mathcal{Y}_{3x}(v, w) + \frac{\beta}{5}\mathcal{Y}_{t_3}(v) = 0 \tag{61}$$

$$\alpha\mathcal{Y}_{t_5}(v) - \frac{3}{2}\mathcal{Y}_{5x}(v, w) + \frac{3}{2}\beta\mathcal{Y}_{2x,t_3}(v, w) = 0. \tag{62}$$

Their compatibility is subject to that of the corresponding linear equations:

$$\beta\psi_{t_3} = -5\psi_{3x} - 15Q_{2x}\psi_x \tag{63}$$

$$\alpha\psi_{t_5} = 9\psi_{5x} + 45Q_{2x}\psi_{3x} + 45Q_{3x}\psi_{2x}$$
$$+ (30Q_{4x} + 45Q_{2x}^2 - 3\beta Q_{xt_3})\psi_x, \tag{64}$$

i.e., to a condition on Q which is easily seen to coincide with

$$\partial_x \left[P_{6x}(Q) + \alpha P_{x,t_5}(Q) + \beta P_{3x,t_3}(Q) - \frac{\beta^2}{5}P_{2t_3}(Q) \right] = 0 \tag{65}$$

Setting now $\alpha = 9$ and $\beta = -5$ we conclude that the system:

$$\psi_{t_3} = \psi_{3x} + 3Q_{2x}\psi_x \tag{66}$$

$$\psi_{t_5} = \psi_{5x} + 5Q_{2x}\psi_{3x} + 5Q_{3x}\psi_{2x} + \left(\frac{10}{3}Q_{4x} + 5Q_{2x}^2 + \frac{5}{3}Q_{xt_3} \right)\psi_x \tag{67}$$

provides us with a zero curvature formulation of a distinguished member of the family (65), the primary version of which is known as the BKP equation [11]:

$$E_{BKP}(Q) \equiv P_{6x}(Q) + 9P_{x,t_5}(Q) - 5P_{3x,t_3}(Q) - 5P_{2t_3}(Q) = 0. \tag{68}$$

A t_3-reduction of Eq. (68) produces a primary version of the Sawada–Kotera equation [13]:

$$E_{SK}(Q) \equiv P_{6x}(Q) + 9P_{x,t_5}(Q) = 0, \tag{69}$$

for which a Lax pair is obtained from Eqs. (66), (67) by setting $Q_{t_3} = 0$ and $\psi_{t_3} = \lambda\psi$.

A t_5-reduction of Eq. (68) produces a primary version of the Ramani equation:

$$E_{Ram}(Q) \equiv P_{6x}(Q) - 5P_{3x,t_3}(Q) - 5P_{2t_3}(Q) = 0, \tag{70}$$

for which a Lax pair is obtained from eqs. (66, 67) by setting $\psi_{t_5} = \lambda\psi$.

A slightly more subtle reduction of BKP can be obtained by noticing that Eq. (68) corresponds to the $a = -\frac{5}{9}$ member of the following family of

P-equations:

$$\tilde{E}_6(Q;a) \equiv P_{x,t_5}(Q) + a\,P_{2t_3}(Q) - \frac{1}{6}(1+3a)P_{6x}(Q)$$

$$-\frac{1}{6}(5+3a)P_{3x,t_3}(Q) = 0 \tag{71}$$

Each member of this family shares, by construction, the two soliton solutions $Q_2 = 2\ln(1 + e^{\theta_1} + 2^{\theta_2} + A_{12}^{\text{KdV}}e^{\theta_1+\theta_2})$,

$$\theta_i = k_i x + k_i^3 t_3 + k_i^5 t_5 \quad \text{and} \quad A_{12}^{\text{KdV}} \equiv \left(\frac{k_1-k_2}{k_1+k_2}\right)^2, \tag{72}$$

with the primary KdV-Eq. (26), as well as with a $1+1$ dimensional NLPDE for Q_{x,t_5} which may be derived from the *x*-derivative of Eq. (71) through elimination of Q_{x,t_3}, and its derivatives, by means of Eq. (26):

$$Q_{2x,t_5} = Q_{7x} + 10Q_{2x}Q_{5x} + 20Q_{3x}Q_{4x} + 30Q_{2x^2}Q_{3x}. \tag{73}$$

The system (26, 71) therefore provides us with a parameter family of $1+2$ dimensional Hirota representations:

$$[D_x D_{t_3} - D_x^4]f \cdot f = 0 \tag{74}$$

$$\left[D_x D_{t_5} + aD_{t_3}^2 - \frac{1}{6}(1+3a)D_x^6 - \frac{1}{6}(5+3a)D_x^3 D_{t_3}\right]f \cdot f = 0 \tag{75}$$

of a fifth order NLPDE for $u = Q_{2x}$ which is known as the Lax (or KdV$_5$)-equation [1, 13]:

$$u_{t_5} = u_{5x} + 10uu_{3x} + 20u_x u_{2x} + 30u^2 u_x. \tag{76}$$

A zero-curvature formulation of this NLPDE can be obtained straight away from the representation (26, 71) in which $a = -\frac{5}{9}$:

$$E_{KdV}(Q) \equiv P_{x,t_3}(Q) - P_{4x}(Q) = 0 \tag{77}$$

$$E_{BKP}(Q) \equiv 9P_{x,t_5}(Q) - 5P_{2t_3}(Q) + P_{6x}(Q) - 5P_{3x,t_3}(Q) = 0 \tag{78}$$

The first equation of this system gives rise to the HB ansatz (28) which is already known to be satisfied if one imposes the constraints (31,32). These constraints, and the use of the identity

$$\partial_x^2 \mathcal{Y}_x \equiv \mathcal{Y}_{3x} - 3\mathcal{Y}_x \mathcal{Y}_{2x} + 2\mathcal{Y}_x^3, \tag{79}$$

enable us to reduce the HB ansatz for Eq. (78) to the condition:

$$\partial_x\left[9\mathcal{Y}_{t_5} - \frac{3}{2}\mathcal{Y}_{5x} - \frac{15}{2}\mathcal{Y}_{2x,t_3} - \frac{45}{2}\lambda\mathcal{Y}_{3x} - 90\lambda^2\mathcal{Y}_x\right] = 0 \tag{80}$$

It follows that the HB conditions (18) associated with the system (77, 78) are satisfied if one imposes the constraints:

$$\mathcal{Y}_{2x}(v, w) = \lambda \tag{81}$$

$$\mathcal{Y}_{t_3}(v) = \mathcal{Y}_{3x}(v, w) + 3\lambda\mathcal{Y}_x(v) \tag{82}$$

$$\mathcal{Y}_{t_5} = \frac{3}{2}\mathcal{Y}_{5x}(v, w) + \frac{15}{2}\mathcal{Y}_{2x,t_3}(v, w) + \frac{45}{2}\lambda\mathcal{Y}_{3x}(v, w) + 90\lambda^2\mathcal{Y}_x(v) \tag{83}$$

or (setting $w = v + Q$ and $v = \ln\psi$):

$$\psi_{2x} = (\lambda - Q_{2x})\psi \tag{84}$$

$$\psi_{t_3} = \psi_{3x} + 3(Q_{2x} + \lambda)\psi_x \tag{85}$$

$$\psi_{t_5} = \frac{1}{6}\psi_{5x} + \frac{5}{6}\psi_{2x,t_3} + \left(\frac{5}{3}Q_{2x} + \frac{5}{2}\lambda\right)\psi_{3x} + \frac{5}{6}Q_{2x}\psi_{t_3}$$

$$+ \left(\frac{5}{6}Q_{4x} + \frac{5}{2}Q_{2x}^2 + \frac{5}{3}Q_{x,t_3} + \frac{15}{2}\lambda Q_{2x} + 10\lambda^2\right)\psi_x \tag{86}$$

Equations (84) and (85) are known to be compatible if Q solves Eq. (26).

Eliminating derivatives with respect to t_3 by means of Eqs. (26) and (85) we may replace Eq. (86) by:

$$\psi_{t_5} = \psi_{5x} + 5(Q_{2x} + \lambda)\psi_{3x} + 5Q_{3x}\psi_{2x}$$

$$+ (5Q_{4x} + 10Q_{2x}^2 + 10\lambda Q_{2x} + 10\lambda^2)\psi_x, \tag{87}$$

or by (eliminating λ from this last equation by means of Eq. (84)):

$$\psi_{t_5} = 16L_5(Q)\psi, \text{ with}$$

$$L_5(Q) = \partial_x^5 + \frac{5}{2}Q_{2x}\partial_x^3 + \frac{15}{4}Q_{3x}\partial_x^2 + \frac{25}{8}\left(Q_{4x} + \frac{3}{5}Q_{2x}^2\right)\partial_x \tag{88}$$

$$+ \frac{15}{16}(Q_{5x} + 2Q_{2x}Q_{3x})$$

The operator L_5 is the fifth order B-operator obtained by Lax [1] in his construction of isospectral deformations of L_2. The operators L_2 and L_5 are also known [12] to produce a Darboux covariant pair of t-evolutions.

A   APPENDIX

The following substitution formulas follow straight from definition (13) and the identities $w_{4x} \equiv \partial_x^2 w_{2x}$, $v_{3x,t_3} \equiv \partial_x v_{2x,t_3} \equiv \partial_{t_3} v_{3x}$, $v_{5x} \equiv \partial_x^2 v_{3x}$ (notice that expressions $\partial_x^p \mathcal{Y}_x$ with $p \geq 2$ have been avoided by making use of the identity

(79), and that a, b, c, d and e are undetermined constants):

$$w_{xt} = \mathcal{Y}_{x,t} - \mathcal{Y}_x \mathcal{Y}_t$$

$$w_{4x} = a(\partial_x^2 \mathcal{Y}_{2x}) + (1 - a)\mathcal{Y}_{4x} + (2a - 4)\mathcal{Y}_x \mathcal{Y}_{3x} + (12 - 6a)\mathcal{Y}_x^2 \mathcal{Y}_{2x}$$
$$\qquad - (3 - 3a)\mathcal{Y}_{2x}^2 + (2a - 6)\mathcal{Y}_x^4 - 2a(\partial_x \mathcal{Y}_x)^2$$

$$v_{3x,t_3} = b(\partial_{t_3} \mathcal{Y}_{3x}) + c(\partial_x \mathcal{Y}_{2x,t_3}) + (1 - b - c)(\partial_x^3 \mathcal{Y}_{t_3})$$
$$\qquad + (3b + 2c)\mathcal{Y}_x[\mathcal{Y}_x(\partial_x \mathcal{Y}_{t_3}) - (\partial_x \mathcal{Y}_{x,t_3})] - (3b + c)\mathcal{Y}_{2x}(\partial_x \mathcal{Y}_{t_3})$$
$$\qquad + (3b + 4c)\mathcal{Y}_x \mathcal{Y}_{t_3}(\partial_x \mathcal{Y}_x) - 2c\mathcal{Y}_{x,t_3}(\partial_x \mathcal{Y}_x) - c\mathcal{Y}_{t_3}(\partial_x \mathcal{Y}_{2x})$$

$$v_{6x} = d(\partial_x^3 \mathcal{Y}_{3x}) + (1 - d)(\partial_x \mathcal{Y}_{5x}) - 5(1 - d - e)[\mathcal{Y}_{4x}(\partial_x \mathcal{Y}_x) + \mathcal{Y}_x(\partial_x \mathcal{Y}_{4x})]$$
$$\qquad - (7d - 10)\mathcal{Y}_{2x}(\partial_x \mathcal{Y}_{3x}) + (10 - 7d - 5e)[2\mathcal{Y}_x^2(\partial_x \mathcal{Y}_{3x}) + 3\mathcal{Y}_{2x}^2(\partial_x \mathcal{Y}_x)]$$
$$\qquad + (d - 10)[\mathcal{Y}_{3x}(\partial_x \mathcal{Y}_{2x}) - 4\mathcal{Y}_x \mathcal{Y}_{3x}(\partial_x \mathcal{Y}_x)]$$
$$\qquad + (60 - 24d - 30e)\mathcal{Y}_x(\mathcal{Y}_{2x} - \mathcal{Y}_x^2)(\partial_x \mathcal{Y}_{2x}) - (9d + 5e)(\partial_x \mathcal{Y}_x)(\partial_x^2 \mathcal{Y}_{2x}).$$
$$\qquad + (12d + 10e)(\partial_x \mathcal{Y}_x)^3 + (120 - 12d - 10e)\mathcal{Y}_x^4(\partial_x \mathcal{Y}_x)$$
$$\qquad + (36d + 30e - 180)\mathcal{Y}_x^2 \mathcal{Y}_{2x}(\partial_x \mathcal{Y}_x)$$

They enable us to express $\frac{1}{2}C_6(v, w)$ as follows:

$$\frac{1}{2}C_6(v, w) = [\alpha \mathcal{Y}_{t_5} + (1 - d)\mathcal{Y}_{5x} + c\mathcal{Y}_{2x,t_3}]_x + [\gamma \mathcal{Y}_{t_3} + b\mathcal{Y}_{3x}]_{t_3}$$
$$\qquad + [d\mathcal{Y}_{3x} + (\beta - b - c)\mathcal{Y}_{t_3}]_{3x} - \mathcal{Y}_x[5(1 - d - e)\mathcal{Y}_{4x}$$
$$\qquad + (3b + 2c)\mathcal{Y}_{x,t_3}]_x + [(10 + 5d + 5e - 15a)\mathcal{Y}_{4x}$$
$$\qquad + (3\beta - 2c)\mathcal{Y}_{x,t_3}](\partial_x \mathcal{Y}_x) + [(d - 10)\mathcal{Y}_{3x} - c\mathcal{Y}_{t_3}](\partial_x \mathcal{Y}_{2x})$$
$$\qquad + \mathcal{Y}_{2x}[(7d + 5)\mathcal{Y}_{3x} - (3b + c - 3\beta)\mathcal{Y}_{t_3}]_x$$
$$\qquad + \mathcal{Y}_x^2[(5 - 14d - 10e)\mathcal{Y}_{3x} + (3b + 2c - 3\beta)\mathcal{Y}_{t_3}]_x \qquad (89)$$
$$\qquad + \mathcal{Y}_x[(30a - 4d - 20)\mathcal{Y}_{3x} + (3b + 4c - 3\beta)\mathcal{Y}_{t_3}](\partial_x \mathcal{Y}_x)$$
$$\qquad - (3d + 5e)\mathcal{Y}_x(\partial_x^3 \mathcal{Y}_{2x}) - (9d + 5e - 15a)(\partial_x \mathcal{Y}_x)(\partial_x^2 \mathcal{Y}_{2x})$$
$$\qquad + (24d + 30e - 15)\mathcal{Y}_x(\mathcal{Y}_x^2 - \mathcal{Y}_{2x})(\partial_x \mathcal{Y}_{2x})$$
$$\qquad + (45a - 21d - 15e - 15)\mathcal{Y}_{2x}^2(\partial_x \mathcal{Y}_x)$$
$$\qquad + (30a - 12d - 10e - 15)[\mathcal{Y}_x^2(\mathcal{Y}_x^2 - 3\mathcal{Y}_{2x}) - (\partial_x \mathcal{Y}_x)^2](\partial_x \mathcal{Y}_x).$$

A reduction of this expression to the x-derivative of a linear combination of $\mathcal{Y}$-polynomials of weight 5, by means of a homogeneous constraint of weight $r \leq 4$, requires the simultaneous vanishing of the last 5 terms. This vanishing can be obtained if one chooses: $a = 1, d = \frac{5}{2}, e = -\frac{3}{2}$, and $b = -\beta$. This choice reduces $C_6(v, w)$ to the expression given in eq. (60).

REFERENCES

1. Lax, P. D. (1968) Integrals of nonlinear equations of evolution and solitary waves, *Comm. Pure Appl. Math.* **21**, pp. 467–490.
2. Ablowitz, M. J., Kaup, D. J., Newell, A. C., and Segur, H. (1974) *Stud. Appl. Math.* **L III**, pp. 249–315.
3. Hietarinta, J. and Kruskal, M. D. (1992) Hirota forms for the six Painlevé equations from singularity analysis, in: *Painlevé Transcendents—Their Asymptotics and Physical Applications*, eds. D. Levi and P. Winternitz, Plenum Press, New York, pp. 175–186.
4. Hirota, R. (1974) A new form of Bäcklund transformations and its relation to the inverse scattering problem, *Prog. Theor. Phys.* **52**, pp. 498–512.
5. Satsuma, J. and Kaup, D. J. (1977) A Bäcklund transformation for a higher order KdV equation, *J. Phys. Soc. Jpn.* **43**, pp. 692–697.
6. Hirota, R. (1976) Direct method of finding exact solutions of nonlinear evolution equations, in: *Bäcklund Transformations, the Inverse Scattering Method, Solitons and their Applications*, Springer Lecture Notes in Mathematics **515**, ed. R.M. Miura, pp. 40–68.
7. Bell, E. T. (1934) Exponential polynomials, *Ann. Math.* **35**, pp. 258–277.
8. Faá di Bruno (1857) Note sur une nouvelle formule de calcul différentiel, *Quart. J. Pure Appl. Math.* **1**, pp. 359–360.
9. Gilson, C., Lambert, F., Nimmo, J., and Willox, R. (1996) On the combinatorics of the Hirota D-operators, *Proc. Roy. Soc. Lond.* **A431**, pp. 361–639.
10. Lambert, F. and Springael, J. (2001) On a direct procedure for the disclosure of Lax pairs and Bäcklund transformations, *Chaos, Solitons and Fractals.* **12**, pp. 2821–2832.
11. Jimbo, M. and Miwa, T. (1983) Solitons and infinite dimensional Lie algebras, *RIMS Kyoto University.* **19**, pp. 943–1001.
12. Lambert, F., Loris, I., and Springael, J. (2001) Classical Darboux transformations and the KP hierarchy, *Inverse Probl.* **17**, pp. 1067–1074.
13. Sawada, K. and Kotera, T. (1974) A method for finding N-soliton solutions of the KdV equation and KdV-like equations, *Progr. Theor. Phys.* **51**, pp. 1355–1367.

COVARIANT FORMS OF LAX ONE-FIELD OPERATORS: FROM ABELIAN TO NONCOMMUTATIVE

Sergey Leble

Gdańsk University of Technology, ul. Narutowicza 11/12, Gdańsk, Poland

Abstract   Polynomials in differentiation operators are considered. Joint covariance with respect to Darboux transformations of a pair of such polynomials (Lax pair) as a function of one-field is studied. Methodically, the transforms of the coefficients are equalized to Frechèt differential (first term of the Taylor series on prolonged space) to establish the operator forms. In the commutative (Abelian) case, as it was recently proved for the KP-KdV Lax operators, it results in binary Bell (Faa de Bruno) differential polynomials having natural bilinear (Hirota) representation. Now next example of generalized Boussinesq equation with variable coefficients is studied, the dressing chain equations for the pair are derived. For a pair of generalized Zakharov–Shabat problems a set of integrable (noncommutative) potentials and hence nonlinear equations are constructed altogether with explicit dressing formulas. Some non-Abelian special functions are introduced.

1   INTRODUCTION

Investigations of general Darboux transformation (DT) theory in the case of differential operators

$$L = \sum_{k=0}^{n} a_k \partial^k \qquad (1)$$

with noncommutative coefficients was launched by papers of Matveev [1]. The proof of a general covariance of the equation

$$\psi_t = L\psi \qquad (2)$$

with respect to the classic DT (the shorthands $\psi' = \partial\psi = \psi_x$ are used through the paper)

$$\psi[1] = \psi' - \sigma\psi, \qquad (3)$$

161

L. Faddeev et al. (eds.),
Bilinear Integrable Systems: From Classical to Quantum, Continuous to Discrete, 161–173.

incorporates the auxiliary relation

$$\sigma_t = \partial r + [r, \sigma], \quad r = \sum_0^N a_n B_n(\sigma), \qquad (4)$$

where B_n are differential Bell (Faa de Bruno [2]) polynomials [3]. The relation (4) generalizes so-called Miura map and became the identity when $\sigma = \phi'\phi^{-1}$, ϕ is a solution of the Eq. (2).

Such operators (1) are used in the Lax representation constructions for nonlinear problems. It opens the way to produce wide classes of solutions of the nonlinear problem. Examples of discrete, non-Abelian, and nonlocal equations, integrable by DT was considered in [4, 5]. Some of them were reviewed and developed in the book [6] and intensely used nowadays [7]. The approach was recently generalized for a wide class of polynomials of automorphism on a differential ring [8].

A study of jointly covariant combinations introduces extra problems of the appropriate choice of potentials on which the polynomial coefficients depend [9]. This problem was recently discussed in [10], where a method of the conditions account was developed. Covariant combinations of (generalized) derivatives and potentials may be hence classified for linear problems. In two words, having the general statement about covariant form of a linear polynomial differential operator that determines transformation formulas for coefficients (Darboux theorem and its Matveev's generalizations), the consistency between two such formulas yields the special constraints. For example, the second-order scalar differential operator has the only place for a potential and the covariance generate the classic Darboux transformation for it.

In scalar case such one-potential constructions have been studied in [11] and developed for higher KdV and KP equations [12]. It was found that the result is conveniently written via such combinations of differentiation operator and exponential functions of the potential as Binary Bell Polynomials (BBP) [13]. The principle is reproduced and developed in the Section 2.1 of this paper to give more explanations.

The whole construction in general (non-Abelian) case is more complicated, but much more rich and promising. The theory could contain two ingredients.

1. The first one would be non-Abelian Hirota construction in the terms of the mentioned binary Bell polynomials. On the level of general formulation some obstacles appears, e.g., an extension of addition formulas [13]) to the non-Abelian case.
2. The second way relates to some generalized polynomials that could be produced as covariant combinations of operators with a faith that observations from Abelian theory could be generalized. Namely the case we would discuss in this paper.

Even the minimal (first order in the ∂-operator) examples of the ZS problems with operator coefficients contains many interesting integrable models. It is seen already from the point of view of symmetry classification [14]. So, the link to DT covariance approach allows to hope for a realization of the main purpose–construction of covariant functions, their classification, and use in the soliton equations theory.

We would begin from the example, using notations from quantum mechanics to emphasize the non-Abelian nature of the consideration. The operators ρ and H could play the roles of density matrices and Hamiltonians, respectively, but one also can think of them as just some operators without any particular quantum mechanical connotations. The approach establishes the covariance with respect to DT of rather general Lax system for the equation

$$-i\rho_t = [H, h(\rho)],$$

where $h(\rho)$–analytic function, in some sense–"Abelian," i.e., the function to be defined by Taylor series [15]. More exactly it is shown that the following statement takes place:

Theorem 1 *Assume $\langle\chi|$ and $\langle\psi|$ are solutions of the following (direct) equations:*

$$z_v\langle\chi| = \langle\chi|(\rho - vH),$$

$$-i\langle\chi_t| = \frac{1}{v}\langle\chi|h(\rho),$$

and $|\varphi\rangle$ stands for the conjugate pair. Here ρ, H are operators left-acting on a "bra" vectors $\langle\psi|$ associated with an element of a Hilbert space. The transforms $\langle\psi_1|, \rho_1, h(\rho)_1$ are defined by

$$\langle\psi_1| = \langle\psi|\left(1 + \frac{v - \mu}{\mu - \lambda}P\right), \tag{5}$$

$$\rho_1 = T\rho T^{-1}, \qquad h_1(\rho) = Th(\rho)T^{-1}, \qquad T = \left(1 + \frac{\mu - v}{v}P\right). \tag{6}$$

where $P = |\varphi\rangle\langle\chi|\varphi\rangle\langle\chi|$. Then the pairs are covariant:

$$z_\lambda\langle\psi_1| = \langle\psi_1|(\rho_1 - \lambda H), \qquad -i\langle\dot{\psi}_1| = \frac{1}{\lambda}\langle\psi_1|h_1(\rho).$$

complex numbers λ, z_λ are independent of t [15].

The cases $f(\rho) = i\rho^3$ and $f(\rho) = i\rho^{-1}$ were considered in [16], see applications in [17]. A step to further generalizations for essentially non-Abelian functions, e.g., $h(X) = XA + AX, [A, X] \neq 0$, is studied in [18]. The case is the development of the matrix representation of the Euler top model [19]. This example of the theory is more close to the spirit of the Section 3.4, more achievements are demonstrated in [20], where abundant set of integrable equations is listed. The list is in a partial correspondence with [14], and give the usual for the DT technique link to solutions via the iteration procedures or dressing chains. One of the main results, we present in the Section 3.3, is how the "true" non-Abelian functions appear in the context of the covariance conditions application.

2 ONE-FIELD LAX PAIR FOR ABELIAN CASE

2.1 Covariance Equations

First we would reproduce the "Abelian" scheme, generalizing the study of the example of the Boussinesq equation [10]. To start with the search we should fix the number of fields. Let us consider the third-order operator (1) with coefficients $b_k, k = 0, 1, 2, 3$, reserving a_k for the second operator in a Lax pair. Suppose, both operators depend on the only potential function w. The problem we consider now may be formulated as follows: To find restrictions on the coefficients $b_3(t), b_2(x, t), b_1 = b(w, t), b_0 = G(w, t)$ compatible with DT transformations rules of the potential function w induced by DT for b_i. The classic DT for the third-order operator coefficients (Matveev generalization [1]) yields

$$b_2[1] = b_2 + b_3', \tag{7}$$

$$b_1[1] = b_1 + b_2' + 3b_3\sigma', \tag{8}$$

$$b_0[1] = b_0 + b_1' + \sigma b_2' + 3b_3(\sigma\sigma' + \sigma''), \tag{9}$$

having in mind that the "elder" coefficient b_3 does not transform. Note also, that $b_3' = 0$ yields invariance of the coefficient b_2.

The general idea of DT form-invariance may be realized considering the coefficients transforms to be consistent with respect to the fixed transform of w. Generalizing the analysis of the third-order operator transformation [10], one arrives at the equations for the functions $b_2(x, t), b(w, t), G(w)$. The covariance of the spectral equation

$$b_3\psi_{xxx} + b_2(x, t)\psi_{xx} + b(w, t)\psi_x + G(w, t)\psi = \lambda\psi \tag{10}$$

may be considered separately, that leads to the link between b_i only. We, however, study the problem of the (10) in the context of Lax representation for some nonlinear equation, hence the covariance of the second Lax equation is taken into account from the very beginning. We name such principle as the "*principle of joint covariance*" [9]. The second (evolution) equation of the case is

$$\psi_t = a_2(t)\psi_{xx} + a_1(t)\psi_x + w\psi, \tag{11}$$

with the operator in the r.h.s. having again the form of (1). We do not consider here a dependence of a_i, b_i on x for the sake of brevity, leaving this interesting question to the next paper.

If one consider the L and A operators of the form (1), specified in Eqs. (10) and (11). as the Lax pair equations, the DT of w implied by the covariance of (11), should be compatible with DT formulas of both coefficients of (10) depending on the only variable w.

$$a_2[1] = a_2 = a(x, t),$$
$$a_1[1] = a_1(x, t) + Da(x, t)$$
$$a_0[1] = w[1] = w + a_1' + 2a_2\sigma' + \sigma a_2' \tag{12}$$

Next important relations being in fact the identities in the DT transformation theory [3], are the particular cases of the generalized Miura map (4):

$$\sigma_t = [a_2(\sigma^2 + \sigma_x) + a_1\sigma + w]_x \tag{13}$$

for the problem (11) and, for the (10)

$$b_3(\sigma^3 + 3\sigma_x\sigma + \sigma_{xx}) + b_2(\sigma^2 + \sigma_x)$$
$$+ b(w, t)\sigma + G(w) = \text{const}; \tag{14}$$

ϕ is a solution of both Lax equations.

Suppose now that the coefficients of the operators are analytical functions of w together with its derivatives (or integrals) with respect to x (such functions are named functions on prolonged space [21]). For the coefficient $b_0 = G(w, t)$ it means

$$G = G(\partial^{-1}w, w, w_x, \ldots, \partial^{-1}w_t, w_t, w_{tx}, \ldots). \tag{15}$$

The covariance condition is obtained for the Frechêt derivative (FD) of the function G on the prolonged space, or the first terms of multidimensional Taylor series for (15), read

$$G(w + a_1' + 2a_2\sigma' + \sigma a_2') = G(w)$$
$$+ G_{w_x}(a_1' + 2a_2\sigma' + \sigma a_2')' + \cdots \tag{16}$$

We shall show only the terms of further importance.

Quite similar expansion arises for the coefficient $b_1 = b(w, t)$, with which we would start in the analogy with the expressions (8, 16). Equalizing the DT and the expansion one obtains the condition

$$b_2' + 3b_3\sigma' = b_w(a_1' + 2a_2\sigma' + \sigma a_2') + b_{w'}(a_1' + 2a_2\sigma' + \sigma a_2')' \cdots \tag{17}$$

This equation we name the (first) "*joint covariance equation*" that guarantee the consistency between transformations of the coefficients of the Lax pair

(10), (11). In the frame of our choice $a_2' = 0$, the equation simplifies and linear independence of the derivatives $\sigma^{(n)}$ yields two constraints

$$3b_3 = 2b_w a_2,$$
$$b_2' = b_w a_1',$$
(18)

or, solving the second and plugging into the first, results in

$$b_w = 3b_3/2a_2,$$
$$b_2' = 3b_3 a_1'/2a_2.$$
(19)

So, if one wants to save the form of the standard DT for the variable w (potential) the simple comparison of both transformation formulas gives for $b(w)$ the following connection (with arbitrary function $\alpha(t)$):

$$b(w, t) = 3b_3 w/2a_2 + \alpha(t).$$
(20)

Equalizing the expansion (16) with the transform of the $b_0 = G(w, t)$ yields:

$$b_1' + \sigma b_2' + 3b_3(\sigma^2/2 + \sigma')' = G_{w_x}(a_1' + 2a_2\sigma' + \sigma a_2')'$$
$$+ G_{\partial^{-1}w_t}[a_{1t} + 2\partial^{-1}(a_2\sigma_t') + \partial^{-1}(\sigma a_2')_t] + \cdots$$
(21)

This second "*joint covariance equation*" also simplifies when $a_2' = 0$:

$$3b_3 w'/2a_2 + \sigma b_2' + 3b_3(\partial^{-1}\sigma_t - w)'/2a_2 + 3b_3\sigma''/2$$
$$= G_{w_x}(a_1' + 2a_2\sigma')' + G_{\partial^{-1}w}[a_1 + 2a_2\sigma]$$
$$+ G_{\partial^{-1}w_t}[a_{1t} + 2a_2\sigma_t] + \ldots,$$
(22)

when (20) is accounted. Note, that the "Miura" (13) is used in the l.h.s. and linearizes the FD with respect to σ. Therefore, the derivatives of the function G

$$G_{w_x} = 3b_3/4a_2,$$
$$G_{\partial^{-1}w_t} = 3b_3/4a_2^2,$$
$$G_{\partial^{-1}w} = b_2'/2a^2,$$
(23)

are accompanied by the constraint

$$a_{1t} + a_2 a_1'' + a_1 a_1' = 0,$$
(24)

which have got the form of the Burgers equation after (19) account. Finally the integration of the relation (19) gives

$$b_2 = 3b_3 a_1/2a_2 + \beta(t)$$
(25)

and the "lower" coefficient of the third-order operator is expressed by

$$G(w, t) = 3b_3 w_x/2a_2 + 3b_3 a_1' \partial^{-1}w/2(a_2)^2 + 3b_3 \partial^{-1}w_t/2a_2^2.$$
(26)

Statement 2 *The expressions* (11, 10, 20, 26) *define the covariant Lax pair when the constraints* (19, 24) *are valid.*

Remark We cut the Frechêt differential formulas on the level that is necessary for the minimal flows. The account of higher terms leads to the whole hierarchy [12].

2.2 Compatibility Condition

In the case $a_2' = 0$ by which we have restricted ourselves, the Lax system (10, 11) produces the following compatibility conditions:

$$
\begin{aligned}
2a_2 b_3' &= 3b_3 a_2', \\
b_{3t} &= 2a_2 b_2' - 3b_3 a_1'' \\
b_{2t} &= a_2 b_2'' + 2a_2 b_1' + a_1 b_2' - 3b_3 a_1'' - 2b_2 a_1' - 3b_3 a_0' \\
b_{1t} &= a_2 b_1'' + a_1 b_1' - b_3 a_1''' - b_2 a_1'' - b_1 a_1' - 3b_3 a_0'' - 2b_2 a_0' + 2a_2 b_0' \\
b_{0t} &= a_1 b_0' + a_2 b_0'' - b_1 a_0' - b_2 a_0'' - b_3 a_0'''
\end{aligned}
\tag{27}
$$

In the particular case of $a_2 = 0$ we extract at once from the first of the equalities (27) the constraint $b_3' = 0$. The direct corollary of (25) is $b_{3t} = 0$. In the rest of the equations the links (27), (25) are taken into account. Hence (24) in the combination with the expression for b_{2t} produce

$$
\beta_t = -2\beta a_1'. \tag{28}
$$

The last two equations (choice of constants $b_3 = 1$, $a_2 = -1$) become

$$
\begin{aligned}
&\alpha w + \alpha_t + 3a_1'' \partial^{-1} w/2 + (2\beta - 3a_1/2)w' + a_1''' + 3a_1 a_1''/2 = 0 \\
&3\partial^{-1}(w_t + a_1 w)_t/4 = (\alpha - 3w/2)w' - w'''/4 \\
&+ 3a_1 w_t/4 + 3a_1 a_1'' \partial^{-1} w/4 + 3a_1 a_1' w/4 - 3a_1' w'/4 \\
&+ (\beta + 3a_1/4)w''.
\end{aligned}
\tag{29}
$$

In the simplest case of constant coefficients ($b_2' = a'1 = 0$) one goes down to

$$
\begin{aligned}
3b_3(w_t + a_1 w)_t/4a_2^2 &= -[(3b_3 w/2a_2 + \alpha)w' - b_3 w'''/4 \\
&+ 3b_3 a_1 w_t/4a_2^2 + (\beta - 3b_3 a_1/4a_2)w'']'.
\end{aligned}
\tag{30}
$$

This equation reduces to the standard Boussinesq equation when ($b_1 = a_1 = 0$, $b_3 = 1$, $a_2 = -1$) [6].

We would repeat that the results given in the Section 2 are simplified to show more clear the algorithm of the covariant Lax pair derivation. More general study ($a_2' \neq 0$) will be published elsewhere.

2.3   Solutions. Dressing Chains for the Boussinesq Equation

The dressing formula for the zero seed potential (39) is standard and includes the only seed solution ϕ, of the Lax equations with zero potential w.

$$w_s = a_1' + 2a_2\sigma' + \sigma a_2' = a_1' + 2a_2 \log_{xx} \phi(x, t) \qquad (31)$$

A next power tool to obtain solutions of nonlinear system is the dressing chain equation: solitonic, finite-gap, and other important solutions were obtained for the KdV equations reducing such chain [22]. Going to the dressing chain, we use the scheme from [10]. We would restrict ourselves further to the case of $a_2 = -1$, $a_0 = u$, $b_3 = 1$, $b_2 = 0$, $b_1 = b(u, t) = -3u/2 + \alpha$, $G = -3u'/4 + 3\partial^{-1}u_t$ to fit the notations from [10]. The general construction is quite similar.

The Miura equations (13), (14) also simplifies

$$\sigma_t = -(\sigma^2 + \sigma_x)_x + u_x \qquad (32)$$

for the problem (11) and

$$\sigma^3 + 3\sigma_x\sigma + \sigma_{xx} + b\sigma + G = \text{const}, \qquad (33)$$

where $b = 3u/2 + \alpha$, $G = -3\partial^{-1}u_t/4 + 3u_x/4$.

Namely the Eq. (32), (33) together with the n-fold iterated DT formula (5)

$$u_{n+1} = u_n - 2\sigma_n' \qquad (34)$$

form the basis to produce the DT dressing chain equations.

We express the iterated potential w_n from (32)

$$-\sigma_{nt} + (\sigma_n^2 + \sigma_n')' = u_n' \qquad (35)$$

and substitute it into the differentiated relation (34) to get the first dressing chain equation

$$\sigma_{n+1,t} - \sigma_{nt} = (\sigma_{n+1}^2 + \sigma_{n+1}')' - (\sigma_n^2 - \sigma_n')'. \qquad (36)$$

Next chain equation is obtained when one plugs the potential from (35) to the iterated (33)

$$\sigma_n^3 + 3\sigma_n'\sigma_n + \sigma_n'' + (-3u_n/2 + \alpha)\sigma_n + -3u_n'/4 + 3\partial^{-1}u_{nt} = c_n. \qquad (37)$$

3   NON-ABELIAN CASE. ZAKHAROV–SHABAT (ZS) PROBLEM

3.1   Joint Covariance Conditions for General ZS Equations

Let us change notations for the first order ($n = 1$) Eq. (1) with the coefficients from a non-Abelian differential ring A (for details of the mathematical objects definitions see [3]) as follows:

$$\psi_t = (J + u\partial)\psi, \tag{38}$$

where the operator $J \in A$ does not depend on x, y, t and the potential $a_0 \equiv u = u(x, y, t) \in A$ is a function of all variables. The operator $\partial = \partial/\partial x$ may be considered as a general differentiation as in [3]. The transformed potential

$$\tilde{u} = u + [J, \sigma], \tag{39}$$

where the $\sigma = \phi_x \phi^{-1}$, is defined by the same formula as before, but the order of elements is important. The covariance of the operator in (38) follows from general transformations of the coefficients of a polynomial [6]. The coefficient J does not transform.

Suppose the second operator of a Lax pair has the same form, but with different entries and derivatives.

$$\psi_y = (Y + w\partial)\psi, \tag{40}$$

$Y \in A$ where the potential $w = F(u) \in A$ is a function of the potential of the first (38) equation. The principle of joint covariance [9] hence reads

$$\tilde{w} = w + [Y, \sigma] = F(u + [J, \sigma]), \tag{41}$$

with the direct corollary

$$F(u) + [Y, \sigma] = F(u + [J, \sigma]). \tag{42}$$

So, the Eq. (42) defines the function $F(u)$, we shall name this equation as **joint covariance equation**. In the case of Abelian algebra we used the Taylor series (generalized by use of a Frechet derivative) to determine the function. Now some more generalization is necessary. Let us make some general remarks.

An operator-valued function $F(u)$ of an operator u in a Banach space may be considered as a generalized Taylor series with coefficients that are expressed in terms of Frechèt derivatives. The linear in u part of the series approximates (in a sense of the space norm) the function

$$F(u) = F(0) + F'(0)u + \cdots.$$

The representation is not unique and the similar expression

$$F(u) = F(0) + u\hat{F}'(0) + \cdots$$

may be introduced (definitions are given in Appendix). Both expressions however are not Hermitian, hence not suitable for the majority of physical models. It means, that the class or such operator functions is too restrictive. To explain what we have in mind, let us consider examples.

3.2   Important Example

From a point of view of the physical modeling the following Hermitian approximation:

$$F(u) = F(0) + H^+ u + uH + \cdots, \quad u^+ = u,$$

is preferable. Such models could be applied to quantum theories: introduction of this approximation is similar to "phi in quadro" (Landau-Ginzburg) model [18]. Let us study, in which conditions the function

$$w = F(u) = Hu + uH, \tag{43}$$

satisfy the joint covariance condition for the Lax pair (38), (40) By direct calculation in (42) one arrives at the equality

$$[Y, \sigma] = H[J, \sigma] + [J, \sigma]H. \tag{44}$$

The obvious choice for arbitrary σ is $Y = H^2, J = H$.

The compatibility conditions for the pair of Eqs. (38) and (40) yields

$$u_y - Hu_t - u_t H + [u, H]u + u[u, H] + H^2 u_x + Hu_x H + H^2 u_x = 0 \tag{45}$$

If the potential does not depend on t, it is reduced to the next equation:

$$u_y + [u^2, H] + H^2 u_x + Hu_x H + H^2 u_x = 0, \tag{46}$$

and x-independence yields the generalized Euler top equations

$$u_y + [u^2, H] = 0, \tag{47}$$

which Lax pair (38), (40) with $Y = J^2, J = H$ was found by Manakov [19].

3.3   Covariant Combinations of Symmetric Polynomials

The next natural example appears if one examine the link (44).

$$P_2(H, u) = H^2 u + HuH + uH^2$$

The direct substitution in the covariance and compatibility equations leads to covariant constraint that turns to the identity, if $Y = H^3, J = H$.

It is easy to check more general connection $Y = J^n, J = H$ connection that leads to the covariance of the function

$$P_n(H, u) = \sum_{p=0}^{n} H^{n-p} u H^p.$$

Such observation was exhibited in [18]. On the way of a further generalization let us consider

$$f(H, u) = Hu + uH + S^2u + SuS + uS^2 \qquad (48)$$

Plugging (48) as $F(u) = f(H, u)$ into (42), representing $Y = AB + CDE$ yields

$$A[B, \sigma] + [A, \sigma]B + CD[E, \sigma] + C[D, \sigma]E + [C, \sigma]DE$$
$$= H[J, \sigma] + [J, \sigma]H + S^2[J, \sigma] + S[J, \sigma]S + [J, \sigma]S^2.$$

The last expression turns to identity if $A = B = J = H, C = \alpha H, D = \alpha H, D = \alpha H, S = \beta H$, and $[\alpha, H] = 0, [\beta, H] = 0$ with the link $\alpha^3 = \beta^2$.

Statement 3 *Darboux covariance define a class of homogeneous polynomials $P_n(H, u)$, symmetric with respect to cyclic permutations. A linear combination of such polynomials $\sum_{n=1}^{N} \beta_n P_n(H, u)$ with the coefficients commuting with u, H is also covariant, if the element $Y = \sum_{n=1}^{N} \alpha_n H^{n+1}$ and $\alpha_1 = \beta_1 = 1, \alpha_n^{n+2} = \beta^{n+1}, n \neq 1$.*

A proof could be made by induction that is based on homogeneity of the P_n and linearity of the constraints with respect to u. The functions $F_H(u) = \sum_0^\infty a_n P_{Hn}(u)$ satisfy the constraints if the series converges.

4 CONCLUSION

The main result of this paper is the covariant Eq. (42). See, also the example (44). A class of potentials from [20] contains polynomials $P_n(H, u)$ and give alternative expressions for it. The linear combinations, introduced here could better reproduce physical situation of interest. So, we used the compatibility condition to find the form of integrable equation and reduction tracing the simplifications appearing for the subclasses of covariant potentials. While doing this we also check the invariance of the equation and heredity of the constraints.

The work is also supported by KBN grant 5P03B 040 20.

APPENDIX

Right and left Frechêt derivatives: The classic notion of a derivative of an operator by other one is defined in a Banach space B. Two specific features in the case of a operator-function $F(u) \in Bu \in B$ should be taken into account: a norm choice when a limiting procedure is made and the non-Abelian character of expressions while the differential and difference introduced.

Definition 3 *Let a Banach space B have a structure of a differential ring. Let F be the operator from B to B' defined on the open set of B. The operator is named the left-differentiable in $u_0 \in B$ if there exist a linear restricted operator $L(u_0)$, acting also from B to B' with the property*

$$L(u_0 + h) - L(u_0) = L(u_0)h + \alpha(u_0, h), \|h\| \to 0, \qquad (A1)$$

where $\|\alpha(u_0, h)\|/\|h\| \to 0$. The operator $L(u_0) = F'(u_0)$ is referred as the operator of the (strong) left derivative of the function $F(u)$. The right derivative $\hat{F}'(u_0)$ could be defined by the similar expression and conditions, if one changes $Lh \to h\hat{L}$ in the equality (A1).

The addition of the half of the right and left differentials

$$(F'(u_0)h + h\hat{F}'(u_0))/2 \qquad (A2)$$

also approximates the difference $L(u_0 + h) - L(u_0)$ in the sense of the FD definition.

REFERENCES

1. Matveev, V. B. (1979) Darboux transformation and explicit solutions of the Kadomtcev–Petviaschvily equation, depending on functional parameters, *Lett. Math. Phys.* **3** pp. 213–216, Darboux transformation and the explicit solutions of differential-difference and difference-difference evolution equations, pp. 217–222, Some comments on the rational solutions of the Zakharov–Schabat equations, pp. 503–512.
2. Bruno, F. D. (1857) Notes sur une nouvelle formule de calcul différentiel Q.J., *Pure Appl. Math.* **1**, pp. 359–360.
3. Zaitsev, A. A. and Leble, S. B. (2000) Division of differential operators, intertwine relations and Darboux transformations, Preprint 12.01.1999 math-ph/9903005, ROMP, **46**, p. 155.
4. Salle, M. (1982) Darboux transformations for nonabelian and nonlocal Toda-chain-type equations, *Theor. Math. Phys.* **53**, pp. 227–237.
5. Leble, S. and Salle, M. (1985) The Darboux transformations for the discrete analogue of the Silin-Tikhonchuk equation, *Dokl. AN SSSR* **284**, pp. 110–114.
6. Matveev, V. B. and Salle, M. A. (1991) *Darboux Transformations and Solitons*, Springer, Berlin.
7. Roger, C. and Schiff, W. K. (2002) *Backlund and Darboux Transformations*, Cambridge University Press.
8. Matveev, V. B. (2000) Darboux transformations, covariance theorems and integrable systems, *Amer. Math. Soc. Transl.* **201**(2), pp. 179–209.
9. Leble S. (1991) Darboux transforms algebras in $2 + 1$ dimensions, in: *Proceedings of NEEDS-91 Workshop, World Scientific. Singapore*, pp. 53–61.
10. Leble, S. (2001) Covariance of lax pairs and integrability of compatibility condition, nlin.SI/0101028, *Theor. Math. Phys.*, **128**, pp. 890–905.

11. Lambert, F., Leble, S., and Springael, J. (2001) Binary Bell polynomials and Dar-boux covariant lax, pairs, *Glasgow Math. J.* **43A**, pp. 55–63.
12. Lambert, F., Loris, I., and Springael, J. (2001) Classical Darboux transformations and the KP hierarchy, *Inverse Problems* **17**, pp. 1067–1074.
13. Gilson, C., Lambert, F., Nimmo, J., and Willox, R. (1996) *Proc. R. Soc. Land A* **452**, pp. 223–234.
14. Mikhailov, A. and Sokolov, V. (2000) Integrable ordinary differential equations on free associative algebras, *Theor. Math. Phys.* **122**, pp. 72–83.
15. Ustinov, N., Leble, S., Czachor, M., and Kuna, M. (2001) '*Darboux-integration of* $i\rho_t = [H, f(\rho)]$' quant-ph/0005030, *Phys. Lett. A.* **279**, pp. 333–340.
16. Czachor, M., Leble, S., Kuna, M., and Naudts, J. (2000) *Nonlinear von Neumann type equations, Trends in Quantum Mechanics Proceedings of the International symposium, ed. H.-D. Doebner et al* World Sci, pp. 209–226.
17. Minic, A. (2002) "Nambu-type quantum mechanics: a nonlinear generalization of geometric QM." *Phys. Lett. B* **536**, 305–314. hep-th/0202173
18. Leble, S. B. and Czachor, M. (1998) *Darboux-integrable nonlinear Liouville-von Neumann equation* quant-ph/9804052, *Phys. Rev. E.* **58**, p. N6.
19. Manakov, S. V. (1976) A remark on the integration of the Eulerian equations of the dynamics of an n-dimensional rigid body (Russian). *Funktsional'nyi Analiz i Pril.* **10**, pp. 93–94. Adler, M. and van Moerbeke, P. (1980) Completely integrable systems, Euclidean Lie algebras, and curves, *Adv. in Math.* **38**, pp. 267–357.
20. Czachor, M. and Ustinov, N. (2000) New class of integrable nonlinear von Neumann-type equations, arXiv:nlinSI/0011013, *J. Math. Phys.* (in press).
21. Olver, P. G. (1986) Applications of Lie groups to differential equations, *Graduate Texts in Mathematics*, Vol. **107**, Springer, Berlin.
22. Weiss, J. (1986) Periodic fixed points of Bäcklund transformations and the Korteweg-de Vries equation, *J. Math. Phys.* **27**, pp. 2647–2656.
23. Schimming, R. and Rida, S. Z. (1996) Noncommutative Bell polynomials, *Int. J. of Algebra and Computation* **6**, pp. 635–644.
24. Kuna, M., Czachor, M., and Leble, S. (1999) Nonlinear von Noeumann equations: Darboux invariance and spectra, *Phys. Lett. A.* **255**, pp. 42–48.

ON THE DIRICHLET BOUNDARY PROBLEM AND HIROTA EQUATIONS

A. Marshakov

Theory Department, P.N. Lebedev Physics Institute and ITEP, Moscow, Russia

A. Zabrodin

Institute of Biochemical Physics and ITEP, Moscow, Russia

Abstract   We review the integrable structure of the Dirichlet boundary problem in two dimensions. The solution to the Dirichlet boundary problem for simplyconnected case is given through a quasiclassical tau-function, which satisfies the Hirota equations of the dispersionless Toda hierarchy, following from properties of the Dirichlet Green function. We also outline a possible generalization to the case of multiply connected domains related to the multi support solutions of matrix models.

1   INTRODUCTION: GREEN FUNCTION AND HADAMARD FORMULA

Solving the Dirichlet boundary problem [1], one reconstructs a harmonic function in a bounded domain from its values on the boundary. In two dimensions, this is one of standard problems of complex analysis having close relations to string theory and matrix models. Remarkably, it possesses a hidden integrable structure [2]. It turns out that variation of a solution to the Dirichlet problem under variation of the domain is described by an infinite hierarchy of non linear partial differential equations known (in the simply-connected case) as dispersionless Toda hierarchy. It is a particular example of the universal hierarchy of quasiclassical or Whitham equations introduced in [3, 4].

The quasiclassical tau-function (or its logarithm F) is the main new object associated with a family of domains in the plane. Any domain in the complex plane with sufficiently smooth boundary can be parametrized by its harmonic moments and the F-function is a function of the full infinite set of the moments. The first order derivatives of F are then moments of the complementary domain. This gives a formal solution to the inverse potential problem, considered for a simply connected case in [5, 6]. The second order derivatives are coefficients of the Taylor expansion of the Dirichlet Green function and therefore they

L. Faddeev et al. (eds.),
Bilinear Integrable Systems: From Classical to Quantum, Continuous to Discrete, 175–190.
© 2006 *Springer. Printed in the Netherlands.*

solve the Dirichlet boundary problem. These coefficients are constrained by infinite number of universal (i.e. domain independent) relations which, unified in a generating form, just constitute the dispersionless Hirota equations. For the third order derivatives there is a nice "residue formula" which allows one to prove [7] that F obeys the Witten–Dijkgraaf–Verlinde–Verlinde (WDVV) equations [8].

Let us remind the formulation of the Dirichlet problem in planar domains. Let D^c be a domain in the complex plane bounded by one or several non intersecting curves. It will be convenient for us to realize the D^c as a complement of another domain, D (which in general may have more than one connected components), and consider the Dirichlet problem in D^c. The problem is to find a harmonic function $u(z)$ in D^c, such that it is continuous up to the boundary $\partial \mathsf{D}^c$ and equals a given function $u_0(\xi)$ on the boundary, and it can be uniquely solved in terms of the Dirichlet Green function $G(z, \xi)$:

$$u(z) = -\frac{1}{2\pi} \oint_{\partial \mathsf{D}^c} u_0(\xi) \partial_n G(z, \xi) |d\xi| \tag{1}$$

where ∂_n is the normal derivative on the boundary with respect to the second variable, and the normal vector $\vec{n}$ is directed inward D^c, $|d\xi| := dl(\xi)$ is an infinitesimal element of the length of the boundary $\partial \mathsf{D}^c$.

The Dirichlet Green function is uniquely determined by the following properties [1]:

($G1$) The function $G(z, z')$ is symmetric and harmonic everywhere in D^c (including ∞ if $\mathsf{D}^c \ni \infty$) in both arguments except $z = z'$ where $G(z, z') = \log|z - z'| + \cdots$ as $z \to z'$;

($G2$) $G(z, z') = 0$ if any of the variables z, z' belongs to the boundary.

Note that the definition implies that $G(z, z') < 0$ inside D^c. In particular, $\partial_n G(z, \xi)$ is strictly negative for all $\xi \in \partial \mathsf{D}^c$.

If D^c is simply-connected (the boundary has only one component), the Dirichlet problem is equivalent to finding a bijective conformal map from D^c onto the unit disk or any other reference domain (where the Green function is known explicitly) which exists by virtue of the Riemann mapping theorem. Let $w(z)$ be such a bijective conformal map of D^c onto the complement to the unit disk, then

$$G(z, z') = \log \left| \frac{w(z) - w(z')}{w(z)\overline{w(z')} - 1} \right| \tag{2}$$

where bar means complex conjugation. It is this formula which allows one to derive the Hirota equations for the tau-function of the Dirichlet problem in the most economic and transparent way [2]. Indeed, the Green function is shown to admit a representation through the logarithm of the tau-function of the form

$$G(z, z') = \log \left| \frac{1}{z} - \frac{1}{z'} \right| + \frac{1}{2} \nabla(z) \nabla(z') F \tag{3}$$

where $\nabla(z)$ (see (9) below) is certain differential operator with constant coefficients (depending only on the point z as a parameter) in the space of harmonic moments. Taking into account that $G(z, \infty) = -\log |w(z)|$, one excludes the Green function from these relations thus obtaining a closed system of equations for F only.

Our main tool to derive (3) is the Hadamard variational formula [9] which gives variation of the Dirichlet Green function under small deformations of the domain in terms of the Green function itself:

$$\delta G(z, z') = \frac{1}{2\pi} \oint_{\partial D^c} \partial_n G(z, \xi) \partial_n G(z', \xi) \delta n(\xi) |d\xi| . \tag{4}$$

Here $\delta n(\xi)$ is the normal displaycement (with sign) of the boundary under the deformation, counted along the normal vector at the boundary point ξ. It was shown in [2] that this remarkable formula reflects all integrable properties of the Dirichlet problem. An extremely simple "pictorial" derivation of the Hadamard formula is presented in figure 1. Looking at the figure and applying (1), one immediately gets (4).

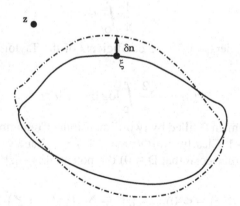

Figure 1. A "pictorial" derivation of the Hadamard formula. We consider a small deformation of the domain, with the new boundary being depicted by the dashed line. According to $(G2)$ the Dirichlet Green function vanishes $G(z, \xi) = 0$ if ξ belongs to the old boundary. Then the variation $\delta G(z, \xi)$ simply equals to the new value, i.e. in the leading order $\delta G(z, \xi) = -\delta n(\xi) \partial_n G(z, \xi)$. Now notice that $\delta G(z, \xi)$ is a harmonic function (the logarithmic singularity cancels since it is the same for both old and new functions) with the boundary value $-\delta_n(\xi) \partial_n G(z, \xi)$. Applying (1) one obtains (4)

2 DIRICHLET PROBLEM FOR SIMPLY-CONNECTED DOMAINS AND DISPERSIONLESS HIROTA EQUATIONS

Let D be a connected domain in the complex plane bounded by a simple analytic curve. We consider the exterior Dirichlet problem in $D^c = C \backslash D$ which is the complement of D in the whole (extended) complex plane. Without loss of generality, we assume that D is compact and contains the point $z = 0$. Then D^c is an unbounded simply-connected domain containing ∞.

2.1 Harmonic Moments and Elementary Deformations

To characterize the shape of the domain D^c we consider its moments with respect to a complete basis of harmonic functions. The simplest basis is $\{z^{-k}\}$, $\{\bar{z}^{-k}\}(k \geq 1)$ and the constant function. Let t_k be the harmonic moments

$$t_k = -\frac{1}{\pi k} \int_{D^c} z^{-k} d^2 z, \quad k = 1, 2, \ldots \tag{5}$$

and $\bar{t}_k$ be the complex conjugated moments. The Stokes formula represents them as contour integrals $t_k = \frac{1}{2\pi i k} \oint_{\partial D} z^{-k} \bar{z} dz$, providing, in particular, a regularization of possibly divergent integrals (5). The moment of constant function is infinite but its variation is always finite and opposite to the variation of the complimentary domain D. Let t_0 be the area (divided by π) of D:

$$t_0 = \frac{1}{\pi} \int_D d^2 z \tag{6}$$

The harmonic moments of D^c are coefficients of the Taylor expansion of the potential

$$\Phi(z) = -\frac{2}{\pi} \int_D \log|z - z'| d^2 z' \tag{7}$$

induced by the domain D filled by two-dimensional Coulomb charges with the uniform density -1. Clearly, $\partial_z \partial_{\bar{z}} \Phi(z) = -1$ if $z \in D$ and vanishes otherwise, so around the origin (recall that $D \ni 0$) the potential is $-|z|^2$ plus a harmonic function:

$$\Phi(z) - \Phi(0) = -|z|^2 + \sum_{k \geq 1} \left(t_k z^k + \bar{t}_k \bar{z}^k \right) \tag{8}$$

A simple calculation shows that t_k are just given by (5).

The basic fact of the theory of deformations of closed analytic curves is that the (in general complex) moments $\{t_k, \bar{t}_k\} \equiv \{t_{\pm k}\}$ supplemented by the real variable t_0 form a set of local coordinates in the "moduli space" of smooth closed curves. This means that under any small deformation of the domain the

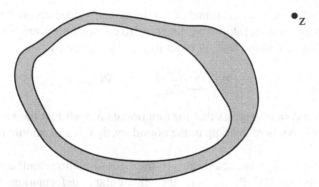

Figure 2. The elementary deformation with the base point z

set $\mathbf{t} = \{t_0, t_{\pm k}\}$ is subject to a small change and vice versa. For more details, see [10, 11, 12]. The differential operators

$$\nabla(z) = \partial_{t_0} + \sum_{k \geq 1} \left(\frac{z^{-k}}{k} \partial_{t_k} + \frac{\bar{z}^{-k}}{k} \partial_{\bar{t}_k} \right) \tag{9}$$

span the complexified tangent space to the space of curves. The operator $\nabla(z)$ has a clear geometrical meaning. To clarify it, we introduce the notion of elementary deformation.

Fix a point $z \in \mathsf{D}^c$ and consider a special infinitesimal deformation of the domain such that the normal displaycement of the boundary is proportional to the gradient of the Green function at the boundary point (Figure 2):

$$\delta n(\xi) = -\frac{\epsilon}{2} \partial_n G(z, \xi) \tag{10}$$

For any sufficiently smooth initial boundary this deformation is well defined as $\epsilon \to 0$. We call infinitesimal deformations from this family, parametrized by $z \in \mathsf{D}^c$, *elementary deformations*. The point z is refered to as the *base point* of the deformation. Note that since $\partial_n G < 0$ (see the remark after the definition of the Green function in the Introduction), δn for the elementary deformations is either strictly positive or strictly negative depending of the sign of the ϵ.

Let δ_z be variation of any quantity under the elementary deformation with the base point z. It is easy to see that $\delta_z t_0 = \epsilon$, $\delta_z t_k = \epsilon z^{-k}/k$. Indeed,

$$\delta_z t_k = \frac{1}{\pi k} \oint \xi^{-k} \delta n(\xi) |d\xi| = -\frac{\epsilon}{2\pi k} \oint \xi^{-k} \partial_n G(z, \xi) |d\xi| = \frac{\epsilon}{k} z^{-k} \tag{11}$$

by virtue of the Dirichlet formula (1).

Let $X = X(\mathbf{t})$ be any functional of our domain that depends on the harmonic moments only (in what follows we are going to consider only such functionals). The variation $\delta_z X$ in the leading order in ϵ is then given by

$$\delta_z X = \sum_k \frac{\partial X}{\partial t_k} \delta_z t_k = \epsilon \nabla(z) X \tag{12}$$

The right hand side suggests that for functionals X such that the series $\nabla(z)X$ converges everywhere in D^c up to the boundary, $\delta_z X$ is a harmonic function of the base point z.

Note that in [2] we used the "bump" deformation and continued it harmonically to D^c. So it was the elementary deformation (11) $\delta_z \propto \oint |d\xi| \partial_n G(z, \xi) \delta^{\text{bump}}(\xi)$ that was really used. The "bump" deformation should be understood as a (carefully taken) limit of δ_z when the point z tends to the boundary.

2.2   The Hadamard Formula as Integrability Condition

Variation of the Green function under small deformations of the domain is known due to Hadamard, see Eq. (4). To find how the Green function changes under small variations of the harmonic moments, we fix three points $a, b, c \in \mathsf{C} \backslash \mathsf{D}$ and compute $\delta_c G(a, b)$ by means of the Hadamard formula (4). Using (12), one can identify the result with the action of the vector field $\nabla(c)$ on the Green function:

$$\nabla(c)G(a, b) = -\frac{1}{4\pi} \oint_{\partial \mathsf{D}} \partial_n G(a, \xi) \partial_n G(b, \xi) \partial_n G(c, \xi) |d\xi| \tag{13}$$

Remarkably, the r.h.s. of (13) is *symmetric* in all three arguments:

$$\nabla(a)G(b, c) = \nabla(b)G(c, a) = \nabla(c)G(a, b) \tag{14}$$

This is the key relation, which allows one to represent the Dirichlet problem as an integrable hierarchy of non linear differential equations [2]. This relation is the integrability condition of the hierarchy.

It follows from (14) (see [2] for details) that there exists a function $F = F(\mathbf{t})$ such that

$$G(z, z') = \log \left| \frac{1}{z} - \frac{1}{z'} \right| + \frac{1}{2} \nabla(z) \nabla(z') F \tag{15}$$

The function F is (logarithm of) the tau-function of the integrable hierarchy. In [13] it was called the tau-function of the (real analytic) curves. Existence of such a representation of the Green function was first conjectured by Takhtajan. This formula was first obtained in [13] (see also [12] for a detailed proof and discussion).

2.3 Dispersionless Hirota Equations for F

Combining (15) and (2), we obtain the relation

$$\log\left|\frac{w(z)-w(z')}{w(z)\overline{w(z')}-1}\right|^2 = \log\left|\frac{1}{z}-\frac{1}{z'}\right|^2 + \nabla(z)\nabla(z')F \qquad (16)$$

which implies an infinite hierarchy of differential equations on the function F. It is convenient to normalize the conformal map $w(z)$ by the conditions that $w(\infty) = \infty$ and $\partial_z w(\infty)$ is real, so that

$$w(z) = \frac{z}{r} + O(1) \quad \text{as } z \to \infty \qquad (17)$$

where the real number $r = \lim_{z\to\infty} dz/dw(z)$ is called the (external) conformal radius of the domain D (equivalently, it can be defined through the Green function as $\log r = \lim_{z\to\infty}(G(z,\infty) + \log|z|)$, see [14]). Then, tending $z' \to \infty$ in (16), one gets

$$\log|w(z)|^2 = \log|z|^2 - \partial_{t_0}\nabla(z)F \qquad (18)$$

The limit $z \to \infty$ of this equality yields a simple formula for the conformal radius:

$$\log r^2 = \partial_{t_0}^2 F \qquad (19)$$

Let us now separate holomorphic and antiholomorphic parts of these equations. To do that it is convenient to introduce holomorphic and antiholomorphic parts of the operator $\nabla(z)$ (9):

$$D(z) = \sum_{k\geq 1}\frac{z^{-k}}{k}\partial_{t_k}, \quad \bar{D}(\bar{z}) = \sum_{k\geq 1}\frac{\bar{z}^{-k}}{k}\partial_{\bar{t}_k}, \qquad (20)$$

Rewrite (16) in the form

$$\log\left(\frac{w(z)-w(z')}{w(z)\overline{w(z')}-1}\right) - \log\left(\frac{1}{z}-\frac{1}{z'}\right) - \left(\frac{1}{2}\partial_{t_0} + D(z)\right)\nabla(z')F$$

$$= -\log\left(\frac{\overline{w(z)}-\overline{w(z')}}{w(z')\overline{w(z)}-1}\right) + \log\left(\frac{1}{\bar{z}}-\frac{1}{\bar{z}'}\right) + \left(\frac{1}{2}\partial_{t_0} + \bar{D}(\bar{z})\right)\nabla(z')F$$

The l.h.s. is a holomorphic function of z while the r.h.s. is antiholomorphic. Therefore, both are equal to a z-independent term which can be found from the limit $z \to \infty$. As a result, we obtain the equation

$$\log\left(\frac{w(z)-w(z')}{w(z)-(\overline{w(z')})^{-1}}\right) = \log\left(1-\frac{z'}{z}\right) + D(z)\nabla(z')F \qquad (21)$$

which, as $z' \to \infty$, turns into the formula for the conformal map $w(z)$:

$$\log w(z) = \log z - \frac{1}{2}\partial_{t_0}^2 F - \partial_{t_0} D(z)F \tag{22}$$

(here we used (19)). Proceeding in a similar way, one can rearrange (21) in order to write it separately for holomorphic and antiholomorphic parts in z':

$$\log \frac{w(z) - w(z')}{z - z'} = -\frac{1}{2}\partial_{t_0}^2 F + D(z)D(z')F \tag{23}$$

$$-\log\left(1 - \frac{1}{w(z)\overline{w(z')}}\right) = D(z)\bar{D}(\bar{z}')F \tag{24}$$

Writing down Eq. (23) for the pairs of points (a, b), (b, c), and (c, a) and summing up the exponentials of the both sides of each equation one arrives at the relation

$$(a - b)e^{D(a)D(b)F} + (b - c)e^{D(b)D(c)F} + (c - a)e^{D(c)D(a)F} = 0 \tag{25}$$

which is the dispersionless Hirota equation (for the KP part of the two-dimensional Toda lattice hierarchy) written in the symmetric form. This equation can be regarded as a very degenerate case of the trisecant Fay identity. It encodes the algebraic relations between the second order derivatives of the function F. As $c \to \infty$, we get these relations in a more explicit but less symmetric form:

$$1 - e^{D(a)D(b)F} = \frac{D(a) - D(b)}{a - b}\partial_{t_1} F \tag{26}$$

which makes it clear that the totality of second derivatives $F_{ij} := \partial_{t_i}\partial_{t_j} F$ are expressed through the derivatives with one of the indices equal to unity.

More general equations of the dispersionless Toda hierarchy obtained in a similar way by combining Eqs. (22–24) include derivatives w.r.t. t_0 and $\bar{t}_k$:

$$(a - b)e^{D(a)D(b)F} = ae^{-\partial_{t_0} D(a)F} - be^{-\partial_{t_0} D(b)F} \tag{27}$$

$$1 - e^{-D(z)\bar{D}(\bar{z})F} = \frac{1}{z\bar{z}}e^{\partial_{t_0}\nabla(z)F} \tag{28}$$

These equations allow one to express the second derivatives ∂_{t_m}, $\partial_{t_n} F$, $\partial_{t_m}\partial_{\bar{t}_n} F$ with $m, n \geq 1$ through the derivatives $\partial_{t_0}\partial_{t_k} F$, $\partial_{t_0}\partial_{\bar{t}_k} F$. In particular, the dispersionless Toda equation,

$$\partial_{t_1}\partial_{\bar{t}_1} F = e^{\partial_{t_0}^2 F} \tag{29}$$

which follows from (28) as $z \to \infty$, expresses $\partial_{t_1}\partial_{\bar{t}_1} F$ through $\partial_{t_0}^2 F$.

For a comprehensive exposition of Hirota equations for dispersionless KP and Toda hierarchies we refer the reader to [15, 16].

2.4   Integral Representation of the Tau-Function

Equation (15) allows one to obtain a representation of the tau-function as a double integral over the domain D. Set $\tilde{\Phi}(z) := \nabla(z)F$. One is able to determine this function via its variation under the elementary deformation:

$$\delta_a \tilde{\Phi}(z) = -2\epsilon \log |a^{-1} - z^{-1}| + 2\epsilon G(a, z) \tag{30}$$

which is read from Eq. (15) by virtue of (12). This allows one to identify $\tilde{\Phi}$ with the "modified potential" $\tilde{\Phi}(z) = \Phi(z) - \Phi(0) + t_0 \log |z|^2$, where Φ is given by (7). Thus we can write

$$\nabla(z)F = \tilde{\Phi}(z) = -\frac{2}{\pi} \int_D \log |z^{-1} - \zeta^{-1}| d^2\zeta = v_0 + 2\mathrm{Re} \sum_{k>0} \frac{v_k}{k} z^{-k} \tag{31}$$

The last equality is to be understood as the Taylor expansion around infinity. The coefficients v_k are moments of the interior domain (the "dual" harmonic moments) defined as

$$v_k = \frac{1}{\pi} \int_D z^k d^2z \quad (k > 0), \qquad v_0 = -\Phi(0) = \frac{2}{\pi} \int_D \log |z| d^2z \tag{32}$$

From (31) it is clear that

$$v_k = \partial_{t_k} F, \quad k \geq 0 \tag{33}$$

In a similar manner, one obtains the integral representation of the tau-function

$$F = -\frac{1}{\pi^2} \int_D \int_D \log |z^{-1} - \zeta^{-1}| d^2z d^2\zeta \tag{34}$$

or

$$F = \frac{1}{2\pi} \int_D \tilde{\Phi}(z) d^2z \tag{35}$$

These formulas remain intact in the multiply-connected case (see below).

3   TOWARDS MULTIPLY-CONNECTED CASE AND GENERALIZED HIROTA EQUATIONS

Now we are going to explain how the above picture can be generalized to the multiply connected case. The details can be found in [17].

Let D_α, $\alpha = 0, 1, \ldots, g$, be a *collection* of $g + 1$ non intersecting bounded connected domains in the complex plane with smooth boundaries ∂D_α. Set $D = \cup_{\alpha=0}^{g} D_\alpha$, so that the complement $D^c = \mathbb{C} \backslash D$ becomes a multiply-connected

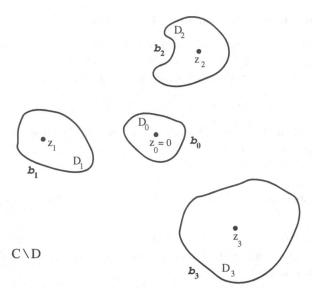

Figure 3. A multiply-connected domain $D^c = C\backslash D$ for $g = 3$. The domain $D = \cup_{\alpha=0}^{3} D_\alpha$ consists of $g + 1 = 4$ disconnected parts D_α with the boundaries b_α. To define the complete set of harmonic moments, we also need the auxiliary points $z_\alpha \in D_\alpha$ which should be always located inside the corresponding domains.

unbounded domain in the complex plane (see Figure 3), b_α being the boundary curves.

It is customary to associate with a planar multiply connected domain its *Schottky double*, a compact Riemann surface without boundary endowed with an antiholomorpic involution, the boundary of the initial domain being the set of fixed points of the involution. The Schottky double of the multiply-connected domain D^c can be thought of as two copies of D^c ("upper" and "lower" sheets of the double) glued along the boundaries $\cup_{\alpha=0}^{g} b_\alpha = \partial D^c$, with points at infinity added (∞ and $\bar{\infty}$). In this set-up the holomorphic coordinate on the upper sheet is z inherited from D^c, while the holomorphic coordinate on the other sheet is $\bar{z}$. The Schottky double of D_c with two infinities added is a compact Riemann surface of genus $g = \#\{D_\alpha\} - 1$.

On the double, one may choose a canonical basis of cycles. The b-cycles are just boundaries of the holes b_α for $\alpha = 1, \ldots, g$. Note that regarded as the oriented boundaries of D^c (not D) they have the *clockwise* orientation. The a_α-cycle connects the α-th hole with the 0-th one. To be more precise, fix points ξ_α on the boundaries, then the a_α cycle starts from ξ_0, goes to ξ_α on the "upper" (holomorphic) sheet of the double and goes back the same way on the "lower" sheet, where the holomorphic coordinate is $\bar{z}$.

3.1 Tau-Function for Algebraic Domains

Comparing to the simply connected case, nothing is changed in posing the standard Dirichlet problem. The definition of the Green function and the formula (1) for the solution of the Dirichlet problem through the Green function are the same too. A difference is in the nature of harmonic functions. Any harmonic function is the real part of an analytic function but in the multiply connected case these analytic funstions are not necessarily single-valued (only their real parts have to be single-valued).

One may still characterize the shape of a multiply connected domain by harmonic moments. However, the set of linearly independent harmonic functions should be extended. The complete basis of harmonic functions in the plane with holes is described in [17].

Here we shall only say a few words about the case which requires the minimal number of additional parameters and minimal modifications of the theory. This is the case of *algebraic domains* (in the sense of [10]), or *quadrature domains* [18, 19], where, roughly speaking, the space of independent harmonic moments is finite-dimensional. For example, one may keep in mind the class of domains with only finite number of non-vanishing moments. It is this class which is directly related to multi-support solutions of matrix models with polynomial potentials. Boundaries of such multiply-connected domains can be explicitly described by algebraic equations [20].

In this case it is enough to incorporate moments with respect to g additional harmonic functions of the form

$$v_\alpha(z) = \log \left| 1 - \frac{z_\alpha}{z} \right|^2, \quad \alpha = 1, \ldots, g$$

where $z_\alpha \in D_\alpha$ are some marked points, one in each hole (see Figure 3). Without loss of generality, it is convenient to put $z_0 = 0$. The "periods" of these functions are: $\oint_{b_\alpha} \partial_n v_\beta(z)|dz| = 4\pi \delta_{\alpha\beta}$. The independent parameters for algebraic domains are:

$$t_0 = \frac{1}{\pi} \int_D d^2z = \frac{\text{Area(D)}}{\pi}$$

$$t_k = -\frac{1}{\pi k} \int_{D^c} z^{-k} d^2z, \quad k \geq 1 \tag{36}$$

$$\phi_\alpha = -\frac{1}{\pi} \int_{D^c} \log \left| 1 - \frac{z_\alpha}{z} \right|^2 d^2z, \quad \alpha = 1, \ldots, g$$

(t_k are complex numbers while t_0 and ϕ_α are real). Instead of ϕ_α it is more convenient to use

$$\Pi_\alpha = \phi_\alpha - 2 \, \text{Re} \sum_{k>0} t_k z_\alpha^k \tag{37}$$

which does not depend on the choice of z_α's. Note that in the case of a finite number of nonvanishing moments the sum is always well defined.

Using the Hadamard formula, one again derives the "exchange relations" (14) which imply the existence of the tau-function and the fundamental relation (15). They have the same form as in the simply connected case. It can be shown that taking derivatives of the tau-function with respect to the additional variables Π_α, one obtains the harmonic measures of boundary components and the period matrix.

The *harmonic measure* $\omega_\alpha(z)$ of the boundary component b_α is the harmonic function in D^c such that it is equal to 1 on b_α and vanishes on the other boundary curves. From the general formula (1) we conclude that

$$\omega_\alpha(z) = -\frac{1}{2\pi} \oint_{b_\alpha} \partial_n G(z, \zeta)|d\zeta|, \quad \alpha = 1, \ldots, g \tag{38}$$

Being harmonic, ω_α can be represented as the real part of a holomorphic function:

$$\omega_\alpha(z) = W_\alpha(z) + \overline{W_\alpha(z)}$$

where $W_\alpha(z)$ are holomorphic multivalued functions in D^c. The differentials dW_α are holomorphic in D^c and purely imaginary on all boundary contours. So they can be extended holomorphically to the lower sheet of the Schottky double as $-d\overline{W_\alpha(z)}$. In fact this is the canonically normalized basis of holomorphic differentials on the double. Indeed, according to the definitions,

$$\oint_{a_\alpha} dW_\beta = 2\mathrm{Re} \int_{\xi_0}^{\xi_\alpha} dW_\beta(z) = \omega_\beta(\xi_\alpha) - \omega_\beta(\xi_0) = \delta_{\alpha\beta}$$

Then the matrix of b-periods of these differentials reads

$$T_{\alpha\beta} = \oint_{b_\alpha} dW_\beta = -\frac{i}{2} \oint_{b_\alpha} \partial_n \omega_\beta dl = i\pi \Omega_{\alpha\beta} \tag{39}$$

The period matrix $T_{\alpha\beta}$ is purely imaginary non degenerate matrix with positively definite imaginary part. In addition to (15), the following relations hold:

$$\omega_\alpha(z) = -\partial_\alpha \nabla(z) F \tag{40}$$

and

$$T_{\alpha\beta} = 2\pi i \partial_\alpha \partial_\beta F \tag{41}$$

where $\partial_\alpha := \partial/\partial\Pi_\alpha$.

In the multiply-connected case, the suitable analog of the conformal map $w(z)$ (or rather of $\log w(z)$) is the embedding of D^c into the g-dimensional complex torus **Jac**, the Jacobi variety of the Schottky double. This embedding is given, up to an overall shift in **Jac**, by the Abel map $z \mapsto \mathbf{W}(z) := (W_1(z), \ldots, W_g(z))$ where

$$W_\alpha(z) = \int_{\xi_0}^z dW_\alpha \tag{42}$$

is the holomorphic part of the harmonic measure ω_α. By virtue of (40), the Abel map is represented through the second order derivatives of the function F:

$$W_\alpha(z) - W_\alpha(\infty) = \int_\infty^z dW_\alpha = -\partial_\alpha D(z)F \tag{43}$$

$$2\mathrm{Re}\, W_\alpha(\infty) = \omega_\alpha(\infty) = -\partial_{t_0} \partial_\alpha F \tag{44}$$

The last formula immediately follows from (40).

3.2 Green Function and Generalized Hirota Equations

The Green function of the Dirichlet boundary problem in the multiply connected case, can be written in terms of the prime form (see [21] for the definition and properties) on the Schottky double (cf. (2)):

$$G(z, \zeta) = \log \left| \frac{E(z, \zeta)}{E(z, \bar\zeta)} \right| \tag{45}$$

Here by $\bar\zeta$ we mean the (holomorphic) coordinate of the "mirror" point on the Schottky double, i.e. the "mirror" of ζ under the antiholomorphic involution. Using (45) together with (15), (40) and (41) one can obtain the following representations of the prime form in terms of the tau-function

$$E(z, \zeta) = (z^{-1} - \zeta^{-1})e^{-\frac{1}{2}(D(z) - D(\zeta))^2 F}$$

$$i E(z, \bar\zeta) = e^{-\frac{1}{2}(\partial_{t_0} + D(z) + \bar D(\bar\zeta))^2 F} \tag{46}$$

$$i E(z, \bar z) = e^{-\frac{1}{2}\nabla^2(z)F}$$

generalizing (23), (24) in the simply-connected case.

This allows us to write the generalized Hirota equations for F in the multiply-connected case. They follow from the Fay identities [21] and (46). In analogy to the simply-conected case, any second order derivative of the function F w.r.t. t_k (and $\bar t_k$), F_{ik}, is expressed through the derivatives $\{F_{\alpha\beta}\}$ where $\alpha, \beta = 0, \ldots, g$ together with $\{F_{\alpha t_i}\}$ and their complex conjugated. To be more precise, one can consider all second derivatives as functions of $\{F_{\alpha\beta}, F_{\alpha k}\}$ modulo certain relations on the latter discussed in [17]; sometimes on this "small phase space" more extra constraints arise, which can be written in the form similar to the Hirota or WDVV equations [22].

For the detailed discussion of the generalized Hirota relations the reader is addressed to [17]. Here we just give the simplest example of such relations, an analog of the dispesrionless Toda equation (2.25) for the tau-function. It reads

$$\partial_{t_1}\partial_{\bar{t}_1}F = \frac{\theta(\omega(\infty)+\mathbf{Z})\theta(\omega(\infty)-\mathbf{Z})}{\theta^2(\mathbf{Z})}e^{\partial_{t_0}^2 F}$$

$$+ \sum_{\alpha,\beta=1}^{g}(\log\theta(\mathbf{Z})),_{\alpha\beta}(\partial_\alpha\partial_{t_1}F)(\partial_\beta\partial_{\bar{t}_1}F) \qquad (47)$$

Here θ is the Riemann theta-function with the period matrix $T_{\alpha\beta}$ and

$$(\log\theta(\mathbf{Z})),_{\alpha\beta} := \partial^2\log\theta(\mathbf{Z})/\partial Z_\alpha\partial Z_\beta$$

The equation holds for any vector-valued parameter $\mathbf{Z}\in\mathbf{Jac}$. It is important to note that the theta-functions are expressed through the second order derivatives of F, so (47) is indeed a partial differential equation for F. For example,

$$\theta(\omega(\infty)) = \sum_{n_\alpha\in\mathbf{Z}}\exp\left(-2\pi^2\sum_{\alpha\beta}n_\alpha n_\beta\partial_{\alpha\beta}^2 F - 2\pi i\sum_\alpha n_\alpha\partial_\alpha\partial_{t_0}F\right)$$

4   CONCLUSION

In these notes we have reviewed the integrable structure of the Dirichlet boundary problem. We have presented the simplest known to us proof of the Hadamard variational formula and derivation of the dispersionless Hirota equations for the simply-connected case.

We have also demonstrated how this approach can be generalized to the case of multiply-connected domains. The main ingredients remain intact, but the conformal map to the reference domain should be substituted by the Abel map into Jacobian of the Schottky double of the multiply-connected domain. Then one can write the generalization of the Hirota equations using the Fay identities, a particular case of which leads to generalization of the dispersionless Toda equation.

Here we have only briefly commented on the properties of the quasiclassical tau-function of the multiply-connected solution. A detailed discussion of this issue and many related problems, including conformal maps in the multiply-connected case, duality transformations on the Schottky double, relation to the multi-support solutions of the matrix models etc, can be found in [17].

ACKNOWLEDGMENTS

We are indebted to I.Krichever and P.Wiegmann for collaboration on different stages of this work and to V.Kazakov, A.Levin, M.Mineev–Weinstein,

S.Natanzon, and L.Takhtajan for illuminating discussions. The work was partialy supported by RFBR under the grant 00-02-16477, by INTAS under the grant 99-0590 and by the Program of support of scientific schools under the grants 1578.2003.2 (A.M.), 1999.2003.2 (A.Z.). The work of A.Z. was also partially supported by the LDRD project 20020006ER "Unstable Fluid/Fluid Interfaces". We are indebted to P. van Moerbeke for encouraging us to write this contribution and to F. Lambert for the warm hospitality on Elba during the conference.

REFERENCES

1. Hurwitz, A. and Courant, R. (1964) *Vorlesungen über allgemeine Funktionentheorie und elliptische Funktionen. Herausgegeben und ergänzt durch einen Abschnitt über geometrische Funktionentheorie*, Springer-Verlag, (Russian translation, adapted by M. A. Evgrafov: *Theory of functions*, Nauka, Moscow, 1968).

2. Marshakov, A., Wiegmann, P., and Zabrodin, A. (2002) *Commun. Math. Phys.* **227**, p. 131, e-print archive: hep-th/0109048.

3. Krichever, I. M. (1989) *Funct. Anal Appl.* **22**, pp. 200–213.

4. Krichever, I. M. (1994) *Commun. Pure. Appl. Math.* **47**, p. 437, e-print archive: hep-th/9205110.

5. Mineev-Weinstein, M., Wiegmann, P. B., and Zabrodin, A. (2000) *Phys. Rev. Lett.* **84**, p. 5106, e-print archive: nlin.SI/0001007.

6. Wiegmann, P. B. and Zabrodin, A. (2000) *Commun. Math. Phys.* **213**, p. 523, e-print archive: hep-th/9909147.

7. Boyarsky, A., Marshakov, A., Ruchayskiy, O., Wiegmann, P., and Zabrodin, A. (2001) *Phys. Lett.* **B515**, pp. 483–492, e-print archive: hep-th/0105260.

8. Witten, E. (1990) *Nucl. Phys.* **B340**, p. 281. Dijkgraaf, R., Verlinde, H. and Verlinde, E. (1991) *Nucl. Phys.* **B352** p. 59.

9. Hadamard, J. (1908) *Mém. présentés par divers savants à l'Acad. sci.*, **33**.

10. Etingof, P. and Varchenko, A. (1992) Why does the boundary of a round drop becomes a curve of order four, *University Lecture Series* 3, Am. Math. Soc., Providence, RI.

11. Krichever, I. (2000) unpublished.

12. Takhtajan, L. (2001) *Lett. Math. Phys.* **56**, pp. 181–228, e-print archive: math.QA/0102164.

13. Kostov, I. K., Krichever, I. M., Mineev-Weinstein, M., Wiegmann, P. B., and Zabrodin, A. (2001) τ-function for analytic curves, in: *Random Matrices and Their Applications*, MSRI publications, Vol. 40, Cambridge Academic Press, Cambridge, e-print archive: hep-th/0005259.

14. Hille, E. (1962) *Analytic function theory*, Vol. II, Ginn and Company.

15. Gibbons, J. and Kodama, Y. (1994) Singular Limits of Dispersive Waves, *Proceedings of NATO ASI* ed. N. Ercolani, London—New York, Plenum; Carroll, R. and Kodama, Y. J. (1995) *J. Phys. A: Math. Gen.* **A28**, p. 6373.

16. Takasaki, K. and Takebe, T. (1995) *Rev. Math. Phys.* **7**, pp. 743–808.
17. Krichever, I., Marshakov, A., and Zabrodin, A. *Integrable Structure of the Dirichlet Boundary Problem in Multiply-Connected Domains*, preprint MPIM-2003–42, ITEP/TH-24/03, FIAN/TD-09/03.
18. Gustafsson, B. (1983) Acta Appl. Math. **1**, pp. 209–240.
19. Aharonov, D. and Shapiro, H. (1976) *J. Anal. Math.* **30**, pp. 39–73
20. Kazakov, V. and Marshakov, A. (2003) *J. Phys. A: Math. Gen.* **36**, pp. 3107–3136, e-print archive: hep-th/0211236.
21. Fay, J. D. (1973) Theta Functions on Riemann Surfaces, Lecture Notes in Mathematics **352**, Springer-Verlag, New York.
22. Marshakov, A., Mironov, A., and Morozov, A. (1996) *Phys. Lett.* **B389**, pp. 43–52, e-print archive: hep-th/9607109.
23. Braden, H. and Marshakov, A. (2002) *Phys. Lett.* **B541**, pp. 376–383, e-print archive: hep-th/0205308.

FUNCTIONAL-DIFFERENCE DEFORMATIONS OF DARBOUX-PÖSHL-TELLER POTENTIALS

Vladimir B. Matveev[1,2,*]
[1]*Université de Bourgogne, Laboratoire Gevrey de Mathématique Physique*
[2]*St-Petersbourg branch of Steklov Mathematical Institut, Fontanka 27,*
127011, St-Petersburg, Russia

Abstract   We consider the functional-difference deformation of the Schrödinger equation. The main goal of this article is to construct some integrable potentials representing a natural difference deformation of the so called two parametric Darboux-Pöshl-Teller model and to describe explicitly the solutions of the related difference Schrödinger equation. In the limit when the difference step tends to zero the related formulas reproduce well known results concerning the Schrödinger operator with DPT potential. We also describe the solutions of the difference KdV (DKdV) equation with the nonsingular "difference DPT" initial data.

1 INTRODUCTION

In 1882 Darboux [1] proved the integrability of the following Sturm–Liouville equation[1]

$$-y'' + \left(\frac{n(n+1)q^2}{\cos^2 qx} + \frac{m(m+1)q^2}{\sin^2 qx} \right) y = \lambda y \qquad (1)$$

Later in 1933 in a frame of study of the quantum theory of the two atomic molecules Pöshl and Teller, (the creator of the US hydrogenous bomb and neutron bomb), rediscovered [3] the hyperbolic version of (1) and proved

---

* The author wishes to thank Max-Planck-Institut fur Mathematik in Bonn, where this work was mainly written, for hospitality and financial support and the organizers of the NATO ARW Workshop "Bilininear Integral Systems: From Classical to Quantum, From Discrete to Continuous" where this work was reported the first time.
[1] Darboux also considered and solved the equation which is an elliptic function generalization of (1):see [2] and the concluding remarks.

L. Faddeev et al. (eds.),
Bilinear Integrable Systems: From Classical to Quantum, Continuous to Discrete, 191–208.
© 2006 *Springer. Printed in the Netherlands.*

independently from Darboux the integrabilty of the equation obtained from
(1) by transformation $q \to iq$. Darboux proposed two independent methods
for solving (1). First, taking $\sin^2 qx$ as a new independent variable he reduced
(1) to Gauss hypergeometric equation and explicitely expressed the solutions
in terms of the hypergeometric functions. Next, exploring the fact that the
functions $y(m, n) := \cos^{n+1} qx \sin^{m+1} qx$ represent the particular solutions of
(1) with $\lambda = q^2(m + n + 2)^2$, namely taking them as generating functions of
some sequence of Darboux transformations he obtained the global solution
of the same equation in terms of the elementary functions. Comparison of
these two solutions provides the nontrivial case of the reduction of the Gauss
hypergeometrical function to the elementary functions. The third representation for the solutions of the same equation in terms of Wronskian determinants
was recently obtained in [4].

In this article[2] we consider the following functional-difference deformation
of the Schrödinger equation

$$f(x + h)v(x, h) + f(x - h) = 2\cosh(ikh) \cdot f(x), \tag{2}$$

where h is a non negative parameter. We can obviously rewrite (2) in a following
form:

$$h^{-2}[v(x, h)f(x + h) + f(x - h) - 2f(x)] = 2h^{-2}(\cosh(ikh) - 1)f(x). \tag{3}$$

Assume now that, when $h \to 0$ the related potential v has the following
asymptotics:

$$v(x, h) = 1 + h^2 g(x) + O(h^3). \tag{4}$$

Under this assumption, when $h \to 0$, the limit of (3) is exactly the Schrödinger
equation with h-independent potential $g(x)$,

$$f'' + g(x)f(x) = -k^2 f(x) \tag{5}$$

In the sequel we use the following abbreviations

$$c(x) := \cosh qx \quad s(x) := \sinh qx$$

Below we consider special class $v_{nm}(x, h, q)$ of the potentials satisfying (4):

$$v_{nm}(x) : c = \frac{c(x + (n + 1)h)c(x - nh)s(x + (m + 1)h)s(x - mh)}{c(x)c(x + h)s(x)s(x + h)} \tag{6}$$

$$= v_{n0}(x)v_{0m}(x) \tag{7}$$

---

[2] Preliminary version of this article first appeared in [5]. Here we improved some typing errors
in [5] and added the references on some later relevant works.

We see that in (6) the additive structure of the original DPT potentials is replaced by the multiplicative structure.

For any C^3 function $l(x)$ a short calculation proves the following asymptotic estimate

$$\frac{l(x + (n + 1)h)l(x - nh)}{l(x)l(x + h)} = 1 + h^2 n(n + 1)\frac{d^2}{dx^2} \log l(x) + 0(h^3).$$

This makes obvious that the potential $v_{nm}(x)$, defined above, when $h \to 0$, has the following asymtotics

$$v_{nm}(x) = 1 + h^2 \left(\frac{n(n + 1)q^2}{\cosh^2 qx} - \frac{m(m + 1)q^2}{\sinh^2 qx} \right) + 0(h^3) \tag{8}$$

Therefore the limit of (3) with $v = v_{nm}(x)$ is just a Schrödinger equation with a hyperbolic Darboux-Pöshl-Teller potential

$$-y''0 + \left(-\frac{n(n + 1)q^2}{\cosh^2 qx} + \frac{m(m + 1)q^2}{\sinh^2 qx} \right) y = k^2 y. \tag{9}$$

As their continuous limits the potentials $v_{nm}(x, h, q)$ are invariant with respect to the transformations $m \to -m - 1$, and $n \to -n - 1$.

It is enough to replace $c(x)$ and $s(x)$ by $\cos qx$ and $\sin qx$ in order to recover from (6) the original trigonometric Darboux potentials (1) in the limit $h \to 0$. Below we show that

$$f(x + h)v_{nm}(x, h) + f(x - h) = 2 \cosh ikh f(x), \tag{10}$$

is an integrable equation and its global solution can be expressed by means of the elementary functions. In the limit $h \to 0$ all known results, concerning the Schrödinger equation, are easily recovered. The cases $n = 0$, or $m = 0$ were studied in the recent papers [6–9] using the different tools and leading to the formulas different from ours having more complicated combinatorial structure. The Lattice specialization of the case $m = 0$, corresponding to the choice $h = 1, x = j, j \in Z$ was studied in [8, 9]. In these articles various interesting connections with q-ultra spherical polynomials, Askey-Wilson polynomials and q-spherical functions of the $m = 0$ case were discovered. Similar connections exist also for general $v_{nm}(x)$ potentials. We expect to discuss them elsewhere.

The article is organized as follows. In the second section we collected a few results concerning the functional-difference version of the Darboux dressing applicable to any equation of the form (2) as well as for the closely related functional difference KdV equation and the related hierarchies. The related results represent a very special case of much more general statements proved in [10]. They can also be considered as a natural functional difference extrapolation of the results of [11], p. 84 concerning the lattice KdV equation. Therefore, in this part all the proofs are omitted. The third section contains the construction of the solutions of (10) with potential $v_{nm}(x)$ defined in (6), based on a proper generalization of the Darboux approach for the Schrödinger equation, combined with the results of the second section. The special solutions corresponding to the products $c^{m+1}(x)s^{n+1}(x)$, used in original Darboux construction [1], are replaced by the appropriate products of the shifts of the same functions c and s, playing the same role in our case. In particular, this leads to especially simple formulas for the discrete eigenvalues and bound states eigenfunctions in the case of v_{n0} potentials. In the last section we explain how to get the same potentials and their eigenfunctions from Casorati determinants by appropriate reduction of the general determinant Darboux dressing formulae. This leads immediately to the solution of the difference KdV equation corresponding to $v_{n0}(x)$ taken as initial data.

In the concluding remarks we mention some possible developments and extensions of the results of this article.

2   FUNCTIONAL DIFFERENCE SCHRÖDINGER AND KdV EQUATIONS AND DARBOUX DRESSING

The functional-difference KdV equation reads

$$\dot{v}(x,t) = v(x)[v(x-h) - v(x+h)]. \tag{11}$$

Its Lax representation can be written as a compatibility condition of the difference Schrödinger equation:

$$v(x)f(x+h,t) + f(x-h,t) = \lambda f(x,t), \tag{12}$$

and of the following evolution equation:

$$\dot{f}(x,t) = -v(x,t)v(x+h,t)f(x+2h,t) \tag{13}$$

This Lax pair is a natural functional interpolation of the lattice Lax pair used in [11], p. 84. For the trivial solution (potential) $v = 1$, two important real solutions of (12) with $\lambda = 2\cosh\beta h$ are

$$f = \cosh(\beta x + \eta), \quad f = \sinh(\beta x + \eta),$$

where η and β are two real parameters. They can be easily extended to the solutions of the system (12)–(13):

$$f = e^{-t\cosh(2\beta h)}\cosh(\beta x - t\sinh(2\beta h) + \eta),$$
$$f = e^{-t\cosh(2\beta h)}\sinh(\beta x - t\sinh(2\beta h) + \eta)$$

The space of the solutions of (12) is infinite dimensional since for any h-periodic function g (i.e., $g(x + h) = g(h)$), $g(x)f(x)$ is again the solution of (12). The same is true for (13) under the assumption that $g(x)$ is a t-independent h periodic function of x.

Multiplying $f(x)$ by any h-antiperiodic function $p(x)$, (i.e., $p(x \pm h) = -p(x)$), for instance taking $p(x) = e^{\pm \frac{i\pi}{h}}$, we see that the product function $\hat{f}(x) := f(x)p(x)$ is the solution of the equation which differs from (12) only by the different sign of λ:

$$v(x)\hat{f}(x + h) + \hat{f}(x - h) = -\lambda \hat{f}(x), \ \hat{f}(x) = p(x)f(x). \qquad (14)$$

Therefore, the spectrum of the difference Schrödinger equation is symmetric for any given $v(x)$.

The difference KdV equation (11) admits an infinite dimensional space of the stationary solutions which are t-independent $2h$-periodic functions of x, which makes again certain difference with its lattice reduction, $x = j \in Z, h = 1$, considered for instance in [11], p. 84.

2.1 Darboux Dressing

Darboux transform of an arbitrary solution $f(x)$ of (12) generated by the fixed solution $f_1(x)$ of the same system with $\lambda = \lambda_1$, is defined by the formula

$$\psi_1(x) = f(x - 2h) - \sigma(x)f(x) \qquad (15)$$

$$= \frac{\begin{vmatrix} f(x - 2h) & f_1(x - 2h) \\ f(x) & f_1(x) \end{vmatrix}}{f_1(x)}, \qquad (16)$$

$$\sigma_1(x) = \frac{f_1(x - 2h)}{f_1(x)}, \qquad (17)$$

where $f_1(x)$ is a fixed solution of (12) with $\lambda = \lambda_1$.

Proposition 1 (Darboux covariance property) *Darboux transform of $f(x)$ represents a general solution of the following equation*

$$v_1(x)\psi_1(x + h) + \psi_1(x - h) = \lambda \psi_1(x), \qquad (18)$$

$$v_1(x) = v(x)\frac{\sigma_1(x)}{\sigma_1(x + h)} = \frac{f_1(x - 2h)f_1(x + h)}{f_1(x)f_1(x - h)}. \qquad (19)$$

Providing that $f_1(x)$ is also a solution of (13) and $v(x)$ is a solution of (11), (hence both depend also on t), $v_1(x, t)$ described by the formula (19) is a new solution of DKdV equation.

In other words, Darboux transform maps (12) into equation of the same form, with the same value of spectral parameter λ, but with a new potential constructed in terms of the initial potential $v(x)$ and a fixed solution $f_1(x)$ of (12). The same transformation maps the given solution of the DKdV equation into a new solution which differs from the old one by the factor explicitly constructed in terms of the fixed solution of the Lax system (12)–(13). One step Darboux dressing of the trivial solution $v = 1$ with the generating function

$$f_1(x, \lambda_1) = e^{-t \cosh(2\beta h)} \cosh(\beta x - t \sinh(2\beta h) + \eta),$$

with $\lambda_1 = 2 \cosh bh$, obviously produces the smooth 1-soliton of the DKdV equation:

$$v = \frac{\cosh(\beta(x - 2h) - t \sinh(2\beta h) + \eta) \cdot \cosh(\beta(x + h) - t \sinh(2\beta h) + \eta)}{\cosh(\beta x - t \sinh(2\beta h) + \eta) \cdot \cosh(\beta(x - h) - t \sinh(2\beta h) + \eta)},$$

representing a wave propagating with constant velocity $\frac{\sinh 2\beta h}{\beta}$.

2.2 Iterated Darboux Transform

In a sequel the notation $\Delta_n(x) = \Delta_n[f_1(x), f_2(x), \ldots, f_n(x)]$ will be used for the following Casorati determinant:

$$\Delta_n(x) = \begin{vmatrix} f_1(x - 2(n-1)h) & f_2(x - 2(n-1)h) & \cdots & f_n(x - 2(n-1)h) \\ f_1(x - 2(n-2)h) & f_2(x - 2(n-2)h) & \cdots & f_n(x - 2(n-2)h) \\ \vdots & \vdots & \vdots & \vdots \\ f_1(x) & f_2(x) & \cdots & f_n(x) \end{vmatrix}$$

We also use below the following notations

$$\psi_n(x) = \frac{\Delta_{n+1}[f(x), f_1(x), f_2(x), \ldots, f_n(x)]}{\Delta_n(x)}, \tag{20}$$

$$F_j(x) = \psi_{j-1}(x)|_{f(x)=f_j(x)}, \quad j = 1, 2, \ldots, n. \tag{21}$$

$$\sigma_j(x) = \frac{F_j(x - 2h)}{F_j(x)}, \tag{22}$$

$$v(x) f_j(x + h) + f_j(x - h) = \lambda_j f_j(x), \tag{23}$$

$$v(x) f(x + h) + f(x - h) = \lambda f(x). \tag{24}$$

Proposition 2 *The function ψ_n represents the general solution of the functional difference equation*

$$v_n(x) \psi_n(x + h) + \psi_n(x - h) = \lambda \psi_n(x), \tag{25}$$

$$v_n(x) = v(x) \frac{\Delta_n(x - 2h) \Delta_n(x + h)}{\Delta_n(x) \Delta_n(x - h)} \tag{26}$$

Providing that $v(x, t)$ *is a solution of* (11), *and* $f_j(x)$ *are also the solutions of* (12)–(13), *the RHS of* (26) *is again the solution of DKdV equation. In particular, setting* $v(x, t) = 1$ *and*

$$f_{2j+1}(x, t) = e^{-t \cosh(2\beta_{2j+1}h)} \cosh(\beta_{2j+1}x - t \sinh(2\beta_{2j+1}h) + \eta_{2j+1}),$$
$$f_{2j} = e^{-t \cosh(2\beta_{2j}h)} \sinh(\beta_{2j}x - t \sinh(2\beta_{2j}h) + \eta_{2j}),$$
$$\eta_j \in R, 0 < \beta_1 < \cdots < \beta_n, \tag{27}$$

we obtain from (26) *the real nonsingular multi solitons solutions of the DKdV equation. While substituting* (27) *into* (26) *the exponential factors in the RHS of* (27) *can be omitted.*

The function $\psi_n(x)$ can be also represented in a following factorized form

$$\psi_n(x) = (T^{-2} - \sigma_n(x)) \dots (T^{-2} - \sigma_1(x)) f(x, \lambda), \tag{28}$$

where T is the shift operator : $T^{\pm k} f(x) = f(x \pm kh)$.

3 DIFFERENCE DPT POTENTIALS AND THE RELATED SOLUTIONS OF THE DIFFERENCE SCHRÖDINGER EQUATION

3.1 Shifted Darboux Dressing

The above version of the Darboux dressing exposed in Section 2 is convienient in a sense that it shows how to construct the new integrable potential (or the new solution of DKdV equation) from the given one, keeping the later as a stable factor entering into the new potential, (or a new solution of DKdV equation). Somehow in the sequel it will be more convienient to use the same formulas corresponding to the result of the n-fold Darboux transform considered at the reference point $x + nh$ rather than at the point x. It is clear from above that the function $\phi_1(x)$ defined by the formula

$$\phi_1(x) = f(x - h) - \kappa_1(x) f(x + h) = \frac{\begin{vmatrix} f(x - h) & f_1(x - h) \\ f(x + h) & f_1(x + h) \end{vmatrix}}{f_1(x + h)} \tag{29}$$

$$\kappa_1(x) = \frac{f_1(x - h)}{f_1(x + h)}, \tag{30}$$

satisfies the equation of the DS type with potential

$$v_1(x) = v_1(x + h) = v(x + h)\frac{\kappa_1(x)}{\kappa_1(x + h)} = v(x + h)\frac{f_1(x - h)f_1(x + 2h)}{f_1(x + h)f_1(x)}. \tag{31}$$

As before we will call the mapping

$$D_{f_1} : f(x) \to \phi_1(x) \tag{32}$$

Darboux transformation of the solution $f(x, \lambda)$ generated by $f_1(x, \lambda_1)$. It transforms the solution of the DS equation with potential $v(x)$ to the solutions of the DS equation with potential $u_1(x)$ with the same value of the spectral parameter λ.

In the sequel the notation $\delta_n(x) = \delta_n[f_1, f_2, \ldots, f_n](x)$ will be used for the following Casorati determinant:

$$\delta_n(x) = \Delta_n(x + (n-1)h)$$

$$= \begin{vmatrix} f_1(x - (n-1)h) & f_2(x - (n-1)h) & \ldots & f_n(x - (n-1)h) \\ f_1(x - (n-3)h) & f_2(x - (n-3)h) & \ldots & f_n(x - (n-3)h) \\ \vdots & \vdots & \vdots & \vdots \\ f_1(x + (n-1)h) & f_2(x + (n-1)h) & \ldots & f_n(x + (n-1)h) \end{vmatrix}.$$

We also use below the following notations

$$\phi_n(x) = \frac{\delta_{n+1}[f, f_1, f_2, \ldots, f_n](x)}{\delta_n(x + h)}, \tag{33}$$

$$\Phi_j(x) = \phi_{j-1}(x)|_{f(x)=f_j(x)}, \quad j = 1, 2, \ldots, n. \tag{34}$$

$$\kappa_j(x) = \frac{\Phi_j(x - h)}{\Phi_j(x + h)}, \tag{35}$$

$$v(x)f_j(x + h) + f_j(x - h) = \lambda_j f_j(x), \tag{36}$$

$$v(x)f(x + h) + f(x - h) = \lambda f(x). \tag{37}$$

Proposition 2a *The function ϕ_n represents the general solution of the functional difference equation*

$$v_n(x)\phi_n(x + h) + \phi_n(x - h) = \lambda \phi_n(x), \tag{38}$$

$$v_n(x) = v(x + nh)\frac{\delta_n(x - h)\delta_n(x + 2h)}{\delta_n(x)\delta_n(x + h)} \tag{39}$$

Providing that $v(x, t)$ is a solution of (11), that $f_j(x)$ are also the solutions of (13) the RHS of (39) is again the solution of DKdV equation.

The function $\phi_n(x)$ can be also represented in a following factorized form

$$\phi_n(x) = (T^{-2} - \kappa_n(x)) \ldots (T^{-2} - \kappa_1(x))f(x, \lambda), \tag{40}$$

where T is the shift operator: $T^{\pm \kappa} f(x) = f(x \pm kh)$.

3.2   Potentials $v_{nm}(x)$ as the Result of the Multiple Darboux Transform

Here we obtain the main result of this article–global solution of the difference Schrödinger equation

$$v_{mn}(x)\psi(x+h) + \psi(x-h) = \lambda\psi(x), \tag{41}$$

where $v_{nm}(x)$, is defined by (6) and n, m are two nonnegative integers. We suppose also that $n \geq m$. The case $n < m$ can be treated in a same way without any difficulties and we less this case as an easy exercise for the reader.

First we prove the following statement.

Proposition 3   *The following* 2 *functions*

$$F_1(x, n) := \prod_{k=0}^{n} c(x - kh), \quad n \geq 0 \tag{42}$$

$$F_2(x, n) := \left(\prod_{k=1}^{n} c(x + kh)\right)^{-1}, \quad n \geq 1 \tag{43}$$

are the solutions of (41) *with* $m = 0$, $\lambda = \lambda_j$, $j = 1, 2$

$$\lambda_1 := 2\cosh qh(n + 1), \quad \lambda_2 := 2\cosh qhn,$$

respectively.

Proof   Substituting $\psi(x) = F_j(x, n)$ in (41) and removing the commune factors in the LHS and in the RHS of the obtained relation, we reduce the proof to checking the following 2 identities:

$$c(x + (n + 1)h) + c(x - (n + 1)h)) = 2\cosh qh(n + 1)c(x),$$
$$c(x - nh) + c(x + nh) = 2\cosh qhn\, c(x)$$

Replacing n by $n - 1$ we transform the first identity to the second one, which can be trivially checked. This completes the proof.   □

Remark   Replacing $c(x)$ by $s(x)$ in the proposition proved above we obtain the similar statement concerning the potential v_{0n}.

Proposition 4   *The following* 4 *functions*

$$F_1(x, n, m) = \Pi_{k=0}^{n}c(x - kh)\Pi_{j=0}^{m}s(x - jh),$$
$$F_2(x, n, m) = \left(\Pi_{k=1}^{n}c(x + kh)\Pi_{j=1}^{m}s(x + jh)\right)^{-1}, \quad m, n \geq 1,$$
$$F_3(x, n, m) = \Pi_{k=0}^{m}s(x - kh)\left(\Pi_{j=1}^{n}c(x + jh)\right)^{-1}, \quad n \geq 1,$$
$$F_4(x, n, m) = \Pi_{k=0}^{n}c(x - kh)\left(\Pi_{j=1}^{m}s(x + jh)\right)^{-1}, \quad m \geq 1,$$

are the solutions of (41) *with* $\lambda = \lambda_j$, $j = 1, 2, 3, 4$

$$\lambda_1 := 2\cosh qh(n + m + 2), \quad \lambda_2 := 2\cosh qh(n + m),$$
$$\lambda_3 := 2\cosh qh(n - m - 1), \quad \lambda_4 := 2\cosh qh(m - n - 1),$$

respectively.

Proof Substituting $\psi(x) = F_j(x, n, m)$ in (41) and removing the commune factors in the LHS and RHS of the obtained relation, we reduce the proof to checking the following 4 identities:

$$c(x + (n + 1)h)s(x + (m + 1)h) + c(x - (n + 1)h))s(x - (m + 1)h)$$
$$= \lambda_1 c(x)s(x),$$
$$c(x - nh)s(x - mh) + c(x + nh)s(x + mh) = \lambda_2 c(x)s(x)$$
$$c(x - nh)s(x + (m + 1)h) + c(x + nh)(s(x - (m + 1)h) = \lambda_3 c(x)s(x),$$
$$s(x - mh)c(x + (n + 1)h) + s(x + mh)(c(x - (n + 1)h) = \lambda_4 c(x)s(x),$$

Replacing n by $n - 1$ and m by $m - 1$ we transform the first identity to the second one, which can be checked by the direct calculation. The proof of the third and of the fourth identities is also straightforward and completes the proof of Proposition 3. $\qquad\qquad\qquad\qquad\qquad\qquad\qquad\qquad\qquad\qquad\qquad\qquad\quad\Box$

The special solutions listed above become particularly simple in the cases $m = 0$ or $n = 0$, corresponding to the potentials $v_{n0}(x)$ or $v_{0m}(x)$: for these cases they are different from the solutions given by Proposition 2 and they correspond to the different values of spectral parameter λ.

Remark Formulas similar to those of Propositions 3 and 4 can be obtained for the trigonometric version of the potentials v_{nm} replacing cosh by cos in the expressions for λ_j and replacing $c(x)$, $s(x)$ by $\cos qx$ and $\sin qx$ respectively, which corresponds to replace q by iq in the formulas listed above.

Now we are in a position to show that the potentials $v_{nm}(x)$ can be obtained from the trivial starting potential $v(x) = 1$ by the action of the n-fold Darboux dressing assuming that $m \leq n$, (otherwise we should use an appropriately chosen m-fold Darboux dressing). This can be done in two steps. First we perform the (n-m) fold consecutive Darboux dressing using the functions $F_1(x, k, 0)$, $k = 0, \ldots, n - m - 1$, as the generating functions of the consecutive single Darboux transforms, thus generating the potential $v_{(n-m)0}(x)$ together with the general solution of the related DS equation. Next, using $F_1(x, n - m + j, j)$, $j = 0, \ldots, m - 1$ as a generating functions of the consecutive Darboux transforms we obtain the potential $v_{nm}(x)$ together with the general solution of the related Schrödinger equation. Due to the Darboux-covariance statement formulated above we only have to check that the following proposition holds.

Proposition 5   *The potential $v_{j0}(x)$ is mapped to $v_{j+1,0}(x)$ by the Darboux transform (32) with the generating function $F_1(x, j)$ and $v_{jk}(x)$ is mapped to $v_{j+1,k+1}(x)$ by the DT with generating function $F_1(x, j, k)$.*

Proof   The proof obviously reduces to check two identities:

$$v_{j+1,0}(x) = v_{j,0}(x + h)\frac{F_1(x - h, j)F_1(x + 2h, j)}{F_1(x + h, j)F_1(x, j)},$$

$$v_{j+1,k+1}(x) = v_{j,k}(x + h)\frac{F_1(x - h, j, k)F_1(x + 2h, j, k)}{F_1(x + h, j, k)F_1(x, j, k)}, \qquad (44)$$

following immediately from the definition of the potentials $v_{jk}(x)$ and the functions $F_1(x, j)$, $F_1(x, j, k)$ after removing the commune factors in nominators and denominators of the RHS of (44).   □

Now the general solution of (41) in the case $n \geq m$ is given by the formula

$$\phi_n(x) = (T^{-2} - \kappa_n(x)) \ldots (T^{-2} - \kappa_1(x)) f(x, \lambda) \qquad (45)$$

where $\kappa_k(x)$ is defined by the formula

$$\kappa_k(x) = \frac{F_1(x - h, k)}{F_1(x + h, k)}, k = 0, \ldots, n - m - 1, \qquad (46)$$

$$\kappa_k(x) = \frac{F_1(x - h, n - m + k, k)}{F_1(x + h, n - m + k, k)}, k = n - m, \ldots, m - 1. \qquad (47)$$

The special solutions of the form $F_2(x, j)$, $F_3(x, j, k)$ represent smooth bound state eigenfunctions exponentially decreasing at infinity. They can be used to construct the new bound states for the same potentials.

3.3 Bound States Eigenfunctions for Difference Schrödinger Equation with Potential $v_{n0}(x)$

From the results derived above we can immediately deduce the following statement

Proposition 6   *Operator $v_{n0}(x)T + T^{-1}$ has $2n$ discrete eigenvalues:*

$$\pm\lambda_j, \lambda_j = 2\cosh qhj, j = 1, \ldots, n.$$

The eigenfunctions $\psi_{n,j}$ corresponding to the eigenvalues λ_{n-j} are given by the formulas

$$\psi_{n,j} = D_{F_1(x,n-1)} \cdot D_{F_1(x,n-2)} \ldots D_{F_1(x,n-j+1)} F_3(x, n - j, 0),$$
$$j = 2, \ldots, n - 1,$$

where the operations $D_{F_1(x,j)}$ are defined as in subsection 3.1.

$$\psi_{n,0}(x) = F_2(x, n), \quad \psi_{n,1}(x) = F_3(x, n, 0).$$

The infinite dimensional subspace of eigenfunctions corresponding to the same eigenvalue is obtained by multiplying $\psi_{n,j}$ by any h periodic function of x. The eigenfunctions corresponding to the eigenvalues $-\lambda_j$ are obviously obtained from the eigenfunctions, corresponding to the eigenvalues λ_j, multiplying $\psi_{n,j}$ by $\exp(i\pi x/h)$.

Different formulas for the bound states eigenfunctions follows also from the determinant representation of the general solution of (12) with potential $v_{n0}(x)$ given in the last section.

One more form for the bound states eigenfunctions quite different from given here can be found in [6].

4   GETTING v_{n0} POTENTIALS FROM CASORATI DETERMINANTS

In this section we use the following solutions of (12) with

$$v = 1, \quad \lambda = 2, 2\cosh qh, 2\cosh 2qh, \dots, 2\cosh qnh:$$

$$f_1(x) = 1, \ f_2(x) = \sinh q(x + h),$$

$$f_3 = \cosh 2q(x + h), \dots, f_{n+1} = \cosh nq(x + h). \quad (48)$$

$f(x, \lambda)$ as before denotes any solution of (12) with $v = 1$.

The statement formulated below provides the Casorati determinant representation for the potentials $v_{n0}(x)$ and the general solution of the related DSE equation:

Proposition 7   *Potential $v_{n0}(x)$ can be represented in a following determinant form*

$$v_{n0}(x) = \frac{\Delta_{n+1}(x - 2h)\Delta_{n+1}(x + h)}{\Delta_{n+1}(x)\Delta_{n+1}(x - h)}, \ \Delta_{n+1}(x) = \Delta_{n+1}[f_1, \dots, f_{n+1}](x)$$

$$(49)$$

The general solution of the DSE equation with the potential $v_{n0}(x)$ is given by the formula

$$\frac{\Delta_{n+2}[f, f_1, \dots, f_{n+1}](x)}{\Delta_{n+1}[f_1, \dots, f_{n+1}](x)},$$

The solution of DKdV equation with the initial condition $v(x, 0) = v_{n0}(x)$ is given by the formula (49) where the functions $f_j(x, t)$ are defined as follows

$$f_1(x) = 1, \quad f_2(x) = \sinh[q(x + h) + t \sinh qh],$$
$$f_3(x) = \cosh[2q(x + h) + t \sinh 2qh], \ldots, f_{n+1}$$
$$= \cosh[nq(x + h) + t \sinh nqh],$$

if n is even number and ending by $f_{n+1} = \sinh[nq(x + h) + t \sinh nqh]$ if n is an odd number.

Proof It is clear that the formulas (6) represent the special reductions of (26) with $v(x, t) = 1$ corresponding to the special choice of λ_j described above and to the particular selection of the solutions $f_j(x)$ of (12), (with $v(x, t) = 1$) for which F_k defined in (42) coincide with $F_1(x, k)$ up to the x independent factor. Assume for instance that

$$f_1(x) = 1, \quad f_2(x) = \cosh qx, \quad f_3 = \cosh 2qx, \ldots, f_{n+1} = \cosh nqx.$$

In this case the determinant $\Delta_{n+1}(x)$ can be easily computed:

$$\Delta_{n+1}(x) = 2^{\frac{n(n-1)}{2}} \prod_{n+1 \geq j > k \geq 1} (\cosh q(x - 2(n - j)) - \cosh q(x - 2(n - k)))$$

Therefore for the function F_n we get the formula

$$F_n(x) = 2^{n-1} \prod_{k=1}^{n} (\cosh qx - \cosh q(x - 2kh)) \tag{50}$$

$$= 2^{2n-1} \prod_{k=1}^{n} \sinh q(x - kh) \sinh qkh. \tag{51}$$

Taking into account, that $\sinh q(x - i\pi/2q - kh) = i \cosh q(x - kh)$, we realize, that the x-dependent factor in $F_n(x + i\pi/2q)$ is given by the formula:

$$\prod_{k=1}^{n} \cosh q(x - kh) = \prod_{k=1}^{n} c(x - kh) = F_1(x - h, n - 1).$$

In other words taking the functions $f_j(x)$ in the form

$$f_1(x) = 1, \quad f_2(x) = \sinh q(x + h), \quad f_3 = \cosh 2q(x + h), \ldots, f_{n+1}$$
$$= \cosh nq(x + h), \tag{52}$$

if n is even number, and ending by $f_{n+1} = \sinh nq(x + h)$, if n is an odd number, we obtain the relation

$$F_n(x) = c F_1(x, n - 1), \tag{53}$$

where the RHS of (53) is defined by (42) and the LHS of the same formula was defined in the Proposition 2a. Since the LHS and RHS of (53) are the solutions of (12) with the same value of $\lambda = 2 \cosh qnh$ the related potentials are the same. Therefore, it is clear that the related global solution of the discrete Schödinger equation with potential $v_{n0}(x)$ can be written as in the Proposition 7 above. The structure of the global solution and the formula for the solution of the difference KdV equation with the initial condition $v(x, 0) = v_{n0}(x)$, given above, now follows immediately from the Proposition 2.   □

Of course it is also possible to get by the reduction of Casorati determinants the whole family of the potentials $v_{nm}(x)$ as it was done for the DPT potentials in [4]. We postpone the corresponding details to a more detailed publication.

5   CONCLUSION

The difference Schrödinger equation with potential $v_{nm}(x)$, considered above, merits some further investigations along the lines developed for its one parametric reductions [6, 7] and further lattice reductions [8, 9], considered before and corresponding to the potentials $v_{n0}(x)$ and their trigonometric versions or to their lattice restrictions.

Let us emphazise that even in this reduced case our formulas for the solutions of the related DSE equation are different from given in the aforementioned works and have more simple combinatorial structure.

New family of integrable functional-difference deformations of the Schrödinger equation with Darboux-Pöschl-Teller potentials was recently constructed by P. Gaillard following the same strategy as in this article [12]. He considered the functional difference equation with the diagonal term

$$g(x)f(x + 2h) + f(x - 2h) + b(x)f(x) = \lambda f(x). \tag{54}$$

For $g = g_{m,n}(x), \quad b = b_{m,n}(x), m, n \in \mathbf{Z}$,

$$g_{m,n}(x) = \frac{c(x - mh)c(x - (m - 1)h)c(x + (m + 1)h)c(x + (m + 2)h)}{c(x)c(x + 2h)(c(x + h))^2}$$

$$\cdot \frac{s(x - nh)s(x - (n - 1)h)s(x + (n + 1)h)s(x + (n + 2)h)}{s(x)s(x + 2h)(s(x + h))^2},$$

$$b_{m,n}(x) = \frac{2s(mh)s((m + 1)h)c(nh)c((n + 1)h)}{c(x - h)c(x + h)}$$

$$- \frac{2c(mh)c((m + 1)h)s(nh)s((n + 1)h)}{s(x - h)s(x + h)},$$

the general solution of (54) can be obtained along the same lines as above although some calculations become much longer.

We propose to call the model considered in this article by DDPT-I model (Difference Darboux-Pöschl Teller-I model) and the model decsribed by the formulas above by DDPT-II model.

Like in the continuous case of (DPT) potentials and the DDPT1-model, we have the invariance of the DDPT2-model by the transformations $m \to -m - 1$, and $n \to -n - 1$. The potential $g_{m,n}(x)$ is expressed by means of $v_{nm}(x)$ of our article as follows

$$g_{m,n}(x) = v_{nm}(x)v_{nm}(x + h)$$

In particular for $\lambda = 2\cosh(2(m + n + 2)qh)$, special solution of (54) (with $g(x)$ and $b(x)$ decribed above is given by the formula

$$F_1(x, m, n) = \prod_{k=0}^{m} c(x - kh) \cdot \prod_{k=0}^{n} s(x - kh), \tag{55}$$

emphazizing the close link between DDPT-I and DDPT-2 models. The knowledge of this particular solution enables one to construct the general solution of (54) as it was done for DDPT-1 model (see [19] for details)

The potentials $g_{m,n}(x)$ and $b_{m,n}(x)$ defined above have the following asymptotics when $h \to 0$,

$$g_{m,n}(x) = 1 + 2h^2 \left[m(m + 1)(\ln c(x))'' + n(n + 1)(\ln s(x))'' \right] + O(h^3), \tag{56}$$

$$b_{m,n}(x) = 2h^2 \left[m(m + 1)(\ln c(x))'' + n(n + 1)(\ln s(x))'' \right] + O(h^3). \tag{57}$$

Taking into account that (54) can be also written as,

$$h^{-2} \left[g(x)f(x + 2h) + f(x - 2h) - 2f(x) + b(x)f(x) \right] = 2h^{-2} \left[k - 1 \right] f(x), \tag{58}$$

with $k = \cosh(aqh)$, we see that in the limit, when $h \to 0$, the equation (54) with the potential $g(x) = g_{m,n}(x)$ and $b(x) = b_{m,n}(x)$ becomes,

$$-f'' + \left(-\frac{m(m + 1)q^2}{\cosh^2 qx} + \frac{n(n + 1)q^2}{\sinh^2 qx} \right) f = -(aq)^2 f, \tag{59}$$

which proves that (54) with the potentials (55) and (55) is also an integrable deformation of DPT model.

Recently an elliptic analogue: of the potentials $v_{n0}(x)$, representing also the functional difference generalization of the Lamé equation was considered by Ruijenaars, Krichever, Zabrodin, Felder, and Varchenko (see for instance [13, 14]).

The related potential is obtained from $v_{n0}(x)$ simply by setting $c(x) = \theta_1(qx)$, where $\theta_1(x)$ is one of the 4 Jacobi elliptic theta functions:[3]

$$\theta_3(x) = \theta_3(x|\tau) = \sum_{-\infty}^{\infty} e^{i\pi(m^2\tau + 2mx)}, \quad \Im\tau > 0,$$

$$\theta_1(x) = ie^{-i\pi(x-\tau/4)}\theta_3\left(x + \frac{1-\tau}{2}\right),$$

$$\theta_2(x) = e^{-i\pi(x-\tau/4)}\theta_3\left(x - \frac{\tau}{2}\right),$$

$$\theta_4(x) = \theta_3\left(x + \frac{1}{2}\right).$$

There exists the beautiful 4 parametric generalization of the Lamé finite gap elliptic potential, discovered first by Darboux, [2].[4] It was rediscovered later by Trebich and Verdier [17, 18] in a context of study of elliptic reductions of the Its-Matveev formula [19] for the general finite gap potentials. Its difference analogue until now was not discussed in the literature. The degenerate case, considered above, suggests the following

Conjecture One of the posible difference integrable elliptic deformations of the DVT potential has the form

$$v(x, n_1, n_2, n_3, n_4, q, h) = \prod_{j=1}^{4} \frac{c_j(x - n_jh)c_j(x + (n_j + 1)h)}{c_j(x)c_j(x+h)},$$

where

$$c_j(x) = \theta_j(qx), \quad j = 1, 2, 3, 4.$$

When $n_2 = n_3 = n_4 = 0$, $v(x, n_1, n_2, n_3, n_4)$ reduces to the Krichever-Zabrodin-Ruijenaars-Felder-Varchenko difference Lamé potential. In the limit $h \to 0$, $v(x, n_1, n_2, n_3, n_4)$ obviously has the following asymptotics

$$v = 1 + h^2 \sum_{j=1}^{4} n_j(n_j + 1)\frac{d^2}{dx^2} \log\theta_j(qx) + 0(h^3),$$

where the coefficient of h^2 is exactly the famous Darboux-Verdier-Trebich potential. It is clear, from the remarks made in the introduction, that the related difference Schrödinger equation in a same limit transforms to the

---

[3] Here we use the same definition as in [15] slightly different from those of [16]:replacing x by x/π we obtain Jacobi θ-functions defined in [16] from those defined in [17].
[4] Darboux proved its integrability via explicit construction of the related solutions of the Sturm-Liouville equation.

Schrödinger equation with the DVT potential. It is obvious that the potential $v(x, n_1, n_2, n_3, n_4)$ represent a double periodic elliptic function with the periods q^{-1} and $q^{-1}\tau$.

Introducing the notation $v = \exp i\pi\tau$, $v^{\frac{1}{4}} = \exp i\pi\tau/4$ we have well known expansions

$$\theta_1(x) = 2v^{\frac{1}{4}} \sin x\pi - 2v^{\frac{9}{4}} \sin 3x\pi + 2v^{\frac{25}{4}} \sin 5x\pi + \cdots,$$

$$\theta_2(x) = 2v^{\frac{1}{4}} \cos x\pi + 2v^{\frac{9}{4}} \cos 3x\pi + 2v^{\frac{25}{4}} \cos 5x\pi + \cdots,$$

$$\theta_3(x) = 1 + 2v \cos 2x\pi + 2v^4 \cos 4x\pi + 2v^9 \cos 6x\pi + \cdots,$$

$$\theta_4(x) = 1 - 2v \cos 2x\pi + 2v^4 \cos 4x\pi - 2v^9 \cos 6x\pi + \cdots.$$

Using these expansions it is easy to show that, when $Im\tau \to +\infty$, and q is a real parameter, $v(x, n_1, n_2, n_3, n_4)$ tends to the trigonometric version of the potential $v_{nm}(x)$ with $c(x) = \cos qx$, $s(x) = \sin qx$. Replacing q by iq, in a same limit, we obtain hyperbolic potentials $v_{nm}(x)$ defined by (6).

In June 2003 I was informed by Professor Treibich about his article [20], where the elliptic analogue of the model considered above was constructed confirming the conjecture formulated above. In [20] also the elliptic analogues of the DDPT-II model is discussed. The formulas for the elliptic anallogue of b(x) in his article need somehow some further effectivization, precision of the reality conditions etc. It will be interesting to recover our results from his construction although technically it might be much more involved comparing to our direct approach.

REFERENCES

1. Darboux, G. (1915) *Leçons sur la théorie des surfaces*, 2nd ed., Vol. 2, Gauthier-Villars, Paris, pp. 210–215.
2. Darboux, G. (1982) Sur une équation linéaire, T. XCIV, *C. R. Acad. Sci., Ser I, Math.* pp. 1645–1648.
3. Pöschl, G. and Teller, E. (1933) Bemerkungen zur Quantenmechanik des anharmonischen Oszillators *Z. Phys.* **83**, pp. 143–151.
4. Gaillard, P. and Matveev, V. B. (2002) *Wronskian addition formula and its applications*, Max-Planck-Institut für Mathematik, Bonn, preprint MPI 02-31, pp. 1–17.
5. Matveev, V. B. (2002) *Functional Difference Analogues of Darboux-Pöshl-Teller Potentials*, preprint MPI 02-125, Max-Planck-Institut für Mathematik Bonn pp. 1–18.
6. van Diejen, J. F. and Kirillov, A. N. (2000) Formulas for q-spherical functions using inverse scattering theory of reflectionless Jacobi operators, *Commun. Math. Phys.* **210**, pp. 335–369.

7. Ruijenaars, S. N. M. (1999) Generalized Lamé functions 2: Hyperbolic and trigonometric specialisations, *J. Math. Phys.* **40**, pp. 1627–1663.

8. Spiridonov, V. and Zhedanov, A. (1995) Discrete reflectionless potentials, quantum
algebras, and q-orthogonal polynomials, *Ann. Phys.* **237**(1), pp. 126–146.

9. Spiridonov, V. and Zhedanov, A. (1995) Discrete darboux transformations, the
discrete time Toda lattice and the Askey-Wilson polynomials, *Ann. Phys.* **2** (4),
pp. 370–398.

10. Matveev, V. B. (2000) Darboux transformations, covariance theorems and integrable systems, in: *M. A. Semenov–Tyanshanskij*, ed., L. D. Faddeev's Seminar on
Mathematical Physics, *Amer. Math. Soc. Transl.* **201**(2), pp. 179–209.

11. Matveev, V. B. and Salle, M. A. (1991) Darboux Transformations and Solitons,
Series in Nonlinear Dynamics, Springer Verlag, Berlin.

12. Gaillard, P. (2004) *New family of Deformations of Darboux-Pöshl-Teller potentials*
pp. 1–17 (to be published in LMP).

13. Felder, J. and Varchenko, A. (1996) *Algebraic Integrability of the Two-Body Ruijenaars Operator*, preprint q-alg/9610024.

14. Zabrodin, A. (2000) Finite-gap difference operators with elliptic coefficients and
their spectral curves, in: *Physical Combinatorics*, Birkhauser series Progress in
Mathematics **191**, ISBN 0-816-4175-0, Birkhauser-Boston, eds. M. Kashiwara
and T. Miwa, pp. 301–317.

15. Bateman, H. (1955) *Higher Trancedental Functions*, ed. Arthur Erdélyi, Vol. 3,
Mc Graw-Hill book Co., New York-Toronto, London.

16. Whittaker, E. T. and Watson, G. N. (1927) *A Course of Modern Analysis*, Cambridge
University Press.

17. Trebich, A. (1992) Revêtements exceptionnels et somme de quatre nombres triangulaires, *Duke Math. J.*, **68**, pp. 217–236.

18. Verdier, J. L. and Trebich, A. (1990) Revêtements tangentiels et somme de 4
nombres triangulaires, *C. R. Acad. Sci., ser. I, Math.* **311**, pp. 51–54.

19. Its, A. R. and Matveev, V. B. (1975) Hill's operator with finitely many gaps, *Funct.
Anal. i Prilozhen.* **9**, pp. 69–70.

20. Treibich, A. (2003) *Difference analogues of Elliptic KdV Solitons and Schrödinger
operators, Inst. Math. Res. Notices.* (6) 314–360.

MAXWELL EQUATIONS FOR QUANTUM SPACE-TIME

R.M. Mir-Kasimov
Bogoliubov Laboratory of Theoretical Physics, Joint Institute for
Nuclear Research, 141980, Dubna, Moscow region, Russia
and
Department of Mathematics, Izmir Institute of High
Technology, 35437, Gulbahce, Izmir, Turkey

Abstract   The realization of the two-dimensional Poincare algebra in terms of the noncommutative differential calculus on the algebra of functions A is considered. A is the commutative algebra of functions generated by the unitary irreducible representations of the isometry group of the De Sitter **momentum space**. Corresponding space-time carries the noncommutative geometry (NG) [1–14]. The Gauge invariance principle consistent with this NG is considered.

1   INTRODUCTION

The suggestions to consider the noncommutative space-time at small distances are as old as quantum field theory itself. One of the first NG models goes back to H. Snyder. The Snyder coordinates $\hat{x}_\mu$ are proportional to the boost generators of the De Sitter or Anti-De Sitter space (see the footnote to the first Snyder paper with this interpretation given by W. Pauli). The further development of Snyder ideas took place in Former Sovjet Union (Lebedev's Institute of Physics (Moscow), and Joint Institute for Nuclear Research (Dubna)). The key role in this new approach played the physical meaning of $\hat{x}_\mu$ (I.E. Tamm, Yu.A. Golfand). It was stressed, that in fact the Snyder modification of the position operators is a consequence of the modification of the **geometry of momentum space**, when the standard quantum-mechanical position operators $\hat{x}_\mu = i\hbar\frac{\partial}{\partial x^\mu}$, i.e., the generators of the translations of the flat Minkowski momentum space must be naturally substituted by the boosts of the curved (De Sitter or Anti-De Sitter) momentum space

$$\hat{x}_\mu = il_0 \left(p_4 \frac{\partial}{\partial p^\mu} - p_\mu \frac{\partial}{\partial p^4} \right), \qquad \left[\hat{x}_\mu, \hat{x}_\nu \right] = -il_0^2 \hat{M}_{\mu\nu} \qquad (1)$$

209

L. Faddeev et al. (eds.),
Bilinear Integrable Systems: From Classical to Quantum, Continuous to Discrete, 209–217.

The geometries of momentum and configurational spaces are closely connected. As quantum mechanical operators of energy and momentum are the derivation operators in space-time, it is clear that transfer to the noncommutative geometry requires the modification of the geometry of momentum space. And vise versa. It is not obvious which is prime. Choosing the momentum space with geometry different from the standard Minkowsky one we obtain different (in general noncommutative) geometry of space-time. The change of the geometry of the momentum space leads to the modification of the procedure of extension of the S-matrix off the mass shell, i.e., to a different dynamical description (Dubna group, see review article [1] for the list of references to Russian papers on Snyder theory). Establishing the geometry of the momentum space off the mass shell is in fact an additional axiom of quantum field theory (QFT). Actually, in the standard QFT, the axiom that the geometry of the momentum space off the mass shell is the pseudo-euclidean (Minkowsky) is accepted as an evident fact, without saying. We can think that some background interaction exists which modifies the geometry of the momentum space [1]. In consequence of the change of the geometry of the p-space the space-time becomes quantum (noncommutative). We stress that the physical meaning of the geometry and topology of the momentum space has no clear physical interpretation as yet. The space-time groups considered in QFT as covariance groups are the isometry groups of space-time.

The explicit character of Snyder's approach to space-time quantization has a remarkable consequence: we can define the common spectrum of a (complete) set of four operators belonging to the centrum of the universal enveloping algebra of the De Sitter Lie algebra and consider the points of this spectrum as the points ξ_μ of the new quantum (and noncommutative in general) space-time. It can be shown that a formulation of the generalized causality condition and QFT in terms of this new numerical quantum space-time is as comprehensive procedure as it is in the usual QFT with the Minkowskian space-time. In this approach, the structure of the singular field theoretic functions is entirely reconstructed as compared to the standard QFT, and the corresponding perturbation theory is free of ultraviolet divergences [4].

Today it is commonly accepted that the most probable scale for these "small" distances (cf. the proportionality coefficient in (1) having the dimension of length) is that of Grand Unified Theories, i.e., the scale close to the Planck one

$$l_0 = \sqrt{\frac{\hbar G_N}{c^3}} \simeq 10^{-33} \text{cm} \tag{2}$$

One of the most convincing arguments for this is that at this scale the curvature radius of space-time is of the order of the De Broglie wavelength of a test particle. Recently the possibility of noncommutative space to emerge in the framework of the string theory has been indicated first in [6]. More recently Yang–Mills theories on noncommutative spaces have emerged in the context of M-theory compactified on a torus in the presence of constant background three-form field, or as a low-energy limit of open strings in a background B-field describing the fluctuations of the D-brane world volume. (See the review article [7] and references therein.)

It is worth mentioning a series of papers [5–8] where it has been shown that the curved momentum space and the corresponding Snyder-like quantum space naturally arise when considering the $2 + 1$ model of gravity interacting with the scalar field. The canonical momenta belong to the hyperboloid in three-dimensional space, Lobachevsky space.

We would like to draw the attention of the reader to the recent paper [15] where interesting details of the prehistory of the Snyder space quantization are delivered.

2 NONCOMMUTATIVE CONFIGURATIONAL SPACE

Let us consider the two-dimensional momentum space of constant curvature modeled by the two-dimensional surface embedded into the three-dimensional pseudo-Euclidean space

$$p_L p^L = p_\mu p^\mu - p_2^2 = -1 \quad L = 0, 1, 2, \quad \mu = 0, 1 \tag{3}$$

The "plane waves" in this case, i.e., the kernels of the Fourier transform connecting the p-space and the new configurational space and at the same time the state vectors describing the free motion of the particle are the matrix elements of the unitary irreducible representations of the isometry group of the space (3) $\langle \xi \mid p \rangle$. Or in other words, the surface (3) is the uniform space of this group.[1] To describe these matrix elements, we use the hyperspherical coordinates

$$p^0 = \sinh \zeta \quad p^1 = \cosh \zeta \sin \omega \quad p^2 = \cosh \zeta \cos \omega$$
$$-\infty < \zeta < \infty \quad -\pi < \omega < \pi \tag{4}$$

In the correspondence limit

$$p^2 \approx 1, \quad p^0 \approx p^1 \approx 0, \quad \zeta \to 0, \quad \omega \to 0 \tag{5}$$

---

[1] Evidently, dimensional physical constants including the fundamental length l_0 must enter expression (3). We use the unit system $\hbar = c = 1$. All relations of the theory must go over into the standard ones when l_0 can be considered as a small quantity. This will be called the correspondence limit.

the generalized plane waves $\langle \xi \mid p \rangle$ have the form

$$\langle \xi \mid p \rangle = \langle \sigma, n \mid \zeta, \omega \rangle = \frac{2^{\sigma+1}\kappa(\sigma, n)}{\sqrt{2\pi}\,\cosh\zeta}e^{in\omega}P_{n-\frac{1}{2}}^{-(\sigma+\frac{1}{2})}(\xi^0\tanh\zeta) \qquad (6)$$

where $P_\nu^\mu(z)$ are the associated Legendre functions, ξ^0 is the sign of the discrete time n, and

$$\kappa(\sigma, n) = \Gamma\left(\frac{\sigma+n+2}{2}\right)\Gamma\left(\frac{\sigma-n+2}{2}\right) \qquad (7)$$

Let us list some important properties of the generalized plane waves:

1. Behavior at the origin

$$\langle \xi \mid 0 \rangle = 1 \qquad (8)$$

2. For principal series of the irreps

$$\sigma = i\Lambda - \frac{1}{2}, \quad 0 \le \Lambda < \infty, \quad \text{or } \sigma = k = 0, 1, 2, \ldots, \quad n = 0, \pm1, \pm2, \ldots \qquad (9)$$

We call the set $\xi = (\sigma, n)$ a point of quantum space.

3. Orthogonality and completeness

$$\frac{1}{(2\pi)^2}\int d\Omega_p \langle \xi \mid p \rangle\langle p \mid \xi' \rangle = \mu^{-1}(\sigma)\delta_{\xi^0\xi'^0}\delta_{nn'}\delta(\Lambda - \Lambda')$$

$$\frac{1}{(2\pi)^2}\int d\Omega_\xi \langle p \mid \xi \rangle\langle \xi \mid p' \rangle = \delta(p(-)p') = |p^2|\delta(\tilde{p} - \tilde{p}') \qquad (10)$$

$$d\Omega_p = \frac{dp^0 dp^1}{|p^2|},$$

$$\mu(\sigma, n) = \frac{\left(\sigma+\frac{1}{2}\right)\cot\pi\left(\sigma+\frac{1}{2}\right)}{2}$$

$$\times \frac{\Gamma\left(\frac{\sigma+n+2}{2}\right)\Gamma\left(\frac{-\sigma+n+1}{2}\right)}{\Gamma\left(\frac{\sigma+n+1}{2}\right)\Gamma\left(\frac{-\sigma+n}{2}\right)}$$

The volume element $d\Omega_\xi$ symbolizes the integration over the continuous part of the proper time σ and summation over its discrete part as well as the summation over the discrete time n (cf. [4]).

4. The correspondence limit

$$\zeta \simeq p^0, \qquad \omega \simeq p^1, \qquad p^2 \simeq 1$$

$$n \simeq x^1, \qquad \sqrt{\sigma^2 + n^2} \simeq |x^0| \tag{11}$$

$$\langle \xi \mid p \rangle \rightarrow e^{ip_\mu x^\mu}$$

Relations (11) suggest our identification of σ and n as quantum analogs of the two-dimensional space-time coordinates (interval and time).

3 HOLOMORPHIC REALIZATION

Let us consider the holomorphic realization of the plane waves (6). We start with the continuous part of the spectrum of the interval σ. Intoducing the new variables

$$z = i\Lambda + n = \sigma + n + \frac{1}{2} \quad \text{and} \quad \bar{z} = -i\Lambda + n = -\sigma + n - \frac{1}{2} \tag{12}$$

and using the connection between associated Legendre functions and hypergeometric functions [16] we come to the following holomorphic representation for the plane wave:

$$\langle \xi \mid p \rangle = \langle \sigma, n \mid \zeta, \omega \rangle = \langle z, \bar{z} \mid \zeta, \omega \rangle$$

$$= \frac{2^{\frac{z-\bar{z}+1}{2}} e^{i\omega \frac{z+\bar{z}}{2}}}{\sqrt{2\pi \cosh \zeta}} \kappa(z, \bar{z}) P_{\frac{z+\bar{z}-1}{2}}^{-\left(\frac{z-\bar{z}}{2}\right)} \left(\xi^0 \tanh \zeta \right) \tag{13}$$

$$= (\cosh \zeta)^{-\frac{z-\bar{z}+1}{2}} e^{i\omega \frac{z+\bar{z}}{2}} F \left(\frac{z+\frac{1}{2}}{2}, \frac{-\bar{z}+\frac{1}{2}}{2}, \frac{1}{2}; \tanh^2 \zeta \right)$$

$$- 2\xi^0 \tanh \zeta \cdot \zeta(z, \bar{z}) F \left(\frac{z+\frac{3}{2}}{2}, \frac{-\bar{z}+\frac{3}{2}}{2}, \frac{3}{2}; \tanh^2 \zeta \right)$$

where

$$\zeta(z, \bar{z}) = \frac{\Gamma\left(\frac{z+\frac{3}{2}}{2}\right) \Gamma\left(\frac{-\bar{z}+\frac{3}{2}}{2}\right)}{\Gamma\left(\frac{z+\frac{1}{2}}{2}\right) \Gamma\left(\frac{-\bar{z}+\frac{1}{2}}{2}\right)} \tag{14}$$

As the hypergeometric function is the entire function of its first two parameters we can easily prove that the plane wave $\langle z, \bar{z} \mid \zeta, \omega \rangle$ is an analytic function in its first argument z and antianalytic in its second argument $\bar{z}$. This function obeys as well as $\zeta(z, \bar{z})$ the symmetry condition

$$\bar{f}(z, \bar{z}) = f(\bar{z}, z) \tag{15}$$

At the same time it is well-known fact that hypergeometric functions don't obey any differential relation in its parameters, but obey the recurrence relations [16]. This makes them the subjects to the noncommutative differential calculus. The holomorphic representation leads to the simplest form of such a calculus. We introduce this calculus starting with basic relations

$$[z, dz] = dz, \qquad [\bar{z}, d\bar{z}] = d\bar{z}, \qquad [\bar{z}, dz] = 0,$$
$$[z, d\bar{z}] = 0, \qquad [z, \bar{z}] = 0 \tag{16}$$

in which in contrast with the standard (commutative) calculus the coordinates z, and $\bar{z}$ commute between themselves but don't commute with corresponding differentials. It can be shown that the comprehensive differential calculus based on the relations (16) exists [1]. We can introduce the generalized interior derivatives right $\overrightarrow{\partial}$ and left $\overleftarrow{\partial}$ of the function $f(z, \bar{z})$ as

$$d_z f = [dz, f] = \overrightarrow{\partial_z} f \, dz = dz \, \overleftarrow{\partial_z} f \tag{17}$$

and similar formulae for the noncommutative differentiation in $\bar{z}$. The Leibnitz rule is fulfilled for the exterior differentiations (17)

$$d_z(fg) = (d_z f)g + f(d_z g) \tag{18}$$

and in modified form for the interior derivatives

$$\overrightarrow{\partial_z}(fg) = (\overrightarrow{\partial_z} f)g + f(\overrightarrow{\partial_z} g) + (\overrightarrow{\partial_z} f)(\overrightarrow{\partial_z} g)$$
$$\overleftarrow{\partial_z}(fg) = (\overleftarrow{\partial_z} f)g + f(\overleftarrow{\partial_z} g) + (\overleftarrow{\partial_z} f)(\overleftarrow{\partial_z} g) \tag{19}$$

We refer the reader for the detailed theory of noncommutative differential forms on the commutative algebra of functions to [1, 7, 14] and deliver here the only information necessary for introducing the physical operators. Using right and left noncommutative Hodge operators $\overrightarrow{*}$ and $\overleftarrow{*}$ [1] we define the symmetrized interior derivatives in the form

$$\partial_z^s = \frac{1}{2}(\overrightarrow{*} + \overleftarrow{*})dz \qquad \partial_z^c = \frac{1}{2}(\overrightarrow{*} - \overleftarrow{*})dz + 1 \tag{20}$$

It can be proved that noncommutative interior derivatives in our case can be expressed as the finite-difference derivatives

$$\overrightarrow{\partial} f = \left(e^{\frac{\partial}{\partial z}} - 1\right) f(z, \bar{z}), \qquad \overleftarrow{\partial} f = \left(1 - e^{-\frac{\partial}{\partial z}}\right) f(z, \bar{z}) \tag{21}$$

momentum operators

$$
p^+ = \frac{1}{(z-\bar{z})}\left\{\left(z+\frac{1}{2}\right)e^{2\frac{\partial}{\partial z}} - \left(\bar{z}+\frac{1}{2}\right)e^{2\frac{\partial}{\partial \bar{z}}}\right\}
$$

$$
p^- = \frac{1}{(z-\bar{z})}\left\{\left(z-\frac{1}{2}\right)e^{-2\frac{\partial}{\partial z}} - \left(\bar{z}-\frac{1}{2}\right)e^{-2\frac{\partial}{\partial \bar{z}}}\right\} \tag{22}
$$

$$
p^0 = -\xi^0\frac{4}{(z-\bar{z})}\zeta(z,\bar{z})\sinh\left(\frac{\partial}{\partial z} - \frac{\partial}{\partial \bar{z}}\right)
$$

$$
p^+ = p^2 + ip^1 \qquad p^- = p^2 - ip^1
$$

Operators p^L mutually commute. Their common eigenfunctions are the plane waves (6) with eigenvalues (4). Momentum operators obey the hermiticity condition

$$
\mu^{-1}(p^+)^\dagger\mu = p^- \qquad \mu^{-1}(p^0)^\dagger\mu = p^0 \tag{23}
$$

Now we write down the three generators

$$
M^{12} = \frac{z+\bar{z}}{2}, \qquad M^{10} = 2\xi^0\zeta(z,\bar{z})\partial_{\frac{z+\bar{z}}{2}}^{(s)}, \qquad M^{20} = 2i\xi^0\zeta(z,\bar{z})\partial_{\frac{z+\bar{z}}{2}}^{(c)} \tag{24}
$$

which are as well as (22) the noncommutative differential operators and complete the set (22) up to the Lie algebra of the inhomogeneous two-dimensional De Sitter group.

For $\partial_{\frac{z+\bar{z}}{2}}^{(s)}$ and $\partial_{\frac{z+\bar{z}}{2}}^{(c)}$ the following relations:

$$
\partial_{\frac{z+\bar{z}}{2}}^{(s)} = \partial_z^{(s)}\partial_{\bar{z}}^{(c)} \pm \partial_z^{(c)}\partial_{\bar{z}}^{(s)} \qquad \partial_{\frac{z+\bar{z}}{2}}^{(c)} = \partial_z^{(c)}\partial_{\bar{z}}^{(c)} \pm \partial_z^{(s)}\partial_{\bar{z}}^{(s)} \tag{25}
$$

are fullfilled.

Let us consider the gauge transformation localized in noncommutative space-time

$$
\psi'(\xi) = \Omega(\xi)\psi(\xi) \qquad \Omega(\xi)^\dagger = \Omega(\xi)^{-1} \tag{26}
$$

Unlike the usual theory the gauge transformation entangles the components of momenta:

$$
\Omega^{-1}(\xi)P^L\Omega(\xi) = C^L{}_K(\xi)P^K \tag{27}
$$

For the sake of simplicity we write down the $C^L{}_K(z,\bar{z})$-matrix separately for the cases when it depends on only one of the variables $\frac{z-\bar{z}}{2}$ or $\frac{z+\bar{z}}{2}$. For

$\Omega = \Omega(\frac{z-\bar{z}}{2})$ we have

$$C^L{}_K\left(\frac{z-\bar{z}}{2}\right) = \left[\left(\partial^{(c)}_{\frac{z-\bar{z}}{2}}\Omega\left(\frac{z-\bar{z}}{2}\right)\right) - \frac{2i}{z-\bar{z}}\left(\partial^{(s)}_{\frac{z-\bar{z}}{2}}\Omega\left(\frac{z-\bar{z}}{2}\right)\right)\hat{\Sigma}\right]^L_K$$

(28)

where

$$\hat{\Sigma} = \begin{pmatrix} \frac{i}{2} & -M^{10} & -M^{20} \\ -M^{10} & \frac{i}{2} & M^{12} \\ -M^{20} & -M^{12} & \frac{i}{2} \end{pmatrix}$$

(29)

For the case $\Omega = \Omega(\frac{z+\bar{z}}{2}) = e^{i\lambda\left(\frac{z+\bar{z}}{2}\right)}$, $\quad \lambda = -i\ln\Omega$

$$C^L{}_K(\frac{z+\bar{z}}{2}) = e^{i\left(\partial^{(c)}_{\frac{z+\bar{z}}{2}}\lambda(\frac{z+\bar{z}}{2})\right)}$$

$$\times \begin{pmatrix} e^{i\left(\left(\partial^c_{\frac{z+\bar{z}}{2}} - 1\right)\lambda\right)} & 0 & 0 \\ 0 & \cos\left(\partial^s_{\frac{z+\bar{z}}{2}}\lambda\right) & -\sin\left(\partial^s_{\frac{z+\bar{z}}{2}}\lambda\right) \\ 0 & \sin\left(\partial^s_{\frac{z+\bar{z}}{2}}\lambda\right) & \cos\left(\partial^s_{\frac{z+\bar{z}}{2}}\lambda\right) \end{pmatrix}$$

(30)

Also the following the following relations are true:

$$\Omega^{-1}(z,\bar{z})P_L\Omega(z,\bar{z}) = P_K C^{\dagger K}_L(z,\bar{z}) \quad C^{\dagger} = C^{-1}$$

(31)

It is easily seen that in a consequence of (27) and (31) the De Sitter condition (3) is invariant in respect to the gauge transformations (26). It can be easily shown [4] that to make the theory gauge invariant we must introduce the complex De Sitter vector $\hat{A}^L$ of electromagnetic field and require that it transforms similarly to (27) and (31):

$$\Omega^{-1}(z,\bar{z})\hat{A}^L\Omega(z,\bar{z}) = C^L{}_K(z,\bar{z})\hat{A}^K$$
$$\Omega^{-1}(z,\bar{z})\hat{A}^{\dagger}_L\Omega(z,\bar{z}) = \hat{A}^{\dagger}_K C^{\dagger K}_L(z,\bar{z})$$

(32)

It follows from (32) that the components of electromagnetic field do not commute with the gauge function and in a consequence do not commute between themselves. Introducing the covariant derivatives

$$\hat{D}^L = -i\left(p^L - \hat{A}^L\right)$$

(33)

we obtain the De Sitter invariant equations for the matter fields. The tensor of electromagnetic field is given as

$$\hat{F}^{KL} = [\hat{D}^L, \hat{D}^K] = [\hat{p}^L, \hat{A}^K] - [\hat{p}^K, \hat{A}^L] - [\hat{A}^L, \hat{A}^K] \qquad (34)$$

The action of the electromagnetic field is

$$S = Tr \int \hat{F}^{KL} \hat{F}_{KL} \, d\Omega_\xi \qquad (35)$$

and the noncommutative analog of the D'Alembert equation takes the form

$$[\hat{D}_K, \hat{F}^{KL}] = 0 \qquad (36)$$

REFERENCES

1. Mir-Kasimov, R. M. (2000) *Phys. Part. Nucl.* **31**(1), p. 44.
2. Kadyshevsky, V. G. (1988) *Nucl. Phys.* **B. 141**, p. 477.
3. Kadyshevsky, V. G. and Fursaev, D. V. (1990) *Theor. and Math. Phys.* **83**, p. 197.
4. Mir-Kasimov, R. M. (1991) *Phys. Lett.* **B. 259**, p. 79; (1991) *J. Phys.* **A24**, p. 4283; (1996) *Phys. Lett.* **B. 378**, p. 181; (1997) *Turk. J. Phys.* **21**, p. 472; (1997) *Int. J. Mod. Phys.* **12**, N1, p. 24; Yadernaya Fizika, (1998) (*Phys. Atom. Nucl.*) **61**, N11, 1951.
5. Deser, S., Jackiw R., and 't Hooft, G. (1988) *Ann. Phys.* **117**, p. 685.
6. Witten, E. (1988) *Nucl. Phys.* **B311**, p. 46.
7. Castellani, L. (xxxx) Non-commutative geometry and physics: a review of selected recent results, hep-th/0005210.
8. 't Hooft, G. (1996) *Class. Quant. Grav.* **13**, p. 1023.
9. Woronowich, S. L. (1989) *Commun. Math. Phys.* **122**, p. 125.
10. Wess, J. and Zumino, B. (1990) *Nucl. Phys.* (*Proc. Suppl.*) **B. 18**, p. 302.
11. Dubois-Violette, M., Kerner, R., and Madore, J. (1990) *J. Math. Phys.* **31**, p. 323.
12. Connes, A. (1994) Non-Commutative Geometry, Academic Press.
13. Madore, J. (1995) *An Introduction to non-Commutative Geometry and Its Physical Applications*, Cambridge University Press.
14. Dimakis, A. and Müller-Hoissen, F. (1998) *Math. Phys.* **40**, p. 1518.
15. Jackiw, R, hep-th/0212146.
16. Higher transcendental functions, (1953) *Bateman H. Manuscript Project*, Mc-Graw Hill Co. Inc., New York, Toronto, London.
17. Manin, Yu. I. (1988) *Les publications du Centre de Recherches Mathématique*, Université de Montreal.
18. Macfarlane, A. J. (1989) *J. Phys. A.* **22**, p. 4581; Biedenharn, L.C. (1989) *ibid* **22**, p. L873.
19. Witten, E. (1986) *Nucl. Phys.* **B268**, p. 253; Seiberg, N. and Witten, E. *JHEP* **9909**, p. 032, hep-th/9908142.

A SOLVABLE MODEL OF INTERACTING PHOTONS

Jan Naudts
Departement Natuurkunde, Universiteit Antwerpen,
Universiteitsplein 1, 2610 Antwerpen, Belgium

1 INTRODUCTION

It is a general belief that the relevant models of quantum field theory cannot be solved analytically. It is even not clear how to formulate these models in a mathematically consistent manner. Only free-field models are well understood. These form the basis for the perturbative approach to interacting fields. But, as is well known, convergence of the perturbation series is problematic. This is the context in which to situate the present attempt to introduce interactions, while keeping a solvable model.

Starting point is a Hilbert space H containing so-called classical wave functions. It determines a C^*-algebra of canonical commutation relations [1], denoted $\overline{\Delta(H)}$. A class of quantum fields, which can be handled analytically, is made up by the quasi-free states of $\overline{\Delta(H)}$. A subclass of these are the Fock states. Their special property is that there exists a representation in a Hilbert space $\mathcal{H}$, together with field operators $\hat{A}(\phi)$, one for each classical wave function ϕ. Creation and annihilation operators $\hat{A}_\pm(\phi)$ are defined in terms of the field operators by

$$\hat{A}_\pm(\phi) = \frac{1}{2}\left(\hat{A}(\phi) \pm i\,\hat{A}(i\phi)\right).$$

Moreover, there exists a vacuum vector Ω in $\mathcal{H}$ which is annihilated by the annihilation operators

$$\hat{A}_-(\phi)\Omega = 0. \tag{1}$$

Hence, the Hilbert space $\mathcal{H}$ has the structure of a Fock space.

Intuitively, it is obvious to consider quasi-free states as (trivial) examples of integrable quantum fields. In the context of conformal field theory the notion of integrability is usually associated with the existence of infinitely many local conservation laws (see [2] for a recent account). These conservation laws are then used to construct the state of the system. Here, the state of the system

L. Faddeev et al. (eds.),
Bilinear Integrable Systems: From Classical to Quantum, Continuous to Discrete, 219–224.
© 2006 *Springer. Printed in the Netherlands.*

is given in an explicit form. The open question is then whether a dynamical context exists in which conservation laws of this state do exist.

As their name suggests, quasi-free states describe rather trivial physics. Interactions between fields are needed to produce non trivial theories. The prototype of such a theory is quantum electrodynamics (QED). It describes the interaction between photon and electron fields using the S-matrix formalism. The model, discussed in the present paper, describes the interaction of a photon field with a kind of background medium. It has been introduced by Czachor et al. [3–5]. In this original version, it imagines a space-time filled with harmonic oscillators carrying the electromagnetic field. In the reformulation of the model, found in [6], the photons interact with a scalar boson field which is described in a semi classical manner. The scalar field can have several possible interpretations, one of which is that of quantized fluctuations of spacetime.

The model has some peculiar features. The representation of the photon field is *not* irreducible, as one assumes usually. In particular, the commutation relations differ from the canonical expression

$$\left[\hat{A}(\phi), \hat{A}(\psi)\right] = -2i \, \text{Im}\langle\phi|\psi\rangle.$$

They become

$$\left[\hat{A}(\phi), \hat{A}(\psi)\right] = i\hat{s}(\phi|\psi), \tag{2}$$

where $\hat{s}(\phi|\psi)$ is an operator belonging to the center of the representation. This kind of commutation relations has been studied before, in the context of generalized free fields – see e.g., Section 12.5 of [7].

Another feature is that the vacuum vector of the photon field depends on the state of the scalar field. In particular, the vacuum is not invariant under Poincaré transformations. Experimentally, there is a strong evidence that vacuum is locally Poincaré invariant. However, recently more and more research papers investigate the possibility of breaking of Poincaré symmetry at very small length scales, e.g. at the scale of Planck's length. Therefore the scalar bosons of the present model, if one would like to interpret them as part of physical reality, should be active at small length scales, or, equivalently, at very large wavevectors.

2   CORRELATION FUNCTION APPROACH

The vacuum-to-vacuum correlation functions

$$\langle\Omega|\hat{A}(\phi_1)\hat{A}(\phi_2)\ldots\hat{A}(\phi_n)\Omega\rangle$$

determine the vacuum state of the electromagnetic field [8]. It is sufficient [1, 9] to know the functions

$$\mathcal{F}(\phi, \psi) = \langle \hat{W}(\psi)^* \Omega | \hat{W}(\phi)^* \Omega \rangle,$$

with Weyl operators defined by $\hat{W}(\phi) = \exp(i\hat{A}(\phi))$. For the free electromagnetic field is [10, 11]

$$\mathcal{F}(\phi; \psi) = \exp(-i\mathrm{Im}\langle\psi|\phi\rangle)\exp\left(-\frac{1}{2}\langle\psi - \phi|\psi - \phi\rangle\right) \tag{3}$$

with the (degenerate) scalar product given by

$$\langle\psi|\phi\rangle = -\int_{\mathbf{R}^3} d\mathbf{k} \frac{1}{2|\mathbf{k}|} \overline{\psi^\mu(\mathbf{k})} \phi_\mu(\mathbf{k}). \tag{4}$$

Positivity of the scalar product follows if one assumes that all classical wave functions satisfy the Lorentz gauge condition

$$|\mathbf{k}|\phi_0(\mathbf{k}) = \sum_{\alpha=1}^{3} k_\alpha \phi_\alpha(\mathbf{k}).$$

Next consider a scalar boson described in the standard way by state vectors in a Fock space. Select in this space a normalized element χ of the form

$$\chi = \chi^{(0)} \oplus \chi^{(1)} \oplus \frac{1}{\sqrt{2!}} \chi^{(2)} \otimes \chi^{(2)} \oplus \cdots. \tag{5}$$

This state vector will be fixed throughout the paper. Instead of creation and annihilation operators consider in this Fock space observables which are functions of momenta. Such observables are denoted $\hat{f}$ and act on the state vector in the following way

$$\hat{f}\chi^{(n)}(\mathbf{k}_1, \mathbf{k}_2, \ldots, \mathbf{k}_n) = f^{(n)}(\mathbf{k}_1, \mathbf{k}_2, \ldots, \mathbf{k}_n)\chi^{(n)}(\mathbf{k}_1, \mathbf{k}_2, \ldots, \mathbf{k}_n). \tag{6}$$

Clearly, by only allowing a state vector of the form (5), and by restricting operators to functions of momenta, a semi-classical description of the boson field is obtained. The only information that can be calculated are the quantum expectation values

$$\langle\hat{f}\rangle_\chi = \langle\chi|\hat{f}\chi\rangle$$

$$= \sum_{n=0}^{\infty} \frac{1}{n!} \left[\prod_{j=1}^{n} \int_{\mathbf{R}^3} d\mathbf{k}_j \frac{1}{2|\mathbf{k}_j|} |\chi^{(n)}(\mathbf{k}_j)|^2\right] f^{(n)}(\mathbf{k}_1, \mathbf{k}_2, \ldots, \mathbf{k}_n).$$

Finally, the interaction between the photon field and the scalar boson field is introduced. The conventional way to do so is by specification of a Lagrangian or a Hamiltonian. Then a difficult, if not impossible, calculation is needed to obtain the correlation functions. Here an alternative path is followed. Correlation functions of the form

$$\mathcal{F}_\chi(f;\phi;\psi) = \langle \hat{W}(\psi)^*\Omega | \hat{f} \hat{W}(\phi)^*\Omega \rangle \qquad (7)$$

are specified explicitly. Next, the properties of the state, determined by these correlation functions, are studied in order to find out the kind of interactions they describe.

The specific *ansatz* is [6]

$$\mathcal{F}_\chi(f;\phi;\psi) = f^{(0)}|\chi^{(0)}|^2$$

$$+ \sum_{n=1}^\infty \frac{1}{n!} \left[\prod_{j=1}^n \int_{\mathbb{R}^3} \mathrm{d}\mathbf{k}_j \frac{1}{2|\mathbf{k}_j|} |\chi^{(n)}(\mathbf{k}_j)|^2 \right.$$

$$\times \exp\left(\frac{1}{2n} \overline{(\psi^\mu(\mathbf{k}_j) - \phi^\mu(\mathbf{k}_j))}(\psi_\mu(\mathbf{k}_j) - \phi_\mu(\mathbf{k}_j)) \right) \Bigg]$$

$$\times e^{is^{(n)}(\phi,\psi)(\mathbf{k}_1,\mathbf{k}_2,\dots,\mathbf{k}_n)/2} f^{(n)}(\mathbf{k}_1, \mathbf{k}_2, \dots, \mathbf{k}_n). \qquad (8)$$

with

$$s^{(n)}(\phi, \psi)(\mathbf{k}_1, \mathbf{k}_2, \dots, \mathbf{k}_n) = \frac{2}{n} \sum_{j=1}^n \mathrm{Im}\, \overline{\psi^\mu(\mathbf{k}_j)} \phi_\mu(\mathbf{k}_j).$$

It is straightforward to verify that the sum converges. Standard techniques from quasi-free state theory can be used to show that all properties, needed for being the correlation functions of a state, are satisfied. A rather tedious calculation shows that these correlation functions are exactly those of the model introduced in [3–5].

3   PROPERTIES OF THE INTERACTING STATE

The generalized GNS-theorem [9] implies the existence of a Hilbert space representation of the correlation functions (8). More precisely, there exists a vacuum vector Ω in a Hilbert space $\mathcal{H}$, and operators $\hat{W}(\phi)$ and $\hat{f}$ in $\mathcal{H}$, such that (7) holds. The properties of these operators can then be deduced from the explicit form of the correlation functions (8).

First notice that the operators $\hat{W}(\phi)$ satisfy the Weyl form of commutation relations, modified to

$$\hat{W}(\phi)\hat{W}(\psi) = \hat{W}(\phi + \psi)e^{i\hat{s}(\phi,\psi)/2}. \qquad (9)$$

Next, standard arguments are used to show that the operators $\hat{W}(\phi)$ can be written into the form $\hat{W}(\phi) = \exp(i\hat{A}(\phi))$, and that the field operators $\hat{A}(\phi)$ are real linear functions of ϕ. Expansion of (9) for small values of ϕ and ψ then yields the generalized commutation relations (2). Finally one shows that the relation $\hat{A}(\phi)\Omega = i\hat{A}(i\phi)\Omega$ holds. This relation is needed to conclude that the vacuum vector is annihilated by the annihilation operators $\hat{A}_-(\phi)\Omega = 0$. One concludes that the Hilbert space has the graded structure of a Fock space.

So far, the differences with the free field situation are small. In order to see further differences, let us now calculate the vacuum fluctuations of the electric field. For the free photon field is

$$\langle \hat{E}_\alpha(q)\hat{E}_\beta(q')\rangle_\Omega = \frac{1}{(2\pi)^4} \int d\mathbf{k}\frac{1}{2|\mathbf{k}|}\rho(\mathbf{k}) \left(\delta_{\alpha\beta}|\mathbf{k}|^2 - \mathbf{k}_\alpha\mathbf{k}_\beta\right) e^{i(q-q')^\mu k_\mu},$$

with the usual convention $k = (|\mathbf{k}|, \mathbf{k})$, and with the spectral function $\rho(\mathbf{k})$ identically equal to 1. For the model described by the correlation functions (8) the spectral function is given by

$$\rho(\mathbf{k}) = \sum_{n=1}^\infty \frac{1}{n!}|\chi^{(n)}(\mathbf{k})|^2 \left[\int d\mathbf{k}'\frac{1}{2|\mathbf{k}'|}|\chi^{(n)}(\mathbf{k}')|^2\right]^{n-1}.$$

Note that

$$|\chi^{(0)}|^2 + \int d\mathbf{k}\frac{1}{2|\mathbf{k}|}\rho(\mathbf{k}) = 1$$

because of normalization of the state vector χ. The contribution to the vacuum energy density in the point q is

$$\frac{1}{2}\sum_\alpha \langle \hat{E}_\alpha(q)^2\rangle_\Omega = \frac{1}{2}\frac{1}{(2\pi)^4}\int d\mathbf{k}|\mathbf{k}|\rho(\mathbf{k}).$$

For free photons this integral diverges. For the present model it converges under mild conditions on the state vector χ. This shows that interactions can make the energy density of the electromagnetic vacuum finite.

4   DISCUSSION

The model, described by the correlation functions (8), appears to describe genuine interactions of the photon field with a background field of semiclassical bosons. The model has not been derived from a Lagrangian or Hamiltonian function. Therefore it is not easy to grasp the nature of these interactions. It is e.g. not clear whether these interactions are local.

At first sight it might seem easy to generalize the present model, e.g., to allow for a general scalar boson field instead of its semiclassical description. However, the correlation functions (8) must satisfy conditions of positivity and covariance. Both conditions are far from trivial. In this sense the existence of the present model is comparable to the existence of soliton-like solutions of nonlinear equations.

The interacting photon field is better behaved than the free field. One feature has been stressed here: finite energy density of the vacuum. Other interesting properties can be found in [3, 4, 5, 6]. A further study of this and similar models of interacting fields might one day indicate how to build a non perturbative theory of QED, free of divergences.

REFERENCES

1. Petz, D. (1990) *An invitation to the algebra of canonical commutation relations*, Leuven University Press.
2. Babujian, H. and Karowski, M. (2002) Towards the construction of Wightman functions of integrable quantum field theories, arXiv:hep-th/0301088, in: *Proceedings of the '6th International Workshop on Conformal Field Theories and Integrable Models'*, Chernologka, September 2002.
3. Czachor, M. 2000 Non-canonical quantum optics, *J. Phys. A: Math. Gen.* **33**(45), 8081–8103.
4. Czachor, M. and Syty, M. A toy model of bosonic non-canonical quantum field, hep-th/0112011.
5. Czachor, M. Non-canonical quantization of electromagnetic fields and the meaning of Z_3, hep-th/0201090.
6. Naudts, J., Kuna, M., and De Roeck, W. (2002) Photon fields in a fluctuating spacetime, hep-th/0210188.
7. Bogoliubov, N. N., Logunov, A. A. and Todorov, I.T. (1975) Introduction to axiomatic quantum field theory, The Benjamin/Cummings Publishing Company.
8. Wightman, A. S. (1956) Quantum field theory in terms of vacuum expectation values, *Phys. Rev.* **101**, pp. 860–866.
9. Naudts, J. and Kuna, M. (2001) Covariance systems, *J. Phys. A: Math. Gen.* **34**, pp. 9265–9280.
10. Carey, A. L., Gaffney, J.M., and Hurst, C. A. (1977) A C^*-algebra formulation of the quantization of the electromagnetic field, *J. Math. Phys.* **18**, pp. 629–640.
11. Kuna, M. and Naudts, J. (2002) *Covariance approach to the free photon field, in: Probing the structure of Quantum Mechanics: nonlinearity, nonlocality, computation and axiomatics*, eds. D. Aerts, M. Czachor, and T. Durt, World Scientific, Singapore, 368–393.

DISCRETIZATION OF A SINE-GORDON TYPE EQUATION

Y. Ohta
Information Engineering, Graduate School of Engineering, Hiroshima University
1-4-1 Kagamiyama, Higashi-Hiroshima 739-8527, Japan

Abstract   An integrable modification of the double sine-Gordon equation is discretized by using Hirota's bilinear theory. The soliton solution is given in terms of the discrete Gram type determinant and the bilinear equations are reduced to the Jacobi formula for determinant.

1 INTRODUCTION

Hirota's direct method is one of the most powerful tools to construct both the integrable systems and their solutions. For instance, for a given integrable equation, the bilinear method provides a simple and direct way to derive the solutions through the bilinear form. For a given continuous integrable equation, it also enables us to construct integrable discrete analogues which share the common solutions with the original continuous equation. For a given function which has desirable properties, by using the direct method we can also generate integrable systems which allow that function as a solution.

In this paper, we shall demonstrate how the direct method works for the purpose of constructing integrable system and discretizing it taking the double sine-Gordon equation as an example. The double sine-Gordon equation is one of the famous NON integrable equations.[1] It allows a double kink solution as a traveling wave solution, but two double kinks do not have elastic collision. In this sense, the double sine-Gordon equation is not a soliton equation. First, we shall give an integrable modification of the double sine-Gordon equation so that it allows multi double kink solution. Next we will show the integrable discretization of the modified double sine-Gordon equation by using the bilinear theory. The solution is written in terms of the discrete Gram type determinant.

L. Faddeev et al. (eds.),
Bilinear Integrable Systems: From Classical to Quantum, Continuous to Discrete, 225–230.
© 2006 *Springer. Printed in the Netherlands.*

2   DOUBLE SINE-GORDON EQUATION

The double sine-Gordon equation has the following form

$$u_{xt} = \sin u + \frac{A}{2} \sin 2u$$

or equivalently

$$u_{xt} = \sin u + A \sin u \cos u \tag{1}$$

where A is a nonzero constant. The above equation possesses a double kink solution which is written by log of a rational function of exponential function. The double sine-Gordon Eq. (1) itself does not have multi double kink solution because the collision of two double kinks is not elastic. On the other hand, for soliton equations, the τ functions of soliton solutions are given in terms of determinant whose components are polynomials of exponential function. Now we regard that the double kink solution of (1) is derived from a special case of determinant (in fact 1×1 determinant) and by generalizing it to the determinant of arbitrary size, we obtain the bilinear equations satisfied by the determinant which will lead to the integrable modification of the original double sine-Gordon equation. By using this procedure, we can integrablize the non integrable Eq. (1).

3   MODIFIED DOUBLE SINE-GORDON EQUATION

The modified double sine-Gordon equation is given by

$$u_{xt} = \sin u + A \int^x (\sin u)_t \, dx \cos u \tag{2}$$

whose difference with (1) is just an integration by x and differentiation by t of a term $\sin u$. It is clear that the traveling wave solutions of (2) and (1) are identical. Moreover (2) allows multi double kink solution.

By using the dependent variable transformation

$$u = \mathrm{i} \log \frac{f^* \bar{f}}{f \, \bar{f}^*}$$

the modified double sine-Gordon Eq. (2) is bilinearized into the following bilinear form

$$2 D_x D_t f \cdot f = f^2 - \bar{f} \underline{f}$$

$$(\frac{2}{c} D_x - 1) \bar{f}^* \cdot f + \bar{f} f^* = 0$$

$$(2c D_t - 1) f^* \cdot f + \bar{f}^* \underline{f} = 0$$

where f is an auxiliary variable, $*$ means the complex conjugate and $A = c^2$. The N double kink solution is given in the following Gram determinant form

$$f = \det\left(\delta_{ij} + i\frac{P_i - c}{P_i + P_j}e^{\xi_i}\right)_{1 \le i, j \le N}$$

$$\bar{f} = \det\left(\delta_{ij} - i\frac{P_i + c}{P_i + P_j}e^{\xi_i}\right)_{1 \le i, j \le N} \tag{3}$$

$$\xi_i = P_i x + \frac{P_i}{P_i^2 - c^2}t + \xi_i^{(0)}$$

where P_i and $\xi_i^{(0)}$ are the wave number and phase constant of i-th double kink, respectively. The Jacobi formula for the Gram determinants (3) reduces to the above bilinear equations.

4 DISCRETE MODIFIED DOUBLE SINE-GORDON EQUATION

We apply the usual discretization procedure to the above modified double sine-Gordon equation based on the bilinear theory. By introducing discrete independent variables with keeping the structure of determinant solution, the discrete analogues of bilinear equations automatically follow from the same algebraic identities as continuous case. Let us denote the τ function of discrete case as f_{kl} instead of the continuous one $f(x, t)$. Guided by the case of the sine-Gordon equation and its discrete analogue (see appendix), we obtain the following result.

In the discrete case, the τ functions are given as

$$f_{kl} = \det\left(\delta_{ij} + i\frac{P_i - c}{P_i + P_j}e^{\xi_i}\right)_{1 \le i, j \le N}$$

$$\bar{f}_{kl} = \det\left(\delta_{ij} - i\frac{P_i + c}{P_i + P_j}e^{\xi_i}\right)_{1 \le i, j \le N}$$

which are completely same with the continuous ones (3). The difference between the continuous and discrete appears in the exponent ξ_i only

$$\xi_i = k \log\frac{1 + a(P_i - c)}{1 - a(P_i + c)} + l \log\frac{1 + b/(P_i - c)}{1 - b/(P_i + c)} + \xi_i^{(0)}$$

where a and b are the difference intervals for k and l, respectively. This means that the space of solution is common for both continuous and discrete and the compatible flows of continuous and discrete evolutions are introduced in the space of solution. The discrete bilinear equations satisfied by the above τ

functions are

$$(1 - ab)f_{k+1,l+1}f_{kl} = f_{k+1,l}f_{k,l+1} - ab\bar{f}_{k+1,l}\underline{f}_{k,l+1}$$

$$(1 - ac)\bar{f}^*_{k+1,l}f_{kl} - \bar{f}^*_{kl}f_{k+1,l} + ac\bar{f}_{k+1,l}f^*_{kl} = 0$$

$$(c - b)f^*_{k,l+1}f_{kl} - cf^*_{kl}f_{k,l+1} + b\bar{f}^*_{kl}\underline{f}_{k,l+1} = 0$$

By using the dependent variable transformation

$$u_{kl} = \frac{1}{2i} \log \frac{f^*_{kl}\bar{f}_{k,l-1}}{f_{kl}\bar{f}^*_{k,l-1}}$$

$$v_{kl} = \frac{1}{2i} \log \frac{f^*_{kl}\bar{f}_{k+1,l}}{f_{kl}\bar{f}^*_{k+1,l}}$$

we obtain the integrable discrete analogue of the integrable modification of double sine-Gordon Eq. (2)

$$\sin(u_{k+1,l+1} - u_{k+1,l} - u_{k,l+1} + u_{kl} + \theta_{kl})$$

$$= a\frac{b - c}{1 - ac}(\cos \varphi_{kl} \sin(u_{k+1,l+1} + u_{k,l+1}) + \cos \varphi_{k,l+1} \sin(u_{k+1,l} + u_{kl}))$$

$$u_{k+1,l} + u_{kl} = v_{kl} + v_{k,l-1} + \frac{1}{2} \sum_{j=-\infty}^{k} (\theta_{jl} - \theta_{j-1,l-1})$$

where θ and φ are defined in term of u and v by

$$\theta_{kl} = \arg \frac{1 + ac(e^{-2iv_{kl}} - 1)}{1 + ac(e^{2iv_{kl}} - 1)}$$

$$\varphi_{kl} = u_{k+1,l} + u_{kl} + \sum_{j=-\infty}^{l-1} (-1)^{l-j}(2u_{k+1,j} + 2u_{k,j+1} - \theta_{kj})$$

5   CONCLUDING REMARKS

The integrablization of the double sine-Gordon equation and its solution are given by using Hirota's direct method in soliton theory. We also gave the discretization of the modified double sine-Gordon equation and its solution in terms of the discrete Gram type determinant. For a given non integrable system, starting from a special solution and embedding it into a certain space of solution of integrable equations, we can construct an integrablization of the non integrable system. By introducing compatible discrete flows of evolution on the space of solution, integrable discrete analogues of the original continuous

equation can be derived. We expect that these integrablization and discretization techniques are applicable for interesting and important systems in various
fields.

APPENDIX

The sine-Gordon equation

$$u_{xt} = \sin u$$

is bilinearized into the bilinear form

$$2D_x D_t f \cdot f = f^2 - f^{*2}$$

through the dependent variable transformation

$$u = 2i \log \frac{f^*}{f}$$

The N kink solution is given by

$$f = \det \left(\delta_{ij} + \frac{i}{p_i + p_j} e^{\xi_i} \right)_{1 \leq i,j \leq N}$$

$$\xi_i = p_i x + \frac{1}{p_i} t + \xi_i^{(0)}$$

where p_i and $\xi_i^{(0)}$ are the wave number and phase constant of i-th kink, respectively.

   Hirota proposed the integrable discrete analogue of the sine-Gordon equation[2]

$$\sin(u_{k+1,l+1} - u_{k+1,l} - u_{k,l+1} + u_{kl})$$
$$= ab \sin(u_{k+1,l+1} + u_{k+1,l} + u_{k,l+1} + u_{kl})$$

which is transformed into the bilinear form

$$(1 - ab) f_{k+1,l+1} f_{kl} = f_{k+1,l} f_{k,l+1} - ab f_{k+1,l}^* f_{k,l+1}^*$$

through the dependent variable transformation

$$u_{kl} = \frac{1}{2i} \log \frac{f_{kl}^*}{f_{kl}}$$

where a and b are the difference intervals for k and l, respectively. The N kink solution is given by

$$f_{kl} = \det\left(\delta_{ij} + \frac{i}{p_i + p_j}e^{\xi_i}\right)_{1 \le i,j \le N}$$

$$\xi_i = k\log\frac{1 + ap_i}{1 - ap_i} + l\log\frac{1 + b/p_i}{1 - b/p_i} + \xi_i^{(0)}$$

The structure of the determinant solutions is quite same for continuous and discrete cases. Only the dispersion relations for the independent variables in the exponent ξ_i are different between (x, t) and (k, l).

REFERENCES

1. Bullough, R. K., Caudrey, P. J., and Gibbs, H. M. (1980) The double sine-Gordon equations: A physically applicable system of equations, in: *Solitons*, ed. R. K. Bullough and P. J. Caudrey, Springer-Verlag, Berlin, Heidelberg, pp. 107–141.
2. Hirota, R. (1977) Nonlinear Partial Difference Equations. III. Discrete Sine-Gordon Equation, *J. Phys. Soc. Jpn.* **43**, pp. 2079–2086.

HIERARCHY OF QUANTUM EXPLICITLY SOLVABLE AND INTEGRABLE MODELS

A.K. Pogrebkov
Steklov Mathematical Institute, Moscow, Russia

Abstract   Realizing bosonic field $v(x)$ as current of massless (chiral) fermions we derive hierarchy of quantum polynomial interactions of the field $v(x)$ that are completely integrable and lead to linear evolutions for the fermionic field. It is proved that in the classical limit this hierarchy reduces to the dispersionless KdV hierarchy. Application of our construction to quantization of generic completely integrable interaction is demonstrated by example of the mKdV equation.

Keywords: Quantum integrable models, fermionization, dispresionless KdV hierarchy

1   INTRODUCTION

Special quantum fields that first appeared in the literature (see, e.g., [1]) under the name "massless two-dimensional fermionic fields," are known for decades to be useful tool of investigation of completely integrable models in quantum (fermionization procedure [2–4]) and in classical (symmetry approach to KP hierarchy [5]) cases. Already in [2] it was shown that when bosonic field of the quantum version of some integrable model is considered as a composition of fermions, the most nonlinear parts of the quantum bosonic Hamiltonian becomes bilinear in terms of these Fermi fields. In [6–8] the same property was proved for the nonlinear Schrödinger equation and some integrable models of statistical physics, where fermionic fields naturally appeared in the so-called limit of the infinite interaction, i.e., again as describing the most nonlinear part of the Hamiltonian. Quantization of the KdV equation is based on analogy of the Gardner–Zakharov–Faddeev (GZF) [9] and Magri [10] Poisson brackets with the current and Virasoro algebras [4, 11, 12]. In [4] we proved that quantization of any of these brackets for the KdV equation by means of fermionization procedure can be performed on the entire x-axis and the Hamiltonian is given as sum of two terms, bilinear with respect to either fermionic or current operators.

L. Faddeev et al. (eds.),
Bilinear Integrable Systems: From Classical to Quantum, Continuous to Discrete, 231–244.
© 2006 *Springer. Printed in the Netherlands.*

We also proved that the quantum dispersionless KdV equation generates linear evolution equation for the Fermi field. Thus this equation is explicitly and uniquely solvable for any instant of time (in contrast to the classical case).

In this article we construct hierarchy of nonlinear interactions for the real bosonic quantum field $v(x)$ that obeys quantized version of the GZF bracket, i.e., commutator relation (2.9) below. The hierarchy itself is determined by the following conditions:

- All equations of this hierarchy are completely integrable in the sense that they have infinite set of local, polynomial (with respect to v and its derivatives) commuting integrals of motion.
- All equations of this hierarchy are explicitly solvable in the following sense. Let v be realized as current of fermionic field ψ. Then all these nonlinear equations for v lead to linear evolution equations for ψ.

We prove that these conditions uniquely determine hierarchy and that in the limit $\hbar \to 0$ this hierarchy reduces to the dispersionless KdV hierarchy. The paper is as follows. In Section 2 we present some well-known results on the "two-dimensional massless" fermions. In the Section 3 the hierarchy is derived and its properties are studied. In Section 4 we demonstrate by means of the modified KdV equation that results of our construction can be applied to quantization of the generic integrable models. Discussion of the classical limit of the hierarchy and some concluding remarks are given in the Section 5. Preliminary version of this article see in [13], more detailed presentation will be given in [14].

2 MASSLESS TWO-DIMENSIONAL FERMIONS

Here we introduce notations and list some standard properties of the massless Fermi fields (see, e.g., [1]). Let $\mathcal{H}$ denote the fermionic Fock space generated by operators $\psi(k)$ and $\psi^*(k)$, where $*$ means Hermitian conjugation, and that obey canonical anticommutation relations,

$$\{\psi^*(k), \psi(p)\}_+ = \delta(k - p), \qquad \{\psi(k), \psi(p)\}_+ = 0. \tag{1}$$

Let $\Omega \in \mathcal{H}$ denote vacuum vector and $\psi(k < 0)$ and $\psi^*(k > 0)$ be annihilation operators,

$$\psi(k)\Omega\Big|_{k<0} = 0, \qquad \psi^*(k)\Omega\Big|_{k>0} = 0, \tag{2}$$

whereas $\psi(k > 0)$ and $\psi^*(k < 0)$ are creation operators. Fermionic field is the Fourier transform,

$$\psi(x) = \frac{1}{\sqrt{2\pi}} \int dk \, e^{ikx} \psi(k), \tag{3}$$

and obeys relations

$$\{\psi^*(x), \psi(y)\}_+ = \delta(x - y), \qquad \{\psi(x), \psi(y)\}_+ = 0, \qquad (4)$$

$$(\Omega, \psi(x)\psi^*(y)\Omega) = (\Omega, \psi^*(x)\psi(y)\Omega) = \frac{-i\varepsilon^2}{x - y - i0}, \qquad (5)$$

where we denoted $\varepsilon = (\sqrt{2\pi})^{-1}$. This notation is convenient as in order to restore the Plank constant $\hbar$ we need not only to substitute all commutators and anticommutators $[\cdot, \cdot] \to [\cdot, \cdot]\hbar^{-1}$, but also put

$$\varepsilon = \sqrt{\frac{\hbar}{2\pi}}. \qquad (6)$$

The current of the massless two-dimensional fermionic field is given by the bilinear combination

$$v(x) = \varepsilon^{-1} : \psi^* \psi : (x), \qquad (7)$$

where the sign $: \ldots :$ denotes the Wick ordering with respect to the fermionic creation-annihilation operators, for example, $: \psi^*(x)\psi(y) : = \psi^*(x)\psi(y) - (\Omega, \psi^*(x)\psi(x)\Omega)$ and $: \psi^*\psi : (x) = \lim_{y \to x} : \psi^*(x)\psi(y):$, etc. Current is a self-adjoint operator-valued distribution in the space $\mathcal{H}$ obeying the following commutation relations:

$$[\psi(x), v(y)] = \varepsilon^{-1}\delta(x - y)\psi(x), \qquad (8)$$

$$[v(x), v(y)] = i\delta'(x - y). \qquad (9)$$

The charge of the fermionic field, $\Lambda = \int dx \, v(x)$, is self-adjoint operator with spectrum $\sqrt{2\pi\hbar}\mathbb{Z}$.

Commutation relation (29) suggests interpretation of $v(x)$ as bosonic field that obeys quantized version of the GZF bracket ([9], see also (58) below). In what follows, we use the decomposition

$$v(x) = v^+(x) + v^-(x) \qquad (10)$$

of this field, where positive and negative parts equal

$$v^\pm(x) = \frac{\pm 1}{2\pi i} \int \frac{dy \, v(y)}{y - x \mp i0} \qquad (11)$$

and admit analytic continuation in the upper and bottom half-planes of variable x, correspondingly. They are mutually conjugate and

$$v^-(x)\Omega = 0. \qquad (12)$$

Let

$$v(k) = \int dx \, e^{-ikx} v(x), \qquad (13)$$

so that $v^\pm(x) = (2\pi)^{-1} \int dk\, e^{ikx}\theta(\pm k)v(k)$, where $\theta(k)$ is step function. Then

$$v^*(k) = v(-k), \qquad v(k)\Omega\big|_{k<0} = 0. \tag{14}$$

Thus $v(k > 0)$ and $v(k < 0)$ are bosonic creation and annihilation operators, correspondingly, that are bilinear with respect to fermionic ones. One can introduce the bosonic Wick ordering for the products of currents, which we denote by the symbol $\vdots \ldots \vdots$, that means that all positive components of the currents are placed to the left from the negative components, for instance,

$$\vdots v(x)v(y)\vdots = v^+(x)v^+(y) + v^+(x)v^-(y) + v^+(y)v^-(x) + v^-(x)v^-(y) \tag{15}$$

and again $\vdots v^2\vdots(x) = \lim_{y\to x} \vdots v(x)v(y)\vdots$. We can also use equality

$$\vdots v(x)v(y)\vdots = v(x)v(y) - (\Omega, v(x)v(y)\Omega), \tag{16}$$

where

$$(\Omega, v(x)v(y)\Omega) = \left(\frac{i\varepsilon}{x - y - i0}\right)^2. \tag{17}$$

Fermionization procedure is essentially based on the relation between these two normal orderings. The bosonic ordering $\vdots \ldots \vdots$ can be extended for expressions that include fermionic field:

$$\vdots v(x)\psi(y)\vdots = v^+(x)\psi(y) + \psi(y)v^-(x). \tag{18}$$

Then by (8) and (11),

$$\vdots v(x)\psi(y)\vdots =: v(x)\psi(y): + i\varepsilon\frac{\psi(x) - \psi(y)}{x - y}, \tag{19}$$

so that this expression as well as its derivatives w.r.t. x and y are well-defined in the limit $y \to x$. In this limit one uses the obvious fact that under the sign of the fermionic normal product any expression of the kind $: \ldots \psi(x) \ldots \psi(x) \ldots :$ equals to zero. In particular, we get relation

$$\vdots v\psi\vdots(x) = i\varepsilon\psi_x(x), \tag{20}$$

that results in the bosonization of fermions [2–3]. More exactly, one can integrate this equality and write (at least formally) that

$$\psi(x) = \vdots e^{-i\varepsilon^{-1}\int^x v(x)\,dx}\vdots \equiv e^{-i\varepsilon^{-1}\int^x v^+(x)\,dx}e^{-i\varepsilon^{-1}\int^x v^-(x)\,dx}, \tag{21}$$

where in the second equality definition of the bosonic normal product was used for the exponent. Relation (21) needs special infrared regularization of the primitive of the current, $\int^x v(x)\,dx$, and its positive and negative components.

This procedure can be performed, say, like in [3], and it leads to a special constant operator conjugated to the charge Λ, that must be included in the r.h.s. of (21).

3 HIERARCHY OF EXPLICITLY SOLVABLE MODELS

Problems of interpretation of Eq. (21) do not appear if we deal with bilinear combinations of fermionic fields of the type (7). In this case neither infrared regularization, nor the above-mentioned auxiliary operator are needed and product of Fermi fields is given directly in terms of the current. An analog of such relation is known in the literature on the symmetries of the KP and KdV hierarchies (see [5]) in the sense of formal series. Let us denote

$$F(x, y) = \varepsilon^{-1} : \psi^*(x + y)\psi(x - y) :. \tag{22}$$

In [13, 14] it is proved that in the sense of operator-valued distribution with respect to x we have equality

$$F(x, y) = \varepsilon \frac{: \exp\left(i\varepsilon^{-1} \int_{x-y}^{x+y} dx'\, v(x')\right) : - 1}{2iy}, \tag{23}$$

where both sides are smooth, infinitely differentiable functions of y. In particular,

$$F(x, 0) = v(x). \tag{24}$$

Let us introduce

$$F_n(x) \equiv \left(\frac{\varepsilon \partial_y}{2i}\right)^n F(x, y)\Big|_{y=0} = \frac{\varepsilon^{n-1}}{(2i)^n} D^n(: \psi^* \cdot \psi :)(x), \tag{25}$$

where in the second equality we used notation for the Hirota derivative [5], that in the generic case of two functions $f(x)$ and $g(x)$ reads as

$$D^n(f \cdot g)(x) = \lim_{y \to 0} \frac{\partial^n}{\partial y^n} f(x + y)g(x - y), \quad n = 1, 2, \ldots. \tag{26}$$

In particular, by (24) we get that

$$F_0(x) = v(x), \tag{27}$$

$$F_1(x) = \frac{1}{2i} D(: \psi^* \cdot \psi :)(x), \tag{28}$$

that are current and energy–momentum density of the massless fermi-field, correspondingly. Thus Eqs. (22) and (23) give relation of the Hirota derivatives of the fermionic fields with polynomials of the current and its derivatives. All

$F_n(x)$ by (25) are self-adjoint operator-valued distributions on the Fock space $\mathcal{H}$ and by (23) we get recursion relations

$$F_{2n+1}(x) = \frac{1}{2n+2} \sum_{m=0}^{n} \frac{(-i\varepsilon/2)^{2m}(2n+1)!}{(2m)!(2(n-m))!} :v^{(2m)}(x)F_{2(n-m)}(x):,$$

$$n = 0, 1, 2, \ldots, \tag{29}$$

and

$$F_{2n}(x) = \frac{1}{2n+1} \sum_{m=0}^{n-1} \frac{(-i\varepsilon/2)^{2m}(2n)!}{(2m)!(2(n-m)-1)!} :v^{(2m)}(x)F_{2(n-m)-1}(x):$$

$$+ \left(\frac{\varepsilon}{2i}\right)^n \frac{v^{(2n)}(x)}{2n+1}, \quad n = 1, 2, 3, \ldots, \tag{30}$$

where F_0 is given in (27). The lowest simplest examples are as follows:

$$F_1(x) = \frac{1}{2}:v^2:(x), \tag{31}$$

$$F_2(x) = \frac{1}{3}:v^3:(x) - \frac{\varepsilon^2 v_{xx}(x)}{12}, \tag{32}$$

$$F_3(x) = \frac{1}{4}:v^4:(x) - \frac{\varepsilon^2}{4}:v(x)v_{xx}(x):, \tag{33}$$

$$F_4(x) = \frac{1}{5}:v^5:(x) - \frac{\varepsilon^2}{2}:v^2(x)v_{xx}(x): + \frac{\varepsilon^4 v_{xxxx}(x)}{80}. \tag{34}$$

By definition (22) operator $F(x, y)$ obeys commutation relation

$$[F(x, y), F(x', y')] = -\varepsilon^{-1}\delta(x - x' + y + y')F(x + y', y + y') \tag{35}$$
$$+ \varepsilon^{-1}\delta(x - x' - y - y')F(x' + y, y + y')$$
$$+ i\frac{\delta(x - x' + y + y') - \delta(x - x' - y - y')}{y + y'},$$

that generates corresponding commutation relations for F_m (closely related with a representation of the gl_∞-algebra). Only the lowest terms, F_0 and F_1, form closed subalgebras:

$$[F_0(x), F_0(x')] = i\delta'(x - x'), \tag{36}$$

$$[F_0(x), F_1(x')] = i\delta'(x - x')F_0(x'), \tag{37}$$

$$[F_1(x), F_1(x')] = i\{F_1(x) + F_1(x')\}\delta'(x - x') - \frac{i\varepsilon^2}{12}\delta'''(x - x') \tag{38}$$

while commutators of the type $[F_m, F_n]$ include F_j's till F_{m+n-1}.

Operator $F(x, y)$ admits integration with respect to x along the entire axis and result of integration is well-defined operator in the fermionic Fock space

$\mathcal{H}$. Indeed, by (3) and (22)

$$\int dx\, F(x,\, y) = \frac{1}{\varepsilon} \int_{0}^{\infty} dk (e^{2iky} \psi^*(-k)\psi(-k) - e^{-2iky}\psi(k)\psi^*(k)), \quad (39)$$

where expression in the r.h.s. is normally ordered and has creation × annihilation form, so that thanks to (2)

$$\int dx\, F(x,\, y)\Omega = 0 \qquad (40)$$

for any y. From here we derive that all operators

$$H_n \equiv \int dx\, F_n(x) = \frac{1}{\varepsilon} \int_{0}^{\infty} dk (\varepsilon k)^n (\psi^*(-k)\psi(-k) - (-1)^n \psi(k)\psi^*(k))$$

$$(41)$$

are well-defined and self-adjoint. For odd n they are positively defined. At the same time by (35) we get

$$\left[\int dx\, F(x,\, y), \int dx'\, F(x',\, y') \right] = 0 \qquad (42)$$

for any y and y'. This means in particular that all

$$[H_m,\, H_n] = 0, \quad m,\, n = 0,\, 1,\, \dots \qquad (43)$$

In other words, these operators define commuting flows on the space $\mathcal{H}$ and we can introduce hierarchy of integrable time evolutions by means of commutation relation

$$v_{t_m}(x) = i[H_m,\, v(x)], \quad m = 0,\, 1,\, \dots, \qquad (44)$$

so that by (43): $(\partial_{t_m}\partial_{t_n} - \partial_{t_n}\partial_{t_m})v(x) = 0$ for any m and n (we do not indicate the time dependence in all cases where it is not necessary). On the other side, by (35)

$$\left[\int dx\, F(x,\, y), v(x') \right] = \varepsilon^{-1}[F(x'+y,\, y) - F(x'-y,\, y)] \qquad (45)$$

$$\equiv \frac{1}{2iy} \left\{ :\exp\left(i\varepsilon^{-1} \int_{x'}^{x'+2y} d\xi\, v(\xi) \right): - :\exp\left(i\varepsilon^{-1} \int_{x'-2y}^{x'} d\xi\, v(\xi) \right): \right\},$$

that leads to highly nonlinear (polynomial) dynamic equations for $v(x)$ in all cases with exception to t_0 and t_1. Thanks to (25), (41), and (44) we have

$$v_{t_0}(x) = 0, \qquad (46)$$

$$v_{t_1}(x) = v_x(x), \qquad (47)$$

and in the generic situation

$$v_{t_n}(x) = \frac{\partial}{\partial x} \sum_{m=0}^{\left[\frac{n-1}{2}\right]} \frac{(i\varepsilon/2)^{2m} n! \, \partial_x^{2m} F_{n-2m-1}(x)}{(n-2m-1)!(2m+1)!}, \quad n = 1, 2, \ldots \tag{48}$$

The simplest examples are as follows:

$$v_{t_2}(x) = \partial_x :v^2:(x), \tag{49}$$

$$v_{t_3}(x) = \partial_x \left(:v^3:(x) - \frac{\varepsilon^2}{2} v_{xx}(x) \right), \tag{50}$$

$$v_{t_4}(x) = \partial_x \left(:v^4:(x) - 2\varepsilon^2 :vv_{xx}:(x) - \varepsilon^2 :v_x^2:(x) \right). \tag{51}$$

These polynomial interactions are closely related to the KdV hierarchy: the second evolution is just dispersionless quantum KdV (cf. [4]), the third evolution coincide with the modified KdV equation for some specific value of the interaction constant, and so on. In the next section we discuss the case of mKdV equation in more detail. Here we emphasize that in spite of the highly nonlinear form of all these equations in terms of the field v, all of them give linear evolutions for fermions. Indeed, introducing the time dependence of $\psi(x)$ in analogy with (44) as $\psi_{t_m} = i[H_m, \psi]$, we get by (41)

$$\psi_{t_m}(x) = \frac{1}{i\varepsilon} (i\varepsilon \partial_x)^m \psi(x), \tag{52}$$

or by (3) $\psi_{t_m}(k) = (i\varepsilon)^{-1} (-\varepsilon k)^m \psi(k)$. Let now $\psi(t_m, x)$, $v(t_m, x)$, and $F(t_m, x, y)$ be operators with time evolution given by some H_m and determined by the condition that at $t_m = 0$ they equal to $\psi(x)$, $v(x)$, and $F(x, y)$, correspondingly. Thanks to (40) the definitions of the both normal products do not depend on time. This means that these operators are related at arbitrary value of t_m by means of the same Eqs. (7), (22), (23), and (27) as at $t_m = 0$. In particular, by (22)

$$F(t_m, x, y) = \frac{1}{\varepsilon} : \psi^*(t_m, x + y)\psi(t_m, x - y) : . \tag{53}$$

Then, thanks to (3), (22), and (52) we get *explicit* expression for $F(t_m, x, y)$ in terms of its initial value $F(x, y)$:

$$F(t_m, x, y) = \frac{2}{(2\pi)^2} \int dx' \int dy' \int dk \int dp F(x - x', y - y') \tag{54}$$
$$\times \exp(i(k - p)x' + i(k + p)y' + i\varepsilon^{m-1}(k^m - p^m)t_m).$$

Thanks to (25) and (27) we obtain for $y = 0$:

$$v(t_m, x) = \frac{2}{(2\pi)^2} \int dx' \int dy' \int dk \int dp F(x - x', y') \tag{55}$$
$$\times \exp(i(k - p)x' - i(k + p)y' + i\varepsilon^{m-1}(k^m - p^m)t_m).$$

Substituting here $F(x, y)$ by means of (23) we get solution of the m's equation of the hierarchy (3.23) in terms of the initial data $v(x)$:

$$v(t_m, x) = \frac{1}{(2\pi)^2} \int dx' \int dy' \int dk \int dp \frac{:\exp\left(i\varepsilon^{-1} \int_{x-x'-\varepsilon y'}^{x-x'+\varepsilon y'} dx'' v(x'')\right): - 1}{2iy'}$$
$$\times \exp\left(ikx' - ipy' + i\frac{t_m}{2^m\varepsilon}[(p + \varepsilon k)^m - (p - \varepsilon k)^m)]\right). \tag{56}$$

Generalization to the case where time evolution is determined by a linear combination of Hamiltonians H_m is straightforward.

Thus we see, that all these models are not only completely integrable, but also explicitly solvable in the fermionic Fock space $\mathcal{H}$. On the other side, taking into account that thanks to (43) and (46) the charge operator $\Lambda = H_0/\sqrt{2\pi}$ commutes with all Hamiltonians and $v(x)$, one can reduce bosonic equations to the zero (or any other, fixed) charge sector of $\mathcal{H}$, that is exactly the standard bosonic Fock space. In that case all relations of the type (23) and (55) remain valid and give explicit solution of the hierarchy (44) in the bosonic Fock space.

4 THE MODIFIED KdV EQUATION

The modified Korteweg–de Vries (mKdV) equation

$$v_t = \partial_x \left(gv^3 - \frac{v_{xx}}{2}\right) \tag{57}$$

for the real function $v(t, x)$ is well-known example of the completely integrable differential equation. If $v(x)$ is a smooth real function that decays rapidly enough when $|x| \to \infty$, the inverse spectral transform (IST) method (see [15, 16] and references therein) is applicable to Eq. (57). Constant g in this equation is an arbitrary real parameter and properties of solutions essentially depend on its sign. In particular, the soliton solutions exist only if $g < 0$.

The mKdV equation is Hamiltonian system with respect to the GZF bracket [9],

$$\{v(x), v(y)\} = \delta'(x - y), \tag{58}$$

so that Eq. (57) can be written in the form $v_t = -\{H, v\}$, where Hamiltonian

$$H = \frac{1}{4} \int dx(g v^4(x) + v_x^2(x)) \tag{59}$$

The direct quantization of the mKdV equation on the whole axis requires some regularization (e.g., space cut-off) of the Hamiltonian in order to supply it with operator meaning. Any such regularization is incompatible with the IST already in the classical case: the continuous and discrete spectra of corresponding linear (Zakharov–Shabat) problem become mixed and the most interesting, soliton solutions cease to exist.

Here we show that realizing $v(x)$ as in (7), i.e., as a composition of fermionic fields we can avoid any cut-off procedure in (59), because the Hamiltonian becomes well-defined in the fermionic Fock space $\mathcal{H}$.

We choose the quantum Hamiltonian to be bosonically ordered expression (59),

$$H = \frac{1}{4} \int dx {:} g v^4(x) + v_x^2(x){:}. \tag{60}$$

Then, thanks to (33) we get

$$H = g H_3 + \frac{1 - g\varepsilon^2}{4} \int dx {:} v_x^2{:}(x), \tag{61}$$

where (41) for $n = 3$ was used. Thus, in analogy with the KdV case (see [4]), the most singular part of the Hamiltonian (60) that was of the fourth order with respect to bosonic operators is only of the second order with respect to fermions. Taking into account that by (13)

$$\int dx {:} v_x^2{:}(x) = 2 \int_0^\infty dk \, k^2 v(k)v(-k) \tag{62}$$

we get that both terms in (61) are bilinear in either fermionic, or bosonic creation–annihilation operators, they are normally ordered and have a diagonal form, i.e., they include "creation×annihilation" terms only. Correspondingly, both these terms are well-defined self-adjoint operators in $\mathcal{H}$ and under our quantization procedure no any regularization of the Hamiltonian is needed. In particular, by (14) and (40)

$$H\Omega = 0 \tag{63}$$

and by (41) and (62) the Hamiltonian (61) is positively defined when $\varepsilon^{-2} \geq g \geq 0$.

It is clear that time evolution given by the Hamiltonian (60),

$$v_t = i[H, v] \equiv \partial_x \left(g : v^3 : - \frac{v_{xx}}{2} \right), \tag{64}$$

is exactly the quantum version of the Eq. (57) normally ordered with respect to the bosonic operators. Thanks to (32) we can exclude the v^3-term and get the quantum **bilinear** form of the mKdV equation in terms of the fermionic fields:

$$v_t(x) = \frac{\partial}{\partial x} \left(3g F_2(x) + \frac{g\varepsilon^2 - 2}{4} v_{xx}(x) \right), \tag{65}$$

that can be considered as a quantum Hirota form of the mKdV equation.

In order to derive time evolution of the fermionic field ψ it is reasonable to rewrite the second term of (61) by means of the fermionic normal ordering. Omitting details we get by definitions of the both normal orderings and Eqs. (7) and (22) the equality

$$:v(x)v(y): =: v(x)v(y) : +\varepsilon \frac{F\left(\frac{x+y}{2}, \frac{x-y}{2}\right) - F\left(\frac{x+y}{2}, \frac{y-x}{2}\right)}{i(x - y)}, \tag{66}$$

that after differentiation gives in the limit $y \to x$

$$:v_x^2:(x) =: v_x^2 : (x) + \frac{1}{2}\partial_x^2 F_1(x) + \frac{2}{3\varepsilon^2} F_3(x), \tag{67}$$

where (25) was used and where by (7) $: v_x^2 : (x) = 2\varepsilon^{-2} : \psi_x^* \psi^* \psi_x \psi :$. Thus we can write (61) as

$$H = \frac{5g + \varepsilon^{-2}}{6} H_3 + \frac{\varepsilon^{-2} - g}{2} \int dx : \psi_x^* \psi^* \psi_x \psi : (x), \tag{68}$$

and thus time evolution of the fermionic field, $\psi_t = i[H, \psi]$ is given by equation

$$\psi_t(x) = -\frac{5g\varepsilon^2 + 1}{6} \psi_{xxx}(x) + \frac{g\varepsilon^2 - 1}{2i\varepsilon} : v_{xx} \psi : (x), \tag{69}$$

that is, of course, nonlinear when $g \neq \varepsilon^{-2}$.

Investigation of the spectrum of the quantum Hamiltonian deserves the separate studying. But like in the [4] it can be shown that in the fermionic Fock space $\mathcal{H}$ for $g < 0$ there exists one-soliton state, i.e., such state that the average of the field v with respect to it equals to the classical one-soliton solution at least at zero (or any fixed) instant of time. This state does not belong to the zero charge sector of $\mathcal{H}$, so it cannot exist in the standard (bosonic) quantization of the mKdV equation. Again, like in [4] it can be shown that existence of this state implies quantization of the soliton action variable.

5 CONCLUSION

We derived hierarchy of nonlinear integrable and at the same time solvable evolutions of the bosonic field $v(x)$ realized as composition of the fermionic fields–current. By (27) this means that $F_0(x)$ was chosen to be a dynamical variable. But the closed subalgebra of commutation relations (36)–(37) is given also by $F_0(x)$ and $F_1(x)$. Moreover, the linear combination

$$\widetilde{F}(x) = F_1(x) + a\partial_x F_0(x) \tag{70}$$

with real constant coefficient a also obeys closed commutation relation,

$$[\widetilde{F}(x), \widetilde{F}(x')] = i\{\widetilde{F}(x) + \widetilde{F}(x')\}\delta'(x - x') - i\left(a^2 + \frac{\varepsilon^2}{12}\right)\delta'''(x - x'), \tag{71}$$

as follows from (36)–(37). This means that $\widetilde{F}(x)$ gives another possible choice of a dynamical variable. In [4] we proved that the dispersionless KdV in this case is also solvable, while – in contrast to the above – it was $v(x)$ that evolved linearly. It is natural to expect that the same property is valid for the entire hierarchy (48) generated by the quantum version (71) of the Magri bracket.

Coefficients of the R.H.S. of the bosonic equations of motion (47)–(51) are uniquely (up to a common factor) fixed by recursion relations (29)–(30). Indeed, transformation

$$v(x) \rightarrow av(ax), \tag{72}$$

is the only canonical scaling transformation that is unitary implemented in $\mathcal{H}$. Here constant $a > 0$ in order to preserve definition (11) of positive and negative parts of v. This transformation generates

$$\psi(x) \rightarrow \sqrt{a}\psi(ax), \qquad F(x, y) \rightarrow aF(ax, ay), \qquad F_n(x) \rightarrow a^n F_n(ax), \tag{73}$$

that is compatible with (27)–(30). Thus by (41) $H_n \rightarrow a^{n-1} H_n$, and thanks to (48) transformation (72) can be compensated by rescaling of times: $t_n \rightarrow a^{1-n} t_n$.

Flows given in (27)–(34) are close to the flows of the KdV hierarchy [15]: they are polynomial with respect to $v(x)$ and its derivatives and have the same leading terms. On the other side, the lowest nontrivial example (32) shows that some essential terms that are involved in the KdV case are absent in (48). In fact, as it was natural to expect by [4], Eq. (32) is the dispersionless KdV equation: the term $v_{xxx}(x)$ is absent. The higher equations, like (33), (34), and so on already include terms with derivatives, so these equations are not the dispersionless ones. On the other side, coefficients of all such terms of all commuting flows introduced in Section 3 are proportional to powers of ε^2, i.e., of $\hbar$ by (6). Thanks to (27) and (29), (30) it is easy to see that in the limit $\hbar \to 0$

$$F_m(x) \to \frac{v^{m+1}(x)}{m+1}, \tag{74}$$

so that by (48) we get in the classical limit equations

$$\partial_{t_m} v(t_m, x) = m v^{m-1}(t_m, x) v_x(t_m, x), \tag{75}$$

i.e., the dispersionless KdV hierarchy. Solution of the initial problem for the mth equation can be written in the parametric form as

$$x = s - m t_m v^{m-1}(s), \qquad v(t_m, x) = v(s), \tag{76}$$

where $v(x)$ is initial data. This solution is known to describe overturn of the front, so the initial problem for the Eqs. (75) has no global solution. On the other side, Eq. (56) gives global solution of the quantum hierarchy (48). It is easy to see that in the limit $\varepsilon \to 0$ (i.e., $\hbar \to 0$) we get from (56)

$$v(t_m, x) = \frac{1}{(2\pi)^2 i} \int dx' \int dy' \int dk \int dp \frac{e^{iy'v(x-x')} - 1}{y'} e^{ikx' - ipy' + imt_m kp^{m-1}},$$

so that for the classical limit of (56) we get representation

$$v(t_m, x) = \int dp [\theta(v(x + m t_m p^{m-1}) - p) - \theta(-p)], \tag{77}$$

where $\theta(p)$ denotes the step function. It is easy to check that (77) coincides with the solution (76) of the classical equation (75) before the first overturn of the front.

Summarizing, it is natural to call the hierarchy introduced in the Section 3 **the quantum dispersionless KdV hierarchy**. Dispersionless limits of integrable hierarchies attract now essential attention in the literature, see [17, 18].

Our construction here is essentially based on the equalities (22) and (23) valid for the standard massless fermionic fields. Thanks to this relations we got description of the quantum dispersionless KdV hierarchy. It is natural to hypothesize that anyonic generalization [19] of the fermions leads to more generic integrable bozonic systems.

ACKNOWLEDGMENT

The author thanks Organizers of the NATO ARW "Bilinear Integrable Systems: from Classical to Quantum, Continuous to Discrete" (Elba, 2002) for kind hospitality and support of his participation. He also thanks Profs. P.P. Kulish and S.P. Novikov for fruitful discussions.

This work is supported in part by the Russian Foundation for Basic Research (grants # 02-01-00484 and 00-15-96046) and by the INTAS (grant # 99-1782).

REFERENCES

1. Wightman, A. (1964) *Introduction to Some Aspects of the Relativistic Dynamics of Quantized Fields*, Princeton University Press Princeton-New Jersey.
2. Coleman, S. (1975) *Phys. Rev.* **D11**, p. 2088; Mandelstam, S. (1975) *Phys. Rev.* **D11**, p. 3026.
3. Pogrebkov, A. K. and Sushko, V. N. (1975) *Theor. Math. Phys.* **24**, p. 937; (1976) **26**, p. 286.
4. Pogrebkov, A. K. (2001) *Theor. Math. Phys.* **129**, p. 1586
5. Miwa, T., Jimbo, M., and Date, E. *Solitons: Differential Equations, Symmetries and Infinite Dimensional Algebras*, Cambridge University Press Cambridge.
6. Colomo, F., Izergin, A. G., Korepin, V. E., and Tognetti V. (1933) *Theor. Math. Phys.* **94**, p. 11.
7. Bogoliubov, N. M., Izergin, A. G., and Korepin, V. E. (1989) *Quantum Inverse Scattering Method and Correlation Functions*, Cambridge University Press Cambridge.
8. Slavnov, N. A. (1996) *Theor. Math. Phys.* **108**, p. 993.
9. Gardner, C. (1971) *Math. Phys.* **12**, p. 1548; Zakharov, V. E. and Faddeev, L. D. (1971) *Funct. Anal. Appl.* **5**, p. 280.
10. Magri, F. (1980) A geometrical approach to the nonlinear solvable equations, in: *Nonlinear Evolution Equations and Dynamical Systems*, Lectures Notes in Physics, Vol. 120, eds. M. Boiti, F. Pempinelli, and G. Soliani, Springer, Berlin, p. 233.
11. Bazhanov, V. V., Lukyanov, S. L., and Zamolodchikov, A. B. (1996) *Commun. Math. Phys.* **177**, p. 381; (1997) **190**, p. 247; (1999) **200**, p. 297.
12. Di Francesco, P., Mathieu, P., and Sénéchal, D. (1992) *Mod Phys. Lett.* **A7**, p. 701.
13. Pogrebkov, A. K. nlin. (2002) SI/0202043.
14. Pogrebkov, A. K. (2003) *Russ. Math. Surv.*, (in press).
15. Novikov, S. P., Manakov, S. V., Pitaevsky, L. P., and Zakharov V. E. (1984) *Theory of Solitons. The Inverse Scattering Method*, New York.
16. Calogero, F. and Degasperis, A. (1982) *Spectral Transform and Solitons*, Vol. 1, Amsterdam, North-Holland.
17. Krichever, I. (1992) *Commun. Math. Phys.* **143**, p. 415; (1992) *Commun. Pure Appl. Math.* **47**, p. 437.
18. Boyarsky, A., Marshakov, A., Ruchayskiy, O., Wiegmann, P., and Zabrodin, A. (2001) *Phys. Lett.* **B515**, p. 483.
19. Ilieva, N. and Thirring, W. (1999) *Eur. Phys. J.* **C6**, p. 705; (1999) *Theor. Math. Phys.* **121**, p. 1294.

A TWO-PARAMETER ELLIPTIC EXTENSION OF THE LATTICE KdV SYSTEM

S.E. Puttock and F.W. Nijhoff
Department of Applied Mathematics, University of Leeds, Leeds
LS2 9JT, United Kingdom

1   INTRODUCTION

In [1] we presented a novel integrable lattice system given by the following coupled system of equations:

$$\left(a + b + u - \hat{\tilde{u}}\right)(a - b + \hat{u} - \tilde{u}) = a^2 - b^2 + f(\tilde{s} - \hat{s})\left(\hat{\tilde{s}} - s\right) \tag{1a}$$

$$\left(\hat{\tilde{s}} - s\right)(\tilde{w} - \hat{w}) = [(a + u)\tilde{s} - (b + u)\hat{s}]\hat{\tilde{s}} - \left[(a - \hat{u})\hat{s} - (b - \hat{u})\tilde{s}\right]s \tag{1b}$$

$$(\hat{s} - \tilde{s})\left(\hat{\tilde{w}} - w\right) = \left[(a - \tilde{u})s + (b + \tilde{u})\hat{\tilde{s}}\right]\hat{s} - \left[(a + \hat{u})\hat{\tilde{s}} + (b - \hat{u})s\right]\tilde{s} \tag{1c}$$

$$\left(a + u - \frac{\tilde{w}}{\tilde{s}}\right)\left(a - \tilde{u} + \frac{w}{s}\right) = a^2 - P(s\tilde{s}) \tag{1d}$$

$$\left(b + u - \frac{\hat{w}}{\hat{s}}\right)\left(b - \hat{u} + \frac{w}{s}\right) = b^2 - P(s\hat{s}) \tag{1e}$$

in which

$$P(x) \equiv \frac{1}{x} + 3e + fx,$$

with e and f being fixed parameters. We consider this system to be an "elliptic" extension of the lattice KdV equation by virtue of the fact that it is naturally associated with the elliptic curve $y^2 = P(x)$, where e and f are the moduli. When the curve degenerates, i.e., when $f = 0$, one immediately notes that the first equation (1a) decouples and we recover the lattice (potential) KdV equation for the variable u. To explain the notation used in (1), we mention that $u = u_{n,m}$, $w = w_{n,m}$, $s = s_{n,m}$ are the dependent variables, depending on the lattice variables $n, m \in Z$, and that the symbols $\tilde{\ }$ and $\hat{\ }$ denote lattice shifts in the n, m directions respectively, i.e., $\tilde{u} = u_{n+1,m}$, $\hat{u} = u_{n,m+1}$, $\hat{\tilde{u}} = u_{n+1,m+1}$ as indicated in Figure 1. Furthermore in (1) a and b denote lattice parameters, i.e., parameters associated with the lattice variables n and m respectively, in contrast to the parameters e and f (associated with the elliptic curve) which are fixed.

245

L. Faddeev et al. (eds.),
Bilinear Integrable Systems: From Classical to Quantum, Continuous to Discrete, 245–251.
© 2006 Springer. Printed in the Netherlands.

Figure 1. Elementary quadrilateral on which the lattice equation is defined

To explain the distinction between lattice parameters and fixed parameters, we recall that the integrability of lattice equations such as the lattice KdV equation can be understood in the following way: the integrability seems to entirely reside in a simple but deep combinatorial property, first described in the paper [12]. This property amounts to the fact that these integrable two-dimensional lattice equations should really be viewed as parameter-*families* (relative to the lattice parameters) of compatible equations which can be consistently embedded in a multidimensional lattice, on each two-dimensional sublattice of which a copy of the lattice equation can be defined. As was shown in [2, 3], cf. also [14], this property is powerful enough to derive subsequently Lax pairs for the lattice equations, which can then be used to study the analytic properties of solutions. A full classification of lattice equations of the type involving variables around elementary plaquettes was recently given in [5]. The richest equation in this classification is a lattice equation, first derived in [6], involving lattice parameters on an elliptic curve, forming the natural discrete analogue of the Krichever–Novikov equation, cf. [7].

In [1] we took a different position towards deriving latice systems associated with elliptic curves, in that we aimed at starting from an underlying structure expressed by means of an infinite matrix system. This is in the spirit of earlier publications [8, 9], where similar structures were exhibited in connection with the lattice KdV equation and other discrete systems. The extension of this construction to the elliptic case, which involves the use of an elliptic Cauchy kernel, was presented in [1]. We will not repeat the details here, but restrict ourselves to highlighting the main results.

2   ELLIPTIC LATTICE SYSTEM

We will exhibit here a number of key properties of the system (1).

2.1   Lax Pair

The Lax pair, which customarily is considered to be a clear indication of the integrability of the model, is given by the overdetermined set of linear discrete equations:

$$(a - k)\tilde{\phi} = L(K)\phi \qquad (2a)$$
$$(b - k)\hat{\phi} = M(K)\phi \qquad (2b)$$

in which the Lax matrices L and M are given by:

$$L(K) = \begin{pmatrix} a - \tilde{u} + \frac{f}{K}\tilde{s}w & 1 - \frac{f}{K}\tilde{s}s \\ K + 3e - a^2 + f\tilde{s}s & a + u - \frac{f}{K}\tilde{w}s \\ +(a - \tilde{u})(a + u) + \frac{f}{K}\tilde{w}w & \end{pmatrix} \qquad (3a)$$

$$M(K) = \begin{pmatrix} b - \hat{u} + \frac{f}{K}\hat{s}w & 1 - \frac{f}{K}\hat{s}s \\ K + 3e - b^2 + f\hat{s}s & b + u - \frac{f}{K}\hat{w}s \\ +(b - \hat{u})(b + u) + \frac{f}{K}\hat{w}w & \end{pmatrix} \qquad (3b)$$

with (k, K) on the elliptic curve representing the spectral parameter. It is straightforward to show that the discrete Lax equation arising from the compatibility condition of the linear system (2a), (2b),

$$\tilde{L}M = \hat{M}L, \qquad (4)$$

gives rise to the set of equations (1a)–(1e). We observe that the matrices L and M depend rationally on K only, and thus we have a rational dependence on the spectral variable. Nevertheless, the solutions seem to depend essentially on the elliptic curve as is apparet from the soliton type solutions presented in the next subsection.

2.2 Soliton Type Solutions

It is relatively straightforward from the structure exhibited in [1] to construct soliton solutions.

Introducing the $N \times N$ matrix M with entries

$$M_{ij} = \frac{1 - f/(K_i K_j)}{k_i + k_j} r_i, \qquad (i, j = 1, \ldots, N) \qquad (5)$$

where the parameters of the solution (k_i, K_i) are points on the elliptic curve:

$$k^2 = K + 3e + \frac{f}{K} \qquad (6)$$

and the column vector $r = (r_i)_{i=1,\ldots,N}$ with components

$$r_i = \left(\frac{a + k_i}{a - k_i}\right)^n \left(\frac{b + k_i}{b - k_i}\right)^m r_i^0, \qquad (7)$$

where the coefficients r_i^0 are independent of n, m.

We note that although the dynamics itself (encoded in the wave factors r_i) does not involve the elliptic curve, the soliton solutions essentially depend on the variables on the curve. In fact, it is easily verified by direct calculation that the formulae (5) provide a solution to the lattice system (1) if and only if the elliptic curve relation (6) holds between the parameters k_i and the parameters K_i.

To present the soliton type solutions of (1) we introduce an "elliptic" matrix U with entries $U_{i,j}$ where we have to distinguish between even and odd entries in the following way:

$$U_{2i,2j} = e \cdot K^i \cdot (1 + M)^{-1} \cdot K^j \cdot r \tag{8a}$$

$$U_{2i+1,2j} = e \cdot K^i \cdot k \cdot (1 + M)^{-1} \cdot K^j \cdot r \tag{8b}$$

$$U_{2i,2j+1} = e \cdot K^i \cdot (1 + M)^{-1} \cdot K^j \cdot k \cdot r \tag{8c}$$

$$U_{2i+1,2j+1} = e \cdot K^i \cdot k \cdot (1 + M)^{-1} \cdot K^j \cdot k \cdot r \tag{8d}$$

in which we have employed the row vector $e = (1, 1, \ldots, 1)$ and the diagonal matrices

$$K = \mathrm{diag}\,(K_1, K_2, \ldots, K_N), \quad k = \mathrm{diag}\,(k_1, k_2, \ldots, k_N)\,.$$

where (k_i, K_i) are points on the elliptic curve (6). We note that the formulae (8) can be thought of as introducing a quasi-gradation on the matrix U. Although it is not manifest, it can be easily shown that the matrix U is symmetric.

If we now select the following entries:

$$u = U_{0,0}, \quad s = U_{-2,0}, \quad h = U_{-2,-2}$$

$$v = 1 - U_{-1,0}, \quad w = 1 + U_{-2,1},$$

it can be shown that they obey the closed-form system of partial difference equations (1a)–(1e) in terms of u, s, and w. Alternatively, we could just as easily derive a lattice system in terms of h, s, and v which is equivalent to our lattice system (1a)–(1e).

2.3 Consistency of the Lattice System

We now address the question of how to define a well-posed initial value problem (IVP) for the lattice system. Motivated by the work on the lattice KdV equation, cf. [10], it is natural to investigate a *local* iteration scheme is given on "staircases," as in Figure 2, assigning initial values u_i for u and s_i for s on the vertices of this staircase, and to consider the discrete-time shift to be the map $(u_i, s_i) \mapsto (\hat{u}_i, \hat{s}_i)$.

We need in addition one "background" value w_0 at a specific point on the staircase.

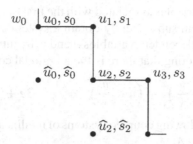

Figure 2. Staircase of initial values on the lattice

Setting the IVP up in this way is just a case of straightforward computation to show that it is well-posed. In fact, from (1a)–(1c) one can solve $\hat{\tilde{u}}$, $\hat{\tilde{s}}$, and $\hat{\tilde{w}}$ in a unique way, given the values of the other variables u, $\tilde{u}$, $\hat{u}$ as well as s, $\tilde{s}$, $\hat{s}$ and w, $\tilde{w}$, $\hat{w}$. Eqs. (1d) and (1e) link the variables $\tilde{w}$ and $\hat{w}$ to w and to the u,s-variables. Thus, it remains to be verified that the shifted forms $(\widehat{1d})$ and $(\widetilde{1e})$ trivialise through back-substitution of $\hat{\tilde{w}}$ which was already obtained. Also it is easily checked that the two ways of calculating $\hat{\tilde{w}}$ from either (1d) followed by $(\widetilde{1e})$, or from (1e) followed by $(\widehat{1d})$ are consistent. Thus, by a simple unambiguous computation the lattice system (1a)–(1e) is shown to be consistent.

2.4 Associated Continuous Systems

As was demonstrated in the past for the lattice systems studied in [8, 11, 12], there exist many compatible continuous systems associated with them. These form, in fact, the continuous symmetries for the lattice systems (whilst the lattice systems constitute the discrete symmetries for the corresponding continuous flows). We will give here a few of the simplest of such associated continuous flows for the purpose of identification of the associated lattice system.

We can derive a hierarchy of partial differential equations which are compatible with the discrete system (1). Thus, in [1] we derived the following set of coupled relations for the first member of this hierarchy:

$$u_t = \frac{1}{4}u_{xxx} + \frac{3}{2}u_x^2 - \frac{3}{2}fs_x^2 \tag{9a}$$

$$s_t = \frac{1}{4}s_{xxx} + \frac{3}{2}s_x u_x - \frac{3}{2}fh_x s_x \tag{9b}$$

$$h_t = \frac{1}{4}h_{xxx} + \frac{3}{2}s_x^2 - \frac{3}{2}fh_x^2 \tag{9c}$$

$$v_t = \frac{1}{4}v_{xxx} + \frac{3}{4}fs_x(h_x + s^2 - fh^2)_x + \frac{3}{2}v_x u_x \tag{9d}$$

$$w_t = \frac{1}{4}w_{xxx} + \frac{3}{4}s_x(u_x + u^2 - fs^2)_x - \frac{3}{2}fw_x h_x. \tag{9e}$$

where x and t are the variables associated with the first two nontrivial time flows in the hierarchy. We can subsequently obtain a system solely in terms of u, s and w, by eliminating dependent variables h and v, by introducing the quantity $A = -u + w/s$, and noting that there is the additional constraint

$$\left(u + \frac{w}{s} \right)_x + \left(u - \frac{w}{s} \right)^2 = \frac{1}{s^2} + 3e + f s^2 \tag{10}$$

Thus, we obtain the following coupled systems of nonlinear evolution equations solely in terms of s and A:

$$s_t = \frac{1}{4} s_{xxx} + \frac{3}{2} s_x \left[\frac{1}{s^2} + 3e + f s^2 - A^2 + A \frac{s_x}{s} - \frac{1}{2} \frac{s_{xx}}{s} \right] \tag{11a}$$

$$A_t = \frac{1}{4} A_{xxx} - \frac{3}{2} A^2 A_x + \frac{3}{2} A_x \left(\frac{1}{s^2} + 3e + f s^2 \right) + \frac{3}{4} \frac{s_x}{s} \left(\frac{1}{s^2} + 3e + f s^2 \right)_x \tag{11b}$$

Alternatively, eqs. (11) can be obtained by a rather subtle continuum limit from the lattice system (1). This system of PDEs is integrable in its own right and admits a Lax pair.

3 DISCUSSION

In this paper we presented an integrable system of partial difference equations associated with an elliptic curve. This system constitutes a *two-parameter deformation* of the lattice KdV system which was investigated in numerous papers e.g., [8, 13, 10, 3, 14], and which is recovered from (1) when the elliptic curve degenerates. The scheme for obtaining the elliptic extension was presented in detail in [1]. We believe this elliptic lattice system serves as a starting point for the derivation of a number of new discrete and continuous systems, which arise from reductions, and which will be the subject of future investigations.

We should mention that there exists an alternative way to extend the lattice systems of KdV type such that there is an underlying elliptic curve. V. Adler discovered in [6] a lattice version of the Krichever–Novikov equation, cf. [7]. The main difference between Adler's equation and the system (1) is that the lattice parameters for Adler's equation are points of the elliptic curve, and the Lax pair for it, presented in [3], has the spectral parameter living also on the elliptic curve. Nonetheless, the formulae for soliton solutions discussed in subsection 2.2, show the presence of the curve through the parameters (k_i, K_i), which seems a clear indication that the elliptic curve is essential in system (1) as well.

REFERENCES

1. Nijhoff, F. W. and Puttock, S. E. On a two-parameter extension of the lattice kdV system associated with an elliptic curve, nlin.SI/0212041, submitted to Intl. *J. Math. Phys.*
2. Nijhoff, F. W. and Walker, A. J. (2001) The discrete and continuous Painlevé VI hierarchy and the Garnier systems, *Glasgow Math. J.* **43A**, pp. 109–123, nlin.SI/0001054.
3. Nijhoff, F. W. (2002) Lax pair for the Adler (lattice Krichever–Novikov) system, *Phys. Lett.* **297A**, pp. 49–58.
4. Bobenko, A. I. and Suris, Yu B. Integrable systems on quad-graphs, nlin.SI/0110004.
5. Adler, V. E., Bobenko, A. I., and Suris, Yu B., Classification of integrable equations on quad-graphs. The consistency approach, nlin.SI/0202024.
6. Adler, V. E. (1998) Bäcklund Transformation for the Krichever-Novikov equation, *Intl. Math. Res. Notices*, **1**, pp. 1–4.
7. Krichever, I. M. and Novikov, S. P. (1979) Holomorphic fiberings and nonlinear equations, *Sov. Math. Dokl.* **20**, pp. 650–654; (1981) Holomorphic bundles over algebraic curves and nonlinear equations, *Russ. Math. Surv.* **35**, pp. 53–79.
8. Nijhoff, F. W., Quispel, G. R. W., and Capel, H. W. (1983) Direct linearization of difference-difference equations, *Phys. Lett.* **97A**, pp. 125–128.
9. Capel, H. W., Nijhoff, F. W., and Papageorgiou, V. G. (1991) Complete integrability of Lagrangian mappings and lattices of KdV type, *Phys. Lett.* **155A**, pp. 377–387.
10. Papageorgiou, V. G., Nijhoff, F. W., and Capel, H. W. (1990) Integrable mappings and nonlinear integrable lattice equations, *Phys. Lett.* **147A**, pp. 106–114.
11. Quispel, G. R. W., Nijhoff, F. W., Capel, H. W., and van der Linden, J. (1984) Linear integral equations and nonlinear difference-difference equations, *Physica* **125A**, pp. 344–380.
12. Nijhoff, F. W. and Capel, H. W. (1995) The discrete Korteweg-de Vries equation, *Acta Applicandae Mathematicae* **39**, pp. 133–158.
13. Quispel, G. R. W., Nijhoff, F. W., Capel, H. W., and van der Linden, J. (1984) Bäcklund transformations and singular integral equations, *Physica* **123A**, pp. 319–359.
14. Nijhoff, F. W., Ramani, A., Grammaticos, B., and Ohta, Y. (2001) On Discrete Painlevé equations associated with the lattice KdV systems and the Painlevé VI equation, *Stud. Appl. Math.* **106**, pp. 261–314.

TRAVELLING WAVES IN A PERTURBED DISCRETE SINE-GORDON EQUATION

Vassilis M. Rothos
School of Mathematical Sciences, Queen Mary, University of London,
Mile End London E1 4NS, UK

Michal Feckan
Department of Mathematical Analysis, Comenius University, Mlynska dolina,
842 48 Bratislava and Mathematical Institute, Slovak Academy of Sciences,
Stefanikova 49, 814 73 Bratislava, Slovakia

Abstract   The existence of traveling waves is studied analytical for discrete sine-Gordon equation with an inter-site potential. The reduced functional differential equation is formulated as an infinite dimensional differential equation which is reduced by a centre manifold method and to a 4-dimensional singular ODE with certain symmetries and with heteroclinic structure. The bifurcations of solutions from heteroclinic ones are investigated for singular perturbed systems.

Keywords: lattice sine-Gordon, center manifold reduction, normal form theory, bifurcations

1   INTRODUCTION

In recent years there has been a flurry of mathematical research arising from condensed matter physics and physical chemistry, namely the study of localised modes in anharmonic molecules and molecular crystals. Using classical approximations, these are described by nonlinear lattice equations. Most nonlinear lattice systems are not integrable even if the PDE model in the continuum limit is; (see [1, 2] and references therein). Prototype models for such nonlinear lattices take the form of various discrete NLS equations or systems, a particularly important class of solutions of which are so called *discrete breathers* which are homoclinic in space and oscillatory in time. Other questions involve the existence and propagation of topological defects or *kinks* which mathematically are heteroclinic connections between a ground and an excited steady state. Prototype models here are discrete version of sine-Gordon equations, also known as known as Frenkel-Kontorova (FK) models. There are many

253

L. Faddeev et al. (eds.),
Bilinear Integrable Systems: From Classical to Quantum, Continuous to Discrete, 253–257.
© 2006 Springer. Printed in the Netherlands.

outstanding issues for such systems relating to the global existence and dynamics of localised modes for general nonlinearities, away from either continuum or anticontinuum limits [3]. The *kinks* solutions have applications to problems such as dislocation and mass transport in solids, charge-density waves, commensurable-incommensurable phase transitions, Josephson transmission lines etc.

In this paper, we consider a perturbed Hamiltonian chain of coupled oscillators with an Hamiltonian

$$
\mathcal{H} = \sum_{n \in \mathbb{Z}} \left(\frac{1}{2} \dot{u}_n^2 + \frac{1}{2\varepsilon^2} (u_{n+1} - u_n)^2 + H(u_n) + \mu G(u_{n+1} - u_n) \right) \tag{1}
$$

where $\varepsilon > 0$, μ are small parameters and $h(x) = H'(x)$ and $g(x) = G'(x)$. $H, G \in C^2(\mathbb{R})$. For $H(x) = G(x) = 1 - \cos x$ we obtain the discrete sine-Gordon equation with inter-site potential as perturbation. The Hamiltonian $\mathcal{H}$ gives the nonlinear lattice eqn:

$$
\ddot{u}_n - \frac{1}{\varepsilon^2} (u_{n+1} - 2u_n + u_{n-1}) + h(u_n) + \mu \{ g(u_n - u_{n-1}) - g(u_{n+1} - u_n) \} = 0 \tag{2}
$$

We suppose for (2) the following conditions

(A1) $h, g \in C^1(\mathbb{R})$ are odd, h is 2π-periodic, $h(x - \pi) = -h(x)$ and g is globally Lipschitz on $\mathbb{R}$.
(A2) $h(0) = h(2\pi) = 0$, $h'(0) = h'(2\pi) = a^2 > 0$ and there is a heteroclinic solution Φ of

$$
\ddot{x} - h(x) = 0 : \Phi(t) = 2\pi - \Phi(-t), \quad \Phi(t) \to 2\pi \quad \text{as } t \to +\infty.
$$

The continuum limit of Eq. (2) for $\mu = 0$

$$
u_{tt} - u_{xx} + h(u) = 0
$$

admits travelling wave solutions

$$
u(x, t) = \Phi \left(\frac{x - vt}{\sqrt{1 - v^2}} \right), \quad 0 < v < 1.
$$

We consider for Eq. (2) travelling wave solutions of stationary profile in a moving reference with constant velocity v/ε. One can write

$$
u_n(t) = V \left(n - \frac{v}{\varepsilon} t \right) \equiv V(z), \quad z = n - \frac{v}{\varepsilon} t, \quad 0 < v < 1.
$$

Equation (2) is reduced to the following functional differential equation:

$$v^2 V''(z) - V(z+1) + 2V(z) - V(z-1) + \varepsilon^2 h(V(z))$$
$$+ \varepsilon^2 \mu \left(g(V(z) - V(z-1)) - g(V(z+1) - V(z)) \right) = 0 \quad (3)$$

where $'$ represents differentiation with respect to z. In this paper, we review the analytical results about the existence of solutions of Eq. (3) near Φ and the relationship between traveling wave solutions of (2) and continuum sine-Gordon for $\varepsilon > 0$, μ small.

2   PERIODIC TRAVELLING WAVES-BIFURCATION ANALYSIS

We apply center manifold theory to the study of existence of travelling wave solution of Eq. (1.8) with small amplitude oscillations on infinite nonlinear lattice.

We introduce a new variable $v \in [-1, 1]$ and functions $X(t, v) = x(t + v)$. The notation $U(t)(v) = (x(t), \xi(t), X(t, v))$ indicates our intention to construct V as a map from $\mathbb{R}$ into some function space living on the v-interval $[-1, 1]$. We introduce the Banach spaces $\mathbb{H}$ and $\mathbb{D}$ for $U(v) = (x, \xi, X(v))$

$$\mathbb{H} = \mathbb{R}^2 \times C[-1, 1], \quad \mathbb{D} = \{U \in \mathbb{R}^2 \times C^1[-1, 1] \mid X(0) = x\}$$

with the usual maximum norms. Then $L \in \mathcal{L}(\mathbb{D}, \mathbb{H})$ and $M \in C^1(\mathbb{D}, \mathbb{D})$. Eqn (3) can be written as follows [4]

$$U_t = LU + \frac{\varepsilon^2}{v^2} M(U) \qquad (4)$$

where

$$L = \begin{pmatrix} 0 & 1 & 0 \\ -\dfrac{2}{v^2} & 0 & \dfrac{1}{v^2}\delta^1 + \dfrac{1}{v^2}\delta^{-1} \\ 0 & 0 & \partial_v \end{pmatrix}, \quad \delta^{\pm 1} X(v) = X(\pm 1)$$

$$M(U) = \left(0, h(x) - \mu \left\{ g(x - \delta^{-1} X(v)) - g(\delta^1 X(v) - x) \right\}, 0 \right)$$

The spectrum $\sigma(L)$ is given by the resolvent equation $(\lambda I - L)U = F$, $F \in \mathbb{H}$, $\lambda \in \mathbb{C}$, $U \in \mathbb{D}$. The resolvent equation is solvable if and only if $N(\lambda) := \lambda^2 + \frac{2}{v^2}(1 - \cosh \lambda) = 0$. Clearly, $\sigma(L)$ is invariant under $\lambda \to \bar{\lambda}$ and $\lambda \to -\lambda$. The central part $\sigma_0(L) = \sigma(L) \cap \iota \mathbb{R}$ is determined by the equation $q^2 + \frac{2}{v^2}(\cos q - 1) = 0$, $q \in \mathbb{R}$. We assume that $v_1 < v < 1$ where $v = v_1$ is the first value from the left of 1 for which the equations

$$\lambda^2 + \frac{2}{v^2}(\cos \lambda - 1) = 0, \quad \lambda - \frac{1}{v^2}\sin \lambda = 0$$

have a common nonzero solution $\lambda \neq 0$. Then equation $N(\iota q) = 0$ has the double root 0 and simple roots $\pm q$. Hence we have $\sigma_0(L) = \{0, \pm \iota q\}$.

The linear operator on the 4th-dimensional central subspace $\mathbb{H}_c$ has the form $L_c = L/\mathbb{H}_c$ in the basis $(\xi_1, \xi_2, \xi_3, \xi_4)$ which satisfies $L\xi_1 = 0$, $L\xi_2 = \xi_1$, $L\xi_3 = -q\xi_4$, $L\xi_4 = q\xi_3$. The projection $P_c\mathbb{H} \to \mathbb{H}_c$ is given by $P_c(U) = P_1(U)\xi_1 + P_2(U)\xi_2 + P_3(U)\xi_3 + P_4(U)\xi_4$ [5]. Condition (A1) implies that M is globally Lipschitz. We can apply the procedure of a center manifold method to get for ε, μ small the reduced equation of (4) over $\mathbb{H}_c$ given by

$$\dot{u}_c = L_c u_c + \frac{\varepsilon^2}{v^2} P_c(M(u_c)) + O(\varepsilon^4), \tag{5}$$

where $u_c = u_1\xi_1 + u_2\xi_2 + u_3\xi_3 + u_4\xi_4$. Introducing the appropriate scaling, we consider the singularly perturbed system of the form:

$$\ddot{x} + h(x) = f_1(x, \dot{x}, y, \varepsilon \dot{y}, \varepsilon), \quad \varepsilon^2 \ddot{y} + y = \varepsilon^2 g_1(x, \dot{x}, y, \varepsilon \dot{y}, \varepsilon) \tag{6}$$

Theorem 1 [5] *For any $k_0 \in \mathbb{N}$ there is an $\varepsilon_0 > 0$ such that for any $0 < \varepsilon < \varepsilon_0$, $|\mu| \leq \varepsilon_0 \varepsilon^{1/4}$ and $T = \varepsilon(k[\varepsilon^{-3/2}]\pi + \tau)$ with $k \in \mathbb{N}$, $k \leq k_0$, $\tau \in [\pi/3, \pi/6]$, system (6) has a 4T-periodic solution $(x_{T,\varepsilon,1}(t), y_{T,\varepsilon,1}(t))$ near $(\phi(t), 0)$, $-T \leq t \leq T$ and has a solution $(x_{T,\varepsilon,2}(t), y_{T,\varepsilon,2}(t))$ on $\mathbb{R}$ near $(\phi(t), 0)$, $-T \leq t \leq T$, such that $x_{T,\varepsilon,\iota}$, $y_{T,\varepsilon,\iota}$ are odd functions and*

$$x_{T,\varepsilon,\iota}(t + 2T) = (-1)^\iota x_{T,\varepsilon,1}(t) + 2\pi(\iota - 1),$$
$$y_{T,\varepsilon,\iota}(t + 2T) = (-1)^\iota y_{T,\varepsilon,1}(t), \quad \iota = 1, 2$$

Theorem 2 [5] *If h, g satisfy the assumptions $(A1 - A2)$ then traveling wave solution $u(x, t) = \Phi(\frac{x - vt}{\sqrt{1 - v^2}})$ for $0 < v_1 < v < 1$ of sine-Gordon can be approximated by the both rotational and librational travelling wave solutions of (2) with very large periods and with the velocity v for $\mu = o(\varepsilon^{1/4})$ small.*

The central part of the spectrum $\sigma(L)$ is $\{0, \pm iq\}$, where 0 has multiplicity two. We can perform a polynomial change of coordinates close to identity, analytically depending on the parameter $\tilde{\mu}$, $u_c = Y + \Phi(Y, \tilde{\mu})$ such that the reduced system (5) is equivalent in a neighborhood of the origin to

$$\frac{dY}{dt} = N(Y, \tilde{\mu}) + R(Y, \tilde{\mu}) \tag{7}$$

where N is the normal form of order 2 and R represents the new terms of order greater or equal to 3, $Y = (y_1, y_2, y_3, y_4)$ and the system (7) has the following symmetry properties:

$$SN(Y, \tilde{\mu}) = -N(SY, \tilde{\mu}), SR(Y, \tilde{\mu}) = -R(SY, \tilde{\mu})$$

with $S(y_1, y_2, y_3, y_4) = (y_1, -y_2, y_3, -y_4)$.

For studying the dynamics of the initial system near the origin, we perform a polynomial change of coordinates for which the "linear and quadratic" part N is as simple as possible. Next, we analyze the truncated system

$$\frac{dY}{dt} = N(Y, \tilde{\mu}),$$

its heteroclinic orbits close to the origin. We focus on the problem of the persistence for the full system of the heteroclinic connections obtained for the truncated system and emphasize the case of solutions tending to exponentially small oscillations at infinity, (see 6).

ACKNOWLEDGEMENTS

VR was partly supported by the EPSRC-UK GrantNo. GR/R02702/01 and LMS. MF was partly supported by Grant GA-SAV 2/1140/21.

REFERENCES

1. Flach, S. and Willis, C. R. (1998) Discrete breathers, *Phys. Rep.* **295**, pp. 181–264.
2. Aubry, S. and MacKay, R. S. (1994) Proof of existence of breathers for time-reversible or Hamiltonian networks of weakly coupled oscillators, *Nonlinearity* **6**, pp. 1623–1643.
3. Iooss, G. and Kirchgassner, K. (2000) Traveling waves in a chain of coupled non-linear oscillators, *Comm. Math. Phys.* **211**(2), pp. 439–464.
4. Feckan, M. and Rothos, V. M. (2002) Bifurcations of periodics from homoclinics in singular o.d.e.: Applications to discretizations of traveling waves of p.d.e., *Comm. Pure Appl. Anal.* **1**, pp. 475–483.
5. Feckan, M. and Rothos, V. M. (2003) Travelling Waves for Perturbed Spatial Discretizations of Partial Differential Equations (preprint).
6. Rothos, V. M. and Feckan, M. (2003) Global Normal Form for Travelling Waves in Nonlinear Lattices (preprint).
7. Aigner, A. A., Champneys A. R., and Rothos V. M. (2003). A new barrier to the existence of moving kinks in Frenkel-Kontorova lattices (submitted Physica D).
8. Elibeck, J. C. and Flesch, R. (1990) Calculation of families of solitary waves on discrete lattices, *Physics Letters* A **149**, pp. 200–202.

QUANTUM VS CLASSICAL CALOGERO–MOSER SYSTEMS

Ryu Sasaki

Yukawa Institute for Theoretical Physics, Kyoto University, Kyoto 606–8502, Japan

Abstract   Calogero–Moser and Toda systems are best known examples of solvable many-particle dynamics on a line which are based on root systems. At the *classical* level, the former (C–M) is integrable for elliptic potentials (Weierstraß $\wp$ function) and their various degenerations. The latter (Toda) has an exponential potential, which is obtained from the former as a special limit of the elliptic potential. First, we discuss *quantum* C–M systems based on any root system. For the models with degenerate potentials, i.e., the rational with/without the harmonic confining force, the hyperbolic and the trigonometric, we demonstrate the following: (i) Construction of a complete set of quantum conserved quantities in terms of a Lax pair. (ii) Triangularity of the quantum Hamiltonian and the entire discrete spectrum. (iii) Equivalence of the quantum Lax pair method and that of so-called differential-reflection (Dunkl) operators. (iv) Algebraic construction of all excited states in terms of creation operators. Next, we discuss the relationship/contrast between the *quantum* and *classical* integrability as seen in the C–M systems.

1 INTRODUCTION

Calogero–Moser systems [1, 2, 3] are one-dimensional multiparticle dynamics with long-range interactions. They are integrable at both *classical* and *quantum* levels. In this lecture we discuss quantum C–M systems with degenerate potentials, that is the rational with/without harmonic force, the hyperbolic and the trigonometric potentials based on any root system [4–12]. The relationship/contrast between quantum and classical integrability is also discussed in some detail in the second half. The quantum *Calogero* systems having $1/q^2$ potential and a confining q^2 potential and the *Sutherland* systems with $1/\sin^2 q$ potentials have "integer" energy spectra characterized by the root system Δ. We show that the corresponding *classical* data, *e.g.*, minimum energy, frequencies of small oscillations, the eigenvalues of the classical Lax pair matrices, etc. at the equilibrium point of the potential are also "integers," or they appear to be "quantized." To be more precise, these quantities are polynomials in the coupling constant(s)

L. Faddeev et al. (eds.),
Bilinear Integrable Systems: From Classical to Quantum, Continuous to Discrete, 259–289.
© 2006 *Springer. Printed in the Netherlands.*

with integer coefficients [13]. The explanation of the highly organized nature of the energy spectra of the spin exchange models (Haldane–Shastry model and generalizations) [14–16] in terms of the Lax pairs at equilibrium is one of the motivations of the present research [13].

For the A_r models, the Lax pairs, conserved quantities and their involution were discussed by many authors with varied degrees of completeness and rigour, see for example [5, 17–26]. The point (iii) was shown by Wadati and collaborators [24] and point (iv) was initiated by Perelomov [18] and developed by Brink and collaborators [23] and Wadati and collaborators [24]. Various properties of classical A_r Calogero and Sutherland systems at equilibrium were discussed by Calogero and collaborators [27, 28]. A rather different approach by Heckman and Opdam [29, 30] to C–M systems with degenerate potentials based on any root system should also be mentioned in this connection. For physical applications of the C–M systems with various potentials to lower dimensional physics, ranging from solid state to particle physics and supersymmetric Yang–Mills theory, we refer to recent papers [7–9] and references therein.

This lecture is organized as follows. In Section 2 quantum C–M Hamiltonian with degenerate potentials is introduced as a factorized form (4). Connection with root systems and the Coxeter invariance is emphasized. Some rudimentary facts of the root systems and reflections are summarized in Appendix A. A universal Coxeter invariant ground state wavefunction and the ground-state energy are derived as simple consequences of the factorized Hamiltonian. In Section 3 we show that all the excited states are also Coxeter invariant and that the Hamiltonian is *triangular* in certain bases. Complete sets of quantum conserved quantities are derived from quantum Lax operator L in Section 4. Instead of the trace, the *total sum* of L^n is conserved. That is $\mathrm{Ts}(L^n) = \sum_{\mu,\nu \in \mathcal{R}} (L^n)_{\mu\nu}$, in which $\mathcal{R}$ is a set of $\mathbf{R}^r$ vectors invariant under the action of the Coxeter group. They form a single Coxeter orbit. In Section 5 the creation and annihilation operators for the *Calogero* system are derived. In Section 6 the equivalence of the Lax pair operator formalism with the so-called differential-reflection (Dunkl) operators is demonstrated. Another form of the quantum conserved quantities is given in terms of the differential-reflection (Dunkl) operators. In Section 7 an algebraic construction of excited states in terms of the differential-reflection (Dunkl) operators for the *Calogero* system is presented. In Section 8 we discuss the properties of classical equilibrium. We define integrable spin exchange models at the equilibrium of the *Calogero* and *Sutherland* systems for any root system. In Sections 9 and 10 we present the classical data of the *Calogero* and *Sutherland* systems, respectively. Most of them are expressed neatly in terms of roots and weights and provide interesting examples that the classical data of integrable systems are "quantized." The final section is for summary, comments, and outlook.

2 QUANTUM CALOGERO–MOSER SYSTEMS

A Calogero–Moser system is a multiparticle Hamiltonian system associated with a root system Δ of rank r, which is a finite set of vectors in $\mathbf{R}^r$ with its standard inner product. A brief review of the properties of the root systems and the associated reflections together with explicit realisations of all the root systems will be found in the Appendix A.

2.1 Factorized Hamiltonian

The dynamical variables of the quantum C–M system are the coordinates $\{q_j\}$ and their canonically conjugate momenta $\{p_j\}$, with the canonical commutation relations:

$$[q_j, p_k] = i\delta_{jk}, \qquad [q_j, q_k] = [p_j, p_k] = 0, \qquad j, k = 1, \ldots, r. \tag{1}$$

These will be denoted by vectors in $\mathbf{R}^r$:

$$q = (q_1, \ldots, q_r), \qquad p = (p_1, \ldots, p_r). \tag{2}$$

The momentum operator p_j acts as a differential operator $p_j = -i\frac{\partial}{\partial q_j}$, $j = 1, \ldots, r$. As for the interactions we consider only the degenerate potentials, that is the rational (with/without harmonic force), hyperbolic, and trigonometric potentials, $\rho \in \Delta$:

$$V(\rho \cdot q) : 1/(\rho \cdot q)^2, \ a^2/\sinh^2 a(\rho \cdot q), \ a^2/\sin^2 a(\rho \cdot q), \tag{3}$$

in which a is an arbitrary real positive constant, determining the period of the trigonometric potentials. They imply integrability for all of the C–M systems based on the crystallographic root systems. Those models based on the noncrystallographic root systems, the dihedral group $I_2(m)$, H_3, and H_4, are integrable only for the rational potential. The rational potential models are also integrable if a confining harmonic potential $\omega^2 q^2/2$, $\omega > 0$, is added to the Hamiltonian. This case will be called the *Calogero* system and the trigonometric case will be referred to as the *Sutherland* system hereafter.

The Hamiltonian for the *quantum* C–M system can be written in a "factorized form"

$$\mathcal{H} = \frac{1}{2} \sum_{j=1}^{r} \left(p_j - i\frac{\partial W}{\partial q_j} \right) \left(p_j + i\frac{\partial W}{\partial q_j} \right), \tag{4}$$

$$= \frac{1}{2} \sum_{j=1}^{r} \left(p_j^2 + \left(\frac{\partial W}{\partial q_j} \right)^2 \right) + \frac{1}{2} \sum_{j=1}^{r} \frac{\partial^2 W}{\partial q_j^2}. \tag{5}$$

Table 1. Functions appearing in the Lax pair and prepotential

Potential	$w(u)$	$x(u)$	$y(u)$
Rational	u	$1/u$	$-1/u^2$
Hyperbolic	$\sinh au$	$a \coth au$	$-a^2/\sinh^2 au$
Trigonometric	$\sin au$	$a \cot au$	$-a^2/\sin^2 au$

The simplest way to introduce the factorized form is through supersymmetry [10, 31], in which function W is called a *prepotential*:[1]

$$W(q) = \sum_{\rho \in \Delta_+} g_\rho \ln |w(\rho \cdot q)| + \left(-\frac{\omega}{2}q^2\right), \quad g_\rho > 1, \quad \omega > 0. \tag{6}$$

The real *positive* coupling constants g_ρ are defined on orbits of the corresponding Coxeter group, i.e., they are identical for roots in the same orbit. That is, for the simple Lie algebra cases one coupling constant $g_\rho = g$ for all roots in simply laced models and two independent coupling constants, $g_\rho = g_L$ for long roots and $g_\rho = g_S$ for short roots in nonsimply laced models. The potential $V(u)$ (3) and the function $w(u)$ are related by

$$y(u) \equiv \frac{d}{du}x(u), \quad \frac{dw(u)}{du}/w(u) \equiv x(u), \tag{7}$$

$$V(u) = -y(u) = x^2(u) + a^2 \times \begin{cases} 0 & \text{rational,} \\ -1 & \text{hyperbolic,} \\ 1 & \text{trigonometric.} \end{cases} \tag{8}$$

Table 1 gives these functions for each potential, it should be noted that the above factorized Hamiltonian (4) consists of an operator part $\hat{\mathcal{H}}$, which is the Hamiltonian in the usual definition, and a constant $\mathcal{E}_0$ which is the ground-state energy to be discussed later:

$$\mathcal{H} = \hat{\mathcal{H}} - \mathcal{E}_0, \tag{9}$$

$$\hat{\mathcal{H}} = \frac{1}{2}p^2 + \frac{1}{2}\sum_{\rho \in \Delta_+} g_\rho(g_\rho - 1)|\rho|^2 V(\rho \cdot q) + \left(\frac{\omega^2}{2}q^2\right). \tag{10}$$

For proofs that the factorized Hamiltonian (4) actually leads to the quantum Hamiltonian (10) for *any root system and potential* see Refs.[5, 10, 17]. It is easy to verify that for any potential $V(u)$, the Hamiltonian is invariant under reflection of the phase space variables in the hyperplane perpendicular to any root

---

[1] The dynamics of the prepotentials W (6), or rather that of $-W$, has been discussed by Dyson [32] from a different point of view (random matrix model).

$$\mathcal{H}(s_\alpha(p), s_\alpha(q)) = \mathcal{H}(p, q), \quad \forall \alpha \in \Delta \tag{11}$$

with s_α defined by (A2).

The main problem is to find all the eigenvalues $\{\lambda\}$ and eigenfunctions $\{\psi\}$ of the above Hamiltonian:

$$\mathcal{H}\psi = \lambda\psi. \tag{12}$$

For any root system and for any choice of potential (3), the C–M system has a hard repulsive potential $\sim 1/(\alpha q)^2$ near the reflection hyperplane $H_\alpha = \{q \in \mathbf{R}^r, \alpha \cdot q = 0\}$. The strength of the singularity is given by the coupling constant $g_\alpha(g_\alpha - 1)$ which is *independent* of the choice of the normalization of the roots. In other words, (12) is a second-order Fuchsian differential equation with regular singularity at each reflection hyperplane H_α. That is any solution of (12) is regular at all points except those on the union of reflection hyperplanes $\cup_{\alpha \in \Delta_+} H_\alpha$. Near the reflection hyperplane H_α, we choose the solution behaving

$$\psi \sim (\alpha \cdot q)^{g_\alpha}(1 + \text{regular terms}),$$

for the square integrability. This reflects the fact that the repulsive potential is classically and quantum mechanically insurmountable. Thus the motion is always confined within one Weyl chamber, which we choose to be the principal Weyl chamber (Π: set of simple roots, see Appendix A)

$$PW = \{q \in \mathbf{R}^r | \alpha \cdot q > 0, \quad \alpha \in \Pi\}, \tag{13}$$

without loss of generality. For the trigonometric potential, the configuration space is further limited due to the periodicity of the potential to

$$PW_T = \{q \in \mathbf{R}^r | \alpha \cdot q > 0, \quad \alpha \in \Pi, \quad \alpha_h \cdot q < \pi/a\}, \tag{14}$$

where α_h is the highest root.

2.2 Ground State Wavefunction and Energy

One straightforward outcome of the factorized Hamiltonian (4) is the *universal* ground-state wavefunction which is given by $\Phi_0(q) = e^{W(q)}$. It is easy to see that it is an eigenstate of the Hamiltonian (4) with *zero* eigenvalue $\mathcal{H}\Phi_0(q) = 0$, since it satisfies the condition

$$(p_j + i\partial W/\partial q_j)e^{W(q)} = 0, \quad j = 1, \ldots, r. \tag{15}$$

By using the decomposition of the factorized Hamiltonian into the operator
Hamiltonian (10) and a constant, we obtain

$$\hat{\mathcal{H}}e^W \equiv \left(\frac{1}{2}p^2 + \frac{1}{2}\sum_{\rho\in\Delta_+} g_\rho(g_\rho - 1)|\rho|^2 V(\rho \cdot q) + \left(\frac{\omega^2}{2}q^2\right)\right)e^W = \mathcal{E}_0 e^W.$$

(16)

In other words, the above solution $\Phi_0 = e^W$ provides an eigenstate of the
Hamiltonian operator $\hat{\mathcal{H}}$ with energy $\mathcal{E}_0$. The ground-state energy for the rational
potential case is vanishing and

$$\mathcal{E}_0 = \omega\left(r/2 + \sum_{\rho\in\Delta_+} g_\rho\right)$$

(17)

for the *Calogero* system. The same for the hyperbolic and trigonometric potential cases are

$$\mathcal{E}_0 = 2a^2\varrho^2 \times \begin{cases} -1 & \text{hyperbolic,} \\ 1 & \text{trigonometric,} \end{cases}$$

(18)

in which

$$\varrho = \frac{1}{2}\sum_{\rho\in\Delta_+} g_\rho\rho$$

(19)

can be considered as a "deformed Weyl vector" [5, 30]. Again these formulas
are universal. A negative $\mathcal{E}_0$ for the obviously positive Hamiltonian of the hyperbolic potential model indicates that the wavefunction is not square integrable,
since it diverges as $|q| \to \infty$ for the hyperbolic and the rational potential cases.
Obviously we have for the *Calogero* and *Sutherland* systems

$$\int_{PW(PW_T)} e^{2W(q)} \, dq < \infty$$

(20)

in which PW and PW_T denote that the integration is over the regions defined
in (13) and (14). The universal ground-state wavefunction Φ_0 and W are characterized as *Coxeter invariant*:

$$\check{s}_\rho\Phi_0 = \Phi_0, \qquad \check{s}_\rho W = W, \qquad \forall\rho \in \Delta,$$

(21)

in which $\check{s}_\rho$ is the representation of the reflection in the function space.

3   COXETER INVARIANT EXCITED STATES, TRIANGULARITY
AND SPECTRUM

In this section we show that all the excited states wavefunctions are Coxeter
invariant, too. With the knowledge of the ground-state wavefunction e^W, the

other states of the C–M systems can be easily obtained as eigenfunctions of a differential operator $\tilde{\mathcal{H}}$ obtained from $\mathcal{H}$ by a similarity transformation:

$$\tilde{\mathcal{H}} = e^{-W} \mathcal{H} e^{W}, \qquad \tilde{\mathcal{H}} \Psi_\lambda = \lambda \Psi_\lambda \iff \mathcal{H} \Psi_\lambda e^{W} = \lambda \Psi_\lambda e^{W}. \tag{22}$$

The transformed Hamiltonian $\tilde{\mathcal{H}}$ takes a simple form:

$$\tilde{\mathcal{H}} = -\frac{1}{2} \sum_{j=1}^{r} \left(\frac{\partial^2}{\partial q_j^2} + 2 \frac{\partial W}{\partial q_j} \frac{\partial}{\partial q_j} \right), \tag{23}$$

which is also Coxeter invariant. Since all the singularities of the Fuchsian differential equation (12) are contained in the ground-state wavefunction e^{W} *the function Ψ_λ above must be regular at finite q including all the reflection boundaries.*

We introduce proper bases of Fock space consisting of Coxeter invariant functions and show that the above Hamiltonian $\tilde{\mathcal{H}}$ (23) is *triangular* in these bases. This establishes the integrability universally[2] and also gives the entire spectrum of the Hamiltonian, see (28), (29), and (48).

3.1 Rational Potential with Harmonic Force

First, let us determine the structure of the set of eigenfunctions of the transformed Hamiltonian $\tilde{\mathcal{H}}$ for the *Calogero* system:

$$\tilde{\mathcal{H}} = \omega q \cdot \frac{\partial}{\partial q} - \frac{1}{2} \sum_{j=1}^{r} \frac{\partial^2}{\partial q_j^2} - \sum_{\rho \in \Delta_+} \frac{g_\rho}{\rho \cdot q} \rho \cdot \frac{\partial}{\partial q}. \tag{24}$$

As remarked above, the eigenfunctions of $\tilde{\mathcal{H}}$ have no singularities at finite q. Thus we look for polynomial (in q) eigenfunctions $P(q)$:

$$\tilde{\mathcal{H}} P(q) = \lambda P(q). \tag{25}$$

The Coxeter invariance of $\tilde{\mathcal{H}}$ (23) translates into a theorem [6] that *the eigenfunctions are Coxeter invariant polynomials* and that the Hamiltonian $\tilde{\mathcal{H}}$ (24) maps a Coxeter invariant polynomial to another.

An obvious basis in the space of Coxeter invariant polynomials is the homogeneous polynomials of various degrees. This basis has a natural order given by the degree. For a given degree the space of homogeneous Coxeter invariant polynomials is finite-dimensional. The explicit form of $\tilde{\mathcal{H}}$ (24) shows that it is *lower triangular* in this basis and the diagonal elements are $\omega \times degree$ as

---

[2] Triangularity of the A_r-type Hamiltonians was noted in the original papers of Calogero [1] and Sutherland [2]. That of rank two models in the Coxeter invariant bases was shown in [30, 33].

Table 2. The degrees f_j in which independent Coxeter invariant polynomials exist

Δ	$f_j = 1 + e_j$	Δ	$f_j = 1 + e_j$
A_r	$2, 3, 4, \ldots, r + 1$	E_8	$2, 8, 12, 14, 18, 20, 24, 30$
B_r	$2, 4, 6, \ldots, 2r$	F_4	$2, 6, 8, 12$
C_r	$2, 4, 6, \ldots, 2r$	G_2	$2, 6$
D_r	$2, 4, \ldots, 2r - 2; r$	$I_2(m)$	$2, m$
E_6	$2, 5, 6, 8, 9, 12$	H_3	$2, 6, 10$
E_7	$2, 6, 8, 10, 12, 14, 18$	H_4	$2, 12, 20, 30$

given by the first term. Independent Coxeter invariant polynomials exist at the degrees f_j listed in Table 2:

$$f_j = 1 + e_j, \quad j = 1, \ldots, r, \tag{26}$$

in which $\{e_j\}, j = 1, \ldots, r$, are the *exponents* of Δ. Let us denote them by

$$z_1(q), \ldots, z_r(q); \qquad z_j(\kappa q) = \kappa^{f_j} z_j(q). \tag{27}$$

Thus we arrive at the quantum *Calogero* system is algebraically solvable for any (crystallographic and noncrystallographic) root system Δ. (See [34] for algebraic linearizability theorem of the classical C–M system.) The spectrum of the operator Hamiltonian $\hat{\mathcal{H}}$ is

$$\omega N + \mathcal{E}_0, \tag{28}$$

with a nonnegative integer N which can be expressed as

$$N = \sum_{j=1}^{r} n_j f_j, \quad n_j \in \mathbf{Z}_+, \tag{29}$$

and the degeneracy of the above eigenvalue is the number of different solutions of (29) for given N. This is generalization of Calogero's original argument for the A_r model [1] to the models based on any root system. Now let us denote by $\vec{N}$ the set of nonnegative integers in (29), $\vec{N} = (n_1, n_2, \ldots, n_r)$, and by $\phi_{\vec{N}}(q)$ the homogeneous Coxeter invariant polynomial determined by $\vec{N}$ and the above basis $\{z_j\}$ (27):

$$\phi_{\vec{N}}(q) = \prod_{j=1}^{r} z_j^{n_j}(q). \tag{30}$$

As shown above, there exists a unique eigenstate $\psi_{\vec{N}}(q)$ for each $\phi_{\vec{N}}(q)$:

$$\psi_{\vec{N}}(q) = \phi_{\vec{N}}(q) + \sum_{\vec{N}' < \vec{N}} d_{\vec{N}'} \phi_{\vec{N}'}(q), \quad d_{\vec{N}'} : \text{const}, \tag{31}$$

$$\hat{\mathcal{H}} \psi_{\vec{N}}(q) = \omega N \psi_{\vec{N}}(q). \tag{32}$$

It satisfies the orthogonality relation

$$(\psi_{\vec{N}}, \phi_{\vec{N}'}) = 0, \quad \vec{N}' < \vec{N},$$ (33)

with respect to the inner product in PW:

$$(\psi, \varphi) = \int_{PW} \psi^*(q)\varphi(q)e^{2W(q)}dq.$$ (34)

These polynomials $\{\psi_{\vec{N}}(q)\}$ are generalizations of the multivariable Laguerre (Hermite) polynomials [20] known for the $A_r(B_r, D_r)$ root systems to any root system.

The explicit example of the simplest root system of rank one would be illuminating. For $\Delta = A_1$, the Hamiltonian $\tilde{\mathcal{H}}$ can be rewritten in terms of a Coxeter invariant variable $u = \omega q^2$ as

$$\tilde{\mathcal{H}} = \omega q \frac{d}{dq} - \frac{1}{2}\frac{d^2}{dq^2} - \frac{g}{q}\frac{d}{dq} = -2\omega\left\{u\frac{d^2}{du^2} + (g + \frac{1}{2} - u)\frac{d}{du}\right\}.$$ (35)

The Laguerre polynomial satisfying the differential equation

$$\left\{u\frac{d^2}{du^2} + \left(g + \frac{1}{2} - u\right)\frac{d}{du} + n\right\}L_n^{(g-\frac{1}{2})}(u) = 0,$$ (36)

provides an eigenfunction with eigenvalue $2\omega n$, which corresponds to the eigenvalue $2\omega n + \mathcal{E}_0$ of $\hat{\mathcal{H}}$. This is a well-known result.

3.2 Trigonometric Potential

Here we consider those root systems associated with Lie algebras. In order to determine the excited states of the *Sutherland* system, we have to consider the periodicity. The prepotential W and the Hamiltonian $\mathcal{H}$ are invariant under the following translation:

$$W(q') = W(q), \quad \mathcal{H}(p, q') = \mathcal{H}(p, q), \quad q' = q + l^\vee \pi/a,$$ (37)

in which $l^\vee$ is an element of the dual weight lattice, that is

$$l^\vee = \sum_{j=1}^{r} l_j \frac{2}{\alpha_j^2}\lambda_j, \quad l_j \in \mathbf{Z}, \ \alpha_j \in \Pi, \ \alpha_j^\vee \cdot \lambda_k = \delta_{jk}.$$ (38)

Known as the Bloch wavefunctions in quantum mechanics with periodic potentials, the wavefunctions diagonalizing the translation operators are expressed as

$$e^{2ia\mu\cdot q}\left(\sum_{\alpha\in L(\Delta)} b_\alpha e^{2ia\alpha\cdot q}\right)e^W, \quad b_\alpha : \text{const}, \ L(\Delta) : \text{root lattice},$$ (39)

in which a vector $\mu \in \mathbf{R}^r$ is as yet unspecified. In other words, up to the overall phase factor $e^{2ia\mu \cdot q}$, this is a Fourier expansion in terms of the simple roots. As in the *Calogero* case, the eigenfunction of the Hamiltonian is *Coxeter invariant*. This translates into the requirement that the unspecified vector μ in (39) should be an element of the weight lattice

$$\mu \in \Lambda(\Delta), \quad \Lambda(\Delta) : \text{weight lattice.} \tag{40}$$

Let us introduce a basis for the Coxeter invariant functions. Let λ be a dominant weight

$$\lambda = \sum_{j=1}^{r} m_j \lambda_j, \quad m_j \in \mathbf{Z}_+, \tag{41}$$

and W_λ be the orbit of λ by the action of the Weyl group:

$$W_\lambda = \{\mu \in \Lambda(\Delta) | \mu = g(\lambda), \forall g \in G_\Delta\}. \tag{42}$$

We define

$$\phi_\lambda(q) \equiv \sum_{\mu \in W_\lambda} e^{2ia\mu \cdot q}, \tag{43}$$

which is Coxeter invariant. The set of functions $\{\phi_\lambda\}$ has an order $\succ$:

$$|\lambda|^2 > |\lambda'|^2 \Rightarrow \phi_\lambda \succ \phi_{\lambda'}. \tag{44}$$

It is easy to show that the similarity transformed Hamiltonian $\tilde{\mathcal{H}}$

$$\tilde{\mathcal{H}} = -\frac{1}{2} \sum_{j=1}^{r} \frac{\partial^2}{\partial q_j^2} - a \sum_{\rho \in \Delta_+} g_\rho \cot (a\rho \cdot q) \rho \cdot \frac{\partial}{\partial q} \tag{45}$$

is *lower triangular* in this basis: Thus we have demonstrated the triangularity of $\tilde{\mathcal{H}}$:

$$\tilde{\mathcal{H}}\phi_\lambda = 2a^2(\lambda^2 + 2\varrho \cdot \lambda)\phi_\lambda + \sum_{|\lambda'|<|\lambda|} c_{\lambda'}\phi_{\lambda'}. \tag{46}$$

It is an eigenfunction of the initial Hamiltonian $\hat{\mathcal{H}}$

$$\hat{\mathcal{H}}\phi_\lambda e^W = 2a^2(\lambda + \varrho)^2 \phi_\lambda e^W + \sum_{|\lambda'|<|\lambda|} c_{\lambda'}\phi_{\lambda'} e^W, \tag{47}$$

with the eigenvalue

$$2a^2(\lambda + \varrho)^2. \tag{48}$$

In other words, for each dominant weight λ there exists an eigenstate of $\tilde{\mathcal{H}}$ with eigenvalue proportional to $\lambda(\lambda + 2\varrho)$. Let us denote this eigenfunction by $\psi_\lambda(q)$:

$$\psi_\lambda(q) = \phi_\lambda(q) + \sum_{|\lambda'|<|\lambda|} d_{\lambda'}\phi_{\lambda'}(q), \quad d_{\lambda'} : \text{const}, \tag{49}$$

$$\tilde{\mathcal{H}}\psi_\lambda(q) = 2a^2\lambda(\lambda + 2\varrho)\psi_\lambda(q), \tag{50}$$

and call it a *generalized Jack polynomial* [35–38]. It satisfies the orthogonality relation

$$(\psi_\lambda, \phi_{\lambda'}) = 0, \quad |\lambda|' < |\lambda|, \tag{51}$$

with respect to the inner product in PW_T:

$$(\psi, \varphi) = \int_{PW_T} \psi^*(q)\varphi(q)e^{2W(q)}\,dq. \tag{52}$$

In the A_r model, specifying a dominant weight λ is the same as giving a Young diagram which designates a Jack polynomial.

Thus we arrive at the quantum *Sutherland* system is algebraically solvable for any crystallographic root system Δ. The spectrum of the Hamiltonian $\hat{\mathcal{H}}$ is given by (48) in which λ is an arbitrary dominant weight. This is generalization of Sutherland's original argument [2] to the models based on any root system. Some remarks are in order

1. The weights μ appearing in the lower order terms $\{\phi_{\lambda'}\}$'s are those weights contained in the Lie algebra representation belonging to the highest weight λ.
2. Let us consider the well-known case $\Delta = A_1$. By rewriting the Hamiltonian $\tilde{\mathcal{H}}$ in terms of the Coxeter invariant variable $z = \cos(a\rho q)$, we obtain

$$\hat{\mathcal{H}} = -\frac{1}{2}\frac{d^2}{dq^2} - ag\rho\cot(a\rho q)\frac{d}{dq} = -\frac{a^2|\rho|^2}{2}\left\{(1-z^2)\frac{d^2}{dz^2} - (1+2g)z\frac{d}{dz}\right\}.$$

$$\tag{53}$$

The Gegenbauer polynomials [5], a special case of Jacobi polynomials $P_n^{(\alpha,\beta)}$ provide eigenfunctions:

$$P_n^{(g-\frac{1}{2},g-\frac{1}{2})}(\cos(a\rho q)), \qquad \mathcal{E} = a^2|\rho|^2(n+g)^2/2, \quad n \in \mathbf{Z}_+. \tag{54}$$

They form orthogonal polynomials with weight $e^{2W} = |\sin(a\rho q)|^{2g}$ in the interval $q \in [0, \pi/a\rho]$, (14).

4   QUANTUM LAX PAIR AND QUANTUM CONSERVED QUANTITIES

Historically, Lax pairs for C–M systems were presented in terms of Lie algebra representations [3, 5], in particular, the vector representation of the A_r models. However, the invariance of C–M systems is that of Coxeter group but not of the associated Lie algebra. Thus the universal and Coxeter covariant Lax pairs are given in terms of the representations of the Coxeter group.

4.1   General Case

The Lax operators without spectral parameter are [10]

$$L(p, q) = p \cdot \hat{H} + X(q), \qquad X(q) = i \sum_{\rho \in \Delta_+} g_\rho(\dot{\rho} \cdot \hat{H}) x(\rho \cdot q) \hat{s}_\rho, \quad (55)$$

$$M(q) = \frac{i}{2} \sum_{\rho \in \Delta_+} g_\rho |\rho|^2 y(\rho \cdot q)(\hat{s}_\rho - I), \tag{56}$$

in which I is the identity operator and $\{\hat{s}_\alpha, \alpha \in \Delta\}$ are the reflection operators of the root system. They act on a set of $\mathbf{R}^r$ vectors $\mathcal{R} = \{\mu^{(k)} \in \mathbf{R}^r, k = 1, \ldots, D\}$, permuting them under the action of the reflection group. The vectors in $\mathcal{R}$ form a basis for the representation space $\mathbf{V}$ of dimension D. The operator M satisfies the *sum up to zero* relation:

$$\sum_{\mu \in \mathcal{R}} M_{\mu\nu} = \sum_{\nu \in \mathcal{R}} M_{\mu\nu} = 0, \tag{57}$$

which is essential for deriving quantum conserved quantities. The matrix elements of the operators $\{\hat{s}_\alpha, \alpha \in \Delta\}$ and $\{\hat{H}_j, j = 1, \ldots, r\}$ are defined as follows:

$$(\hat{s}_\rho)_{\mu\nu} = \delta_{\mu, s_\rho(\nu)} = \delta_{\nu, s_\rho(\mu)}, \qquad (\hat{H}_j)_{\mu\nu} = \mu_j \delta_{\mu\nu}, \qquad \rho \in \Delta, \ \mu, \nu \in \mathcal{R}. \tag{58}$$

The form of the function x depends on the chosen potential, and the function y are defined by (7), (8).

The underlying idea of the Lax operator L, (55), is quite simple. As seen from (64), L is a "square root" of the Hamiltonian. Thus one part of L contains p which is not associated with roots and another part contains $x(\rho \cdot q)$, a "square root" of the potential $V(\rho \cdot q)$, which being associated with a root ρ is therefore accompanied by the reflection operator $\hat{s}_\rho$. Another explanation is the factorized Hamiltonian $\mathcal{H}$ (4). We obtain, roughly speaking, $L \sim \sqrt{\mathcal{H}} \sim p + i \frac{\partial W}{\partial q} \hat{s}$ and the property of reflection $\hat{s}^2 = 1$ explains the sign change in the first term in (4).

It is straightforward to show that the quantum Lax equation

$$\frac{d}{dt}L = i[\mathcal{H}, L] = [L, M],\tag{59}$$

is equivalent to the quantum equations of motion derived from the Hamiltonian (4). From this it follows:

$$\frac{d}{dt}(L^n)_{\mu\nu} = i[\mathcal{H}, (L^n)_{\mu\nu}] = [L^n, M]_{\mu\nu}$$
$$= \sum_{\lambda\in\mathcal{R}}\left((L^n)_{\mu\lambda}M_{\lambda\nu} - M_{\mu\lambda}(L^n)_{\lambda\nu}\right), \quad n = 1, \ldots.\tag{60}$$

Thanks to the *sum up to zero* property of the M operator (57), we obtain *quantum conserved quantities* as the *total sum* (Ts) of all the matrix elements of L^n:

$$Q_n = \mathrm{Ts}(L^n) \equiv \sum_{\mu,\nu\in\mathcal{R}}(L^n)_{\mu\nu}, \quad [\mathcal{H}, Q_n] = 0, \quad n = 1, \ldots.\tag{61}$$

A universal proof of the involution of the quantum conserved quantities

$$[Q_n, Q_m] = 0, \quad n, m = 1, \ldots,\tag{62}$$

can be found in [6]. See [12] for the classical Liouville integrability of the most general C–M systems with elliptic potentials.

Independent conserved quantities appear at such power n that

$$n = 1 + e_j, \quad e_j : \text{exponent}, \quad j = 1, \ldots, r,\tag{63}$$

of each root system. See I. Thus we have r independent conserved quantities in C–M systems. These are the degrees at which independent Coxeter invariant polynomials exist. In fact, the j-th conserved quantity is a degree $1 + e_j$ polynomial in the momenta p. In particular, the power 2 is universal to all the root systems and the quantum Hamiltonian (4) is given by

$$\mathcal{H} = \frac{1}{2C_{\mathcal{R}}}\mathrm{Ts}(L^2) + \text{const},\tag{64}$$

where the constant $C_{\mathcal{R}}$ is the quadratic Casimir invariant, which depends on the representation. Lax pairs and the quantum conserved quantities Q_n do depend on the chosen representations.

4.2 Rational Potential with Harmonic Force

The quantum Lax pair for the *Calogero* system needs a separate formulation. The canonical equations of motion are equivalent to the following Lax equations for $L^{\pm}$:

$$\frac{d}{dt}L^{\pm} = i[\mathcal{H}, L^{\pm}] = [L^{\pm}, M] \pm i\omega L^{\pm},\tag{65}$$

in which (see Section 4 of [8]) M is the same as before (56), and $L^{\pm}$ and Q are defined by

$$L^{\pm} = L \pm i\omega Q, \quad Q = q \cdot \hat{H}, \tag{66}$$

with L, $\hat{H}$ as earlier (55), (56). They $(L^{\pm})$ are a multiparticle analogue of the creation-annihilation operators of the harmonic oscillator, as we will see shortly. If we define hermitian operators $\mathcal{L}_1$ and $\mathcal{L}_2$ by

$$\mathcal{L}_1 = L^+ L^-, \qquad \mathcal{L}_2 = L^- L^+, \tag{67}$$

they satisfy Lax-like equations

$$\frac{d}{dt}\mathcal{L}_k = i[\mathcal{H}, \mathcal{L}_k] = [\mathcal{L}_k, M], \quad k = 1, 2. \tag{68}$$

From these we can construct conserved quantities

$$\mathrm{Ts}(\mathcal{L}_j^n), \quad j = 1, 2, \ n = 1, 2, \ldots, \tag{69}$$

as before [22, 24]. It is elementary to check that the first conserved quantities give the Hamiltonian

$$\mathcal{H} \propto \mathrm{Ts}(\mathcal{L}_1) = \mathrm{Ts}(\mathcal{L}_2) + \mathrm{const.} \tag{70}$$

5 ALGEBRAIC CONSTRUCTION OF EXCITED STATES I

In this section we show that all the excited states of the *Calogero* systems can be constructed algebraically. Later in Section 7 we show the same results in terms of the ℓ operators to be introduced in Section 6. The main result is surprisingly simple and can be stated universally:

Corresponding to each partition of an integer N which specify the energy level (28) into the sum of the degrees of Coxeter invariant polynomials (29), we have an eigenstate of the Hamiltonian $\hat{\mathcal{H}}$ with eigenvalue $\omega N + \mathcal{E}_0$:

$$\prod_{j=1}^{r} \left(B_{f_j}^+\right)^{n_j} e^W, \quad N = \sum_{j=1}^{r} n_j f_j, \quad n_j \in \mathbf{Z}_+, \tag{71}$$

in which the integers $\{f_j\}$, $j = 1, \ldots, r$ are listed in I. They exhaust all the excited states. In other words the above states give the complete basis of the Fock space. The creation operators $B_{f_j}^+$ and the corresponding annihilation operators[3] $B_{f_j}^-$ are defined in terms of the Lax operators $L^{\pm}$ (66) as follows:

$$B_{f_j}^{\pm} = \mathrm{Ts}(L^{\pm})^{f_j}, \quad j = 1, \ldots, r. \tag{72}$$

---

[3] We adopt the notation by Olshanetsky and Perelomov [5, 18].

They are hermitian conjugate to each other

$$(B_{f_j}^{\pm})^{\dagger} = B_{f_j}^{\mp}. \tag{73}$$

The creation (annihilation) operators commute among themselves:

$$[B_k^+, B_l^+] = [B_k^-, B_l^-] = 0, \quad k, l \in \{f_j | j = 1, \dots, r\}, \tag{74}$$

so that the state (71) does not depend on the order of the creation. The ground state is annihilated by all the annihilation operators

$$B_{f_j}^- e^W = 0, \quad j = 1, \dots, r. \tag{75}$$

Some remarks are in order:

1. Reflecting the universality of the first exponent, $f_1 = 2$, the creation and annihilation operators of the least quanta, 2ω, exist in all the models. They form an $sl(2, \mathbf{R})$ algebra together with the Hamiltonian $\hat{\mathcal{H}}$:

$$[\hat{\mathcal{H}}, b_2^{\pm}] = \pm 2\omega b_2^{\pm}, \quad [b_2^+, b_2^-] = -\omega^{-1}\hat{\mathcal{H}}, \tag{76}$$

in which $b_2^{\pm}$ are normalized forms of $B_2^{\pm}$. The $sl(2, \mathbf{R})$ algebra was discussed by many authors [18, 23, 29, 39]. The states created by $B_2^+(b_2^+)$ only can be expressed by the Laguerre polynomial:

$$(b_2^+)^n e^W = n! L_n^{(\tilde{\mathcal{E}}_0 - 1)}(\omega q^2)? e^W, \quad \tilde{\mathcal{E}}_0 \equiv \mathcal{E}_0/\omega. \tag{77}$$

The Laguerre polynomial wavefunctions appear as "radial" wavefunctions in all the cases [40].

2. The operators $\{Q_n\}$ and $\{B_n^{\pm}\}$ do not form a Lie algebra. They satisfy interesting non-linear relations, for example,

$$[[B_n^+, b_2^-], b_2^+] = nB_n^+, \quad [[B_n^-, b_2^+], b_2^-] = nB_n^-. \tag{78}$$

This tells, for example, that although B_n^+ and b_2^+ create different units of quanta n and 2, they are not independent

$$[B_n^+, b_2^-] \neq 0 \neq [B_n^-, b_2^+].$$

6 ℓ OPERATORS

In this section we will show the universal equivalence of the quantum conserved quantities obtained in the Lax operator formalism of Section 4 and those derived in the "commuting differential operators" formalism initiated by Dunkl [11]. We propose to call the operators in the latter approach simply "ℓ operators," since they are essentially the same as the L operator in the Lax pair formalism and that they are not mutually commuting, as we will show presently, when the interaction potentials are trigonometric (hyperbolic), (88). Although these two formalisms are formally equivalent, the ℓ operator formalism has many advantages over the Lax pair one. Roughly speaking, the "vector-like" objects ℓ_μ's are easier to handle than the matrix $L_{\mu\nu}$.

Let us fix a representation $\mathcal{R}$ of the Coxeter group G_Δ and define for each element $\mu \in \mathcal{R}$ the following *differential-reflection operator*

$$\ell_\mu = \ell \cdot \mu = p \cdot \mu + i \sum_{\rho \in \Delta_+} g_\rho(\rho \cdot \mu)x(\rho \cdot q)\check{s}_\rho, \quad \mu \in \mathcal{R}, \quad \ell_\mu^\dagger = \ell_\mu. \quad (79)$$

The quantum conserved quantities Q_n derived in the previous section (61) can be expressed as polynomials in the ℓ operators as follows:

$$Q_n\psi = \sum_{\mu,\nu \in \mathcal{R}} (L^n)_{\mu\nu}\psi = \left(\sum_{\mu \in \mathcal{R}} \ell_\mu^n\right)\psi, \quad (80)$$

in which ψ is an arbitrary Coxeter invariant state, $\check{s}_\rho\psi = \psi$.

Commutation relations among ℓ operators can be evaluated in a similar manner as those appearing in the Lax pair [8, 10]. We obtain

$$[\ell_\mu, \ell_\nu] = -a^2 \sum_{\rho,\sigma \in \Delta_+} g_\rho g_\sigma(\rho \cdot \mu)(\sigma \cdot \nu)[\check{s}_\rho, \check{s}_\sigma] \times \begin{cases} 0 & \text{rational,} \\ -1 & \text{hyperbolic,} \\ 1 & \text{trigonometric.} \end{cases}$$
$$(81)$$

One important use of the ℓ operators is the proof of involution of quantum conserved quantities. For the rational potential models Heckman [29] gave a universal proof based on the commutation relation (81):

$$[Q_n, Q_m]\psi = \sum_{\mu,\nu \in \mathcal{R}} [\ell_\mu^n, \ell_\nu^m]\psi = 0, \quad \text{rational model.} \quad (82)$$

This was the motivation for the introduction of the commuting differential-reflection operators by Dunkl [11]. In fact, Dunkl's and Heckman's operators were the similarity transformation of ℓ_μ by the ground-state wavefunction e^W:

$$\tilde{\ell}_\mu = e^{-W}\ell_\mu e^W = p \cdot \mu + i \sum_{\rho \in \Delta_+} g_\rho \frac{(\rho \cdot \mu)}{(\rho \cdot q)}(\check{s}_\rho - 1). \quad (83)$$

As for the *Calogero* system, we define $\ell^\pm$ corresponding to $L^\pm$ (66):

$$\ell_\mu^\pm = \ell^\pm \cdot \mu = p \cdot \mu \pm i\omega(q \cdot \mu) + i \sum_{\rho \in \Delta_+} g_\rho \frac{(\rho \cdot \mu)}{(\rho \cdot q)}\check{s}_\rho, \quad \mu \in \mathcal{R}, \quad (\ell_\mu^\pm)^\dagger = \ell_\mu^\mp.$$
$$(84)$$

The conserved quantities are expressed as polynomials in $\ell^\pm$ operators:

$$\mathrm{Ts}(\mathcal{L}_1^n)\psi = \sum_{\mu,\nu\in\mathcal{R}} (L^+L^-)_{\mu\nu}^n\,\psi = \sum_{\mu\in\mathcal{R}} \left(\ell_\mu^+\ell_\mu^-\right)^n\psi, \qquad (85)$$

$$\mathrm{Ts}(\mathcal{L}_2^n)\psi = \sum_{\mu,\nu\in\mathcal{R}} (L^-L^+)_{\mu\nu}^n\,\psi = \sum_{\mu\in\mathcal{R}} \left(\ell_\mu^-\ell_\mu^+\right)^n\psi, \qquad (86)$$

Likewise the creation and annihilation operators $B_n^\pm$ (72) are expressed as

$$B_n^\pm\psi = \mathrm{Ts}(L^\pm)^n\psi = \sum_{\mu,\nu\in\mathcal{R}} (L^\pm)_{\mu\nu}\,\psi = \sum_{\mu\in\mathcal{R}} \left(\ell_\mu^\pm\right)^n\psi. \qquad (87)$$

The commutation relations among $\ell^\pm$ operators are easy to evaluate, since ℓ operators commute in the rational potential models (81):

$$[\ell_\mu^+, \ell_\nu^+] = [\ell_\mu^-, \ell_\nu^-] = 0, \quad [\ell_\mu^-, \ell_\nu^+] = 2\omega\left(\mu\cdot\nu + \sum_{\rho\in\Delta_+} g_\rho(\rho\cdot\mu)(\rho^\vee\cdot\nu)\check{s}_\rho\right).$$

$$(88)$$

From these it follows that the creation (annihilation) operators $B_n^\pm$ do commute among themselves:

$$[B_n^+, B_m^+]\psi = [B_n^-, B_m^-]\psi = 0. \qquad (89)$$

It is also clear that $\ell_\mu^\pm/\sqrt{2\omega}$ are the "deformation" of the creation (annihilation) operators of the ordinary multicomponent harmonic oscillators. In fact we have

$$\ell_\mu^+\,?e^W = 2i\omega(\mu\cdot q)\,?e^W \quad \text{and} \quad \ell_\mu^-\,?e^W = 0. \qquad (90)$$

In the next section we present an alternative scheme of algebraic construction of excited states of the *Calogero* system by pursuing the analogy that $\ell^\pm$ are the creation and annihilation operators of the unit quantum. This method was applied to the A_r models by Brink et al. and others [21, 23, 24].

7 ALGEBRAIC CONSTRUCTION OF EXCITED STATES II

7.1 Operator Solution of the Triangular Hamiltonian

In Subsection 3.1, we have shown that an eigenfunction of $\mathcal{H}$ with eigenvalue $N\omega$ is given by

$$\left(P_N(q) + \tilde{P}_{N-2}(q)\right)e^W, \qquad (91)$$

in which $P_N(q)$ is a Coxeter invariant polynomial in q of homogeneous degree N and $\tilde{P}_{N-2}(q)$ is a Coxeter invariant polynomial in q of degree $N-2$ and lower. The nonleading polynomial $\tilde{P}_{N-2}(q)$ is completely determined by the leading one $P_N(q)$ due to the triangularity. This solution can be written in an operator form as follows.

Suppose $P_N(q)$ is expressed as

$$P_N(q) = \sum_{\{\mu\}} c_{\{\mu\}}(q \cdot \mu_1) \cdots (q \cdot \mu_N), \quad \mu_j \in \mathcal{R}, \quad c_{\{\mu\}} : \text{const.} \quad (92)$$

We obtain a Coxeter invariant polynomial in the creation operators ℓ^+ by replacing $q \cdot \mu$ by $\ell_\mu^+/(2i\omega)$:

$$P_N(q) \Rightarrow \frac{1}{(2i\omega)^N} P_N(\ell^+).$$

This creates the above eigenfunction of $\mathcal{H}$ from the ground state:

$$\frac{1}{(2i\omega)^N} P_N(\ell^+) e^W = \left(P_N(q) + \tilde{P}_{N-2}(q) \right) e^W. \quad (93)$$

The ℓ operator formulas of higher conserved quantities (85) contain extra terms:

$$\text{Ts}(\mathcal{L}_1^n) = \sum_{\mu,\nu \in \mathcal{R}} (L^+ L^-)_{\mu\nu}^n = \sum_{\mu \in \mathcal{R}} (\ell_\mu^+ \ell_\mu^-)^n + VT. \quad (94)$$

Here VT stands for *vanishing terms* when they act on a Coxeter invariant state. The same is true for most formulas derived in Section 6.

8   CLASSICAL EQUILIBRIUM AND SPIN EXCHANGE MODELS

Here we discuss the properties of the classical potential V_C,

$$V_C = \frac{1}{2} \sum_{j=1}^{r} \left(\frac{\partial W}{\partial q_j} \right)^2, \quad (95)$$

which is obtained from (5) by dropping the last term which is a *quantum correction*. (Hereafter we set the constant a determining the period of the trigonometric potential in (3) to be unity $a = 1$, for simplicity.) The classical equilibrium point

$$p = 0, \quad q = \bar{q} \quad (96)$$

can be characterised by two equivalent ways. It is a *minimal* point of the classical potential

$$\left. \frac{\partial V_C}{\partial q_j} \right|_{\bar{q}} = 0, \quad j = 1, \ldots, r, \quad (97)$$

whereas it is a *maximal* point of the prepotential W and of the ground-state wavefunction $\phi_0 = e^W$:

$$\left.\frac{\partial W}{\partial q_j}\right|_{\bar{q}} = 0, \quad j = 1, \ldots, r. \tag{98}$$

Note that the condition (15) $(p + i\partial W/\partial q_j)e^W = 0$ is also satisfied classically at this point. In the Lax representation it is a point at which two Lax matrices commute:

$$0 = [\bar{L}, \bar{M}], \qquad 0 = [\bar{\mathcal{L}}_{(1,2)}, \bar{M}], \tag{99}$$

in which $\bar{L} = L(0, \bar{q})$, $\bar{M} = M(\bar{q})$ etc and $d\bar{L}/dt = 0$, etc at the equilibrium point. The value of a quantity A at the equilibrium is expressed by $\bar{A}$.

By differentiating (95), we obtain

$$\frac{\partial V_C}{\partial q_j} = \sum_{l=1}^{r} \frac{\partial^2 W}{\partial q_j \partial q_l} \frac{\partial W}{\partial q_l}. \tag{100}$$

Since $\partial^2 W/\partial q_j \partial q_k$ is *negative definite* everywhere, the above two conditions (97) and (98) are in fact equivalent. By differentiating (100) again, we obtain

$$\frac{\partial^2 V_C}{\partial q_j \partial q_k} = \sum_{l=1}^{r} \frac{\partial^2 W}{\partial q_j \partial q_l} \frac{\partial^2 W}{\partial q_l \partial q_k} + \sum_{l=1}^{r} \frac{\partial^3 W}{\partial q_j \partial q_k \partial q_l} \frac{\partial W}{\partial q_l}.$$

Thus at the equilibrium point of the classical potential V_C, the following relation holds:

$$\left.\frac{\partial^2 V_C}{\partial q_j \partial q_k}\right|_{\bar{q}} = \sum_{l=1}^{r} \left.\frac{\partial^2 W}{\partial q_j \partial q_l}\right|_{\bar{q}} \left.\frac{\partial^2 W}{\partial q_l \partial q_k}\right|_{\bar{q}}. \tag{101}$$

If we define the following two symmetric $r \times r$ matrices $\tilde{V}$ and $\tilde{W}$,

$$\tilde{V} = \text{Matrix}\left[\left.\frac{\partial^2 V_C}{\partial q_j \partial q_k}\right|_{\bar{q}}\right], \qquad \tilde{W} = \text{Matrix}\left[\left.\frac{\partial^2 W}{\partial q_j \partial q_k}\right|_{\bar{q}}\right], \tag{102}$$

we have

$$\tilde{V} = \tilde{W}^2, \tag{103}$$

and

$$\text{Eigenvalues}(\tilde{V}) = \{w_1^2, \ldots, w_r^2\},$$
$$\text{Eigenvalues}(\tilde{W}) = \{-w_1, \ldots, -w_r\}, \quad w_j > 0, \quad j = 1, \ldots, r. \tag{104}$$

That is $\tilde{V}$ is *positive definite* and the point $\bar{q}$ is actually a minimal point of V_C. As shown in the following two sections, the matrices at equilibrium, $\tilde{W}, \bar{L}, \bar{\mathcal{L}}$, and $\bar{M}$ have "integer" eigenvalues and in most cases with high multiplicities.

Next let us briefly summarize the basic ingredients of the spin exchange models associated with the *Calogero* and *Sutherland* systems based on the root system Δ and with the set of vectors $\mathcal{R}$, [15]. They are defined at the equilibrium points (96) of the corresponding classical systems. Here we call each element μ of $\mathcal{R}$ a *site* to which a dynamical degree of freedom called *spin* is attached. The spin takes a finite set of discrete values. In the simplest, and typical case, they are an up ($\uparrow$) and a down ($\downarrow$). The dynamical state of the spin exchange model is represented by a vector ψ_{Spin} which takes values in the tensor product of $D = \#\mathcal{R}$ copies of a vector space $\mathcal{V}$ whose basis consists of an up ($\uparrow$) and a down ($\downarrow$):

$$\psi_{\text{Spin}} \in \overset{D}{\otimes} \mathcal{V}_\mu. \tag{105}$$

The Hamiltonian of the spin exchange model $\mathcal{H}_{\text{Spin}}$ is

$$\mathcal{H}_{\text{Spin}} = \begin{cases} \dfrac{1}{2} \sum\limits_{\rho \in \Delta_+} g_\rho \rho^2 \dfrac{1}{(\rho \cdot \bar{q})^2} (1 - \hat{P}_\rho), \\[2ex] \dfrac{1}{2} \sum\limits_{\rho \in \Delta_+} g_\rho \rho^2 \dfrac{1}{\sin^2(\rho \cdot \bar{q})} (1 - \hat{P}_\rho), \end{cases} \tag{106}$$

in which $\{\hat{P}_\rho\}$, $\rho \in \Delta_+$ are the dynamical variables called spin exchange operators. The operator $\hat{P}_\rho$ exchanges the spins of sites μ and $s_\rho(\mu)$, $\forall \mu \in \mathcal{R}$. In terms of the operator-valued Lax pairs

$$L_{\text{Spin}} = \begin{cases} i \sum\limits_{\rho \in \Delta_+} g_\rho (\rho \cdot \hat{H}) \dfrac{1}{\rho \cdot \bar{q}} \hat{P}_\rho \hat{s}_\rho, \\[2ex] i \sum\limits_{\rho \in \Delta_+} g_\rho (\rho \cdot \hat{H}) \cot(\rho \cdot \bar{q}) \hat{P}_\rho \hat{s}_\rho, \end{cases} \tag{107}$$

$$M_{\text{Spin}} = \begin{cases} -\dfrac{i}{2} \sum\limits_{\rho \in \Delta_+} g_\rho \rho^2 \dfrac{1}{(\rho \cdot \bar{q})^2} \hat{P}_\rho (\hat{s}_\rho - I), \\[2ex] -\dfrac{i}{2} \sum\limits_{\rho \in \Delta_+} g_\rho \rho^2 \dfrac{1}{\sin^2(\rho \cdot \bar{q})} \hat{P}_\rho (\hat{s}_\rho - I), \end{cases} \tag{108}$$

the Heisenberg equations of motion for the trigonometric spin exchange model can be written in a matrix form

$$i[\mathcal{H}_{\text{Spin}}, L_{\text{Spin}}] = [L_{\text{Spin}}, M_{\text{Spin}}]. \tag{109}$$

Since the M_{Spin} matrix enjoys the *sum up to zero* property,

$$\sum_{\mu \in \mathcal{R}} (M_{\text{Spin}})_{\mu\nu} = \sum_{\nu \in \mathcal{R}} (M_{\text{Spin}})_{\mu\nu} = 0, \tag{110}$$

one obtains conserved quantities via the *total sum* of L_{Spin}^k:

$$\left[\mathcal{H}_{\text{Spin}}, \text{Ts}(L_{\text{Spin}}^k) \right] = 0, \quad \text{Ts}\left(L_{\text{Spin}}^k \right) \equiv \sum_{\mu,\nu \in \mathcal{R}} \left(L_{\text{Spin}}^k \right)_{\mu\nu}, \quad k = 3, \ldots$$

(111)

These are necessary ingredients for complete integrability.

The rational spin exchange model needs some modification similar to those for the *Calogero* systems. We define

$$L_{\text{Spin}}^{\pm} = L_{\text{Spin}} \pm i\omega \bar{Q}, \quad \bar{Q} = \bar{q} \cdot \hat{H},$$

(112)

then the Heisenberg equations of motion in a matrix form read

$$i[\mathcal{H}_{\text{Spin}}, L_{\text{Spin}}^+ L_{\text{Spin}}^-] = [L_{\text{Spin}}^+ L_{\text{Spin}}^-, M_{\text{Spin}}]$$

(113)

and conserved quantities are given by

$$\text{Ts}\left(\left(L_{\text{Spin}}^+ L_{\text{Spin}}^- \right)^k \right) \equiv \sum_{\mu,\nu \in \mathcal{R}} \left(L_{\text{Spin}}^+ L_{\text{Spin}}^- \right)_{\mu\nu}^k, \quad k = 3, \ldots,.$$

Let us emphasize that the current definition of completely integrable spin exchange models is universal, in the sense that it applies to any root system Δ and to an arbitrary choice of the set of vectors $\mathcal{R}$. It contains all the known examples of spin exchange models as subcases. For the A_r root system and for the set of vector weights, $\mathcal{R} = \mathbf{V}$ (vector weights), the trigonometric spin exchange model reduces to the well-known Haldane–Shastry model [14], the rational spin exchange model reduces to the so-called Polychronakos model [21].

As is clear from the formulation, the dynamics of spin exchange models depends on the details of the classical potential V_C or W at the *equilibrium point* and on $\mathcal{R}$. It is quite natural to expect that the highly organized spectra of the known spin exchange models [14, 16] are correlated with the remarkable properties of the $\tilde{W}$ and $\tilde{L}, \tilde{\mathcal{L}}, \tilde{M}$, the Lax matrices at the equilibrium point—the integer eigenvalues and their high multiplicities.

9 CLASSICAL DATA I: CALOGERO SYSTEMS

9.1 Minimum Energy

The equations (97) and (98) determining the classical equilibrium read:

$$\left. \frac{\partial V_C}{\partial q_j} \right|_{\bar{q}} = 0 \Rightarrow \sum_{\rho \in \Delta_+} g_\rho^2 \frac{\rho^2 \rho_j}{(\rho \cdot \bar{q})^3} = \omega^2 \bar{q}_j,$$

(114)

$$j = 1, \ldots, r.$$

$$\left. \frac{\partial W}{\partial q_j} \right|_{\bar{q}} = 0 \Rightarrow \sum_{\rho \in \Delta_+} g_\rho \frac{\rho_j}{(\rho \cdot \bar{q})} = \omega \bar{q}_j,$$

(115)

By multiplying $\bar{q}_j$ to both equations, we obtain the *virial theorem* for the classical potential V_C

$$\sum_{\rho \in \Delta_+} g_\rho^2 \frac{\rho^j}{(\rho \cdot \bar{q})^2} = \omega^2 \bar{q}^2, \tag{116}$$

and a relationship

$$\omega \bar{q}^2 = \sum_{\rho \in \Delta_+} g_\rho \frac{\rho \cdot \bar{q}}{(\rho \cdot \bar{q})} = \sum_{\rho \in \Delta_+} g_\rho. \tag{117}$$

We arrive at the minimal value of the classical potential (95):

$$V_C(\bar{q}) = \omega^2 \bar{q}^2 = \omega \left(\sum_{\rho \in \Delta_+} g_\rho \right) = \tilde{\mathcal{E}}_0, \tag{118}$$

which has the general structure of a *coupling constant*(s) times an *integer*:

$$\tilde{\mathcal{E}}_0 = \begin{cases} \omega g \times \#\Delta/2, & \text{simply laced,} \\ \omega (g_L \times \#\Delta_L + g_s \times \#\Delta_S)/2, & \text{nonsimply laced.} \end{cases} \tag{119}$$

Here, $\#\Delta$ is the total number of roots, $\#\Delta_L(\#\Delta_S)$ is the number of long (short) roots, and $\#\Delta = \#\Delta_L + \#\Delta_S$.

9.2 Equilibrium Point and Eigenvalues of $\widetilde{W}$

9.2.1 A_r

Calogero and collaborators discussed this problem about quarter of a century ago [4, 27]. The Eq. (115) read

$$\sum_{k \neq j}^{r+1} \frac{1}{\bar{q}_j - \bar{q}_k} = \frac{\omega}{g} \bar{q}_j, \quad j = 1, \ldots, r+1. \tag{120}$$

These determine $\{\bar{x}_j = \sqrt{\frac{\omega}{g}} \bar{q}_j\}$, $j = 1, \ldots, r+1$ to be the zeros of the Hermite polynomial $H_{r+1}(x)$ (Stieltjes) [41]. We obtain

$$A_r : \mathrm{Spec}(\widetilde{W}) = -\omega\{1, 2, \ldots, r+1\}. \tag{121}$$

9.2.2 $B_r(D_r)$

The Eq. (115) read (assuming $\bar{q}_j \neq 0$)

$$\sum_{k \neq j}^{r} \frac{1}{\bar{q}_j^2 - \bar{q}_k^2} + \frac{g_S/2g_L}{\bar{q}_j^2} = \frac{\omega}{2g_L}, \quad j = 1, \ldots, r, \tag{122}$$

and determine $\{\bar{q}_j^2\}$, $j = 1, \ldots, r$, as the zeros of the associated Laguerre polynomial $L_r^{(\alpha)}(cx)$, with $\alpha = g_S/g_L - 1$, $c = \omega/g_L$, [4, 41]. The spectrum of $\widetilde{W}$
for B_r and D_r are

$$B_r : \quad \text{Spec}(\widetilde{W}) = -\omega\{2, 4, 6, \ldots, 2r - 2, 2r\}, \tag{123}$$
$$D_r : \quad \text{Spec}(\widetilde{W}) = -\omega\{2, 4, 6, \ldots, 2r - 2, r\}. \tag{124}$$

9.2.3   Exceptional Root Systems

In each of these cases we have calculated the equilibrium position numerically,
and evaluated the spectrum of $\widetilde{W}$. The results are

$$F_4 : \quad \text{Spec}(\widetilde{W}) = -\omega\{2, 6, 8, 12\}, \tag{125}$$
$$E_6 : \quad \text{Spec}(\widetilde{W}) = -\omega\{2, 5, 6, 8, 9, 12\}, \tag{126}$$
$$E_7 : \quad \text{Spec}(\widetilde{W}) = -\omega\{2, 6, 10, 12, 14, 18\}, \tag{127}$$
$$E_8 : \quad \text{Spec}(\widetilde{W}) = -\omega\{2, 8, 12, 14, 18, 20, 24, 30\}. \tag{128}$$

The eigenvalues of $\widetilde{W}$ are the numbers listed in I, i.e., $1 + exponent$, as expected
from the spectrum (28).

9.2.4   Universal Spectrum of M

Let us denote by $\mathbf{v}_0$ a special vector in $\mathbf{R}^D$ with each element unity:

$$\mathbf{v}_0 = (1, 1, \ldots, 1)^T \in \mathbf{R}^D, \quad D = \#\mathcal{R}, \text{ or } \mathbf{v}_{0\mu} = 1, \quad \forall \mu \in \mathcal{R}. \tag{129}$$

The condition for classical equilibrium (115) and *sum up to zero* conditions
(57) can be expressed neatly in matrix–vector notation as

$$\bar{L}^-\mathbf{v}_0 = 0, \quad \mathbf{v}_0^T \bar{L}^+ = 0, \quad \bar{M}\mathbf{v}_0 = 0, \quad \mathbf{v}_0^T \bar{M} = 0, \tag{130}$$

inspiring the idea that $\mathbf{v}_0$ is the classical (Coxeter invariant) *ground* state of
a matrix counterpart of the Hamiltonian ($\bar{M}$) and that $\bar{L}^-$ is an *annihilation*
operator. The analogy goes further when we evaluate the Lax equation for $L^\pm$
(65) at the classical equilibrium to obtain

$$[\bar{M}, \bar{L}^\pm] = \pm i\omega\bar{L}^\pm. \tag{131}$$

The relation (131) simply means that the eigenvalues of $\bar{M}$ are integer spaced
in units of $i\omega$. We obtain

$$\bar{M}\mathbf{v}_0 = 0, \quad \bar{M}\bar{L}^+\mathbf{v}_0 = i\omega\bar{L}^+\mathbf{v}_0, \ldots, \quad \bar{M}(\bar{L}^+)^n\mathbf{v}_0 = in\omega(\bar{L}^+)^n\mathbf{v}_0,$$
$$\tag{132}$$

implying $\bar{L}^+$ is a corresponding *creation* operator. This also means there is a *universal* formula:

$$\text{Spec}(\bar{M}) = i\omega\{0, 1, 2, \ldots, \},\tag{133}$$

with possible degeneracies. Here is the summary of the spectrum of $\bar{M}$ with [multiplicity] for the classical root systems and choices of $\mathcal{R}$:

$$A_r: \quad \textbf{(V)} \ \text{Spec}(\bar{M}) = i\omega\{0, 1, \ldots, r - 1, r\},\tag{134}$$

$$B_r: \quad (\Delta_S) \ \text{Spec}(\bar{M}) = i\omega\{0, 1, 2, \ldots, 2r - 1\},\tag{135}$$

$$D_r: \quad \textbf{(V)} \ \text{Spec}(\bar{M}) = i\omega\{0, 1, 2, \ldots, r - 1[2], \ldots, 2r - 2\}, \tag{136}$$

$$D_4: \quad \textbf{(S)} \ \text{Spec}(\bar{M}) = i\omega\{0, 1, 2, 3\,[2], 4, 5, 6\}.\tag{137}$$

The above results and those for E_6 with **27** and E_7 with **56**, the eigenvalue with [multiplicity] can be neatly expressed as the *heights* of the vectors in $\mathcal{R}$:

$$\text{Spec}(\bar{M}) = i\omega\{\delta \cdot \mu + h_{\max} | \mu \in \mathcal{R}\}, \quad h_{\max} \equiv \max(\delta \cdot \mathcal{R}), \tag{138}$$

in which δ is the *Weyl* vector $\delta = \sum_{\rho \in \Delta_+} \rho/2$ as obtained from (19) by setting all the coupling constant(s) to unity $g_\rho = 1$. The eigenvalues and multiplicities of $\bar{M}$ in the root type Lax pairs of simply laced crystallographic root systems can also be understood as the *height* and multiplicities of Δ:

$$\text{Spec}(\bar{M}) = \left\{ \begin{array}{ll} \delta \cdot \alpha + h_{\max}, & \text{for } \delta \cdot \alpha < 0 \\ \delta \cdot \alpha + h_{\max} - 1, & \text{for } \delta \cdot \alpha > 0 \end{array} \middle| \alpha \in \Delta(\Delta_L) \right\}. \tag{139}$$

10   CLASSICAL DATA II: SUTHERLAND SYSTEMS

10.1   Minimum Energy

The classical minimum energy of the *Sutherland* system, $2\varrho^2$ (18) is, in fact, "quantised." If all the coupling constants are unity $g_\rho = 1 \Rightarrow \varrho = \delta$, the Freudenthal-de Vries ("strange") formula leads to

$$2\varrho^2 = \dim(\mathfrak{g}_\Delta)\rho_h^2 h^\vee/12.\tag{140}$$

10.2   Equilibrium Point and Eigenvalues of $\widetilde{W}$

The equations determining the equilibrium position (98) read

$$\sum_{\rho \in \Delta_+} g_\rho \cot(\rho \cdot \bar{q})\rho_j = 0, \quad j = 1, \ldots, r,\tag{141}$$

and they can be expressed in terms of the L, M matrices at equilibrium:

$$\bar{L}\mathbf{v}_0 = 0 = \mathbf{v}_0^T \bar{L}, \qquad \bar{M}\mathbf{v}_0 = 0 = \mathbf{v}_0^T \bar{M}.\tag{142}$$

10.2.1 A_r

In this case the equilibrium position is *"equally-spaced"* $\bar{q} = \pi(0, 1, \ldots, r - 1, r)/(r + 1) + \xi v_0$, $\xi \in \mathbf{R}$. This is the reason why the Haldane–Shastry model is better understood than the other spin exchange models. We have

$$A_r : \mathrm{Spec}(\widetilde{W}) = -2g\{r, \ldots, (r + 1 - j)j, \ldots, r\} \propto \mathrm{Spec}(\bar{M})(\mathbf{V}), \quad (143)$$

$$A_r(\mathbf{V}) : \mathrm{Spec}(\bar{L}) = g \left\{ \begin{array}{ll} 0[2], \pm 2, \pm 4, \ldots, \pm(r - 1) & r: \text{odd} \\ 0, \quad \pm 1, \pm 3, \ldots, \pm(r - 1) & r: \text{even} \end{array} \right\}. \quad (144)$$

10.2.2 BC_r and D_r

In terms of $\bar{x}_j \equiv \cos 2\bar{q}_j$, the equation determining equilibrium (141) read

$$\sum_{k \neq j}^{r} \frac{1}{\bar{x}_j - \bar{x}_k} + \frac{g_S + g_L}{2g_M} \frac{1}{\bar{x}_j - 1} + \frac{g_L}{2g_M} \frac{1}{\bar{x}_j + 1} = 0, \quad j = 1, \ldots, r,$$
$$(145)$$

which are the equations satisfied by the zeros $\{\bar{x}_j\}$ of Jacobi polynomial $P_r^{(\alpha, \beta)}(x)$ [41] with $\alpha = (g_L + g_S)/g_M - 1$, $\beta = g_L/g_M - 1$.

The problem of finding the maximal point of the D_r prepotential W is the same as the classical problem of maximizing the van der Monde determinant

$$V\,dM(x_1, \ldots, x_r) = \prod_{j<k}^{r}(x_j - x_k), \quad (146)$$

under the boundary conditions, $1 = x_1 > x_2 > \cdots > x_{r-1} > x_r = -1$.

The spectrum of $\widetilde{W}$ and $\bar{M}$ are

$$D_r : \mathrm{Spec}(\widetilde{W}) = -g\{4(r - 1), 4(2r - 3), \ldots, 2j(2r - 1 - j), \ldots,$$
$$2(r - 2)(r + 1), r(r - 1)[2]\}, \quad (147)$$

$$D_r(\mathbf{V}) : \mathrm{Spec}(\bar{M}) = ig\{0, 4(r - 1)[2], \ldots, 2j(2r - 1 - j)[2], \ldots,$$
$$2(r - 2)(r + 1)[2], r(r - 1)[2], 2r(r - 1)\}. \quad (148)$$

For more results on the other root systems including the exceptional ones, see [13].

11   SUMMARY, COMMENTS AND OUTLOOK

Various issues related to *quantum* vs *classical* integrability of C–M systems based on any root system are presented. These are construction of involutive quantum conserved quantities, the relationship between the Lax pair and the differential-reflection (Dunkl) operator formalisms, construction of excited states by creation operators, properties of the classical potentials and Lax pair operators at equilibrium, etc. They are mainly generalizations of the results known for the models based on A-type root systems. Integrability of the models based on other classical root systems and the exceptional ones including the noncrystallographic models are also discussed in [42, 43–46].

Among the interesting recent developments of the related subjects which could not be covered in this lecture are the quadratic algebras [47, 48] for the superintegrable systems [19] with the rational potential and quantum Inozemtsev systems [49] as multiparticle Quasi-Exactly Solvable systems and multivariable $\mathcal{N}$-fold supersymmetry.

There are still many interesting problems to be addressed to: The structure and properties of the eigenfunctions of the trigonometric potential models, which are generalizations of the Jack polynomials [35–38]. Comprehensive treatment of Liouville integrability of the *Calogero* systems. Understanding the roles of supersymmetry and shape invariance in C–M systems [47, 50]. Formulation of various aspects of quantum C–M systems with elliptic potentials; Lax pair, the differential-reflection operators [51, 52], conserved quantities, supersymmetry, and excited states wavefunctions.

APPENDIX A: ROOT SYSTEMS

In this appendix we recapitulate the rudimentary facts of the root systems and reflections to be used in the main text. The set of roots Δ is invariant under reflections in the hyperplane perpendicular to each vector in Δ. In other words,

$$s_\alpha(\beta) \in \Delta, \quad \forall \alpha, \beta \in \Delta, \tag{A1}$$

where

$$s_\alpha(\beta) = \beta - (\alpha^\vee \cdot \beta)\alpha, \quad \alpha^\vee \equiv 2\alpha/|\alpha|^2. \tag{A2}$$

The set of reflections $\{s_\alpha, \alpha \in \Delta\}$ generates a group G_Δ, known as a Coxeter group, or finite reflection group. The orbit of $\beta \in \Delta$ is the set of root vectors resulting from the action of the Coxeter group on it. The set of positive roots Δ_+ may be defined in terms of a vector $U \in \mathbf{R}^r$, with $\alpha \cdot U \neq 0, \forall \alpha \in \Delta$, as those roots $\alpha \in \Delta$ such that $\alpha \cdot U > 0$. Given Δ_+, there is a unique set of r simple roots $\Pi = \{\alpha_j, j = 1, \ldots, r\}$ defined such that they span the root space and the coefficients $\{a_j\}$ in $\beta = \sum_{j=1}^r a_j\alpha_j$ for $\beta \in \Delta_+$ are all nonnegative. The highest root α_h, for which $\sum_{j=1}^r a_j$ is maximal, is then also determined uniquely. The subset of reflections $\{s_\alpha, \alpha \in \Pi\}$ in fact generates the Coxeter group G_Δ. The products of s_α, with $\alpha \in \Pi$, are subject solely to the relations

$$(s_\alpha s_\beta)^{m(\alpha,\beta)} = 1, \qquad \alpha, \beta \in \Pi. \tag{A3}$$

The interpretation is that $s_\alpha s_\beta$ is a rotation in some plane by $2\pi/m(\alpha, \beta)$. The set of positive integers $m(\alpha, \beta)$ (with $m(\alpha, \alpha) = 1$, $\forall \alpha \in \Pi$) uniquely specify the Coxeter group. The weight lattice $\Lambda(\Delta)$ is defined as the $\mathbf{Z}$-span of the fundamental weights $\{\lambda_j\}$, $j = 1, \ldots, r$, defined by

$$\alpha_j^\vee \cdot \lambda_k = \delta_{jk}, \quad \alpha_j \in \Pi. \tag{A4}$$

The root systems for finite reflection groups may be divided into two types: crystallographic and noncrystallographic. Crystallographic root systems satisfy the additional condition

$$\alpha^\vee \cdot \beta \in \mathbf{Z}, \quad \forall \alpha, \beta \in \Delta, \tag{A5}$$

which implies that the $\mathbf{Z}$-span of Π is a lattice in $\mathbf{R}^r$ and contains all roots in Δ. We call this the root lattice, which is denoted by $L(\Delta)$. These root systems are associated with simple Lie algebras: $\{A_r, r \geq 1\}$, $\{B_r, r \geq 2\}$, $\{C_r, r \geq 2\}$, $\{D_r, r \geq 4\}$, E_6, E_7, E_8, F_4, and G_2. The Coxeter groups for these root systems are called Weyl groups. The remaining noncrystallographic root systems are H_3, H_4, whose Coxeter groups are the symmetry groups of the icosahedron and four-dimensional 600-cell, respectively, and the dihedral group of order $2m$, $\{I_2(m), m \geq 4\}$.

Here we give the explicit examples of root systems. In all cases but the A_r, $\{e_j\}$ denotes an orthonormal basis in $\mathbf{R}^r$, $e_j \in \mathbf{R}^r$, $e_j \cdot e_k = \delta_{jk}$. The crystallographic root systems are:

$$A_r : \Delta = \{\pm(e_j - e_k), \ j \neq k = 1, \ldots, r+1 | e_j \in \mathbf{R}^{r+1}, e_j \cdot e_k = \delta_{jk}\},$$
$$\Pi = \{e_j - e_{j+1}, \ j = 1, \ldots, r\}, \tag{A6}$$

$$B_r : \Delta = \{\pm e_j \pm e_k, \pm e_j, \ j \neq k = 1, \ldots, r\},$$
$$\Pi = \{e_j - e_{j+1}, \ j = 1, \ldots, r-1\} \cup \{e_r\}, \tag{A7}$$

$$C_r : \Delta = \{\pm e_j \pm e_k, \pm 2e_j, \ j, k = 1, \ldots, r\},$$
$$\Pi = \{e_j - e_{j+1}, \ j = 1, \ldots, r-1\} \cup \{2e_r\}, \tag{A8}$$

$$D_r : \Delta = \{\pm e_j \pm e_k, \ j \neq k = 1, \ldots, r\},$$
$$\Pi = \{e_j - e_{j+1}, \ j = 1, \ldots, r-1\} \cup \{e_{r-1} + e_r\}, \tag{A9}$$

$$E_6 : \Delta = \{\pm e_j \pm e_k, \ j \neq k = 1, \ldots, 5\}$$
$$\cup \left\{ \frac{1}{2}(\pm e_1 \ldots \pm e_5 \pm \sqrt{3}e_6), \text{(even+)} \right\},$$

$$\Pi = \left\{ \frac{1}{2}(e_1 - e_2 - e_3 - e_4 + e_5 - \sqrt{3}e_6), e_4 - e_5, e_3 - e_4, e_4 + e_5, \right.$$
$$\left. \frac{1}{2}(e_1 - e_2 - e_3 - e_4 - e_5 + \sqrt{3}e_6), e_2 - e_3 \right\}, \tag{A10}$$

$E_7 : \Delta = \{\pm e_j \pm e_k, j \neq k = 1, \dots, 6\} \cup \{\pm\sqrt{2}e_7\}$

$$\cup \left\{ \frac{1}{2}(\pm e_1 \cdots \pm e_6 \pm \sqrt{2}e_7), (\text{even} +) \right\}, \qquad (A11)$$

$\Pi = \Big\{ e_2 - e_3, e_3 - e_4, e_4 - e_5, e_5 - e_6,$

$$\frac{1}{2}(e_1 - e_2 - e_3 - e_4 - e_5 + e_6 - \sqrt{2}e_7), \sqrt{2}e_7, e_5 + e_6 \Big\}, \quad (A12)$$

$E_8 : \Delta = \{\pm e_j \pm e_k, j \neq k = 1, \dots, 8\} \cup \left\{ \frac{1}{2}(\pm e_1 \cdots \pm e_8), (\text{even} +) \right\},$

$$\Pi = \left\{ \frac{1}{2}(e_1 - e_2 - e_3 - e_4 - e_5 - e_6 - e_7 + e_8), e_7 + e_8 \right\}$$

$$\cup \{e_j - e_{j+1}, \ j = 2, \dots, 7\}, \qquad (A13)$$

$$F_4 : \Delta = \left\{ \pm e_j \pm e_k, +e_j, \frac{1}{2}(\pm e_1 \cdots \pm e_4), j \neq k = 1, \dots, 4 \right\},$$

$$\Pi = \left\{ e_2 - e_3, e_3 - e_4, e_4, \frac{1}{2}(e_1 - e_2 - e_3 - e_4) \right\}, \qquad (A14)$$

$$G_2 : \Delta = \left\{ (\pm\sqrt{2}, 0), \left(\pm\sqrt{\frac{3}{2}}, \pm\frac{1}{\sqrt{2}} \right), \left(0, \pm\sqrt{\frac{2}{3}} \right), \left(\pm\frac{1}{\sqrt{2}}, \pm\frac{1}{\sqrt{6}} \right) \right\},$$

$$\Pi = \left\{ (\sqrt{2}, 0), \left(-\frac{1}{\sqrt{2}}, \frac{1}{\sqrt{6}} \right) \right\}. \qquad (A15)$$

The noncrystallographic root systems are:

1. $I_2(m)$: This is a symmetry group of a regular m-gon. For odd m Δ consists of a single orbit, whereas for even m it has two orbits. In both cases we have a representation in which all the roots have length unity

$$\Delta = \{(\cos((j-1)\pi/m), \sin((j-1)\pi/m)), j = 1, \dots, m\},$$
$$\Pi = \{(1, 0), (\cos((m-1)\pi/m), \sin((m-1)\pi/m)))\} \qquad (A16)$$

2. H_4: Define $a \equiv \cos\pi/5 = (1 + \sqrt{5})/4, b \equiv \cos 2\pi/5 = (-1 + \sqrt{5})/4$. Then the H_4 roots are generated by the following simple roots [53]:

$$\alpha_1 = \left(a, -\frac{1}{2}, b, 0 \right), \quad \alpha_2 = \left(-a, \frac{1}{2}, b, 0 \right). \qquad (A17)$$

$$\alpha_3 = \left(\frac{1}{2}, b, -a, 0 \right), \quad \alpha_4 = \left(-\frac{1}{2}, -a, 0, b \right).$$

The full set of roots of H_4 in this basis may be obtained from $(1,0,0,0)$, $(\frac{1}{2}, \frac{1}{2}, \frac{1}{2}, \frac{1}{2})$, and $(a, \frac{1}{2}, b, 0)$ by even permutations and arbitrary sign changes of coordinates. These 120 roots form a single orbit.

3. H_3: A subset of (A17), $\{\alpha_1, \alpha_2, \alpha_3\}$ is a choice of simple roots for the H_3 root system. In this basis, the full set of roots for H_3 results from even permutations and arbitrary sign changes of $(1,0,0)$ and $(a, \frac{1}{2}, b)$. These 30 roots also form a single orbit.

REFERENCES

1. Calogero, F. (1971) *J. Math. Phys.* **12**, pp. 419–436.
2. Sutherland, B. (1972) *Phys. Rev.* **A5**, pp. 1372–1376.
3. Moser, J. (1975) *Adv. Math.* **16**, pp. 197–220; Lecture Notes in Physics **38**, (1975), Springer-Verlag; Calogero, F., Marchioro C., and Ragnisco, O. (1975) *Lett. Nuovo Cim.* **13**, pp. 383–387; Calogero, F. (1975) *Lett. Nuovo Cim.* **13**, pp. 411–416.
4. Olshanetsky, M. A. and Perelomov, A. M. (1976) *Inventions Math.* **37**, pp. 93–108.
5. Olshanetsky, M. A. and Perelomov, A. M. (1983) *Phys. Rep.* **94**, pp. 313–404.
6. Khastgir, S. P., Pocklington, A. J., and Sasaki, R. (2000) *J. Phys.* **A33**, pp. 9033–9064.
7. Bordner, A. J., Corrigan, E., and Sasaki, R. (1998) *Prog. Theor. Phys.* **100**, pp. 1107–1129; Bordner, A. J., Sasaki, R., and Takasaki, K. (1999) *Prog. Theor. Phys.* **101**, pp. 487–518; Bordner, A. J. and Sasaki, R. (1999) *Prog. Theor. Phys.* **101**, pp. 799–829; Khastgir, S. P., Sasaki R., and Takasaki, K. (1999) *Prog. Theor. Phys.* **102**, pp. 749–776.
8. Bordner, A. J., Corrigan, E., and Sasaki, R. (1999) *Prog. Theor. Phys.* **102**, pp. 499–529.
9. D'Hoker, E. and Phong, D. H. (1998) *Nucl. Phys.* **B530**, pp. 537–610.
10. Bordner, A. J., Manton, N. S., and Sasaki, R. (2000) *Prog. Theor. Phys.* **103**, pp. 463–487.
11. Dunkl, C. F. (1989) *Trans. Amer. Math. Soc.* **311**, pp. 167–183; Buchstaber, V. M., Felder, G., and Veselov, A. P. (1994) *Duke Math. J.* **76**, pp. 885–911.
12. Khastgir, S. P. and Sasaki, R. (2001) *Phys. Lett.* **A279**, pp. 189–193; Hurtubise, J. C. and Markman, E. (2001) *Comm. Math. Phys.* **223**, pp. 533–552.
13. Corrigan, E. and Sasaki, R. (2002) *J. Phys.* **A35**, pp. 7017–7061.
14. Haldane, F. D. M. (1988) *Phys. Rev. Lett.* **60**, pp. 635–638; Shastry, B. S. (1988) *ibid* **60**, pp. 639–642.
15. Inozemtsev, V. I. and Sasaki, R. (2001) *J. Phys.* **A34**, pp. 7621–7632.
16. Inozemtsev, V. I. and Sasaki, R. (2001) *Nucl. Phys.* **B618**, pp. 689–698.
17. Olshanetsky, M. A. and Perelomov, A. M. (1977) *Funct. Anal. Appl.* **12**, pp. 121–128.
18. Perelomov, A. M. (1971) *Theor. Math. Phys.* **6**, pp. 263–282.
19. Wojciechowski, S. (1976) *Lett. Nuouv. Cim.* **18**, pp. 103–107; *Phys. Lett.* **A95**, (1983) pp. 279–281.

20. Lassalle, M. (1991) *Acad. C. R. Sci. Paris, t. Sér. I Math.* **312**, pp. 425–428, 725–728, **313**, pp. 579–582.

21. Polykronakos, A. P. (1992) *Phys. Rev. Lett.* **69**, pp. 703–705.

22. Shastry, B. S. and Sutherland, B. (1993) *Phys. Rev. Lett.* **70**, pp. 4029–4033.

23. Brink, L., Hansson, T. H., and Vasiliev, M. A. (1992) *Phys. Lett.* **B286**, pp. 109–111; Brink, L., Hansson, T. H., Konstein, S., and Vasiliev, M. A. (1993) *Nucl. Phys.* **B401**, pp. 591–612; Brink, L., Turbiner, A., and Wyllard, N. (1998) *J. Math. Phys.* **39**, pp. 1285–1315.

24. Ujino, H., Wadati, M., and Hikami, K. (1993) *J. Phys. Soc. Jpn.* **62**, pp. 3035–3043; Ujino, H. and Wadati, M. (1996) *J. Phys. Soc. Jpn.* **65**, pp. 2423–2439; Ujino, H. (1995) *J. Phys. Soc. Jpn* **64**, pp. 2703–2706; Nishino, A., Ujino, H., and Wadati, M. Chaos Solitons Fractals (2000) **11**, pp. 657–674.

25. Sogo, K. (1996) *J. Phys. Soc. Jpn* **65**, pp. 3097–3099; Gurappa, N. and Panigrahi, P. K. (1999) *Phys. Rev.* **B59**, pp. R2490–R2493.

26. Ruijsenaars, S. N. M. (1999) CRM Series in *Math. Phys.* **1**, pp. 251–352, Springer.

27. Calogero, F. (1977) *Lett. Nuovo Cim.* **19**, pp. 505–507; *Lett. Nuovo Cim.* **22**, (1977) pp. 251–253; *Lett. Nuovo Cim.* **24**, (1979) pp. 601–604; *J. Math. Phys.* **22**, (1981) pp. 919–934.

28. Calogero, F. and Perelomov, A. M. (1978) *Commun. Math. Phys.* **59**, pp. 109–116.

29. Heckman, G. J. (1991) in Birkhäuser, Barker, W. and Sally, P. (eds.) Basel; (1991) *Inv. Math.* **103**, pp. 341–350.

30. Heckman, G. J. and Opdam, E. M. (1987) *Comp. Math.* **64**, pp. 329–352; Heckman, G. J. (1987) *Comp. Math.* **64**, pp. 353–373; Opdam, E. M. (1988) *Comp. Math.* **67**, pp. 21–49, 191–209.

31. Freedman, D. Z. and Mende, P. F. (1990) *Nucl. Phys.* **344**, pp. 317–343.

32. Dyson, F. J. (1962) *J. Math. Phys.* **3**, pp. 140–156, 157–165, 166–175; *J. Math. Phys.* **3**, (1962) pp. 1191–1198.

33. Rühl, W. and Turbiner, A. (1995) *Mod. Phys. Lett.* **A10**, pp. 2213–2222; Haschke, O. and Rühl, W. (2000) Lecture Notes in Physics **539**, pp. 118–140, Springer, Berlin.

34. Caseiro, R., Françoise, J. P., and Sasaki, R. (2000) *J. Math. Phys.* **41**, pp. 4679–4689.

35. Stanley, R. (1989) *Adv. Math.* **77**, pp. 76–115; Macdonald, I. G. *Symmetric Functions and Hall Polynomials*, 2nd ed., Oxford University Press.

36. Lapointe, L. and Vinet, L. (1997) *Adv. Math.* **130**, pp. 261–279; *Commun. Math. Phys.* **178**, (1996), pp. 425–452.

37. Baker, T. H. and Forrester, P. J. (1997) *Commun. Math. Phys.* **188**, pp. 175–216.

38. Awata, H., Matsuo, Y., Odake, S. and Shiraishi, J. (1995) *Nucl. Phys.* **B449**, pp. 347–374.

39. Gambardella, P. J. (1975) *J. Math. Phys.* **16**, pp. 1172–1187.

40. Calogero, F. (1969) *J. Math. Phys.* **10**, pp. 2191–2196; *J. Math. Phys.* **10**, pp. 2197–2200.

41. Szegö, G. (1939) *Orthogonal Polynomials*, American Mathematical Society, New York.

42. Wolfes, J. (1974) *J. Math. Phys.* **15**, pp. 1420–1424; Calogero, F. and Marchioro, C. (1974) *J. Math. Phys.* **15**, pp. 1425–1430.
43. Rosenbaum, M., Turbiner, A., and Capella, A. (1998) *Int. J. Mod. Phys.* **A13**, pp. 3885–3904; Gurappa, N., Khare, A., and Panigrahi, P. K. (1998) *Phys. Lett.* **A244**, pp. 467–472.
44. Haschke, O. and Rühl, W. (1998) *Mod. Phys. Lett.* **A13**, pp. 3109–3122; *Modern Phys. Lett.* **A14**, (1999) pp. 937–949.
45. Kakei, S. (1996) *J. Phys.* **A29**, pp. L619–L624; *J. Phys.* **A30**, (1997) pp. L535–L541.
46. Ghosh, P. K., Khare, A., and Sivakumar, M. (1998) *Phys. Rev.* **A58**, pp. 821–825; Sukhatme, U. and Khare, A. quant-ph/9902072.
47. Kuanetsov, V. B. (1996) *Phys. Lett.* **A218**, pp. 212–222.
48. Caseiro, R., Francoise J.-P., and Sasaki, R. (2001) *J. Math. Phys.* **42**, pp. 5329–5340.
49. Sasaki, R. and Takasaki, K. (2001) *J. Phys.* **A34**, pp. 9533–9553, 10335.
50. Efthimiou, C. and Spector, D. (1997) *Phys. Rev.* **A56**, pp. 208–219.
51. Cherednik, I. V. (1995) *Comm. Math. Phys.* **169**, pp. 441–461.
52. Oshima, T. and Sekiguchi, H. (1995) *J. Math. Sci. Univ. Tokyo* **2**, pp. 1–75.
53. Humphreys, J. E. Cambridge Univ. Press, Cambridge 1990.
54. Krichever, I. M. (1980) *Funct. Anal. Appl.* **14**, pp. 282–289.

GEOMETRICAL DYNAMICS OF AN INTEGRABLE PIECEWISE-LINEAR MAPPING

Daisuke Takahashi and Masataka Iwao
Department of Mathematical Sciences, Waseda University,
3-4-1 Ohkubo, Shinjuku-ku, Tokyo 169-8555, Japan

Abstract   A special type of piecewise-linear mapping is discussed. It is obtained by ultradiscretizing the Quispel–Robert–Thompson system. In a special case of a parameter, it becomes a periodic mapping with a constant period for any initial data. In a general case, it becomes an integrable mapping and a period of solution is constant for each solution orbit. We show a structure of solutions discussing the dynamics in a phase plane from a viewpoint of the integrable system theory.

1   INTRODUCTION

There have been many studies using a piecewise-linear mapping in the area of dynamical system theory [1]. The standard form of one-dimensional mapping is

$$x_{n+1} = f(x_n)$$

where $f(x)$ is linear in each local region of x. For example, the tent map $f(x) = 2x \, (0 \le x \le 1/2), 2(1 - x)(1/2 < x \le 1)$ and the Bernoulli shift $f(x) = 2x \, (0 \le x \le 1/2), 2x - 1(1/2 < x \le 1)$ are often used in the chaotic system theory to explain the typical dynamics of chaos. One of the advantages to study a piecewise-linear mapping is that we can analyze its dynamics exactly utilizing the local linearity.

Recently, piecewise-linear mappings appear together with the ultradiscretizing method in the integrable system theory [2]. For example, consider the discrete Painlevé equation [3],

$$x_{n+1} = (1 + \alpha \lambda^n x_n)/x_{n-1},$$

which is integrable because it has a conserved quantity. If we use a transformation of variable x_n and constants α, λ,

$$x_n = e^{X_n/\varepsilon}, \quad \alpha = e^{A/\varepsilon}, \quad \lambda = e^{L/\varepsilon},$$

291

L. Faddeev et al. (eds.),
Bilinear Integrable Systems: From Classical to Quantum, Continuous to Discrete, 291–300.
© 2006 *Springer. Printed in the Netherlands.*

and take a limit $\varepsilon \to +0$, we obtain an ultradiscrete Painlevé equation

$$X_{n+1} = \max(0, X_n + A + L) - X_{n-1}. \tag{1}$$

Note that the max function is defined by

$$\max(A, B) = \begin{cases} A & (A \geq B) \\ B & (A < B), \end{cases}$$

and we use the following formula in the derivation,

$$\lim_{\varepsilon \to +0} \varepsilon \log(e^{A/\varepsilon} + e^{B/\varepsilon} + \cdots) = \max(A, B, \ldots).$$

The remarkable features of (1) are (i) It is also integrable, that is, it has a conserved quantity, (ii) X can be discrete, that is, X_n is always integer if A, L and initial values of X_n are all integer.

In this paper, we discuss a structure of solutions to an integrable piecewise-linear mapping from a viewpoint of the integrable system theory. The mapping is obtained by ultradiscretizing an integrable difference system, the Quispel–Robert–Thompson (QRT) system. The general form of the QRT system gives a wide range of difference equations [4]. However, when we ultradiscretize the equations, positivity of solution is necessary. Therefore, we restrict its form to the following special one in this paper,

$$x_{n+1} = \frac{1 + a x_n}{x_n^\sigma x_{n-1}}, \tag{2}$$

where a is a constant and $\sigma = 0$, 1 or 2. Using transformations, $x_n = e^{X_n/\varepsilon}$ and $a = e^{A/\varepsilon}$, and taking a limit $\varepsilon \to +0$, we obtain

$$X_{n+1} = \max(0, X_n + A) - \sigma X_n - X_{n-1}, \tag{3}$$

from the above equation[3]. Note that (3) with $\sigma = 0$ is equivalent to (1) with $L = 0$.

First we consider a case of $A = 0$. We show (3) is linearizable in that case by a transformation of variable and its solutions are obtained in an explicit form. Second we consider a case of $A \neq 0$ and discuss a structure of solutions.

2   PERIODIC CASE

In this section, we assume $A = 0$ in (3) (or a is positive definite in (2)). Then we obtain

$$X_{n+1} = \max(0, X_n) - \sigma X_n - X_{n-1}. \tag{4}$$

We can easily show that any solution to this equation is always periodic with a constant period. For example, in the case of $\sigma = 0$, $X_2 \sim X_6$ are expressed by initial values X_0 and X_1 as follows:

$$X_2 = \max(0, X_1) - X_0, \quad X_3 = \max(0, X_0, X_1) - X_0 - X_1,$$
$$X_4 = \max(0, X_0) - X_1, \quad X_5 = X_0, \quad X_6 = X_1, \tag{5}$$

where $\max(A, B, C, \ldots)$ denotes the maximum value among $A, B, C, \ldots$. We use the following formulae on max function in the derivation of the above solution,

$$\max(A, B) = \max(B, A),$$
$$\max(A, \max(B, C)) = \max(\max(A, B), C) = \max(A, B, C),$$
$$\max(A, B) + X = \max(A + X, B + X).$$

For example, X_3 is expressed by X_0 and X_1 through the following calculation,

$$X_3 = \max(0, X_2) - X_1 = \max(0, \max(0, X_1) - X_0) - X_1$$
$$= \max(X_0, \max(0, X_1)) - X_0 - X_1 = \max(0, X_0, X_1) - X_0 - X_1.$$

Since (5) gives $X_5 = X_0$ and $X_6 = X_1$ and (4) is of the second order, any solution from arbitrary X_0 and X_1 other than $X_0 = X_1 = 0$ is always periodic with period 5. The case of $X_0 = X_1 = 0$ is exceptional and X_n is always 0 in that case. Similarly, any solution is periodic with period 7 and 8 in the case of $\sigma = 1$ and 2 respectively.

Equation (4) is derived from (2) through the ultradiscretization. If we assume $\sigma = 0$ and $a = 1$ in (2), solutions to (2) are also periodic with a constant period [5]. We obtain the following pattern of solution,

$$x_2 = \frac{1 + x_1}{x_0}, \quad x_3 = \frac{1 + (1 + x_1)/x_0}{x_1} = \frac{1 + x_0 + x_1}{x_0 x_1},$$

$$x_4 = \frac{1 + (1 + x_0 + x_1)/x_0 x_1}{(1 + x_1)/x_0} = \frac{(1 + x_0)(1 + x_1)}{(1 + x_1)x_1} = \frac{1 + x_0}{x_1},$$

$$x_5 = \frac{1 + (1 + x_0)/x_1}{(1 + x_0 + x_1)/x_0 x_1} = \frac{(1 + x_0 + x_1)x_0}{1 + x_0 + x_1} = x_0,$$

$$x_6 = \frac{1 + x_0}{(1 + x_0)/x_1} = x_1. \tag{6}$$

Therefore, a solution from any positive x_0 and x_1 is always periodic with period 5. Moreover, every solution in (6) is transformed to that in (5) through the above ultradiscretization. It means that both the difference equation and its solution can be transformed consistently through the ultradiscretization.

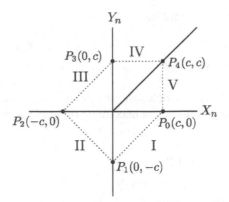

Figure 1. 5 fan areas in the phase plane

3 LINEARIZABILITY OF PERIODIC PIECEWISE-LINEAR MAPPING

Equation (4) can be rewritten by the following piecewise-linear mapping,

$$\begin{cases} X_{n+1} = Y_n \\ Y_{n+1} = \max(0, Y_n) - \sigma Y_n - X_n. \end{cases} \tag{7}$$

The only nonlinearity of this mapping is the term $\max(0, Y_n)$. Therefore a different type of linear mappings are applied to the upper and the lower half plane in a phase plane (X_n, Y_n),

$$\begin{pmatrix} X_{n+1} \\ Y_{n+1} \end{pmatrix} = \begin{cases} \begin{pmatrix} 0 & 1 \\ -1 & 1-\sigma \end{pmatrix} \begin{pmatrix} X_n \\ Y_n \end{pmatrix} & (Y_n \geq 0) \\ \begin{pmatrix} 0 & 1 \\ -1 & -\sigma \end{pmatrix} \begin{pmatrix} X_n \\ Y_n \end{pmatrix} & (Y_n < 0). \end{cases}$$

We can easily see the periodicity of this mapping by the following geometric dynamics in a phase plane. In the case of $\sigma = 0$, let us consider a sequence of mappings of a point $P_0(c, 0)$ $(c > 0)$ in the phase plane (X_n, Y_n). Then, we obtain a periodic sequence of points,

$$P_0(c, 0) \rightarrow P_1(0, -c) \rightarrow P_2(-c, 0) \rightarrow P_3(0, c) \rightarrow P_4(c, c) \rightarrow P_0 \rightarrow \cdots.$$

Since the parameter c is an arbitrary positive number, the phase plane is divided into 5 local "fan" areas as shown in Figure 1. Each area is linearly mapped each other in the following order,

$$\text{I} \rightarrow \text{II} \rightarrow \text{III} \rightarrow \text{IV} \rightarrow \text{V} \rightarrow \text{I} \rightarrow \cdots,$$

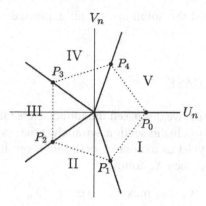

Figure 2. 5 fan areas mapped by a rotation by $-2\pi/5$

and segments $P_j P_{j+1}$ are mapped as follows,

$$P_0 P_1 \to P_1 P_2 \to P_2 P_3 \to P_3 P_4 \to P_4 P_0 \to P_0 P_1 \to \cdots.$$

Though this mapping is nonlinear, it is equivalent to a linear mapping defined by a rotation by an angle $-2\pi/5$, through a combination of local affine transformations. Figure 2 shows corresponding 5 fan areas mapped by this linear mapping in a phase plane (U_n, V_n).

The transformation from (U_n, V_n) to (X_n, Y_n) is again expressed by the max function as follows,

$$X_n = \max\left(\sin\frac{2\pi}{5} \cdot U_n + \left(1 - \cos\frac{2\pi}{5}\right) \cdot V_n,\ \sin\frac{\pi}{5} \cdot U_n + \cos\frac{\pi}{5} \cdot V_n, \right.$$

$$\left. \sin\frac{2\pi}{5} \cdot U_n + \cos\frac{2\pi}{5} \cdot V_n \right). \tag{8}$$

Note that we omit an expression of Y_n since $Y_n = X_{n+1}$. Since $U_n = r_0 \cos(\theta_0 - 2n\pi/5)$ and $V_n = r_0 \sin(\theta_0 - 2n\pi/5)$, we can get a general solution of X_n as follows,

$$X_n = r \cdot \max\left(-\sin\left(\theta_0 - \frac{2n+2}{5}\pi\right) + \sin\left(\theta_0 - \frac{2n}{5}\pi\right), \right.$$

$$\left. -\sin\left(\theta_0 \frac{2n+4}{5}\pi\right),\ \sin\left(\theta_0 - \frac{2n+8}{8}\pi\right) \right),$$

where $r (> 0)$ and θ_0 are arbitrary constants.

We can obtain a general solution of (4) for $\sigma = 1$ and 2 similarly. Thus we show that the mapping (7) equivalent to (4) is a linearizable mapping and solutions are obtained by the linearizability. Note that the mapping, the transformation and the solutions are all expressed by the max function consistently.

4 INTEGRABLE CASE

In the previous sections, we discussed the ultradiscrete QRT system (3) with a special parameter $A = 0$. In this section, we analyze the system with a parameter $A \neq 0$. For simplicity, let us assume $\sigma = 0$. Moreover, if we use a scaling of variable $|A|X_n \rightarrow X_n$, then X_n follows

$$X_{n+1} = \max(0, X_n \pm 1) - X_{n-1}, \tag{9}$$

where $X_n + 1$ is chosen when $A > 0$ and $X_n - 1$ when $A < 0$. Therefore, solutions to (3) for $A = \pm 1$ and 0 give those for general A through the scaling. Below we consider only the case of $A = +1$,

$$X_{n+1} = \max(0, X_n + 1) - X_{n-1},$$

or

$$\begin{cases} X_{n+1} = Y_n \\ Y_{n+1} = \max(0, Y_n + 1) - X_n. \end{cases} \tag{10}$$

It is a well known fact that there exists a conserved quantity for (2). In the case of $\sigma = 0$, the quantity is

$$h = \frac{1}{x_n x_{n+1}}(a + (1 + a^2)(x_n + x_{n+1}) + a(x_n^2 + x_{n+1}^2)$$
$$+ x_n x_{n+1}(x_n + x_{n+1})).$$

Using transformations $x_n = e^{X_n/\varepsilon}$ and $a = e^{1/\varepsilon}$ and defining H by $\lim_{\varepsilon \to +0} \varepsilon \log h$, we obtain a conserved quantity for (10),

$$H = \max(1 - X_n - Y_n, 2 - X_n, 2 - Y_n,$$
$$1 + X_n - Y_n, 1 - X_n + Y_n, X_n, Y_n).$$

Orbits of solutions in the phase plane (X_n, Y_n) are given by contour lines obtained by $H = \text{const}$. Figure 3 shows some contour lines of H. The point $P(1, 1)$ is a fixed point of the mapping, that is, $X_n = Y_n = 1$ for any n if $X_0 = Y_0 = 1$. Positions of vertices of the hexagon Γ are $(3, 3)$, $(3, 1)$, $(1, -1)$, $(-1, -1)$, $(-1, 1)$, and $(1, 3)$.

In an inner region of Γ, any solution other than the fixed point P is always periodic with period 6. Since $X_n \geq -1$ and $Y_n \geq -1$ in that region, the mapping (10) becomes a linear mapping and the periodicity is due to this linearity.

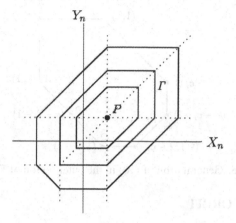

Figure 3. Contour lines of H

In the outer region of Γ, behavior of solutions becomes more complicated. Solution is still periodic but its period depends on an orbit. Figure 4 shows a solution from $(X_0, Y_0) = (4, 4)$ which is periodic with period 17. Since all segments connecting two neighboring P_j's are included in a region defined by $Y_n \geq -1$ or $Y_n \leq -1$, the segment $P_0 P_6$ is mapped linearly in the following sequence,

$$P_0 P_6 \rightarrow P_1 P_7 \rightarrow \cdots \rightarrow P_{10} P_{16} \rightarrow P_{11} P_0 \rightarrow \cdots \rightarrow P_{16} P_5 \rightarrow P_0 P_6 \rightarrow \cdots .$$

It means that a solution from any point on the polygon shown in Figure 4 is always periodic with period 17. However, if we change the orbit, the period becomes different. For example, the period of a solution from $(X_0, Y_0) = (5, 5)$ is 11 and that from $(9/2, 9/2)$ is 39.

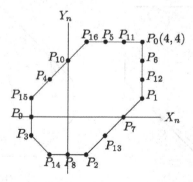

Figure 4. Solution from $(X_0, Y_0) = (4, 4)$

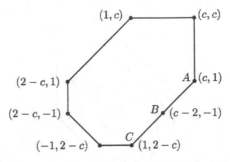

Figure 5. General orbit of (10) in the outer region of $\Gamma(c > 3)$

5 PERIOD OF ORBIT

Next we discuss a relation between the period of a solution to (10) and its orbit.
Figure 5 shows a general orbit in the outer region of $\Gamma(c > 3)$. Every point on
AC comes back to AC after a certain times of mapping. Any point on AB comes
back to AC after 6 mappings and that on BC after 5 mappings. Figure 6 (a)
shows a typical mapping of the former and (b) the latter. Assume that k counts
the number of cycles of mapping and P_k denotes a point on AC at the k-th cycle
of a solution from an initial point P_0. Moreover, define r_k by

$$r_k = AP_k/AC.$$

By this definition, $0 < r_k < 1$ holds for any k. Moreover, r_k satisfies the following recurrence formula,

$$r_{k+1} = \begin{cases} r_k + 1 - \dfrac{2}{c-1} & (r_k < \dfrac{2}{c-1}) \\ r_k - \dfrac{2}{c-1} & \text{(otherwise)}. \end{cases}$$

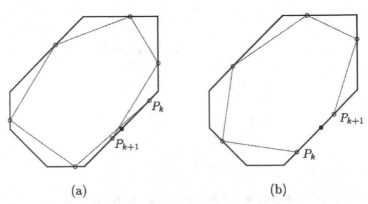

        (a)   (b)

Figure 6. A sequence of mappings of a point on (a) AB, (b) BC

This is a simple one-dimensional dynamical system and we can easily see the solution is

$$r_k = \left\{ r_0 - \frac{2}{c-1}k \right\},$$

where $\{x\}$ denotes a fractional part of x. If $r_k = r_0$, that is, $P_k = P_0$, k must satisfy

$$\left\{ \frac{2}{c-1}k \right\} = 0 \quad \Leftrightarrow \quad \frac{2}{c-1}k \text{ is an integer.}$$

Therefore, if c is a rational number, k satisfying the above condition exists and the solution from P_0 becomes periodic. If not, $r_k \neq r_0 \, (P_k \neq P_0)$ holds for any k.

Moreover, we can derive a period of solution from a value of c. If c is irrational, the period is ∞ according to the above discussion. If c is rational and is expressed by p/q where p and q are relatively prime integers, the period of solution is

$$\begin{cases} (5p - 3q)/2 & (p \equiv q \bmod 2) \\ 5p - 3q & \text{(otherwise)}. \end{cases}$$

Similar results can be obtained for other cases, (9) with $A = -1$ and (3) with $\sigma = 1$ and 2. A period of solution is decided by each orbit and does not depend on the initial position of solution on the orbit. Solutions to (3) with $\sigma = 2$ are reported in the reference [5]. They are derived by ultradiscretizing the solutions to the original QRT system (2) including an elliptic function and the function taking fractional part also appears. Comparing with our results suggests there is a strong relation between geometric piecewise-linear dynamics and elliptic functions through ultradiscretization.

6   CONCLUDING REMARKS

We studied integrable piecewise-linear mappings (3) obtained by ultradiscretizing the QRT system. In the case of $A = 0$, all solutions have the same period other than the fixed point. The mapping is expressed by a max function and is linearizable through the transformation of variables including a max function. Explicit solutions are also expressed by a max function using this linearizability.

In the case of $A \neq 0$, we showed a period of any solution on the same orbit is the same and it depends on the orbit. We can calculate the period from a parameter of the orbit by the function taking fractional part.

Finally we propose the following future problems. (i) Does a general class exist for linearizable piecewise-linear mappings? (ii) Can we obtain such a

class by ultradiscretization of difference mappings? (iii) Is there an integrable
piecewise-linear mapping with different periods depending on initial points on
the same orbit?

REFERENCES

1. Devaney, R. L. (1989) *An Introduction to Chaotic Dynamical Systems* (2nd Ed.),
   Addison-Wesley.
2. Tokihiro, T., Takahashi, D., Matsukidaira, J., and Satsuma, J. (1996) From soliton
   equations to integrable cellular automata through a limiting procedure, *Phys. Rev.
   Lett.* **76**, pp. 3247–3250.
3. Grammaticos, B., Ohta, Y., Ramani, A., Takahashi, D., and Tamizhmani, K. M.
   (1997) Cellular automata and ultra-discrete Painleve equations, *Phys. Lett. A* **226**,
   pp. 53–58.
4. Quispel, G. R. W., Robert, J. A. G., and Thompson, C. T. (1989) Integrable mappings
   and soliton equations II, *Physica D* **34**, pp. 183–192.
5. Takahashi, D., Tokihiro, T., Grammaticos, B., Ohta, Y., and Ramani, A. (1997)
   Constructing solutions to the ultra-discrete Painleve equations, *J. Phys. A: Math.
   Gen.* **30**, pp. 7953–7966.
6. Hirota, R. and Yahagi, H. (2002) Recurrence equations, an integrable system, *J. Phys.
   Soc. Jpn.* **71**, pp. 2867–2872.

FREE BOSONS AND DISPERSIONLESS LIMIT OF HIROTA TAU-FUNCTION

Leon A. Takhtajan

Department of Mathematics, Suny at Stony Brook, Stony Brook, NY 11794-3651, USA

1 INTRODUCTION

1.1 The Tau-Function

Let

$$\mathcal{F} = \bigoplus_{p \in \mathbb{Z}} \mathcal{F}_p$$

be the Fock space of charged fermions, $\psi(z)$ and $\bar{\psi}(z)$ be the fermion fields, and

$$J(z) =: \bar{\psi}(z)\psi(z) := \sum_{n \in \mathbb{Z}} J_n z^{-n-1} dz$$

be the fermion current operator.

For every element $\tilde{U}$ of the charge zero sector in the principal $\mathbb{C}^*$-bundle $\widetilde{UGM}$ over the universal Grassmannian manifold UGM, Hirota's τ-function is defined by

$$\tau(\mathbb{T}, \tilde{U}) = \langle 0 | e^{H(\mathbb{T})} | \tilde{U} \rangle,$$

where $|\tilde{U}\rangle$ is the image of $\tilde{U}$ in $\mathcal{F}_0$ under the Plücker embedding, and

$$H(\mathbb{T}) = \sum_{n=1}^{\infty} t_n J_n.$$

Corresponding wave functions

$$\Psi(z, \mathbb{T}, U) = \frac{\langle -1 | e^{H(\mathbb{T})} \psi(z) | \tilde{U} \rangle}{\tau(\mathbb{T}, \tilde{U})},$$

$$\bar{\Psi}(z, \mathbb{T}, U) = \frac{\langle 1 | e^{H(\mathbb{T})} \bar{\psi}(z) | \tilde{U} \rangle}{\tau(\mathbb{T}, \tilde{U})},$$

where $U \in UGM$, satisfy the bilinear relation

$$\mathrm{Res}_{z=\infty} \Psi(z, \mathbb{T}, U) \bar{\Psi}(z, \mathbb{T}', U) = 0.$$

L. Faddeev et al. (eds.),
Bilinear Integrable Systems: From Classical to Quantum, Continuous to Discrete, 301–311.
© 2006 *Springer. Printed in the Netherlands.*

The Hirota bilinear equation for the τ-function is a direct consequence of the bilinear relation. The Hirota equation characterizes that τ-function is associated with an element in $\widetilde{UGM}$, and it is equivalent to the KP hierarchy (see [1–4], and the exposition in [5]).

The difference analog of the KP hierarchy is the $2D$ Toda hierarchy. Corresponding τ-function depends on times $t_0, t_n, \bar{t}_n, n \in \mathbb{N}$, and satisfies Hirota equations [6]

$$z_1 e^{(\partial_0 - D(z_1))}\tau \cdot e^{-D(z_2)}\tau - z_2 e^{(\partial_0 - D(z_2))}\tau \cdot e^{-D(z_1)}\tau$$
$$= (z_1 - z_2)e^{-(D(z_1)+D(z_2))}\tau \cdot e^{\partial_0}\tau,$$

and

$$z_1\bar{z}_2 e^{-D(z_1)}\tau \cdot e^{-\bar{D}(\bar{z}_2)}\tau - e^{(\bar{D}(\bar{z}_2)+D(z_1))}\tau \cdot \tau$$
$$= e^{-(\partial_0 + D(z_1))}\tau \cdot e^{(\partial_0 + \bar{D}(\bar{z}_2))}\tau,$$

where $\partial_0 = \partial/\partial t_0$ and

$$D(z) = \sum_{n=1}^{\infty} \frac{z^{-n}}{n} \frac{\partial}{\partial t_n}, \quad \bar{D}(\bar{z}) = \sum_{n=1}^{\infty} \frac{\bar{z}^{-n}}{n} \frac{\partial}{\partial \bar{t}_n}.$$

1.2 Dispersionless Limit

Introducing parameter $\hbar$—the lattice spacing, and rescaling

$$t_0 \mapsto t_0/\hbar, \quad t_n \mapsto t_n/\hbar, \quad \bar{t}_n \mapsto \bar{t}_n/\hbar, \quad \tau \mapsto \tau_\hbar,$$

one obtains the τ-function of dispersionless $2D$ Toda hierarchy [7]

$$F = \log \tau = \lim_{\hbar \to 0} \hbar^2 \log \tau_\hbar,$$

which is a special case of Krichever's universal Whitham hierarchy [8, 9]. The τ-function satisfies dispersionless Hirota equations

$$(z_1 - z_2)e^{D(z_1)D(z_2)F}$$
$$= z_1 e^{-\partial_0 D(z_1)F} - z_2 e^{-\partial_0 D(z_2)F},$$

and

$$z_1\bar{z}_2 \left(1 - e^{D(z_1)\bar{D}(\bar{z}_2)F}\right) = e^{\partial_0(\partial_0 + D(z_1) + \bar{D}(\bar{z}_2))F}$$

—a semiclassical limit of differential Fay identity [10]. Dispersionless Hirota equations imply that the free energy F satisfies WDVV equations [11].

2 SPACES OF CONTOURS

2.1 Definitions

Let $\mathrm{Diff}_+(S^1)$ be the group of orientation preserving diffeomorphisms of S^1, let S^1 be the subgroup consisting of rigid rotations and let $\mathrm{Diff}_+(S^1)/S^1$ be the corresponding homogeneous space. It is an infinite-dimensional complex Frechet manifold isomorphic to the Frechét manifold of univalent functions on the unit disk D which are smooth up to the boundary and normalized by the conditions $f(0) = 0$ and $f'(0) = 1$ (see [12]). It is also isomorphic to the space of closed smooth curves on the complex plane $\mathbb{C}$ of conformal radius 1 encompassing 0.

Let $\mathcal{C}_1$ be the space of closed smooth curves on the complex plane $\mathbb{C}$ of Euclidean area 1 encompassing 0. It is an infinite-dimensional complex manifold with complex coordinates given by classical harmonic moments of the exterior of the contour $C \in \mathcal{C}_1$,

$$t_n = \frac{1}{2\pi i n} \int_C \bar{z}^{-n} dz, \quad n \in \mathbb{N}$$

(see [13] and references therein).

Let $T(1) = \mathrm{Homeo}_{qs}(S^1)/\mathrm{M\ddot{o}b}(S^1)$ be Bers' universal Teichmüller space—the space of normalized "fractal" contours of conformal radius 1. It is an infinite-dimensional complex Banach manifold and the inclusion map $\mathrm{Diff}_+(S^1)/\mathrm{M\ddot{o}b}(S^1) \hookrightarrow T(1)$ is holomorphic (see [14, 15]).

2.2 Deformation Theory

Let Ω be simply connected domain in $\mathbb{C}$ containing 0 and bounded by a smooth contour C, and let G be the conformal map $G : \mathbb{C}\backslash\Omega \to \mathbb{C}\backslash D$, normalized by $G(\infty) = \infty$ and $G'(\infty) > 0$. The Faber polynomials associated with G are defined by the Laurent expansion at $z = \infty$

$$\frac{zG'(z)}{G(z) - w} = \sum_{n=0}^{\infty} F_n(w)z^{-n}, \quad |G(z)| > |w|,$$

obtained by substituting Laurent series for $G(z)$

$$G(z) = b_{-1}z + b_0 + \frac{b_1}{z} + \cdots$$

into the geometric series for $(G(z) - w)^{-1}$. In terms of the inverse map $g = G^{-1}$,

$$F_n(w) = [g^n(w)]_+,$$

and Faber polynomials are uniquely characterized by the property

$$F_n(G(z)) = z^n + O(z^{-1}) \text{ as } z \to \infty.$$

The deformation theory describes the tangent vector space $T_C \mathcal{C}$ to the manifold $\mathcal{C}$ at the contour C in terms of the data associated with C.

Deformation of the contour C is a smooth family of contours $\{C_t\}_{t \in (-\epsilon, \epsilon)}$ such that $C_0 = C$; i.e., $C_t = \{z(\sigma, t), \sigma \in \mathbb{R}/2\pi \mathbb{Z}\}$ for every $|t| < \epsilon$, where $z(\sigma, t) \in C^\infty(\mathbb{R}/2\pi \mathbb{Z} \times (-\epsilon, \epsilon))$. Corresponding infinitesimal deformation is the vector field $v = \dot{z}(\sigma) d/d\sigma$ along C, where dot stands for $d/dt|_{t=0}$. A trivial deformation C_t consists of reparameterizations of the contour C, so that the vector field v is tangential to C. The tangent vector space $T_C \mathcal{C}$ is a real vector space of normal vector fields to C.

With every infinitesimal deformation there is associated 1-form on C

$$\dot{\omega}_C = \bar{\dot{z}} dz - \dot{z} d\bar{z}$$

—a restriction to C of a d^{-1} of the Lie derivative L_v of the standard 2-form $d\bar{z} \wedge dz$ on Ω. It satisfies the "calculus formula."

$$\frac{d}{dt}\bigg|_{t=0} \int_{C_t} f(z, \bar{z}, t) dz = \int_C (\dot{f} \, dz + \frac{\partial f}{\partial \bar{z}} \dot{\omega}_C).$$

In classical terms,

$$\delta n \, ds = \frac{1}{2i} \dot{\omega}_C,$$

where $ds := |z'(\sigma)| d\sigma$, $n(\sigma)$ is the outer normal to C, and $\delta n(\sigma) \in C^\infty(\mathbb{R}/2\pi \mathbb{Z}, \mathbb{R})$ defines the infinitesimal deformation of the contour C.

Theorem 1   ("*Krichever's lemma*", [16, 17])

(i) *Any deformation C_t of C which does not change the area πt_0 of Ω and harmonic moments of exterior t_n is infinitesimally trivial. The parameters $\{t_0 - t_0(C), t_n - t_n(C), \bar{t}_n - \bar{t}_n(C)\}$ are local coordinates on $\mathcal{C}$ near C.*

(ii) *The following 1-forms on C*

$$\dot{\omega}_C^{(n)} = \frac{\partial \bar{z}}{\partial t_n}\bigg|_{t_n=t_n(C)} dz - \frac{\partial z}{\partial t_n}\bigg|_{t_n=t_n(C)} d\bar{z},$$

extend to meromorphic $(1, 0)$-forms on the double $\mathbb{P}^1_C$ of the exterior domain $\mathbb{P}^1 \setminus \Omega$ with a single pole at ∞ of order $n + 1$ if $n \in \mathbb{N}$, and simple poles at ∞ and $\overline{\infty}$ with residues 1 and -1 if $n = 0$. Explicitly,

$$\dot{\omega}_C^{(n)} = d(F_n \circ G), \quad \dot{\omega}_C^{(0)} = d \log G$$

in the domain $\mathbb{P}^1 \setminus \Omega$, and

$$\dot{\omega}_C^{(n)} = d(F_n \circ 1/\overline{G}), \quad \dot{\omega}_C^{(0)} = d \log 1/\overline{G}$$

in the domain $\overline{\mathbb{P}^1 \setminus \Omega}$.

(iii) The 1-forms $\dot{w}_C^{(n)}$ satisfy the property

$$\frac{1}{2\pi i} \int_C \frac{z^{-m}}{m} \dot{w}_C^{(n)} = \delta_{mn},$$

and can be identified with the vector fields $\partial/\partial t_n$. For every $a > 0$ the holomorphic tangent vector space $T'_C C_a$ to C_a at C is canonically identified with the complex vector space $\mathcal{M}^{1,0}(\mathbb{P}^1_C)$ of meromorphic $(1,0)$-forms on $\mathbb{P}^1_C$ with a single pole at ∞ of order ≥ 2.

(iv) The holomorphic cotangent vector space $T'^*_C C_a$ to C_a at C is naturally identified with the complex vector space $\mathcal{H}^{1,0}(\mathbb{P}^1\backslash\Omega)$ of holomorphic $(1,0)$-forms on $\mathbb{P}^1\backslash\Omega$ which are smooth up to the boundary, and the pairing

$$(,)_C : T'_C C_a \otimes T'^*_C C_a \to \mathbb{C}$$

is given by

$$(\omega, u)_C = \frac{1}{2\pi i} \int_C d^{-1} u\omega.$$

Differentials dt_n correspond to $(1, 0)$-forms

$$dt_n(z) = d(z^{-n}/n) = -z^{-n-1} dz.$$

3   BOSONIC PARTITION FUNCTION

For $\varphi \in C^\infty(\mathbb{P}^1, \mathbb{R})$ consider the action

$$S_0(\varphi) = \frac{i}{4} \iint_{\mathbb{P}^1} \partial\varphi \wedge \bar{\partial}\varphi,$$

which describes the standard theory of free bosons on the Riemann sphere $\mathbb{P}^1$. For every $C \in \mathcal{C}$ define the "topological term" by

$$S_{top}(\varphi) = \iint_C (A(\Omega)\delta_0 - \chi_\Omega) \, \varphi d^2 z$$

$$= A(\Omega)\varphi(0) - \iint_\Omega \varphi d^2 z,$$

where χ_Ω is a characteristic function of Ω, and δ_0 is a Dirac delta-function at 0 with respect to the Lebesgue measure $d^2 z$. The total bosonic action

$$S_C(\varphi) = S_0(\varphi) + S_{top}(\varphi)$$

describes the theory of free bosons on $\mathbb{P}^1$ in the presence of a contour C, and defines a family of field theories parameterized by $\mathcal{C}$.

For every $C \in \mathcal{C}$ the partition function of the corresponding quantum field theory is defined by

$$\langle 1 \rangle_C = \int_{C^\infty(\mathbb{P}^1, \mathbb{R})/\mathbb{R}} [\mathcal{D}\varphi] e^{-\frac{1}{\pi} S_C(\varphi)}.$$

Specifically, approximate χ_Ω and δ_0 by smooth functions $\chi_\Omega^{(\epsilon)}$ and $\delta_0^{(\epsilon)}$ with compact supports satisfying

$$\iint_{\mathbb{C}} \left(A(\Omega) \delta_0^{(\epsilon)} - \chi_\Omega^{(\epsilon)} \right) d^2 z = 0,$$

and define

$$\langle 1 \rangle_C = \lim_{\epsilon \to 0} \exp \left\{ \frac{A^2(\Omega)}{\pi^2} \iint_{\mathbb{C}} \iint_{\mathbb{C}} \log |z - w| \delta_0^{(\epsilon)}(z) \right.$$

$$\left. \delta_0^{(\epsilon)}(w) d^2 z d^2 w \right\} \int_{C^\infty(\mathbb{P}_1, \mathbb{R})/\mathbb{R}} [\mathcal{D}\varphi] e^{-\frac{1}{\pi} S_C^{(\epsilon)}(\varphi)},$$

where

$$S_C^{(\epsilon)}(\varphi) = S_0(\varphi) + \iint_{\mathbb{C}} \left(A(\Omega) \delta_0^{(\epsilon)} - \chi_\Omega^{(\epsilon)} \right) \varphi d^2 z.$$

The τ-function $\tau = \tau(C)$ of a smooth contour C is defined as the normalized expectation value of C,

$$\tau = \langle\langle C \rangle\rangle = \frac{\langle 1 \rangle_C}{\langle 1 \rangle_0},$$

where $\langle \, \rangle$ stands for expectation value in the standard theory of free bosons on $\mathbb{P}^1$ with the action functional S_0 (it corresponds to the case $C = \emptyset$—the empty set).

The τ-function is well-defined and

$$\log \tau = -\frac{1}{\pi^2} \iint_\Omega \iint_\Omega \log |z - w| d^2 z d^2 w$$

$$+ \frac{2}{\pi^2} A(\Omega) \iint_\Omega \log |z| d^2 z$$

$$= -\frac{1}{\pi^2} \iint_\Omega \iint_\Omega \log \left| \frac{1}{z} - \frac{1}{w} \right| d^2 z d^2 w,$$

which is $-1/\pi^2$ times a regularized energy of the pseudo-measure $d\mu = d^2 z - A(\Omega)\delta_0$ on the domain Ω [17]. Also, $\tau = \tau_{MWZ}$ (see [13, 18]).

4   CURRENT WARD IDENTITIES

Let $\jmath = \partial\varphi$ and $\bar\jmath = \bar\partial\varphi$ be holomorphic and anti-holomorphic components of the bosonic field current $d\varphi$. By definition,

$$\langle X \rangle = \int_{C^\infty(\mathbb{P}^1,\mathbb{R})/\mathbb{R}} [\mathcal{D}\varphi] X \, e^{-\frac{1}{\pi}S_C(\varphi)},$$

where $X = \jmath(z_1)\ldots\jmath(z_m)\bar\jmath(w_1)\ldots\bar\jmath(w_n)$. Correlation functions for the theory on $\mathbb{P}^1\backslash\Omega$ with action functional

$$S_{ext}(\varphi) = \frac{i}{4}\iint_{\mathbb{P}^1\backslash\Omega} \partial\varphi \wedge \bar\partial\varphi,$$

and Dirichlet boundary condition, are defined similarly and are denoted by $\langle\cdots\rangle_{DBC}$.

4.1   1-Point Correlation Functions

Set

$$\langle\langle \jmath(z) \rangle\rangle = \frac{\langle \jmath(z) \rangle}{\langle 1 \rangle_C}.$$

Then

$$\langle\langle \jmath(z) \rangle\rangle = \frac{\partial\Phi(z)}{\partial z}dz,$$

where $\Phi(z)$ satisfies

$$-\frac{\partial^2\Phi(z)}{\partial z\partial\bar z} = \begin{cases} \chi_\Omega(z) - A(\Omega)\delta_0(z) & \text{if } z \in \Omega, \\ 0 & \text{if } z \in \mathbb{C}\backslash\Omega, \end{cases}$$

is continuous on C and is normalized by $\Phi(\infty) = 0$. The function Φ is a logarithmic potential of the pseudo-measure $d\mu = d^2z - A(\Omega)\delta_0$ on Ω. At $z = \infty$

$$\frac{\partial\Phi(z)}{\partial z} = -\sum_{n=1}^{\infty} v_n z^{-n-1},$$

where

$$v_n = \frac{1}{2\pi i}\int_C z^n \bar z \, dz, \quad n \in \mathbb{N},$$

are the harmonic moments of interior of the contour C.

Using deformation theory (calculus formula and Theorem 1) and an explicit form of S_{top}, one gets the following (cf. [13, 18]).

Theorem 2 *For every $a > 0$ the normalized 1-point current correlation functions of free bosons on $\mathbb{P}^1$ parameterized by $C \in \mathcal{C}_a$ satisfy the Ward identities, given by the following Laurent expansions at $z = \infty$*

$$\langle\langle \jmath(z) \rangle\rangle = -\sum_{n=1}^{\infty} z^{-n-1} \frac{\partial \log \tau}{\partial t_n} dz = \mathbf{d}' \log \tau,$$

and

$$\langle\langle \bar{\jmath}(z) \rangle\rangle = -\sum_{n=1}^{\infty} \bar{z}^{-n-1} \frac{\partial \log \tau}{\partial \bar{t}_n} dz = \mathbf{d}'' \log \tau.$$

Corollary 3 *([13, 16, 18]) The function $\log \tau \in C^{\infty}(\mathcal{C}, \mathbb{R})$ is a generating function for the harmonic moments of interior,*

$$v_0 = \frac{\partial \log \tau}{\partial t_0} \quad and \quad v_n = \frac{\partial \log \tau}{\partial t_n}, \qquad n \in \mathbb{N},$$

where

$$v_0 = \frac{2}{\pi} \iint_{\Omega} \log |z| \bar{z} d^2 z.$$

Corollary 4 *("Explicit formula" for the conformal map G, [13, 18])*

$$\log G(z) = \log z - \frac{1}{2} \frac{\partial^2 \log \tau}{\partial t_0^2} - \sum_{n=1}^{\infty} \frac{z^{-n}}{n} \frac{\partial^2 \log \tau}{\partial t_0 \partial t_n}.$$

4.2   2-Point Correlation Functions

Set

$$\langle\langle \jmath(z) \jmath(w) \rangle\rangle = \frac{\langle \jmath(z) \jmath(w) \rangle}{\langle \mathbf{1} \rangle_C} - \langle\langle \jmath(z) \rangle\rangle \langle\langle \jmath(w) \rangle\rangle$$

and

$$\langle\langle \jmath(z) \jmath(w) \rangle\rangle_{DBC} = \frac{\langle \jmath(z) \jmath(w) \rangle_{DBC}}{\langle \mathbf{1} \rangle_{DBC}}.$$

Clearly,

$$\langle\langle \jmath(z) \jmath(w) \rangle\rangle = -\frac{dz \otimes dw}{(z - w)^2},$$

and

$$\langle\langle \jmath(z) \jmath(w) \rangle\rangle_{DBC} = -\frac{G'(z) G'(w)}{(G(z) - G(w))^2} dz \otimes dw.$$

Using Ward identity for the 1-point function, deformation theory and properties of Faber polynomials, one gets the following (cf. [18]).

Theorem 5 *For every $a > 0$ normalized reduced 2-point current correlation functions for free bosons on $\mathbb{P}^1$ parameterized by $C \in C_a$ satisfy the Ward identities, given by the following Laurent series expansions at $z = w = \infty$*

$$\langle\langle j(z)j(w)\rangle\rangle - \langle\langle j(z)j(w)\rangle\rangle_{DBC}$$

$$= \left(\frac{G'(z)G'(w)}{(G(z) - G(w))^2} - \frac{1}{(z-w)^2} \right) dz \otimes dw$$

$$= \sum_{m,n=1}^{\infty} z^{-m-1} w^{-n-1} \frac{\partial^2 \log \tau}{\partial t_m \partial t_n} dz \otimes dw,$$

and

$$\langle\langle j(z)\bar{j}(w)\rangle\rangle - \langle\langle j(z)\bar{j}(w)\rangle\rangle_{DBC}$$

$$= \frac{G'(z)\overline{G'(w)}}{(1 - G(z)\overline{G(w)})^2} dz \otimes d\bar{w}$$

$$= \sum_{m,n=1}^{\infty} z^{-m-1} \bar{w}^{-n-1} \frac{\partial^2 \log \tau}{\partial t_m \partial \bar{t}_n} dz \otimes d\bar{w}$$

$$= d'd'' \log \tau.$$

All higher reduced multipoint current correlation functions vanish.

Corollary 6 *For every $a > 0$ the Hermitian metric H on C_a defined by*

$$H\left(\frac{\partial}{\partial t_m}, \frac{\partial}{\partial t_n} \right)$$

$$-\frac{1}{(2\pi i)^2} \int_{C_+} \int_{C_+} z^m \bar{w}^n K(z, \bar{w}) dz d\bar{w},$$

where C_+ is an arbitrary contour containing C inside, and K is the Bergman reproducing kernel for the domain $\mathbb{P}^1 \backslash \Omega$, is Kähler with the Kähler potential $\log \tau$.

Corollary 7 ([13, 18])

$$\log \frac{G(z) - G(w)}{z - w} = -\frac{1}{2} \frac{\partial^2 \log \tau}{\partial t_0^2}$$

$$+ \sum_{m,n=1}^{\infty} \frac{z^{-m} w^{-n}}{mn} \frac{\partial^2 \log \tau}{\partial t_m \partial t_n}$$

and

$$\log \left(\frac{G(z)\overline{G(w)}}{G(z)\overline{G(w)} - 1} \right) = \sum_{m,n=1}^{\infty} \frac{z^{-m}\bar{w}^{-n}}{mn} \frac{\partial^2 \log \tau}{\partial t_m \partial \bar{t}_n}.$$

From Corollary 3 one immediately gets dispersionless Hirota equations [19].

REFERENCES

1. Hirota, R. (1980) Direct methods in soliton theory, in: *Solitons*, Springer-Verlag, New York.
2. Date, E., Kashiwara, M., Jimbo, M., and Miwa, T. (1983) Transformation groups for soliton equations, in: *Nonlinear integrable systems—classical theory and quantum theory* (Kyoto, 1981), World Scientific, Singapore, pp. 39–119.
3. Sato, M. and Sato, Y. (1983) Soliton equations as dynamical systems on infinite-dimensional Grassmann manifold, in: *Nonlinear partial differential equations in applied science* (Tokyo, 1982), North-Holland Math. Stud., Vol. 81, North-Holland, Amsterdam, pp. 259–271.
4. Sato, M. (1989) The KP hierarchy and infinite-dimensional Grassmann manifolds, in: *Theta functions—Bowdoin 1987, Part 1* (Brunswick, ME, 1987), *Proc. Sympos. Pure Math.*, Vol. 49, *Amer. Math. Soc.*, Providence, RI, pp. 51–66.
5. Kawamoto, N., Namikawa, Y., Tsuchiya, A., and Yasuhiko Yamada, (1988) Geometric realization of conformal field theory on Riemann surfaces, *Comm. Math. Phys.* **116**(2), pp. 247–308.
6. Hirota, R. (1981) Discrete analogue of a generalized Toda equation, *J. Phys. Soc. Japan* **50**(11), pp. 3785–3791. MR **83e**:58035.
7. Takasaki, K. and Takebe, T. (1995) Integrable hierarchies and dispersionless limit, *Rev. Math. Phys.* **7**(5), pp. 743–808.
8. Krichever, I. M. (1992) The dispersionless Lax equations and topological minimal models, *Comm. Math. Phys.* **143**(2), pp. 415–429.
9. Krichever, I. M. (1994) The τ-function of the universal Whitham hierarchy, matrix models and topological field theories, *Comm. Pure Appl. Math.* **47**(4), 437–475.
10. Adler, M. and van Moerbeke, P. (1992) A matrix integral solution to two-dimensional W_p-gravity, *Comm. Math. Phys.* **147** (1), pp. 25–56.
11. Boyarsky, A., Marshakov, A., Ruchayskiy, O., Wiegmann, P., and Zabrodin, A. (2001) Associativity equations in dispersionless integrable hierarchies, *Phys. Lett. B* **515**(3–4), pp. 483–492.
12. Kirillov, A. A. (1987) Kähler structure on the K-orbits of a group of diffeomorphisms of the circle, *Funktsional. Anal. i Prilozhen.* **21**(2), pp. 42–45.
13. Wiegmann, P. B. and Zabrodin, A. (2000) Conformal maps and integrable hierarchies, *Comm. Math. Phys.* **213**(3), pp. 523–538.
14. Bers, Lipman. (1981) Finite-dimensional Teichmüller spaces and generalizations, *Bull. Amer. Math. Soc. (N.S.)* **5**(2), pp. 131–172.

15. Nag, S. and Verjovsky, A. (1990) Diff(S^1) and the Teichmüller spaces, *Comm. Math. Phys.* **130**(1), pp. 123–138.
16. Krichever, I. M. (2000) unpublished manuscript.
17. Takhtajan, L. A. (2001) Free bosons and tau-functions for compact Riemann surfaces and closed smooth Jordan curves. Current correlation functions, *Lett. Math. Phys.* **56**(3), pp. 181–228, EuroConférence Moshé Flato 2000, Part III (Dijon).
18. Kostov, I. K., Krichever, I., Mineev-Weinstein, M., Wiegmann, P. B., and Zabrodin, A. (2001) The τ-function for analytic curves, in: Random Matrix Models and their Applications, *Math. Sci. Res. Inst. Publ.*, Vol. 40, Cambridge University Press, Cambridge, pp. 285–299.
19. Zabrodin, A. V. (2001) The dispersionless limit of the Hirota equations in some problems of complex analysis, *Teoret. Mat. Fiz.* **129**(2), pp. 239–257.

SIMILARITY REDUCTIONS OF HIROTA BILINEAR EQUATIONS AND PAINLEVÉ EQUATIONS

K.M. Tamizhmani
Department of Mathematics, Pondicherry University, Kalapet,
Pondicherry-605014, India

B. Grammaticos
GMPIB, Université Paris VII, Tour 24-14, 5e étage, case 7021, 75251 Paris, France

A. Ramani
CPT, Ecole Polytechnique, CNRS, UMR 7644, 91128 Palaiseau, France

Y. Ohta
Information Engineering, Graduate School of Engineering, Hiroshima University,
1-4-1 Kagamiyama, Higashi, Hiroshima 739-8527, Japan

T. Tamizhmani
Department of Mathematics, K.M. Centre for Post-Graduate Studies, Lawspet,
Pondicherry-605008, India

Abstract   Using a particular class of symmetries of Hirota bilinear soliton equations we reduce them into bilinear ordinary differential equations. We convert these bilinear equations into nonlinear forms. By this process we obtain a class of higher order equations of Painlevé type.

1 INTRODUCTION

Symmetry analysis is very useful to find a class of particular solutions of linear and nonlinear equations. The underlying invariances of the given partial differential equations (PDEs) are widely used to reduce PDEs of higher to lower dimensions in terms of a new independent variable, called similarity variable [1]. In the case of soliton equations these reduced ordinary differential equations (ODEs) are identified with one of the Painlevé equations [2]. It is well-known that soliton equations can be expressed in terms of Hirota's bilinear forms and their Painlevé properties have also been studied [3, 4]. The symmetries of these equations have been studied for large class of equations in [5].

313

L. Faddeev et al. (eds.),
Bilinear Integrable Systems: From Classical to Quantum, Continuous to Discrete, 313–323.
© 2006 Springer. Printed in the Netherlands.

A natural question arises whether the similarity reductions of these bilinear equations can be identified with bilinear forms of Painlevé equations [6–8]. In an exploratory approach in this paper, we consider only simple symmetries of the bilinear equations and reduce them to bilinear ODEs. Our analysis also includes bilinear soliton equations of higher degree [5]. The similarity reductions of these bilinear equations result in higher order ODEs of Painlevé type. This is important since there is considerable activity around the study of higher order Painlevé equations [9]. We present both bilinear and nonlinear forms of these ODEs. We illustrate the above method with many interesting examples.

2 LIE POINT SYMMETRY APPROACH TO HIROTA BILINEAR EQUATIONS

KdV-type Hirota bilinear equations are given in the following form:

$$A(D)F \cdot F = 0$$

where $A(D)$ is a polynomial in Hirota differential operators and is always even. The symbol D is defined as

$$D_x^m D_y^n \ldots F \cdot G = (\partial_x - \partial_x')^m (\partial_y - \partial_y')^n F(x, y, \ldots) G(x, y, \ldots)|_{x=x', y=y' \ldots}$$

We consider the infinitesimal Lie one-parameter point transformation which is given by [1]

$$\bar{x}^i = x^i + \epsilon \xi^i(x) + O(\epsilon^2) \quad i = 1, 2, \ldots, p$$
$$\bar{F}^\alpha = F^\alpha + \epsilon \eta_\alpha(x, F) + O(\epsilon^2), \quad \alpha = 1, 2, \ldots, q$$

where $x = (x^1, \ldots, x^p)$ and $F = (F^1, \ldots F^q)$ with corresponding infinitesimal generator

$$V = \sum_i^p \xi(x) \frac{\partial}{\partial x^i} + \sum_\alpha^q \eta_\alpha(x, F) \frac{\partial}{\partial F^\alpha}$$

Then the invariant condition becomes

$$PrV^{(n)}[A(D)F \cdot F]|_{A(D)F \cdot F=0} = 0$$

$$PrV^{(n)}V = V + \sum_\alpha \sum_J \eta_\alpha^J(x, F^{(n)}) \frac{\partial}{\partial F_J^\alpha}$$

where $J = (j_1 \cdots j_k)$ with $1 \leq j_k \leq p, 1 \leq k \leq n$. The coefficient of η_α^J of $Pr^{(n)}V$ are given by

$$\eta_\alpha^J(x, F^{(n)}) = D_J \left(\eta_\alpha - \sum_{i=1}^p \xi^i F_i^\alpha \right) + \sum_{i=1}^p \xi^i F_{J,i}^\alpha$$

where $D_J = D_{j_1} \ldots D_{j_k}$ and D_j is a total differential operator. The corresponding Lie equations and characteristic systems are

$$\frac{d\bar{x}^i}{d\epsilon} = \xi^i(\bar{x}), \quad \bar{x}^i\big|_{\epsilon=0} = x^i, \, i = 1, 2, \ldots n$$

$$\frac{dx^1}{\xi^1} = \frac{dx^2}{\xi^2} = \cdots = \frac{dx^n}{\xi^n}$$

3 SIMILARITY REDUCTION OF THE KdV FAMILY

3.1 KdV equation

First we consider KdV equation in bilinear form

$$\left(D_x^4 - 4D_x D_t\right)F \cdot F = 0 \tag{1}$$

Following the approach explained in the above section, we can obtain algorithmically the underlying Lie point symmetries. The infinitesimal generators of the symmetry group are given by

$$\xi = \gamma x/3 + 12\epsilon t + \alpha$$
$$\tau = \gamma t + \beta$$
$$\theta = (\epsilon x^2/2 + \delta x + A(t))F$$

The corresponding vector fields are

$$V_1 = \partial_x$$
$$V_2 = \partial_t$$
$$V_3 = \frac{x}{3}\partial_x + t\partial_t$$
$$V_4 = xF\partial_F$$
$$V_5 = \frac{x^2}{2}F\partial_F + 12t\partial_t$$
$$V_6 = A(t)F\partial_F$$

We take a simple vector field corresponding to the scaling symmetry V_3 and get the similarity variable

$$z = \frac{x}{t^{1/3}} \tag{2}$$

Also expanding the bilinear Eq. (1) we get

$$F_{xxxx}F - 4F_{xxx}F_x + 3F_{xx}F_{xx} - 4F_{xt}F + 4F_xF_t = 0$$

On using the similarity variable (2) this equation reduces to

$$F_{zzzz}F - 4F_{zzz}F_z + 3F_{zz}F_{zz} + 4/3zF_{zz}F + 4/3F_zF - 4/3zF_zF_z = 0$$

Again this can be written in the bilinear form as

$$\left(D_z^4 + 4/3zD_z^2 + 4/3\,\partial_z\right)F \cdot F = 0$$

which is the bilinear form of Painlevé 34. We convert this bilinear equation into nonlinear form by using the Hirota transformation:

$$u = (2\log F)_{zz}$$

and finally we get

$$u_{zz} + 3u^2 + 4/3zu + 4/3(2\log F)_z = 0$$

Differentiating the above equation once and using the dependent variable transformation, we get

$$u_{zzz} + 6uu_z + 4/3zu + 8u = 0$$

We can show this equation is equivalent to Painlevé 34

$$2uu_{zz} - u_z^2 + 4u^3 - 8zu^2 + 16\alpha^2 = 0$$

We should remark that if we consider other invariances we will get other types of reductions.

3.2   Boussinesq Equation

Next we consider the Boussinesq equation in bilinear form

$$\left(D_x^4 + 3D_t^2\right)F \cdot F = 0$$

One can find easily the scaling symmetry corresponding to this bilinear equation which gives the similarity variable

$$z = \frac{x}{t^{1/2}}$$

the corresponding vector field of which is

$$V = 2t\partial_t + x\partial_x$$

The reduced equation in bilinear form becomes

$$\left(D_z^4 + 3/4z^2D_z^2 + 9/4z\,\partial_z\right)F \cdot F = 0$$

By introducing the nonlinear variable

$$u = (2\log F)_{zz}$$

finally we get

$$u_{zz} + 3u^2 + 3/4z^2u + 9/4z(2\log F)_z = 0$$

By using the nonlinear variable

$$u = 2w_z$$

we obtain

$$w_{zzz} + 6w_z^2 + \frac{3}{4}z^2w_z + \frac{9}{4}zw = 0$$

3.3 Sawada–Kotera Equation

Following the same approach described above, for Sawada–Kotera equation

$$\left(D_x^6 + D_x D_y\right)F \cdot F = 0 \tag{3}$$

we find the similarity variable

$$z = \frac{x}{y^{1/5}}$$

which corresponds to the vector field

$$V = 5y\partial_y + x\partial_x$$

Reducing the Sawada-Kotera equation (3) in the bilinear form leads to

$$\left(5D_z^6 - zD_z^2 - 1/2\,\partial_z\right)F \cdot F = 0$$

Substituting

$$u = (2\log F)_{zz}$$

in the bilinear form we get the nonlinear form

$$u_{zzzz} + 15u_{zz}u + 15u^3 - \frac{zu}{5} - \frac{1}{5}(2\log F)_z = 0$$

In order to avoid log term we differentiate once and again use the nonlinear variable transformation. We finally get

$$u_{zzzzz} + 15u_{zzz}u + 15u_{zz}u_z + 45u^2u_z - \frac{z}{5}u_z - \frac{2}{5}u = 0$$

This equation can be identified with the classification of Cosgrove for fifth order nonlinear ODEs of Painlevé type [9].

3.4   5-Reduced BKP Equation

The bilinear form of 5-reduced BKP equation is given as

$$\left(D_x^6 + 5D_x^3 D_t - 5D_t^2\right) F \cdot F = 0$$

By using the similarity variable

$$z = \frac{x}{t^{1/3}}$$

which corresponds to the vector field

$$V = 3t\partial_t + x\partial_x$$

the reduced equation in the bilinear form becomes

$$\left(D_z^6 - 5/3z D_z^4 - 5/9z^2 D_z^2 - 5\,\partial_z D_z^2 - 20/9z\,\partial_z\right) F \cdot F = 0$$

Substituting

$$u = (2\log F)_{zz}$$

in the bilinear form we get the nonlinear form

$$(u_{zzzz} + 15u_{zz}u + 15u^3) - 5/3z(u_{zz} + 3u^2) - 5/9z^2u - (5/F^2\partial_z D_z^2)F \cdot F$$
$$-20/9z(2\log F)_z = 0$$

Some care should be taken for the term $\partial_z D_z^2 F \cdot F$. Now

$$\frac{\partial_z D_z^2 F \cdot F}{F^2} = 2\frac{(F_{zz}F - F_z F_z)_z}{F^2} = 2\left(\frac{F_{zzz}}{F} - \frac{F_z}{F}\frac{F_z}{F}\right)$$
$$= u_z + 4\left(\frac{F_{zz}}{F}\frac{F_z}{F} - \left(\frac{F_z}{F}\right)^3\right) = u_z + 2u\frac{F_z}{F}$$

where we have used the dependent variable transformation. Let $2w_z = u$. In order to avoid a log term we differentiate once again use the nonlinear variable transformation, we finally get

$$w_{zzzzz} + 30w_{zzz}w_z + 60w_z^3 - 5/3z w_{zzz} - 10z w_z^2 - 5w_{zz}$$
$$- 5/9z^2 w_z - 20/9z w = 0$$

3.5   Hietarinta Equation

The bilinear form of the Hietarinta equation is given by

$$\left(D_x^4 + D_t^3 D_x\right) F \cdot F = 0$$

By using the similarity variable

$$z = \frac{x}{t}$$

which corresponds to the vector field

$$V = t\partial_t + x\partial_x$$

the reduced equation in the bilinear form becomes

$$((1 - z^3)D_z^4 - 9z^2\,\partial_z\,D_z^2 - 12zD_z^2 - 3z\,\partial_z^2 - 3\,\partial_z)F \cdot F = 0$$

Substituting

$$u = (2\log F)_{zz}$$

in the bilinear form and using the same analysis as in the previous example with $2w_z = u$, we get the nonlinear form

$$(1 - z^3)(w_{zzz} + 6w_z^2) - 9z^2 w_{zz} - 18z^2 ww_z - 15zw_z - 6zw^2 - 3w = 0$$

3.6 Ito Equation

The bilinear form of Ito equation is

$$\left(D_t^3 + D_x^3\,D_t\right)F \cdot F = 0$$

Using the similarity variable

$$z = \frac{x}{t^{1/3}}$$

which corresponds to the vector field

$$V = 3t\,\partial_t + x\,\partial_x$$

the reduced equation in the bilinear form becomes

$$\left(D_z^4 + \frac{3}{z}\,\partial_z\,D_z^2 - 1/3zD_z^2 - 4/3\,\partial_z\right)F \cdot F = 0$$

Substituting

$$u = (2\log F)_{zz}$$

in the bilinear form and using the same analysis as in the previous example, we get the nonlinear form

$$w_{zzz} + \frac{3}{z}w_{zz} + 6w_z^2 + \frac{6}{z}ww_z - \frac{z}{3}w_z - 8/3w = 0$$

where $2w_z = u$.

4 SIMILARITY REDUCTIONS OF THE mKdV FAMILY

4.1 mKdV Equation

It is straightforward to generalize the similarity analysis implemented above to the case of more than one τ function. In this section, we consider the case of mKdV-type Hirota bilinear Equations

$$A(D)F \cdot G = 0$$
$$B(D)F \cdot G = 0$$

As a simple example, we consider the mKdV bilinear equation in the form of

$$D_x^2 G \cdot F = 0$$
$$(D_x^3 + D_t)G \cdot F = 0$$

This bilinear equation can be transformed into

$$D_x^2 g \cdot f = 0$$
$$\left(D_x^3 + D_t + \frac{\alpha}{t}\right) g \cdot f = 0$$

under the gauge transformation $G = t^\alpha g$, $F = f$. Now in this case the similarity variable is given by

$$z = \frac{x}{t^{1/3}}$$

which corresponds to the vector field

$$V = 3t\, \partial_t + x\, \partial_x$$

Then the reduced bilinear equation becomes

$$D_z^2 g \cdot f = 0$$
$$\left(D_z^3 - \frac{z}{3}D_z + \alpha\right)g \cdot f = 0$$

which is the bilinear equation of P_{II}. The above bilinear equations become

$$(\log(gf))_{zz} + (\log(g/f))_z^2 = 0$$
$$(\log(g/f))_{zzz} + 3(\log(g/f))_z(\log(gf))_{zz} + (\log(g/f))_z^3$$
$$-\frac{z}{3}(\log(g/f))_z + \alpha = 0$$

Eliminating $(\log((gf)))_{zz}$ in the second equation using the first we get

$$(\log(g/f))_{zzz} - 2(\log(g/f))_z^3 - \frac{z}{3}(\log(g/f))_z + \alpha = 0$$

By introducing the nonlinear variable

$$u = (\log(g/f))_z$$

we obtain

$$u'' = 2u^3 + \frac{z}{3}w - \alpha = 0$$

which is P_{II}.

4.2 3-Reduced mKP Equation

The bilinear form of 3-Reduced mKP equation is given by

$$\left(D_x^2 - D_y\right)G \cdot F = 0$$

$$\left(D_x^3 + 3D_x D_y\right)G \cdot F = 0$$

Under the gauge transformation $G = y^\alpha g$, $F = f$ these bilinear equations become

$$\left(D_x^2 - D_y - \frac{\alpha}{y}\right)g \cdot f = 0$$

$$\left(D_x^3 + 3D_x D_y + 3\frac{\alpha}{y}D_x\right)g \cdot f = 0$$

The similarity variable in this case is

$$z = \frac{x}{y^{1/2}}$$

Then the reduced equations assume the form

$$\left(D_z^2 + \frac{z}{2}D_z - \alpha\right)g \cdot f = 0$$

$$\left(D_z^3 - 3/2zD_z^2 - 3/2\,\partial_z + 3\alpha D_z\right)g \cdot f = 0$$

which can easily be identified as the bilinear form of P_{IV}. We introduce the nonlinear variable transformation

$$u = (\log(g/f))_z$$

Then the bilinear equations becomes

$$(\log(g/f))_{zz} + u^2 + \frac{z}{2}u - \alpha = 0$$

$$u_{zz} + 3(\log(gf))_{zz}u + u^3 - 3/2z((\log(gf))_{zz} + u^2) - 3/2(\log(gf))_z + 3\alpha u = 0$$

Differentiate once and eliminate the log terms by using the dependent variable transformation, finally, we get

$$u_{zzz} - 6u^2 u_z - 3zuu_z + 6\alpha u_z + 3/4z^2 u_z + 9/4zu - 3\alpha = 0$$

To integrate this equation we multiply the above equation by u and z and subtract one from the other and integrating, we get

$$(2u - z)u_{zz} - u_z^2 - 3u^4 + 6\alpha u^2 + \frac{9}{4}z^2 u^2 + u_z - 6\alpha zu - \frac{3}{4}z^3 u + \frac{3}{2}\alpha z^2 = \beta$$

Let $u = \dfrac{z}{2} + v$. Finally we get

$$v_{zz} = \frac{v_z}{2v} + 3/2v^3 + (9/8z^2 - 3\alpha)v + 3zv^2 + \frac{\beta - 1/4}{2v}$$

which is P_{IV} equation.

5   CONCLUDING REMARKS

We described the method to derive similarity reductions of PDEs in Hirota bilinear form. As a consequence we have obtained the bilinear form of the reduced ODEs. In certain cases these ODEs are identified directly with Painlevé equations. In many other cases we have presented higher order Painlevé-type equations both in bilinear and nonlinear forms.

REFERENCES

1. Olver, R. J. (1986) *Applications of Lie Groups to Differential Equations*, Springer-Verlag, New York.
2. Ablowitz, M. J., Ramani, A., and Segur, H. (1978) Nonlinear evolution equations and ordinary differential equations of Painlevé type, *Lett. Nuovo Cim.* **23**, pp. 333–338.
3. Hietarinta, J. (1987) A search for bilinear equations passing Hirota's three-soliton condition: I. KdV-type bilinear equations, *J. Math. Phys.* **28**, pp. 1732–1742.
4. Grammaticos, B., Ramani, A., and Hietarinta, J. (1990) A search for integrable bilinear equations: the Painlevé approach, *J. Math. Phys.* **31**, pp. 2572–2578.
5. Tamizhmani, K. M., Ramani, A., and Grammaticos, B. (1991) Lie symmetries of Hirota's bilinear equations, *J. Math. Phys.* **32**, pp. 2635–2659.
6. Hietarinta, J. and Kruskal, M. D. (1992) Hirota forms for the six Painlevé equations from singularity analysis, *Painlevé Transcendents*, eds. D. Levi, and P. Winternitz, Plenum Press, New York, pp. 175–185.

7. Ohta, Y., Ramani, A., Grammaticos, B., and Tamizhmani, K. M., (1996) From discrete to continuous Painlevé equations: a bilinear approach, *Phys. Lett.* **216A**, pp. 255–261.
8. Ohta, Y. and Nakamura, A. (1992) Similarity KP equation and various different pepresentations of its solutions, *J. Phys. Soc. Jpn.* **61**, pp. 4295–4313.
9. Cosgrove, C. M. (2000) *Stud. Appl. Math.* **104**, p. 1; Kudryashov (1998) On new transcendents defined by nonlinear ordinary differential equations, *J. Phys. A.* **31**, pp. L129–L137.

ON FUNDAMENTAL CYCLE OF PERIODIC BOX-BALL SYSTEMS

Tetsuji Tokihiro
Graduate School of Mathematical Sciences, University of Tokyo,
3-8-1 Komaba, Tokyo 153-8914, Japan

Abstract   We review the novel properties of the fundamental cycle of periodic Box-Ball systems (PBBSs). According to integrable nature of the PBBS, the explicit formula for the fundamental cycle exists and its asymptotic behaviour can be estimated when the system size N goes to infinity. The upper and lower bounds for the maximum fundamental cycle is given and almost all fundamental cycle is shown to be of order of $N^{\log N}$.

1   INTRODUCTION

The periodic Box-Ball system (PBBS) is a dynamical system of balls in a one dimensional array of boxes with periodic boundary condition [1, 2]. The PBBS is obtained from the discrete Toda equation [3], which is a well known integrable partial difference equation, with a periodic boundary condition through a limiting procedure called ultradiscretization [4, 5]. Using inverse ultradiscretization, the initial value problem of PBBS is solvable by inverse scattering transform [6]. Hence, the PBBS may be called an *integrable* dynamical system. On the other hand, an important feature of an integrable dynamical system is that its trajectry in the phase space is restricted to a low dimensional subspace determined by the conserved quantities [7]. In particular, it does not have ergodicity. Accordingly its Poincaré section in two-dimensional plane locates on one dimensional curves and quite different from that of non integrable (or chaotic) systems. However, since the PBBS is composed of a finite number of boxes and balls, it can only take on a finite number of patterns. In other words, the phase space of the PBBS consists of only finite number of points. For dynamical systems with such phase spaces, it is not clear to specify the difference between integrable and nonintegrable systems from the trajectory. Recently Yoshihara et al. have obtained the formulae to determine the fundamental cycle, i.e., the shortest period of the discrete periodic motion of the PBBS [8]. Mada and the author examined integrability of the PBBS from its fundamental cycle based on their results [9]. If the PBBS is ergodic, the fundamental cycle T is of order of the volume (number of points) of the phase space. However, it is proved to

L. Faddeev et al. (eds.),
Bilinear Integrable Systems: From Classical to Quantum, Continuous to Discrete, 325–334.

be qualitatively smaller than that of the ergodic system and the PBBS may be regarded as an integrable dynamical system. In this article, we review these recent results about the fundamental cycle of the PBBS.

2 PERIODIC BOX-BALL SYSTEM AND ITS FUNDAMENTAL CYCLE

Let us consider a one-dimensional array of N boxes. To be able to impose a periodic boundary condition, we assume that the Nth box is the adjacent box to the first one. The box capacity is one for all the boxes, and each box is either empty or filled with a ball at any time step. We denote the number of balls by M, such that $M \leq \frac{N}{2}$. The balls are moved according to a deterministic time evolution rule.

1. In each filled box, create a copy of the ball.
2. Move all the copies once according to the following rules.
3. Choose one of the copies and move it to the nearest empty box on the right of it.
4. Choose one of the remaining copies and move it to the nearest empty box on the right of it.
5. Repeat the above procedure until all the copies have moved.
6. Delete all the original balls.

A PBBS has conseved quantities which are characterized by a Young diagram with M boxes. The Young diagram is constructed as follows. We denote an empty box by "0" and a filled box by "1". Then the PBBS is represented as a 0, 1 sequence in which the last entry is regarded as adjacent to the first entry. Let p_1 be the number of the 10 pairs in the sequence. If we eliminate these 10 pairs, we obtain a new 0, 1 sequence. We denote by p_2 the number of 10 pairs in the new sequence. We repeat the above procedure until all the "1" s are eliminated and obtain $p_2, p_3, \ldots, p_l$. Clearly $p_1 \geq p_2 \geq \cdots \geq p_l$ and $\sum_{i=1}^{l} p_i = M$. These $\{p_i\}_{i=1}^{l}$ are conserved in time evolution. Since $\{p_1, p_2, \ldots, p_l\}$ is a weakly decreasing series of positive integers, we can associate it with a Young diagram with p_j boxes in the j-th column $(j = 1, 2, \ldots, l)$. Then the lengths of the rows are also weakly decreasing positive integers, and we denote them

$$\underbrace{\{L_1, L_1, \ldots, L_1,}_{n_1} \underbrace{L_2, L_2, \ldots, L_2,}_{n_2} \cdots, \underbrace{L_s, L_s, \ldots, L_s\}}_{n_s}$$

where $L_1 > L_2 > \cdots > L_s$. The set $\{L_j, n_j\}_{j=1}^{s}$ is an alternative expression of the conserved quantities of the system. In the limit $N \to \infty$, L_j means the length of j-th largest soliton and n_j is the number of solitons with length L_j.

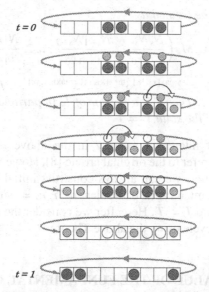

Figure 1. Time evolution rule for PBBS

The following Proposition and Theorem for the fundamental cycle of the PBBS are essential. Let $\ell_0 := N - 2M = N - \sum_{j=1}^{l} 2p_j = N - \sum_{j=1}^{s} 2n_j L_j$, $N_0 := \ell_0$, $L_{s+1} := 0$, and

$$\ell_j := L_j - L_{j+1}, \quad (j = 1, 2, \ldots, s) \tag{1}$$

$$N_j := \ell_0 + 2n_1(L_1 - L_{j+1}) + 2n_2(L_2 - L_{j+1}) + \cdots + 2n_j(L_j - L_{j+1})$$

$$= \ell_0 + \sum_{k=1}^{j} 2n_k(L_k - L_{j+1}). \tag{2}$$

Then, for a fixed number of boxes N and conserved quantities $\{L_j, n_j\}$, the number of possible states of the PBBS $\Omega(N; \{L_j, n_j\})$ is given by the following formula.

Proposition 1

$$\Omega(N; \{L_j, n_j\}) = \frac{N}{\ell_0} \binom{\ell_0 + n_1 - 1}{n_1} \binom{N_1 + n_2 - 1}{n_2} \binom{N_2 + n_3 - 1}{n_3}$$

$$\times \cdots \times \binom{N_{s-1} + n_s - 1}{n_s} \tag{3}$$

The fundamental cycle T is given as:

Theorem 2 *Let $\tilde{T}$ be defined as*

$$\tilde{T} := L.C.M. \left(\frac{N_s N_{s-1}}{\ell_s \ell_0}, \frac{N_{s-1} N_{s-2}}{\ell_{s-1} \ell_0}, \ldots, \frac{N_1 N_0}{\ell_1 \ell_0}, 1 \right), \tag{4}$$

where $L.C.M.(x, y) := 2^{\max[x_2, y_2]} 3^{\max[x_3, y_3]} 5^{\max[x_5, y_5]} \ldots$ for $x = 2^{x_2} 3^{x_3} 5^{x_5} \ldots$ and $y = 2^{y_2} 3^{y_3} 5^{y_5} \ldots$. Then T is a divisor of $\tilde{T}$. In particular, when there is no internal symmetry in the state $T = \tilde{T}$.

The definition of internal symmetry in the above Proposition is rather complicated and we refer to the original article [8]. However, for given number of conserved quantities, we can always construct initial states which do not have any internal symmetry, in particular, if $^\forall i, n_i = 1$ the PBBS never has internal symmetry and $T = \tilde{T}$. Hereafter we consider the asymptotic behavior of the fundamental cycle using the above Theorem.

3   MAXIMUM VALUE OF THE FUNDAMENTAL CYCLE

To take an appropriate limit, we fix the ball densty $\rho := M/N$. The volume of the phase space $V(N; \rho)$ is

$$V(N; \rho) = \binom{N}{M} \sim \frac{1}{\sqrt{2\pi \rho (1 - \rho) N}} R^N, \quad (R := (1 - \rho)^{\rho - 1} \rho^{-\rho}). \tag{5}$$

Thus the volume of the phase space increases exponentially with respect to the system size N. On the other hand, for a given number of balls M, there are P_M different Young diagrams which correspond to conserved quantities. Here P_M is the number of partitions of M. The following estimation of P_M is well known [10].

$$P_M = \frac{\exp[\pi \sqrt{2M/3}]}{4\sqrt{3} M} \left(1 + O \left(\frac{\log M}{M^{1/4}} \right) \right).$$

Since $M = \rho N$, we have $P_M \sim \exp[\pi \sqrt{2\rho/3} \sqrt{N}]/(4\sqrt{3} \rho N)$. The restricted phase space determined by the conserved quantities has the volume $V(N; \rho)/P_M$ in average. This average volume still grows exponentially with respect to the system size and we cannot see an integrable nature of PBBS as a dynamical system from these conserved quantities. So detailed analysis in the fundamental cycle is important to reveal the integrability of the PBBS.

Now we present the estimation of the maximum fundamental cycle $T_{\max} := \max[T]$.

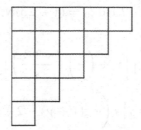

Figure 2. An example of triangular Young diagram

Theorem 3 *For $N \gg 1$ and $M = \rho N (0 < \rho < 1/2)$, the maximum value of the fundamental cycle $T_{\max} \equiv T_{\max}(N; \rho)$ satisfies*

$$\exp\left[2\left(1 - \max\left[\sqrt{2 - 4\rho} - 1, 0\right]\right)\sqrt{N}\left(1 - \frac{c}{\log N}\right)\right]$$

$$< T_{\max} < \exp\left[2\sqrt{2\rho}\sqrt{N}\log N\right]. \tag{6}$$

Here c is a positive integer and $c \sim 0.1$ for $N \geq 10^{16}$.

From the Theorem 3, we find that $\log T_{\max}(N; \rho) \gtrsim \sqrt{N}$. On the other hand $\log V(N; \rho) \sim N$, and we can conclude that the PBBS does not have ergodic property.

Although formula (6) is rather rough estimation for the maximum fundmental cycle. It seems a difficult problem to obtain sharper bound for $T_{\max}$ analytically because of its number theoretical aspects. From the above arguments and numerical calculation however, we expect that the fundamental cycle of the initial state, which has the conserved quantities determined by the triangular Young diagram for the partition $(s, s - 1, s - 2, \ldots, 2, 1)$, is almost of order of $T_{\max}$. In this case, all the solitons have different length and the fundamental cycle is given as

$$T^{(t)}(N, \rho) = \text{L.C.M.}\left(\frac{N_s N_{s-1}}{\ell_0}, \frac{N_{s-1}N_{s-2}}{\ell_0}, \ldots, \frac{N_1 N_0}{\ell_0}, 1\right) \tag{7}$$

where $N_k = \ell_0 + k(k + 1)$ and $\ell_0 = N - 2M = (\rho^{-1} - 2)s(s + 1)/2$.

The number of possible states for the triangular Young diagram $\Omega^{(t)}(N, \rho)$ is given as

$$\Omega^{(t)}(N, \rho) = \prod_{k=1}^{s}(\ell_0 + k(k + 1)), \tag{8}$$

where $M = \rho N = s(s+1)/2$ and $\ell_0 = (1-2\rho)N$. By putting $\gamma := \ell_0/s^2$, we have

$$\Omega^{(t)}(N, \rho) = s^{2s} \prod_{k=1}^{s}\left[\gamma + \left(\frac{k}{s}\right)\left(\frac{k+1}{s}\right)\right]$$

$$\simeq s^{2s} \exp\left[s\left(\log(1+\gamma) - 2 + 2\sqrt{\gamma}\arctan\frac{1}{\sqrt{\gamma}}\right)\right]$$

Since $\gamma = -1 + 1/(2\rho)$, by putting $\alpha(\rho) := \log(1+\gamma) - 2 + 2\sqrt{\gamma}\arctan\frac{1}{\sqrt{\gamma}} + \log(2\rho)$, we have

$$\Omega^{(t)}(N, \rho) \simeq \exp\left[\sqrt{2\rho}\sqrt{N}(\log N + \alpha(\rho))\right]. \qquad (9)$$

Thus $\Omega^{(t)}(N, \rho) \sim e^{(\sqrt{2\rho})\sqrt{N}\log N}$ and

$$\frac{\log \Omega^{(t)}(N, \rho)}{\log V(N, \rho)} \sim \frac{\log N}{\sqrt{N}}.$$

Hence the number of possible states for the triangular Young diagram is much smaller than the volume of the phase space. Figure 3 show the ratio $T^{(t)}(N, \rho)/\Omega^{(t)}(N, \rho)$ obtained numerically. The results show that the fundamental cycle $T^{(t)}$ is much smaller than the number of states $\Omega^{(t)}$. Although the results are not enough to estimate the asymptotic value of $T^{(t)}$, we see in this example that, even if we restrict ourselves to the phase space determined by the conserved quantities, an trajectry does not have ergodicity in the sence that it will never visit most of the states with the same conserved quantities.

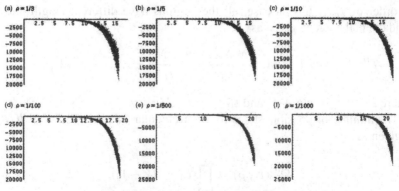

Figure 3. Results of numerically calculated $\log[T^{(t)}(N, \rho)/\Omega^{(t)}(N, \rho)]$

4 ASYMPTOTIC BEHAVIOR OF FUNDAMENTAL CYCLE FOR GENERIC INITIAL STATES

In the preceding section, we have proved that $\log T_{max} \sim \sqrt{N}$. For a generic initial state, however, we expect that its fundamental cycle is qualitatively much smaller. For example, initital states which correspond to rectangular Young diagram have the fundamental cycle less than or equal to the system size N. The number of these initial states grows exponentially with respect to N, while that of the initial states correspond to triangular diagrams grows much slowly like (9).

To examine the asymptotic behavior for a generic initial state, we define the generating function as

$$F(N, K, \ell_0; x) := \frac{N}{\ell_0} \left(\prod_{j=1}^{K} \sum_{n_j=0}^{\infty} \right) \binom{\ell_0 + n_K - 1}{n_K}$$

$$\times \binom{\ell_0 + 2n_K + n_{K-1} - 1}{n_{K-1}} \binom{\ell_0 + 4n_K + 2n_{K-1} + 2n_{K-2} - 1}{n_{K-2}} \cdots$$

$$\times \binom{\ell_0 + \left(\Sigma_{i=2}^{K} 2(i-1)n_i \right) + n_1 - 1}{n_1} x^{\Sigma_{i=1}^{K} i n_i}. \tag{10}$$

From Proposition 1, we find

Proposition 4 *Let N, M and ℓ_0 be the number of boxes of a PBBS, that of balls and $\ell_0 = N - 2M$ respectively. Then the coefficient of x^M of $F(N, K, \ell_0; x)$, $f(N, K; M)$, is the number of initial states whose largest solitons have length less than or equal to K.*

The function $F(N, K, \ell_0; x)$ has the following expression:

Proposition 5

$$F(N, K, \ell_0; x) = \frac{N}{\ell_0} (Y_K(x))^{\ell_0}, \tag{11}$$

where $Y_K(x)$ is recursively defined as

$$X_1(x) := \frac{1}{1-x}$$

$$Y_k(x) := X_1(x)X_2(x) \cdots X_k(x) \quad (k = 1, 2, \ldots)$$

$$X_k(x) := \frac{1}{1 - \{Y_1(x)Y_2(x) \cdots Y_{k-1}(x)\}^2 x^k} \quad (k = 1, 2, \ldots) \tag{12}$$

Now we introduce

$$a_k(x) := \sum_{j=0}^{\left[\frac{k+1}{2}\right]} \binom{k+1-j}{j} (-1)^j x^j \quad (k \geq -1,\ k \in \mathbb{Z}) \tag{13}$$

For polynomials $a_k(x)$, we have the following Lemma.

Lemma 6 *Let $a_k(x)$ be as above, then*

$$a_{k+1}(x) = a_k(x) - x a_{k-1}(x) \quad (k = 0, 1, 2, \ldots). \tag{14}$$

$$a_{k+1}(x) a_{k-1}(x) = a_k(x)^2 - x^{k+1} \quad (k = 0, 1, 2, \ldots) \tag{15}$$

$$a_k(x) = \frac{\alpha(x)^{k+2} - \beta(x)^{k+2}}{\alpha(x) - \beta(x)} \quad (k = 0, 1, 2, \ldots) \tag{16}$$

where $\alpha(x)$ and $\beta(x)$ are two distinct roots of the quadratic equation

$$t^2 - t + x = 0.$$

Note that $\alpha(x)$ and $\beta(x)$ are explicitly given as

$$\alpha = \frac{1 + \sqrt{1 - 4x}}{2}, \quad \beta = \frac{1 - \sqrt{1 - 4x}}{2}, \tag{17}$$

and $\alpha(x)\beta(x) = x$, $\alpha(x) + \beta(x) = 1$.

Proposition 7

$$Y_k(x) = \frac{a_{k-1}(x)}{a_k(x)} \tag{18}$$

$$= \frac{\alpha(x)^{k+1} - \beta(x)^{k+1}}{\alpha(x)^{k+2} - \beta(x)^{k+2}} \quad (k = 1, 2, 3 \ldots) \tag{19}$$

From Propositions 5 and 7, we have an explicit form of the generating function $F(N, K, \ell_0; x)$. Then the coefficient $f(N, K; M)$ is given by the contour integral

$$f(N, K; M) = \frac{1}{2\pi i} \oint_{|z| = \epsilon \ll 1} \frac{F(N, K, \ell_0; z)}{z^{M+1}} dz. \tag{20}$$

Asymptotic behavior of the right hand side of (20) may be estimated with, for example, the method of steepest decent. However, (20) is still complicated and we shall try to obtain a simpler expression.

The following Lemma is easily obtained by induction.

Lemma 8

$$\left(\frac{1}{\alpha(x)}\right)^m = \sum_{r=0}^{\infty} \frac{m \cdot (2r + m - 1)!}{(r + m)! r!} x^r \quad (m = 1, 2, \ldots). \tag{21}$$

Then we obtain an explicit formula for $f(N, K; M)$ defined in Proposition 4 as

Proposition 9

$$f(n, K; M) := \frac{N}{\ell_0} \sum_{j=0,(K+1)j+(K+2)i \leq M}^{\ell_0} \sum_{i=0}^{\infty} \binom{\ell_0}{j}\binom{\ell_0 + i - 1}{i}(-1)^j$$

$$\times \frac{(\ell_0 + 2(K + 1)j + 2(K + 2)i) \cdot (2M + \ell_0 - 1)!}{(M + \ell_0 + (K + 1)j + (K + 2)i)!(M - (K + 1)j - (K + 2)i)!}, \quad (22)$$

where $\ell_0 = N - 2M$.

Finally we obtain

Theorem 10 *The coefficient $f(N, K; M)$ is given by the Cauchy integral*

$$f(N, K; M) = \frac{N}{2\pi i \ell_0} \oint_C \frac{dz}{z^{M+1}} \left(\frac{1 - z^{K+1}}{1 - z^{K+2}}\right)^{\ell_0} (1 + z)^{2M + \ell_0 - 1}(1 - z).$$

$$(23)$$

Here C denotes the contour $|z| = x_0 (< 1)$.

We evaluate (23) by the method of steepest decent and we obtain

Theorem 11 *For sufficiently large K,*

$$f(N, K; M) \sim \frac{N}{\ell_0 \sqrt{2\pi t_0 N}}(1 - t_0)\frac{(1 + t_0)^N}{t_0^M}\left(\frac{1 - t_0^{K+1}}{1 - t_0^{K+2}}\right)^{\ell_0} \quad (N \to +\infty),$$

$$(24)$$

where $\ell_0 = N - 2M$, $M = N\rho$ and $\rho = \frac{t_0}{1+t_0}(0 < t_0 < 1)$.

Utilizing the Theorem 11, the asymptotic behavior of fundamental cycle for generic initial states are given by the theorem

Theorem 12 *Let $\bar{V}(N; \rho)$ be the number of initial states which have the fundamental cycle less than $\exp\left[\frac{2(\log N)^2}{-\log t_0}\right]$ Then*

$$\lim_{N \to \infty} \frac{\bar{V}(N; \rho)}{V(N; \rho)} = 1. \quad (25)$$

5 CONCLUDING REMARKS

We have reviewed integrability of PBBS in terms of asymptotic behavior of its fundamental cycles. As a dynamical system, PBBS is shown to have no ergodicity in the sence that a trajectry does not visit most of the states in the phase space. Although the maximum fundamental cycle $T_{\max} \gtrsim e^{\sqrt{N}}$ (Theorem 3), a generic state has fundamental cycle $T \lesssim e^{(\log N)^2}$ (Theorem 12). To obtain more sharp estimation, we may have to invoke some number theoretical technique, which is a problem we wish to address in the future.

REFERENCES

1. Takahashi, D. (1991) On some soliton systems defined by boxes and balls, In: *Proceedings of the International Symposium on Nonlinear Theory and Its Applications, NOLTA'93*, p. 555.
2. Yura, F. and Tokihrio, T. (2002) *J. Phys. A: Math. Gen.* **35**, p. 3787.
3. Hirota, R., Tsujimoto, S., and Imai, T. (1992) Difference scheme of soliton equations, in *Future Directions of Nonlinear Dynamics in Physics and Biological Systems*, ed. P. L. Christiansen, J. C. Eilbeck, and R. D. Parmentier, Series B: Physics Vol. 312, Plenum, pp. 7–15.
4. Tokihiro, T., Takahashi, D., Matsukidaira, J., and Satsuma, J. (1996) *Phys. Rev. Lett.* **76**, p. 3247.
5. Matsukidaira, J., Satsuma, J., Takahashi, D., Tokihiro, T., and Torii, M. (1997) *Phys. Lett. A* **255**, p. 287.
6. Kimijima, T. and Tokihiro, T. (2002) *Inverse Problems* **18**, p. 1705.
7. Arnold, V. I. (1988) *Mathematical Methods of Classical Mechanics*, Springer-Verlag, New York.
8. Yoshihara, D., Yura, F., and Tokihiro, T. (2003) *J. Phys. A: Math. Gen.* **36**, pp. 99–121.
9. Mada, J. and Tokihrio, T. (2003) J. *Phys. A: Math. Gen.* **36**, pp. 7251–7268.
10. See for example, Newman, D. J. (1998) *Analytic Number Theory*, Springer-Verlag, New York.

COMBINATORICS AND INTEGRABLE GEOMETRY

Pierre van Moerbeke

Department of Mathematics, Université de Louvain, 1348 Louvain-la-Neuve, Belgium and Brandeis University, Waltham, Mass 02454, USA. The author
acknowledges the support of the Clay Mathematics Institute, One Bow Street, Cambridge, MA 02138, USA. E-mail: vanmoerbeke@math.ucl.ac.be and @brandeis.edu. The support of a National Science Foundation grant # DMS-01-00782, European Science Foundation, Nato, FNRS and Francqui Foundation grants is gratefully acknowledged.

1 INTRODUCTION

Since Russel's horse back journey along the canal from Glasgow to Edinburg in 1834, since the birth of the Korteweg-de Vries equation in 1895 and since the remarkable renaissance initiated by M. Kruskal and coworkers in the late 60's, the field of integrable systems has emerged as being at the crossroads of important new developments in the sciences.

Integrable systems typically have many different solutions. Besides the soliton and scattering solutions, other important solutions of KdV have arisen, namely rational and algebro-geometrical solutions. This was the royal road to the infinite-dimensional Grassmannian description of the KP-solutions, leading to the fundamental concept of Sato's τ-function, which enjoys Plücker relations and Hirota bilinear relations. In this way, the τ-function is a far reaching generalization of classical theta functions and is nowadays a unifying theme in mathematics: representation theory, curve theory, symmetric function theory, matrix models, random matrices, combinatorics, topological field theory, the theory of orthogonal polynomials and Painlevé theory all live under the same hat! This general field goes under the somewhat bizarre name of "integrable mathematics."

This lecture illustrates another application of integrable systems, this time, to unitary matrix integrals and ultimately to combinatorics and probability theory. Unitary matrix integrals, with an appropriate set of time parameters inserted to make it a τ function, satisfy a new lattice, the Toeplitz lattice, related to the 2d-Toda lattice for a very special type of initial condition. Besides, it also satisfies constraints, which form a very small subalgebra of the Virasoro algebra (Section 2).

L. Faddeev et al. (eds.),
Bilinear Integrable Systems: From Classical to Quantum, Continuous to Discrete, 335–362.

Along a seemingly different vein, certain unitary matrix integrals, developed in a series with respect to a parameter, have coefficients which contain information concerning random permutations, random words and random walks. Turned around, the generating function for certain probabilities turns out to be a unitary matrix integral (Section 3).

The connection of these combinatorial problems with integrable systems is precious: it enables one to find differential and difference equations for these probabilities! This is explained in Section 4. The purpose of this lecture is to explain these connections. For a more comprehensive account of these results, including the ones on random matrices, see [1].

2   A UNITARY MATRIX INTEGRAL: VIRASORO AND THE TOEPLITZ LATTICE

In this section, we consider integrals over the unitary group $U(n)$ with regard to the invariant measure dM. Since the spectrum $z_1, \ldots, z_n$ of M lies on the circle S^1 and since the integrand only involves traces, it is natural to integrate out the "angular part" of dM and to keep its spectral part[1] $|\Delta_n(z)|^2 dz_1 \ldots dz_n$. For $\varepsilon \in \mathbb{Z}$, define the following integrals, depending on formal time parameters $t = (t_1, t_2, \cdots)$ and $s = (s_1, s_2, \cdots)$, with $\tau_0 = 1$,

$$
\begin{aligned}
\tau_n^\varepsilon(t, s) &= \int_{U(n)} (\det M)^\varepsilon e^{\sum_1^\infty \mathrm{Tr}(t_j M^j - s_j \bar{M}^j)} dM \\
&= \frac{1}{n!} \int_{(S^1)^n} |\Delta_n(z)|^2 \prod_{k=1}^n \left(z_k^\varepsilon e^{\sum_1^\infty (t_j z_k^j - s_j z_k^{-j})} \frac{dz_k}{2\pi i z_k} \right) \\
&= \det \left(\oint_{S^1} \frac{dz}{2\pi i z} z^{\ell-m+\varepsilon} e^{\sum_1^\infty (t_j z^j - s_j z^{-j})} \right)_{1 \le \ell, m \le n},
\end{aligned}
\tag{1}
$$

the latter being a Toeplitz determinant. The last equality follows from the fact that the product of two Vandermonde's can be expressed as sum of determinants:

$$
\Delta_n(u) \Delta_n(v) = \sum_{\sigma \in S_n} \det \left(u_{\sigma(k)}^{\ell-1} v_{\sigma(k)}^{k-1} \right)_{1 \le \ell, k \le n},
\tag{2}
$$

and from distributing the factors in the product (in (1)) over the columns of the matrix, appearing in the last formula of (1). Now, the main point is that the matrix integrals above satisfy **two distinct systems of equations**. These equations will be useful for the combinatorial problems discussed in Section 3.

---

[1] with the Vandermonde determinant $\Delta_n(z) = \Pi_{1 \le i < j \le n}(z_i - z_j)$.

2.1   Unitary Matrix Integrals and the Virasoro Algebra

Proposition 1   *(See [2]) The integrals* (1) *satisfy the Virasoro constraints,*

$$\mathbb{V}_k^\varepsilon(t, s, n)\tau_n^\varepsilon(t, s) = 0, \quad \text{for} \quad k = -1, 0, 1 \tag{3}$$

where $\mathbb{V}_k^\varepsilon := \mathbb{V}_k^\varepsilon(t, s, n)$ *are the operators*

$$\mathbb{V}_{-1}^\varepsilon = \sum_{i\geq 1}(i + 1)t_{i+1}\frac{\partial}{\partial t_i} - \sum_{i\geq 2}(i - 1)s_{i-1}\frac{\partial}{\partial s_i} + nt_1 + (n - \varepsilon)\frac{\partial}{\partial s_1}$$

$$\mathbb{V}_0^\varepsilon = \sum_{i\geq 1}\left(it_i\frac{\partial}{\partial t_i} - is_i\frac{\partial}{\partial s_i}\right) + \varepsilon n = 0 \tag{4}$$

$$\mathbb{V}_1^\varepsilon = -\sum_{i\geq 1}(i + 1)s_{i+1}\frac{\partial}{\partial s_i} + \sum_{i\geq 2}(i - 1)t_{i-1}\frac{\partial}{\partial t_i} + ns_1 + (n + \varepsilon)\frac{\partial}{\partial t_1}.$$

Remark   Note that the generators $\mathbb{V}_k^\varepsilon$ are part of an ∞-dimensional Virasoro algebra; the claim here is that the integrals above satisfy only these three constraints, unlike the case of Hermitian matrix integrals, which satisfy a large subalgebra of constraints!

Proof   For the exponent $\varepsilon \neq 0$, the proof is a slight modification of the case $\varepsilon = 0$; so, we stick to the case $\varepsilon = 0$. The Virasoro operators $\mathbb{V}_k := \mathbb{V}_k^\varepsilon\big|_{\varepsilon=0}$ are generated by the following *vertex operator*[2]

$$\mathbb{X}(t, s; u) := \Lambda^\top e^{\sum_1^\infty (t_i u^i - s_i u^{-i})} e^{-\sum_1^\infty \left(\frac{u^{-i}}{i}\frac{\partial}{\partial t_i} - \frac{u^i}{i}\frac{\partial}{\partial s_i}\right)}. \tag{5}$$

This means they are a commutator realization of differentiation:

$$\frac{\partial}{\partial u}u^{k+1}\frac{\mathbb{X}(t, s; u)}{u} = \left[\mathbb{V}_k(t, s), \frac{\mathbb{X}(t, s; u)}{u}\right]. \tag{6}$$

Then the following operator, obtained by integrating the vertex operator (5),

$$\mathbb{Y}(t, s) = \oint_{S^1}\frac{du}{2\pi i u}\mathbb{X}(t, s; u, u^{-1}) \tag{7}$$

has, using (6), the commutation property

$$[\mathbb{Y}, \mathbb{V}_k] = 0.$$

---

[2] The operator Λ is the semi-infinite shift matrix, with zeroes everywhere, except for 1's just above the diagonal, i.e., $(\Lambda v)_n = v_{n+1}$ and $(\Lambda^\top v)_n = v_{n-1}$.

Then one checks that the integrals $I_n = n! \tau_n^{(0)}$ in (1) (for $n \geq 1$) are fixed points for $\mathbb{Y}(t, s)$; namely, taking into account the shift $\Lambda^\top$ in (5), one computes

$$
\mathbb{Y}(t, s) I_n(t, s) = \oint_{S^1} \frac{du}{2\pi i u} e^{\Sigma_1^\infty (t_i u^i - s_i u^{-i})} e^{-\Sigma_1^\infty \left(\frac{u^{-i}}{i} \frac{\partial}{\partial t_i} - \frac{u^i}{i} \frac{\partial}{\partial s_i} \right)}
$$

$$
\int_{(S^1)^{n-1}} \Delta_{n-1}(z) \Delta_{n-1}(\bar{z}) \prod_{k=1}^{n-1} e^{\Sigma_1^\infty (t_i z_k^i - s_i z_k^{-i})} \frac{dz_k}{2\pi i z_k}
$$

$$
= \oint_{S^1} \frac{du}{2\pi i u} e^{\Sigma_1^\infty (t_i u^i - s_i u^{-i})} \int_{(S^1)^{n-1}} \Delta_{n-1}(z) \Delta_{n-1}(\bar{z})
$$

$$
\times \prod_{k=1}^{n-1} \left(1 - \frac{z_k}{u} \right) \left(1 - \frac{u}{z_k} \right) e^{\Sigma_1^\infty (t_i z_k^i - s_i z_k^{-i})} \frac{dz_k}{2\pi i z_k}
$$

$$
= \int_{(S^1)^n} |\Delta_n(z)|^2 \prod_{k=1}^{n} \left(e^{\Sigma_1^\infty (t_i z_k^i - s_i z_k^{-i})} \frac{dz_k}{2\pi i z_k} \right) = I_n(t, s)
$$

Using this fixed point property and the fact that $(\Lambda^\top)^n I_n = I_0$, we have for $\mathbb{Y} := \mathbb{Y}(t, s)$,

$$
\begin{aligned}
0 &= [\mathbb{V}_k, \mathbb{Y}^n] I_n \\
&= \mathbb{V}_k \mathbb{Y}^n I_n - \mathbb{Y}^n \mathbb{V}_k I_n \\
&= \mathbb{V}_k I_n - \mathbb{Y}^n \mathbb{V}_k I_n. \\
&= \mathbb{V}_k I_n - \oint_{S_1} \frac{du}{2\pi i u} e^{\Sigma_1^\infty (t_i u^i - s_i u^{-i})} e^{-\Sigma_1^\infty \left(\frac{u^{-i}}{i} \frac{\partial}{\partial t_i} - \frac{u^i}{i} \frac{\partial}{\partial s_i} \right)} \\
&\cdots \oint_{S_1} \frac{du}{2\pi i u} e^{\Sigma_1^\infty (t_i u^i - s_i u^{-i})} e^{-\Sigma_1^\infty \left(\frac{u^{-i}}{i} \frac{\partial}{\partial t_i} - \frac{u^i}{i} \frac{\partial}{\partial s_i} \right)} \mathbb{V}_k I_0.
\end{aligned}
$$

Now one checks visually that for $I_0 = 1$,

$$
\mathbb{V}_k I_0 = 0 \quad \text{for} \quad k = -1, 0, 1,
$$

ending the proof of Proposition 1. The details of the proof can be found in Adler-van Moerbeke [2].   ∎

2.2   The Toeplitz Lattice

Considering the integral $\tau_n^\varepsilon(t, s)$, as in (1), and setting, for short,

$$
\tau_n := \tau_n^{(0)}, \quad \tau_n^\pm := \tau_n^{\pm 1},
$$

define the ratios

$$
x_n(t, s) = (-1)^n \frac{\tau_n^+(t, s)}{\tau_n(t, s)} \quad \text{and} \quad y_n(t, s) := (-1)^n \frac{\tau_n^-(t, s)}{\tau_n(t, s)}, \tag{8}
$$

and the semi infinite matrices (they are not "rank 2," but try to be!)

$$L_1 := \begin{pmatrix} -x_1 y_0 & 1 - x_1 y_1 & 0 & 0 \\ -x_2 y_0 & -x_2 y_1 & 1 - x_2 y_2 & 0 \\ -x_3 y_0 & -x_3 y_1 & -x_3 y_2 & 1 - x_3 y_3 \\ -x_4 y_0 & -x_4 y_1 & -x_4 y_2 & -x_4 y_3 \\ & & & & \ddots \end{pmatrix}$$

and

$$L_2 := \begin{pmatrix} -x_0 y_1 & -x_0 y_2 & -x_0 y_3 & -x_0 y_4 \\ 1 - x_1 y_1 & -x_1 y_2 & -x_1 y_3 & -x_1 y_4 \\ 0 & 1 - x_2 y_2 & -x_2 y_3 & -x_2 y_4 \\ 0 & 0 & 1 - x_3 y_3 & -x_3 y_4 \\ & & & & \ddots \end{pmatrix} \tag{9}$$

Throughout the paper, set[3]

$$h_n = \frac{\tau_{n+1}}{\tau_n} \quad \text{and} \quad v_n := 1 - x_n y_n \stackrel{*}{=} \frac{h_n}{h_{n-1}} = \frac{\tau_{n+1} \tau_{n-1}}{\tau_n^2}. \tag{10}$$

One checks that the quantities x_n and y_n satisfy the following commuting Hamiltonian vector fields, introduced by Adler and van Moerbeke in [2],

$$\frac{\partial x_n}{\partial t_i} = (1 - x_n y_n)\frac{\partial G_i}{\partial y_n} \quad \frac{\partial y_n}{\partial t_i} = -(1 - x_n y_n)\frac{\partial G_i}{\partial x_n}$$

$$\frac{\partial x_n}{\partial s_i} = (1 - x_n y_n)\frac{\partial H_i}{\partial y_n} \quad \frac{\partial y_n}{\partial s_i} = -(1 - x_n y_n)\frac{\partial H_i}{\partial x_n}, \tag{11}$$

$$\textbf{(Toeplitz lattice)}$$

with Hamiltonians

$$G_i = -\frac{1}{i}\text{Tr}\, L_1^i, \quad H_i = -\frac{1}{i}\text{Tr}\, L_2^i, \quad i = 1, 2, 3, \ldots \tag{12}$$

and symplectic structure

$$\omega := \sum_1^\infty \frac{dx_k \wedge dy_k}{1 - x_k y_k}.$$

One imposes initial conditions $x_n(0, 0) = y_n(0, 0) = 0$ for $n \geq 1$ and boundary conditions $x_0(t, s) = y_0(t, s) = 1$. The G_i and F_i are functions in involution with regard to the Hamiltonian vector fields (11). Setting $h :=$ diagonal

---

[3] The proof of equality $\stackrel{*}{=}$ hinges on associated bi-orthogonal polynomials on the circle, introduced later.

$(h_0, h_1, \ldots)$, with h_i as in (10), we conjugate L_1 with a diagonal matrix so as to have 1's in the first superdiagonal:

$$\hat{L}_1 := hL_1h^{-1} \quad \text{and} \quad \hat{L}_2 := L_2.$$

The Hamiltonian vector fields (11) imply the 2-Toda lattice equations for the matrices $\hat{L}_1$ and $\hat{L}_2$,

$$\frac{\partial \hat{L}_i}{\partial t_n} = \left[(\hat{L}_1^n)_+ , \hat{L}_i \right] \quad \text{and} \quad \frac{\partial \hat{L}_i}{\partial s_n} = \left[(\hat{L}_2^n)_- , \hat{L}_i \right]$$

$$i = 1, 2 \quad \text{and} \quad n = 1, 2, \ldots .$$

$$\textbf{(two-Toda Lattice)} \tag{13}$$

Thus the particular structure of L_1 and L_2 is preserved by the 2-Toda Lattice equations. In particular, this implies that the τ_n's satisfy the KP-hierarchy.

Other equations for the τ_n's are obtained by noting that the expressions formed by means of the matrix integrals (1) above[4]

$$p_n^{(1)}(t, s; z) = z^n \frac{\tau_n(t - [z^{-1}], s)}{\tau_n(t, s)} \quad \text{and} \quad p_n^{(2)}(t, s; z) = z^n \frac{\tau_n(t, s + [z^{-1}])}{\tau_n(t, s)}$$

are actually polynomials in z, with coefficients depending on t, s; moreover, they are *bi-orthogonal polynomials* on the circle for the following (t, s)-dependent inner product[5],

$$\langle f(z), g(z) \rangle_{t,s} := \oint_{S^1} \frac{dz}{2\pi i z} f(z) g(z^{-1}) e^{\sum_1^\infty (t_i z^i - s_i z^{-i})}. \tag{15}$$

Using bi-orthogonality one shows that the variables x_n and y_n, defined in (8), equal the z^0-term of the bi-orthogonal polynomials,

$$x_n(t, s) = p_n^{(1)}(t, s; 0) \quad \text{and} \quad y_n(t, s) = p_n^{(2)}(t, s; 0) . \tag{16}$$

(i) This fact implies the following identity for the h_n's:

$$\left(1 - \frac{h_{n+1}}{h_n} \right) \left(1 - \frac{h_n}{h_{n-1}} \right) = -\frac{\partial}{\partial t_1} \log h_n \frac{\partial}{\partial s_1} \log h_n . \tag{17}$$

(ii) The mere fact that L_1 and L_2 satisfy the two-Toda lattice implies that the integrals $\tau_n(t, s)$ satisfy, besides the *KP-hierarchy* in t and s (separately), the following equations, combining (t, s)-partials and nearest

---

[4] For $\alpha \in \mathbb{C}$, define $[\alpha] := \left(\alpha, \frac{1}{2}\alpha^2, \frac{1}{3}\alpha^3, \ldots \right) \in \mathbb{C}^\infty$.
[5] For this inner-product, we have $(z^k)^\top = z^{-k}$, i.e.,

$$\langle z^k f(z), g(z) \rangle_{t,s} = \langle f(z), z^{-k} g(z) \rangle_{t,s} . \tag{14}$$

neighbors $\tau_{n\pm1}$,

$$\frac{\partial^2}{\partial s_1 \partial t_1} \log \tau_n = -\frac{\tau_{n-1}\tau_{n+1}}{\tau_n^2},$$

$$\frac{\partial^2}{\partial s_2 \partial t_1} \log \tau_n = -2\frac{\partial}{\partial s_1} \log \frac{\tau_n}{\tau_{n-1}} \cdot \frac{\partial^2}{\partial s_1 \partial t_1} \log \tau_n - \frac{\partial^3}{\partial s_1^2 \partial t_1} \log \tau_n. \quad (18)$$

3 MATRIX INTEGRALS AND COMBINATORICS

3.1 Largest Increasing Sequences in Random Permutations and Words

Consider the group of *permutations of length k*

$$S_k = \{\text{permutaions } \pi \text{ of } \{1, \dots, k\}\}$$

$$= \left\{ \pi_k = \pi = \begin{pmatrix} 1 & \cdots & k \\ \pi(1) & \cdots & \pi(k) \end{pmatrix}, \quad \text{for distinct } 1 \leq \pi(j) \leq k \right\},$$

equipped with the uniform probability distribution

$$P_k(\pi_k) = 1/k!. \quad (19)$$

Also consider *words of length k, taken from an alphabet* $1, \dots, p$,

$$S_k^p = \{\text{words } \sigma \text{ of length } k \text{ from an alphabet } \{1, \dots, p\}\}$$

$$= \left\{ \sigma = \sigma_k = \begin{pmatrix} 1 & 2 & \cdots & k \\ \sigma(1) & \sigma(2) & \cdots & \sigma(k) \end{pmatrix}, \quad \text{for arbitrary } 1 \leq \sigma(j) \leq p \right\}$$

$$(20)$$

and uniform probability $P_k^p(\sigma) = 1/k^p$ on S_k^p.

An *increasing subsequence* of $\pi_k \in S_k$ or $\sigma_k \in S_k^p$ is a sequence[6] $1 \leq j_1 < \cdots < j_\alpha \leq k$, such that $\pi(j_1) \leq \cdots \leq \pi(j_\alpha)$. Define

$$\left. \begin{matrix} L_k(\pi_k) \\ L_k(\sigma_k) \end{matrix} \right\} = \text{length of the longest increasing subsequence of} \begin{cases} \pi_k \\ \sigma_k \end{cases} \quad (21)$$

We shall be interested in the probabilities

$$P_k(L_k(\pi) \leq n, \pi \in S_k) \quad \text{and} \quad P_k^p(L_k(\sigma) \leq n, \sigma \in S_k^p).$$

$$\textit{Examples}: \begin{cases} \text{for } \pi_7 = (\underline{3}, 1, \underline{4}, 2, \underline{6}, \underline{7}, 5) \in S_7, \text{ we have } L_7(\pi_7) = 4. \\ \text{for } \pi_5 = (5, \underline{1}, \underline{4}, 3, 2) \in S_5, \text{ we have } L_5(\pi_5) = 2. \\ \text{for } \sigma_7 = (2, \underline{1}, 3, 2, \underline{1}, \underline{1}, \underline{2}) \in S_7^3, \text{ we have } L_7(\sigma_7) = 4. \end{cases}$$

---

[6] For permutations one automatically has strict inequalities $\pi(j_1) < \cdots < \pi(j_\alpha)$.

In 1990, Gessel [3] considered the generating function (22) below and showed that it equals a Toeplitz determinant (determinant of a matrix, whose (i, j)th entry depends on $i - j$ only). By now, Theorem 2 below has many different proofs; at the end of Section 3.3, we sketch a proof based on integrable ideas. See also Section 4.2.

Theorem 2 *(Gessel [3]) The following generating function has an expression in terms of a $U(n)$-matrix integral[7]*

$$\sum_{k=0}^{\infty} \frac{\xi^k}{k!} P_k(L_k(\pi) \leq n) = \int_{U(n)} e^{\sqrt{\xi}\, \text{Tr}(M+\bar{M})} dM$$

$$= \frac{1}{n!} \oint_{(S^1)^n} |\Delta_n(z)|^2 \prod_{k=1}^{n} \left(e^{\sqrt{\xi}(z_k+\bar{z}_k)} \frac{dz_k}{2\pi i z_k} \right)$$

$$= \det \left(\oint_{S^1} \frac{dz}{2\pi i z} z^{\ell-m} e^{\sqrt{\xi}(z+z^{-1})} \right)_{1 \leq \ell, k \leq n} \qquad (22)$$

Theorem 3 *(Tracy-Widom [4]) We also have[8]*

$$\sum_{k=0}^{\infty} \frac{(p\xi)^k}{k!} P_k^p(L_k(\sigma) \leq n) = \int_{U(n)} e^{\xi Tr\bar{M}} \det(I + M)^p dM$$

$$= \det \left(\oint_{S^1} \frac{dz}{2\pi i z} z^{k-\ell} e^{\xi z^{-1}} (1+z)^p \right)_{1 \leq k, \ell \leq n}.$$

Consider instead the subgroups of odd permutations, with $2^k k!$ elements, the hyperoctahedral group,

$$S_{2k}^{\text{odd}} = \left\{ \begin{array}{l} \pi_{2k} \in S_{2k}, \pi_{2k} : (-k, \ldots, -1, 1, \ldots, k)\circlearrowleft \\ \text{with } \pi_{2k}(-j) = -\pi_{2k}(j), \text{ for all } j \end{array} \right\} \subset S_{2k}$$

$$S_{2k+1}^{\text{odd}} = \left\{ \begin{array}{l} \pi_{2k+1} \in S_{2k+1}, \pi_{2k} : (-k, \ldots, -1, 0, 1, \ldots, k)\circlearrowleft \\ \text{with } \pi_{2k+1}(-j) = -\pi_{2k+1}(j), \text{ for all } j \end{array} \right\} \subset S_{2k}$$

Then, according to Rains [5] and Tracy-Widom [4], the following generating functions, again involving the length of the longest increasing sequence, are related to matrix integrals:

---

[7] The expression (22) is a determinant of Bessel functions, since $J_n(u)$ is defined by $e^{u(t-t^{-1})} = \Sigma_{-\infty}^{\infty} t^n J_n(2u)$ and thus

$$e^{\sqrt{\xi}(z+z^{-1})} = e^{\sqrt{-\xi}((-iz)-(-iz)^{-1})} = \sum (-iz)^n J_n(2\sqrt{-\xi}).$$

[8] The functions appearing in the contour integration are confluent hypergeometric functions $_1F_1$.

Theorem 4   *For* $\pi_{2k} \in S_{2k}^{odd}$ *and* $\pi_{2k+1} S_{2k}^{odd}$, *one has the following generating functions:*

$$\sum_0^\infty \frac{(2\xi)^k}{k!} P(L(\pi_{2k}) \leq n \text{ for } \pi_{2k} \in S_{2k}^{odd}) = \int_{U(n)} e^{\sqrt{\xi} \, \text{Tr}(M^2+\bar{M}^2)} dM$$

$$\sum_0^\infty \frac{(2\xi)^k}{k!} P(L(\pi_{2k+1}) \leq n \text{ for } \pi_{2k+1} \in S_{2k}^{odd})$$

$$= \frac{1}{4} \left(\frac{\partial}{\partial t}\right)^2 \int_{U(n)} dM \left(e^{\text{Tr}(t(M+\bar{M})+\sqrt{\xi}(M^2+\bar{M}^2)} + e^{\text{Tr}(t(M+\bar{M})-\sqrt{\xi}(M^2+\bar{M}^2))} \right) \Bigg|_{t=0}$$

Generating functions for other combinatorial quantities related to integrals over the Grassmannian $\text{Gr}(p, \mathbb{R}^n)$ and $\text{Gr}(p, \mathbb{C}^n)$ of p-planes in $\mathbb{R}^n$ or $\mathbb{C}^n$ have been investigated by Adler-van Moerbeke [6].

3.2   Combinatorial Background

The reader is reminded of a few basic facts in combinatorics. Standard references to this subject are MacDonald, Sagan, Stanley, Stanton and White [7–10].

- A *partition* λ of n (noted $\lambda \vdash n$) or a *Young diagram* λ of weight n is represented by a sequence of integers $\lambda_1 \geq \lambda_2 \geq \cdots \geq \lambda_\ell \geq 0$, such that $n = |\lambda| := \lambda_1 + \cdots + \lambda_\ell; n = |\lambda|$ is called the weight. A *dual Young diagram* $\lambda^\top = (\lambda_1^\top \geq \lambda_2^\top \geq \cdots)$ is the diagram obtained by flipping the diagram λ about its diagonal; clearly $|\lambda| = |\lambda^\top|$. Define $\mathbb{Y}_n := \{$all partitions λ with $|\lambda| = n\}$.

  A *skew-partition* or *skew Young diagram* $\lambda \backslash \mu$, for $\lambda \supset \mu$, is defined as the shape obtained by removing the diagram μ from λ.

- The *Schur polynomial* $\mathbf{s}_\lambda(t)$ associated with a Young diagram $\lambda \vdash n$, is defined by

$$\mathbf{s}_\lambda(t_1, t_2, \ldots) = \det(\mathbf{s}_{\lambda_i - i + j}(t))_{1 \leq i, j \leq \ell}$$

in terms of elementary Schur polynomials $\mathbf{s}_i(t)$, defined by

$$e^{\sum_1^\infty t_i z^i} =: \sum_{i \geq 0} \mathbf{s}_i(t) z^i, \quad \text{and} \quad \mathbf{s}_i(t) = 0 \text{ for } i < 0.$$

The *skew Schur polynomial* $\mathbf{s}_{\lambda \backslash \mu}(t)$, associated with a skew Young diagram $\lambda \backslash \mu$, is defined by

$$\mathbf{s}_{\lambda \backslash \mu}(t) := \det(\mathbf{s}_{\lambda_i - i - \mu_j + j}(t))_{1 \leq i, j \leq n}. \tag{23}$$

The $\mathbf{s}_\lambda$'s form a basis of the space of symmetric functions in $x_1, x_2, \ldots$, via the map $k t_k = \sum_{i \geq 1} x_i^k$.

- A *standard Young tableau* P of shape $\lambda \vdash n$ is an array of integers $1, \ldots, n$ placed in the Young diagram, which are strictly increasing from left to right *and* from top to bottom. A *standard skew Young tableau* of shape $\lambda \backslash \mu \vdash n$ is defined in a similar way. Then, it is well-known that

$$f^{\lambda} := \# \left\{ \begin{array}{l} \text{standard tableaux of shape } \lambda \vdash n \\ \text{filled with integers } 1, \ldots, n \end{array} \right\} = \left. \frac{|\lambda|!}{u^{|\lambda|}} \mathbf{s}_{\lambda}(t) \right|_{t_i = u \delta_{i1}}$$

$$f^{\lambda \backslash \mu} := \# \left\{ \begin{array}{l} \text{standard skew tableaux of shape} \\ \lambda \backslash \mu \vdash n \text{ filled with integers } 1, \ldots, n \end{array} \right\} = \left. \frac{|\lambda \backslash \mu|!}{u^{|\lambda \backslash \mu|}} \mathbf{s}_{\lambda \backslash \mu}(t) \right|_{t_i = u \delta_{i1}}$$

$$(24)$$

- A *semi standard Young tableau* of shape $\lambda \vdash n$ is an array of integers $1, \ldots, p$ placed in the Young diagram λ, which are non-decreasing from left to right *and* strictly increasing from top to bottom. The *number of semi-standard Young tableaux* of a given shape $\lambda \vdash n$, filled with integers 1 to p for $p \geq \lambda_1^{\top}$, has the following expression in terms of Schur polynomials:

$$\# \left\{ \begin{array}{l} \text{semi standard tableaux of shape } \lambda \\ \text{filled with numbers from } 1 \text{ to } p \end{array} \right\} = \mathbf{s}_{\lambda} \left(p, \frac{p}{2}, \frac{p}{3}, \ldots \right). \quad (25)$$

- *Robinson-Schensted-Knuth (RSK) correspondence*: There is a 1-1 correspondence

$$S_k \longleftrightarrow \left\{ \begin{array}{l} \text{pairs of standard Young tableaux } (P, Q), \\ \text{both of same arbitrary shape } \lambda, \text{ with} \\ |\lambda| = k, \text{ filled with integers } 1, \ldots, k \end{array} \right\} \quad (26)$$

Given a permutation $\pi = (i_1, \ldots, i_k)$, the RSK correspondence constructs two standard Young tableaux P, Q having the same shape λ. This construction is inductive. Namely, having obtained two equally shaped Young diagrams P_j, Q_j from $i_1, \ldots, i_j$, with the numbers $(i_1, \ldots, i_j)$ in the boxes of P_j and the numbers $(1, \ldots, j)$ in the boxes of Q_j, one creates a new diagram Q_{j+1}, by putting the *next number* i_{j+1} *in the first row of* P, according to the rules:

(i) if $i_{j+1} \geq$ all numbers appearing in the first row of P_j, then one creates a new box containing i_{j+1} to the right of the first column,

(ii) if not, place i_{j+1} in the box (of the first row) with the smallest higher number. That number then gets pushed down to the second row of P_j according to the rules (i) and (ii), as if the first row had been removed.

The diagram Q is a bookkeeping device; namely, add a box (with the number $j + 1$ in it) to Q_j exactly at the place, where the new box has been added to P_j. This produces a new diagram Q_{j+1} of same shape as P_{j+1}.

The inverse of this map is constructed by reversing the steps above. The Robinson-Schensted-Knuth correspondence has the following properties:

- length (longest increasing subsequence of π) = # (columns in P)
- length (longest decreasing subsequence of π) = # (rows in P)
- $\pi \mapsto (P, Q)$, then $\pi^{-1} \mapsto (Q, P)$ (27)

So-called **Plancherel measure** $\tilde{P}_k$ on $\mathbb{Y}_k$ is the probability induced from the uniform probability P_k on S_k (see (19)), via the RSK map (26). For an arbitrary partition $\lambda \vdash k$, it is computed as follows:

$$\tilde{P}_k(\lambda) := P_k(\text{permutations } \pi \in S_k \text{ leading to } \lambda \in \mathbb{Y}_k \text{ by RSK})$$

$$= \frac{\#\{\text{ permutations leading to } \lambda \in \mathbb{Y}_k \text{ by RSK}\}}{k!}$$

$$= \frac{\#\left\{\begin{array}{l}\text{pairs of standard tableaux } (P, Q), \text{ both} \\ \text{of shape } \lambda, \text{ filled with numbers } 1, \ldots, k\end{array}\right\}}{k!}$$

$$= \frac{(f^\lambda)^2}{k!}, \quad \text{using}(24).$$

Note that, by the first property in (27), we have

$$L_k(\pi) \le n \Longleftrightarrow (P, Q) \text{ has shape } \lambda \text{ with } |\lambda| = k \text{ and } \lambda_1 \le n.$$

These facts prove the following Proposition:

Proposition 5 *Let P_k be uniform probability on the permutations in S_k and $\tilde{P}_k$ Plancherel measure on $\mathbb{Y}_k := \{\text{partitions } \lambda \vdash k\}$. Then:*

$$P_k(L_k(\pi) \le n) = \frac{1}{k!}\#\left\{\begin{array}{l}\text{pairs of standard Young tableaux } (P, Q), \text{ both of} \\ \text{same arbitrary shape } \lambda, \text{ with} |\lambda| = k \text{ and } \lambda_1 \le n\end{array}\right\}$$

$$= \frac{1}{k!}\sum_{\substack{|\lambda|=k \\ \lambda_1 \le n}}(f^\lambda)^2$$

$$= \tilde{P}_k(\lambda_1 \le n). \tag{28}$$

From a slight extension of the RSK correspondence for "words," we have

$$S_k^p \longleftrightarrow \left\{\begin{array}{l}\text{semi-standard and standard Young tableaux} \\ (P, Q) \text{ of same shape } \lambda \text{ and } |\lambda| = k, \text{ filled} \\ \text{resp., with integers } (1, \ldots, p) \text{ and } (1, \ldots, k),\end{array}\right\},$$

and thus the uniform probability P_k^p on S_k^p induces a probability measure $\tilde{P}_k^p$ on

$$\mathbb{Y}_k^p = \{\text{partitions } \lambda \text{ such that } |\lambda| = k, \ \lambda_1^\top \le p\},$$

namely

$$\tilde{P}_k^p(\lambda) = P_k^p \{\text{words } \sigma \in S_k^p \text{ leading to } \lambda \in \mathbb{Y}_k^p \text{ by RSK}\}$$

$$= \frac{f^\lambda s_\lambda \left(p, \frac{p}{2}, \frac{p}{3}, \dots\right)}{p^k}, \quad \lambda \in \mathbb{Y}_k^p.$$

Proposition 6 *Let P_k be uniform probability on words in S_k^p and $\tilde{P}_k^p$ the induced measure on $\mathbb{Y}_k^p$. Then:*

$$P_k^p(L(\sigma) \le n) = \frac{1}{p^k} \# \left\{ \begin{array}{l} \text{semi-standard and standard Young tableaux } (P,Q) \\ \text{of same shape } \lambda, \text{ with} |\lambda| = k \text{ and } \lambda_1 \le n, \text{filled} \\ \text{resp., with integers } (1, \dots, p) \text{ and } (1, \dots, k), \end{array} \right\}$$

$$= \frac{1}{p^k} \sum_{\substack{|\lambda|=k \\ \lambda_1 \le n}} f^\lambda s_\lambda \left(p, \frac{p}{2}, \frac{p}{3}, \dots\right)$$

$$= \tilde{P}_k^p(\lambda_1 \le n). \tag{29}$$

Example For permutation $\pi = \begin{pmatrix} 1 & 2 & 3 & 4 & 5 \\ 5 & \underline{1} & 4 & \underline{3} & 2 \end{pmatrix} \in S_5$, the RSK algorithm gives

$$
\begin{array}{cccccc}
P \Rightarrow & 5 & 1 & 1\ 4 & 1\ 3 & 1\ 2 \\
 & & 5 & 5 & 4 & 3 \\
 & & & & 5 & 4 \\
 & & & & & 5
\end{array}
$$

$$
\begin{array}{cccccc}
Q \Rightarrow & 1 & 1 & 1\ 3 & 1\ 3 & 1\ 3 \\
 & & 2 & 2 & 2 & 2 \\
 & & & & 4 & 4 \\
 & & & & & 5
\end{array}
$$

Hence

$$\pi \mapsto (P, Q) = \left(\left(\overbrace{\begin{array}{|c|c|} \hline 1 & 2 \\ \hline 3 \\ \hline 4 \\ \hline 5 \\ \hline \end{array}}^{2} \right), \left(\begin{array}{|c|c|} \hline 1 & 3 \\ \hline 2 \\ \hline 4 \\ \hline 5 \\ \hline \end{array} \right) \right).$$

$$\text{standard} \qquad\qquad \text{standard}$$

Note that the sequence 1,3, underlined in the permutation above is a longest increasing sequence, and so $L_5(\pi) = 2$; of course, we also have

$$L_5(\pi) = 2 = \#\{\text{columns of } P \text{ or } Q\}.$$

3.3 A Probability on Partitions and Toeplitz Determinants

Define yet another "probability measure" on the set $\mathbb{Y}$ of Young diagrams

$$\mathbb{P}(\lambda) = Z^{-1} s_\lambda(t) s_\lambda(s), \quad Z = e^{\sum_{i \geq 1} i t_i s_i}. \tag{30}$$

Cauchy's identity[9] guarantees that $\mathbb{P}(\lambda)$ is a probability measure, in the sense

$$\sum_{\lambda \in \mathbb{Y}} \mathbb{P}(\lambda) = 1,$$

without necessarily $0 \leq \mathbb{P}(\lambda) \leq 1$. This probability measure has been introduced and extensively studied by Borodin, Okounkov, Olshanski and others; see [11, 12] and references within. In the following Proposition, the Toeplitz determinants appearing in (1) acquire a probabilistic meaning in terms of the new probability $\mathbb{P}$:

Proposition 7 *Given the probability (30), the following holds*

$$\mathbb{P}(\lambda \text{ with } \lambda_1 \leq n) = Z^{-1} \det \left(\oint_{S^1} \frac{dz}{2\pi i z} z^{k-\ell} e^{-\sum_1^\infty (t_i z^i + s_i z^{-i})} \right)_{1 \leq k, \ell \leq n} \tag{31}$$

and

$$\mathbb{P}(\lambda \text{ with } \lambda_1^\top \leq n) = Z^{-1} \det \left(\oint_{S^1} \frac{dz}{2\pi i z} z^{k-\ell} e^{\sum_1^\infty (t_i z^i + s_i z^{-i})} \right)_{1 \leq k, \ell \leq n}$$

with Z given by (31).

Proof Consider the semi infinite Toeplitz matrix

$$m_\infty(t, s) = (\mu_{k\ell})_{k, \ell \geq 0}, \quad \text{with } \mu_{k\ell}(t, s) = \oint_{S^1} z^{k-\ell} e^{\sum_1^\infty (t_j z^j - s_j z^{-j})} \frac{dz}{2\pi i z}.$$

Note that

$$\frac{\partial \mu_{k\ell}}{\partial t_i} = \oint_{S^1} z^{k-\ell+i} e^{\sum_1^\infty (t_j z^j - s_j z^{-j})} \frac{dz}{2\pi i z} = \mu_{k+i, \ell}$$

$$\frac{\partial \mu_{k\ell}}{\partial s_i} = -\oint_{S^1} z^{k-\ell-i} e^{\sum (t_j z^j - s_j z^{-j})} \frac{dz}{2\pi i z} = -\mu_{k, l+i} \tag{32}$$

---

[9] Cauchy's identity takes on the following form in the t and s variables:

$$\sum_{\lambda \in \mathbb{Y}} s_\lambda(t) s_\lambda(s) = e^{\sum_1^\infty i t_i s_i}.$$

with initial condition $\mu_{k\ell}(0,0) = \delta_{k\ell}$. In matrix notation, this amounts to the system of differential equations[10]

$$\frac{\partial m_\infty}{\partial t_i} = \Lambda^i m_\infty \quad \text{and} \quad \frac{\partial m_\infty}{\partial s_i} = -m_\infty (\Lambda^\top)^i,$$
$$\text{with initial condition } m_\infty(0,0) = I_\infty. \tag{33}$$

The solution to this initial value problem is given by the following two expressions:

(i) $$m_\infty(t,s) = (\mu_{k\ell}(t,s))_{k,\ell \geq 0}, \tag{34}$$

as follows from the differential equation (32), and

(ii) $$m_\infty(t,s) = e^{\sum_1^\infty t_i \Lambda^i} m_\infty(0,0) e^{-\sum_1^\infty s_i \Lambda^{\top i}}, \tag{35}$$

upon using $(\partial/\partial t_k)e^{\sum_1^\infty t_i \Lambda^i} = \Lambda^k e^{\sum_1^\infty t_i \Lambda^i}$. Then, by the uniqueness of solutions of ode's, the two solutions coincide, and in particular the $n \times n$ upper-left blocks of (34) and (35), namely

$$m_n(t,s) = E_n(t)m_\infty(0,0)E_n^\top(-s), \tag{36}$$

where

$$E_n(t) = \begin{pmatrix} 1 & s_1(t) & s_2(t) & s_3(t) & \cdots & s_{n-1}(t) & \cdots \\ 0 & 1 & s_1(t) & s_2(t) & \cdots & s_{n-2}(t) & \cdots \\ \vdots & & & & & & \\ & & & & & s_1(t) & \cdots \\ 0 & & \cdots & & 0 & 1 & \cdots \end{pmatrix} = (s_{j-i}(t))_{\substack{1 \leq i < n \\ 1 \leq j < \infty}}$$

is the $n \times n$ upper-left blocks of

$$e^{\sum_1^\infty t_i \Lambda^i} = \sum_0^\infty s_i(t)\Lambda^i = \begin{pmatrix} 1 & s_1(t) & s_2(t) & s_3(t) & \cdots \\ 0 & 1 & s_1(t) & s_2(t) & \cdots \\ 0 & 0 & 1 & s_1(t) & \cdots \\ 0 & 0 & 0 & 1 & \\ \vdots & \vdots & \vdots & \vdots & \end{pmatrix} = (s_{j-i}(t))_{\substack{1 \leq i < \infty \\ 1 \leq j < \infty}}.$$

Therefore the determinants of the matrices (36) coincide:

$$\det m_n(t,s) = \det(E_n(t)m_\infty(0,0)E_n^\top(-s)). \tag{37}$$

---

[10] The operator Λ is the semi-infinite shift matrix defined in footnote 2. Also I_∞ is the semi-infinite identity matrix.

Moreover, from the **Cauchy-Binet formula**[11], applied twice, one proves the following: given an arbitrary semi-infinite initial condition $m_\infty(0,0)$, the expression below admits an expansion in Schur polynomials,

$$\det(E_n(t)m_\infty(0,0)E_n^\top(-s)) = \sum_{\substack{\lambda,\nu \\ \lambda_1^\top,\nu_1^\top \leq n}} \det(m^{\lambda,\nu})s_\lambda(t)s_\nu(-s), \quad \text{for } n > 0,$$

$$(38)$$

where the sum is taken over all Young diagrams λ and ν, with first columns $\leq n$ (i.e., $\lambda_1^\top$ and $\nu_1^\top \leq n$) and where $m^{\lambda,\nu}$ is the matrix

$$m^{\lambda,\nu} := \left(\mu_{\lambda_i-i+n,\nu_j-j+n}(0,0)\right)_{1\leq i,j\leq n}. \quad (39)$$

Applying formula (39) to $m_\infty(0,0) = I_\infty$, we have

$$\det m^{\lambda,\nu} = \det\left(\mu_{\lambda_i-i+n,\nu_j-j+n}\right)_{1\leq i,j\leq n} \neq 0 \text{ if and only if } \lambda = \nu, \quad (40)$$

in which case $\det m^{\lambda,\lambda} = 1$. Therefore,

$$\sum_{\substack{\lambda\in Y \\ \lambda_1^\top \leq n}} s_\lambda(t)s_\lambda(-s) = \det\left(\oint_{S^1}\frac{dz}{2\pi i z}z^{k-\ell}e^{\sum_1^\infty(t_i z^i - s_i z^{-i})}\right)_{1\leq k,\ell\leq n}. \quad (41)$$

But, we also have, using the probability $\mathbb{P}$, defined in (30), that

$$\mathbb{P}(\lambda \text{ with } \lambda_1^\top \leq n) = Z^{-1}\sum_{\substack{\lambda\in Y \\ \lambda_1^\top \leq n}} s_\lambda(t)s_\lambda(s) \quad (42)$$

Comparing the two formulas (41) and (42) and changing $s \mapsto -s$ in (41), yield

$$\mathbb{P}(\lambda \text{ with } \lambda_1^\top \leq n) = Z^{-1}\det\left(\oint_{S^1}\frac{dz}{2\pi i z}z^{k-\ell}e^{\sum_1^\infty(t_i z^i + s_i z^{-i})}\right)_{1\leq k,\ell\leq n}$$

$$= Z^{-1}\sum_{\substack{\lambda\in Y \\ \lambda_1^\top \leq n}} s_\lambda(t)s_\lambda(s). \quad (43)$$

---

[11] Given two matrices $\underset{(m,n)}{A}$, $\underset{(n,m)}{B}$, for n large $\geq m$

$$\det(AB) = \det\left(\sum_i a_{\ell i}b_{ik}\right)_{1\leq k,\ell\leq m}$$

$$= \sum_{1\leq i_1<\ldots<i_m\leq n} \det(a_{k,i_\ell})_{1\leq k,\ell\leq m}\det(b_{i_k},\ell)_{1\leq k,\ell\leq m}.$$

Using $s_\lambda(-t) = (-1)^{|\lambda|} \mathbf{s}_\lambda \mathsf{T}(t)$, one easily checks

$$\mathbb{P}(\lambda \text{ with } \lambda_1 \leq n) = Z^{-1} \sum_{\substack{\lambda \in \mathbb{Y} \\ \lambda_1 \leq n}} \mathbf{s}_\lambda(t) \mathbf{s}_\lambda(s), \quad \text{by definition}$$

$$= Z^{-1} \sum_{\substack{\lambda \in \mathbb{Y} \\ \lambda_1^\top \leq n}} \mathbf{s}_\lambda \mathsf{T}(t) \mathbf{s}_\lambda \mathsf{T}(s)$$

$$= Z^{-1} \sum_{\substack{\lambda \in \mathbb{Y} \\ \lambda_1^\top \leq n}} \mathbf{s}_\lambda(-t) \mathbf{s}_\lambda(-s)$$

$$= Z^{-1} \det \left(\oint_{S^1} \frac{dz}{2\pi i z} z^{k-\ell} e^{-\sum_1^\infty (t_i z^i + s_i z^{-i})} \right)_{1 \leq k, \ell \leq n},$$

using (43) in the last equality, with Z as in (30). This establishes Proposition 7.   ∎

Proof of Theorem 2   For real $\xi > 0$, consider the locus

$$\mathcal{L}_1 = \{\text{all } s_k = t_k = 0, \text{ except } t_1 = s_1 = \sqrt{\xi}\} \tag{44}$$

Indeed, for an arbitrary $\lambda \in \mathbb{Y}$, the probability (30) evaluated along $\mathcal{L}_1$ reads:

$$\mathbb{P}(\lambda)\Big|_{\mathcal{L}_1} = e^{-\sum_{k \geq 1} k t_k s_k} \mathbf{s}_\lambda(t) \mathbf{s}_\lambda(s)\Big|_{\substack{t_i = \sqrt{\xi} \delta_{i1} \\ s_i = \sqrt{\xi} \delta_{i1}}}$$

$$= e^{-\xi} \xi^{|\lambda|/2} \frac{f^\lambda}{|\lambda|!} \xi^{|\lambda|/2} \frac{f^\lambda}{|\lambda|!}, \quad \text{using (24)},$$

$$= e^{-\xi} \frac{\xi^{|\lambda|}}{|\lambda|!} \frac{(f^\lambda)^2}{|\lambda|!}.$$

Therefore

$$\mathbb{P}(\lambda_1 \leq n)|_{\mathcal{L}_1} = \sum_{\substack{\lambda \in \mathbb{Y} \\ \lambda_1 \leq n}} e^{-\xi} \frac{\xi^{|\lambda|}}{|\lambda|!} \frac{(f^\lambda)^2}{|\lambda|!}$$

$$= e^{-\xi} \sum_0^\infty \frac{\xi^k}{k!} \sum_{\substack{|\lambda| = k \\ \lambda_1 \leq n}} \frac{(f^\lambda)^2}{k!}$$

$$= e^{-\xi} \sum_0^\infty \frac{\xi^k}{k!} P_k(L_k(\pi) \leq n), \quad \text{by Proposition 3.4.} \tag{45}$$

The next step is to evaluate (31) in Proposition 3.5 along the locus $\mathcal{L}_1$,

$$\mathbb{P}(\lambda_1 \leq n)\Big|_{\mathcal{L}_1} = e^{-\sum_{i\geq 1} i t_i s_i} \det\left(\oint_{S^1} \frac{dz}{2\pi i z} z^{k-\ell} e^{-\sum_1^\infty (t_i z^i + s_i z^{-i})}\right)_{1\leq k,\ell\leq n}\Bigg|_{\mathcal{L}_1}$$

$$= e^{-\xi} \det\left(\oint_{S^1} \frac{dz}{2\pi i z} z^{k-\ell} e^{-\sqrt{\xi}(z+z^{-1})}\right)_{1\leq k,\ell\leq n}$$

$$= e^{-\xi} \det\left(\oint_{S^1} \frac{dz}{2\pi i z} z^{k-\ell} e^{\sqrt{\xi}(z+z^{-1})}\right)_{1\leq k,\ell\leq n}, \qquad (46)$$

by changing $z \mapsto -z$. Finally, comparing (45) and (46) yields (22), ending the proof of Theorem 2. ∎

Proof of Theorem 3 The proof of this theorem goes along the same lines, except one uses Proposition 3.4 and one evaluates (31) along the locus

$$\mathcal{L}_2 = \{t_k = \delta_{k1}\xi \text{ and } k s_k = p\},$$

instead of $\mathcal{L}_1$; then one makes the change of variable $z \mapsto -z^{-1}$ in the integral. ∎

3.4 Non Intersecting Random Walks

Consider n walkers in $\mathbb{Z}$, walking from $x = (x_1 < x_2 < \cdots < x_n)$ to $y = (y_1 < y_2 < \cdots < y_n)$, such that, at each moment, only one walker moves either one step to the left, or one step to the right, with all possible moves equally likely. This section deals with a generating function for the probability

$$P(k, x, y) := P\left(\begin{array}{l}\text{that } n \text{ walkers in } \mathbb{Z} \text{ go from } x_1, \ldots, x_n \text{ to} \\ y_1, \ldots, y_n \text{ in } k \text{ steps, and do not intersect}\end{array}\right) = \frac{b_{xy}^{(k)}}{(2n)^k}$$

We now state a Theorem which generalizes Theorem 3.1; the latter can be recovered by assuming close packing $x = y = (0, 1, \ldots, n-1)$. In Section 4.3 discrete equations will be found for $P(k; x, y)$.

Theorem 8 *(Adler-van Moerbeke [13]) The generating function for the $P(k; x, y)$ above has the following matrix integral representation:*

$$\sum_{k\geq 0} \frac{(2nz)^k}{k!} P(k; x, y) = \int_{U(n)} s_\lambda(M) s_\mu(\bar{M}) e^{z\,\mathrm{Tr}(M+\bar{M})} dM =: a_{\lambda\mu}(z)$$

$$= \det\left(\oint_{S^1} \frac{du}{2\pi i u} u^{\lambda_\ell - \ell - \mu_k + k} e^{z(u+u^{-1})}\right)_{1\leq k,\ell\leq n},$$

where $\mathbf{s}_\lambda$ *and* $\mathbf{s}_\mu$ *are Schur polynomials*[12] *with regard to the partitions* λ *and* μ, *themselves determined by the initial and final positions* x *and* y,

$$\lambda_{n-i+1} := x_i - i + 1, \quad \mu_{n-i+1} := y_i - i + 1. \quad for \ i = 1, \ldots, n. \quad (47)$$

Remark   The partitions λ and μ measure the discrepancy of x and y from close packing $0, 1, \ldots, n - 1$!

Remark   Connections of random walks with Young diagrams have been known in various situations in the combinatorial literature; see R. Stanley [9] (p. 313), P. Forrester [14], D. Grabiner & P. Magyar [15, 16] and J. Baik [17].

Proof   Consider the locus

$$\mathcal{L}_1 = \{\text{all } t_k = s_k = 0, \text{except } t_1 = z, s_1 = -z\}.$$

Then, since

$$e^{\sum_1^\infty (t_i u^i - s_i u^{-i})}\Big|_{\mathcal{L}_1} = e^{z(u+u^{-1})},$$

we have, combining (38) and (37),

$$\int_{U(n)} e^{z\text{Tr}(M+\bar{M})} e^{\sum_1^\infty \text{Tr}(t_i M^i - s_i \bar{M}^i)} dM = \sum_{\substack{\lambda,\mu \text{ such that} \\ \lambda_1^\top, \mu_1^\top \leq n}} a_{\lambda\mu}(z) s_\lambda(t) s_\mu(-s), \quad (48)$$

with (for definitions and formulas for skew Schur polynomials and tableaux, see (23) and (24))

$$a_{\lambda\mu}(z) \overset{(i)}{=} \det\left(\oint_{S^1} u^{\lambda_\ell - \ell - \mu_k + k} e^{z(u+u^{-1})} \frac{du}{2\pi i u}\right)_{1 \leq \ell, k \leq n}$$

$$\overset{(ii)}{=} \int_{U(n)} s_\lambda(M) S_\mu(\bar{M}) e^{z\text{Tr}(M+\bar{M})} dM$$

$$\overset{(iii)}{=} \sum_{\substack{\nu \text{ with} \\ \nu_1^\top \leq n}} S_{\nu\backslash\lambda}(t) s_{\nu\backslash\mu}(-s)\Big|_{\mathcal{L}_1}$$

$$\overset{(iv)}{=} \sum_{\substack{\nu \text{ with } \nu \supset \lambda, \mu \\ \nu_1^\top \leq n}} \frac{z^{|\nu\backslash\lambda|}}{|\nu\backslash\lambda|!} f^{\nu\backslash\lambda} \frac{z^{|\nu\backslash\mu|}}{|\nu\backslash\mu|!} f^{\nu\backslash\mu}$$

---

[12] Given a unitary matrix M, the notation $s_\lambda(M)$ denotes a symmetric function of the eigenvalues $x_1, \ldots, x_n$ of the unitary matrix M and thus in the notation of the present paper $s_\lambda(M) := s_\lambda(\text{Tr } M, \frac{1}{2}\text{Tr } M^2, \frac{1}{3}\text{Tr } M^3, \ldots)$.

$$\overset{(v)}{=} \sum_{k=0}^{\infty} \frac{z^k}{k!} \frac{k!}{k_1! \, k_2!} \sum_{\substack{v \text{ with } v \supset \lambda, \mu \\ |v\backslash\lambda|=k_1 \\ |v\backslash\mu|=k_2 \\ v_1^T \le n}} f^{v\backslash\lambda} f^{v\backslash\mu},$$

$$\text{where } k_{\left\{\begin{smallmatrix}1\\2\end{smallmatrix}\right\}} = \frac{1}{2}(k \mp |\lambda| \pm |\mu|),$$

$$= \sum_{k\ge 0} \frac{z^k}{k!} \# \left\{ \begin{array}{l} \text{ways that } n \text{ non intersecting} \\ \text{walkers in } \mathbb{Z} \text{ move in } k \text{ steps} \\ \text{from } x_1 < x_2 < \cdots < x_n \\ \text{to } y_1 < y_2 < \cdots < y_n \end{array} \right\} = \sum_{k\ge 0} \frac{(2nz)^k}{k!} P(k; x, y).$$

Equality (i) follows from (40) and (38). The Fourier coefficients $a_{\lambda\mu}(z)$ of (48) can be obtained by taking the inner-product[13] of the sum (48) with $s_\alpha(t) s_\beta(-s)$. Equality (iii) is the analogue of (43) for skew-partitions and also follows from the Cauchy-Binet formula. Equality (iv) follows from formula (24) for skew-partitions. Equality (v) follows immediately from (iv), whereas the last equality follows from an analogue of RSK as is now explained.

Consider, as in the picture below, the two skew-tableaux P and Q of shapes $v\backslash\lambda$ and $\mu\backslash\lambda$, with integers $1, \ldots, |v\backslash\lambda|$ and $1, \ldots, |v\backslash\mu|$ inserted respectively (strictly increasing from left to right and from top to bottom). The integers c_{ij} in the tableau P provide the instants of left move for the corresponding walker (indicated on the left), assuming they all depart from $(x_1, \ldots, x_n)$, which itself is specified by v. This construction implies that, at each instant, only one walker moves and they never intersect. That takes an amount of time $|v\backslash\lambda| = \frac{1}{2}(k - |\lambda| + |\mu|) = k_1$, at which they end up at a position specified by v. At the next stage and from that position, they start moving right at the instants $k - c'_{ij}$, where the c'_{ij} are given by the second skew tableau and forced to end up at positions $(y_1, \ldots, y_n)$, itself specified by μ; see (47). Again the construction implies here that they never intersect and only one walker moves at the time. The time (of right move) elapsed is $|v\backslash\mu| = \frac{1}{2}(k + |\lambda| - |\mu|) = k_2$. So, the total time elapsed is $k_1 + k_2 = k$.

The final argument hinges on the fact that any motion, where exactly one walker moves either left or right during time k can be transformed (in a canonical way) into a motion where the walkers first move left during time k_1 and then move right during time k_2. The precise construction is based on an idea of Forrester [14]. This map is many-to-one: there are precisely $\frac{k!}{k_1! k_2!}$ walks leading

---

13

$$\langle s_\alpha, s_\lambda \rangle := s_\alpha \left(\frac{\partial}{\partial t_1}, \frac{1}{2} \frac{\partial}{\partial t_2}, \ldots \right) s_\lambda(t) \Big|_{t=0}$$

to a walk where walkers first move left and then right.

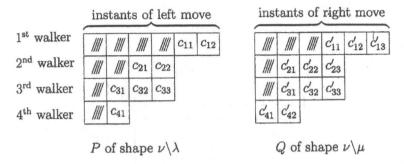

$$P \text{ of shape } \nu\backslash\lambda \qquad\qquad Q \text{ of shape } \nu\backslash\mu$$

This sketches the proof of Theorem 8.   ∎

4 WHAT DO INTEGRABLE SYSTEMS TELL US ABOUT COMBINATORICS?

The fact that the matrix integrals are related to the Virasoro constraints and the Toeplitz lattice will lead to various statements about the various combinatorial problems considered in Section 3.

4.1 Recursion Relations for Unitary Matrix Integrals

Motivated by the integrals appearing in Theorems 2, 3, and 4, consider the integrals, for $\varepsilon = 0, \pm$, (different from the I_n introduced before)

$$I_n^\varepsilon := \frac{1}{n!} \int_{(S^1)^n} |\Delta_n(z)|^2 \prod_{k=1}^n z_k^\varepsilon e^{\sum_{j=1}^N \frac{u_j}{j}(z_k^j + z_k^{-j})} \frac{dz_k}{2\pi i z_k}. \tag{49}$$

They enjoy the following property:

Theorem 9 *The integral $I_n := I_n^0$ can be expressed as a polynomial in I_1 and the expressions $x_1, \ldots, x_{n-1}$,*

$$I_n = (I_1)^n \prod_1^{n-1}(1 - x_k^2)^{n-k}, \tag{50}$$

with the x_k's satisfying rational $2N + 1$-step recursion relations in terms of prior x_i's; to be precise

$$\left(\left(\sum_1^N u_i L_1^i \right)_{k+1,k+1} + \left(\sum_1^N u_i L_1^i \right)_{k,k} - 2 \left(\sum_1^N u_i L_1^{i-1} \right)_{k+1,k} \right) = \frac{k x_k^2}{1 - x_k^2}, \tag{51}$$

where L_1 is the matrix[14] *defined in* (9) *and the u_i's appear in the integral* (49). *The left hand side of this expression is polynomial in the* $x_{k-N}, \ldots, x_k, \ldots, x_{k+N}$ *and linear in x_{k+N} and the parameters $u_1, \ldots, u_N$. This implies the recursion relation*

$$x_{k+N} = F(x_{k+N-1}, \ldots, x_k, \ldots, x_{k-N}; u_1, \ldots, u_N),$$

with F rational in all arguments.

Remark   Note the x_n's are the same ratios as in (8) but for the integrals (4.1.1), i.e.,

$$x_n = (-1)^n \frac{I_n^+}{I_n}, \text{ with } I_n := I_n^\varepsilon \Big|_{\varepsilon=0} \text{ and } I_n^+ := I_n^\varepsilon \Big|_{\varepsilon=+1},$$

Example 1   Symbol $e^{t(z+z^{-1})}$.

This concerns the integral in Theorem 2, expressing the generating function for the *probabilities of the length of longest increasing sequences in random permutations.* Setting $u_1 = u$, $u_i = 0$ for $i \geq 2$ in the equation (51), one finds that

$$x_n = \frac{\int_{(S^1)^n} |\Delta_n(z)|^2 \prod_{k=1}^n z_k e^{u\left(z_k + z_k^{-1}\right)} \dfrac{dz_k}{2\pi i z_k}}{\int_{(S^1)^n} |\Delta_n(z)|^2 \prod_{k=1}^n e^{u\left(z_k + z_k^{-1}\right)} \dfrac{dz_k}{2\pi i z_k}} \tag{52}$$

satisfies the simple three-step rational relation,

$$u(x_{k+1} + x_{k-1}) = \frac{k x_k}{x_k^2 - 1}. \tag{53}$$

This so-called MacMillan equation [18] for x_n was first derived by Borodin [19] and Baik [20], using Riemann-Hilbert methods. In [13], we show this is part of the much larger system of equations (51), closely related to the Toeplitz lattice. This map (53) is the simplest instance of a family of area-preserving maps of the plane, having an invariant, as found by McMillan, and extended by Suris [21] to maps of the form $\partial_n^2 x(n) = f(x(n))$, having an analytic invariant of two variables $\Phi(\beta, \gamma)$. The invariant in the case of the maps (53) is

$$\Phi(\beta, \gamma) = t(1 - \beta^2)(1 - \gamma^2) - n\beta\gamma,$$

which means that for all n,

$$\Phi(x_{n+1}, x_n) = \Phi(x_n, x_{n-1}).$$

---

[14] Note in the case of an integral the type (49), we have $x_n = y_n$, and thus $L_2 = L_1^\top$.

For more on this matter, see the review by B. Grammaticos, F. Nijhoff, A. Ramani [21].

Example 2 Symbol $e^{t(z+z^{-1})+u(z^2+z^{-2})}$.

These symbols appear in the longest increasing sequence problem for the *hyperoctahedral group*; see Theorem 3.3. Here we set $u_1 = t$, $u_2 = u$, $u_i = 0$ for $i \geq 3$ in the equation (51); one finds

$$x_n = \frac{\int_{(S^1)^n} |\Delta_n(z)|^2 \prod_{k=1}^{n} z_k e^{t(z_k+z_k^{-1})+u(z^2+z^{-2})} \frac{dz_k}{2\pi i z_k}}{\int_{(S^1)^n} |\Delta_n(z)|^2 \prod_{k=1}^{n} e^{t(z_k+z_k^{-1})+u(z^2+z^{-2})} \frac{dz_k}{2\pi i z_k}} \tag{54}$$

satisfies the five-step rational relation, $(v_n := 1 - x_n^2)$

$$0 = n x_n + t v_n (x_{n-1} + x_{n+1}) + 2 u v_n$$
$$\times (x_{n+2} v_{n+1} + x_{n-2} v_{n-1} - x_n (x_{n+1} + x_{n-1})^2). \tag{55}$$

Also here the map has a polynomial invariant

$$\Phi(\alpha, \beta, \gamma, \delta) = \big(t + 2u(\alpha(\delta - \beta) - \gamma(\delta + \beta))\big)(1 - \beta^2)(1 - \gamma^2) - n\beta\gamma;$$

that is for all n,

$$\Phi(x_{n-1}, x_n, x_{n+1}, x_{n+2}) = \Phi(x_{n-2}, x_{n-1}, x_n, x_{n+1}).$$

Proof of Theorem 9 Formula (50) follows straightforwardly from the identity (10). Moreover Proposition 2.1 implies the integrals

$$\tau_n^\varepsilon(t, s) = \frac{1}{n!} \int_{(S^1)^n} |\Delta_n(z)|^2 \prod_{k=1}^{n} z_k^\varepsilon e^{\sum_1^\infty \left(t_i z_k^i - s_i z_k^{-i}\right)} \frac{dz_k}{2\pi i z_k} \tag{56}$$

satisfy the Virasoro constraints (3). Thus, setting $\mathbb{V}_n := \mathbb{V}_n^\varepsilon\big|_{\varepsilon=0}$ and $\mathbb{V}_n^+ := \mathbb{V}_n^\varepsilon\big|_{\varepsilon=1}$, we have

$$0 = \frac{\mathbb{V}_0^+ \tau_n^+}{\tau_n^\varepsilon} - \frac{\mathbb{V}_0 \tau_n}{\tau_n}$$

$$= \sum_{i \geq 1} \left(i t_i \frac{\partial}{\partial t_i} - i s_i \frac{\partial}{\partial s_i} \right) \log x_n + n, \quad \text{where } x_n = (-1)^n \frac{\tau_n^+}{\tau_n}$$

$$= \frac{1 - x_n^2}{x_n} \frac{\partial}{\partial x_n} \sum_{i \geq 1} (i t_i G_i - i s_i H_i) + n, \quad \text{using (2.2.4)}$$

$$= \frac{1 - x_n^2}{x_n^2} \sum_{i \geq 1} \left\{ \begin{array}{l} i t_i (-(L_1^i)_{n+1,n+1} + (L_1^{i-1})_{n+1,n}) \\ + i s_i ((L_2^i)_{nn} - (L_2^{i-1})_{n,n+1}) \end{array} \right\} + n.$$

Setting

$$it_i = -is_i = \begin{cases} u_i & \text{for} \quad 1 \le i \le N \\ 0 & \text{for} \quad i > N, \end{cases}$$

leads to the claim (51). Relations (53) and (55) are obtained by specialization.   ∎

4.2 The Painlevé V Equation for the Longest Increasing Sequence Problem

The statement of Theorem 2 can now be completed by the following Theorem,
due to Tracy-Widom [23]. The integrable method explained below captures
many other situations, like longest increasing sequences in involutions and
words; see Adler-van Moerbeke [2].

Theorem 10 *For every $n \ge 0$, the generating function (22) for the probability
of the longest increasing sequence can be expressed in terms of a specific
solution of the Painlevé V equation:*

$$\sum_{k=0}^{\infty} \frac{\xi^k}{k!} P_k(L_k(\pi) \le n) = \exp \int_0^{\xi} \log\left(\frac{\xi}{u}\right) g_n(u)du; \qquad (57)$$

*the function $g_n = g$ is the unique solution to the **Painlevé V equation**, with the
following initial condition:*

$$\begin{cases} g'' - \frac{g'^2}{2}\left(\frac{1}{g-1} + \frac{1}{g}\right) + \frac{g'}{u} + \frac{2}{u}g(g-1) - \frac{n^2}{2u^2}\frac{g-1}{g} = 0 \\ \text{with } g_n(u) = 1 - \frac{u^n}{(n!)^2} + O(u^{n+1}), \quad \text{near } u = 0. \end{cases} \qquad (58)$$

Proof For the sake of this proof, consider the locus

$$\mathcal{L} = \{\text{all } t_i = s_i = 0, \text{ except } t_1, s_1 \ne 0\}.$$

From (4), we have on $\mathcal{L}$,

$$0 = \frac{\mathbb{V}_0 \tau_n}{\tau_n}\bigg|_{\mathcal{L}} = \left(t_1\frac{\partial}{\partial t_1} - s_1\frac{\partial}{\partial s_1}\right) \log \tau_n\bigg|_{\mathcal{L}}$$

$$0 = \frac{\mathbb{V}_0 \tau_n}{\tau_n} - \frac{\mathbb{V}_0 \tau_{n-1}}{\tau_{n-1}}\bigg|_{\mathcal{L}} = \left(t_1\frac{\partial}{\partial t_1} - s_1\frac{\partial}{\partial s_1}\right) \log \frac{\tau_n}{\tau_{n-1}}\bigg|_{\mathcal{L}}$$

$$0 = \frac{\partial}{\partial t_1}\frac{\mathbb{V}_{-1}\tau_n}{\tau_n}\bigg|_{\mathcal{L}} = \left(-s_1\frac{\partial^2}{\partial s_2 \partial t_1} + n\frac{\partial^2}{\partial t_1 \partial s_1}\right) \log \tau_n\bigg|_{\mathcal{L}} + n.$$

Then combining with identities (18) and (17), one finds after some computations that

$$g_n(x) = -\frac{\partial^2}{\partial t_1 \partial s_1} \log \tau_n(t, s)\Big|_{\mathcal{L}} = \frac{d}{dx} x \frac{d}{dx} \left(\log \tau_n(t, s)\Big|_{\substack{t_i = \delta_{i0}\sqrt{x} \\ s_i = -\delta_{i0}\sqrt{x}}}\right) \quad (59)$$

satisfies equation (58). The initial condition follows from the combinatorics. ∎

4.3   Backward and Forward Equation for Nonintersecting Random Walks

Consider the n random walkers, walking in k steps from $x = (x_1 < x_2 < \cdots < x_n)$ to $y = (y_1 < y_2 < \cdots < y_n)$, as introduced in Section 3.4. These data define difference operators[15] for $k, n \in \mathbb{Z}_+, x, y \in \mathbb{Z}$,

$$\mathcal{A}_1 := \sum_{i=1}^{n} \left(\frac{k}{2n}\Lambda_k^{-1}\partial_{2y_i}^+ + x_i\partial_{x_i}^- + \partial_{y_i}^+ y_i - (x_i - y_i)\right)$$

$$\mathcal{A}_2 := \sum_{i=1}^{n} \left(\frac{k}{2n}\Lambda_k^{-1}\partial_{2x_i}^+ + y_i\partial_{y_i}^- + \partial_{x_i}^+ x_i - (y_i - x_i)\right) \quad (61)$$

With these definitions, we have

Theorem 11   [13] *The probability*

$$P(k; x, y) = \frac{b_{xy}^{(k)}}{(2n)^k} = P \begin{pmatrix} \text{that n non-intersecting walkers in } \mathbb{Z} \text{ move during} \\ k \text{ instants from } x_1 < x_2 < \cdots < x_n \text{ to } y_1 < y_2 \\ < \cdots < y_n, \text{ where at each instant exactly one} \\ \text{walker moves either one step to the left, or one} \\ \text{step to the right} \end{pmatrix}$$

$$(62)$$

satisfies both a forward and backward random walk equation,

$$\mathcal{A}_i P(k, x, y) = 0, \quad (63)$$

Remark   "Forward and backward," because $\mathcal{A}_1$ essentially involves the end points y, whereas $\mathcal{A}_2$ involves the initial points x.

---

[15] in terms of difference operators, acting on functions $f(k, x, y)$, with $k \in \mathbb{Z}_+, x, y \in \mathbb{Z}$:

$$\partial_{\alpha x_i}^+ f := f(k, x + \alpha e_i, y) - f(k, x, y)$$
$$\partial_{\alpha x_i}^- f := f(k, x, y) - f(k, x - \alpha e_i, y)$$
$$\Lambda_k^{-1} f := f(k - 1, x, y). \quad (60)$$

Proof The unitary integral below is obtained from the integral $\tau_n^0(t, s)$, appearing in (1), by means of the shifts $t_1 \mapsto t_1 + z, s_1 \mapsto s_1 - z$. Thus it satisfies the Virasoro constraints for $k = -1, 0, 1$, with the same shifts inserted. This integral has a double Fourier expansion in Schur polynomials; see (48). So we have, with $\mathbb{V}_k$ defined in (4),

$$
0 = \mathbb{V}_k \Big|_{\substack{t_1 \mapsto t_1 + z \\ s_1 \mapsto s_1 - z}} \int_{U(n)} e^{z \operatorname{Tr}(M + \bar{M})} e^{\sum_1^\infty \operatorname{Tr}(t_i M^i - s_i \bar{M}^i)} dM
$$

$$
= \mathbb{V}_k \Big|_{\substack{t_1 \mapsto t_1 + z \\ s_1 \mapsto s_1 - z}} \sum_{\substack{\lambda, \mu \text{ such that} \\ \lambda_1^\top, \mu_1^\top \le n}} a_{\lambda\mu}(z) s_\lambda(t) s_\mu(-s)
$$

$$
\overset{*}{=} \sum_{\substack{\lambda_1^\top \le n \\ \mu_1^\top \le n}} s_\lambda(t) s_\mu(-s) \mathcal{L}(a_{\lambda\mu}(z)),
$$

To explain the equality $\overset{*}{=}$ above, notice the Virasoro constraints $\mathbb{V}_k$ act on the terms $s_\lambda(t)s_\mu(-s)$ in the expansion. Since the constraints (4) decouple as a sum of a t-part and an s-part, it suffices to show $V_k(t)s_\lambda(t)$ can be expanded in a Fourier series in $s_\mu(t)$'s; this is done below. Therefore $\mathbb{V}_k s_\lambda(t)s_\mu(-s)$ can again be expanded in double Fourier series, yielding new coefficients $\mathcal{L}(a_{\lambda\mu}(z))$, depending linearly on the old ones $a_{\lambda\mu}(z)$. Thus we must compute $V_k(t)s_\lambda(t)$ for

$$
V_k(t) = \frac{1}{2} \sum_{i+j=k} \frac{\partial^2}{\partial t_i \partial t_j} + \sum_{-i+j=k} i t_i \frac{\partial}{\partial t_j} + \frac{1}{2} \sum_{-i-j=k} (i t_i)(j t_j). \tag{64}
$$

This will generalize the Murnaghan-Nakayama rules,

$$
n t_n \, s_\lambda(t) = \sum_{\substack{\mu \\ \mu \backslash \lambda \in B(n)}} (-1)^{\operatorname{ht}(\mu \backslash \lambda)} s_\mu(t)
$$

$$
\frac{\partial}{\partial t_n} s_\lambda(t) = \sum_{\substack{\mu \\ \lambda \backslash \mu \in B(n)}} (-1)^{\operatorname{ht}(\lambda \backslash \mu)} s_\mu(t). \tag{65}
$$

To explain the notation, $b \in B(i)$ denotes a border-strip (i.e., a connected skew-shape $\lambda \backslash \mu$ containing i boxes, with no 2×2 square) and the height ht b of a border strip b is defined as

$$
\operatorname{ht} b := \#\{\text{rows in } b\} - 1. \tag{66}
$$

Indeed in [14] it is shown that

$$
V_{-n} s_\lambda(t) = \sum_{\substack{\mu \\ \mu \backslash \lambda \in B(n)}} d_{\lambda\mu}^{(n)} s_\mu(t) \quad \text{and} \quad V_n s_\lambda(t) = \sum_{\substack{\mu \\ \lambda \backslash \mu \in B(n)}} d_{\mu\lambda}^{(n)} s_\mu(t) \tag{67}
$$

with the same precise sum, except the coefficients are different: $(n \geq 1)$

$$
d_{\lambda\mu}^{(n)} = \sum_{i \geq 1} \sum_{\substack{\nu \text{ such that} \\ \left\{\begin{array}{l} \lambda\backslash\nu \in B(i) \\ \mu\backslash\nu \in B(n+i) \\ \lambda\backslash\nu \subset \mu\backslash\nu \end{array}\right\}}} (-1)^{\mathrm{ht}(\lambda\backslash\nu)+\mathrm{ht}(\mu\backslash\nu)}
$$

$$
+ \frac{1}{2} \sum_{i=1}^{n} \sum_{\substack{\nu \text{ such that} \\ \left\{\begin{array}{l} \nu\backslash\lambda \in B(i) \\ \mu\backslash\nu \in B(n-i) \end{array}\right\}}} (-1)^{\mathrm{ht}(\nu\backslash\lambda)+\mathrm{ht}(\mu\backslash\nu)}. \tag{68}
$$

In view of the infinite sum in the Virasoro generators (64), one would expect $V_n s_\lambda$ to be expressible as an *infinite sum of Schur polynomials*. This is not so: acting with Virasoro V_n leads to the same precise sum as acting with $n t_n$ (resp. $\partial/\partial t_n$), except the coefficients in (67) are different from the ones in (65). This is to say *the two operators have the same band structure or locality*! Then setting

$$
a_{\lambda\mu}(z) = \sum_{k \geq 0} b_{xy}^{(k)} \frac{z^k}{K!},
$$

leads to the result (63), upon remembering the relation (47) between the λ, μ's and the x,y's.   ∎

REFERENCES

1. van Moerbeke, P. (2001) Integrable lattices: random matrices and random permutations, in: *Random Matrices and their Applications*, Mathematical Sciences research Institute Publications #40, Cambridge University Press, pp. 321–406. http://www.msri.org/publications/books/Book40/files/moerbeke.pdf
2. Adler, M. and van Moerbeke, P. (2000) Integrals over classical groups, random permutations, Toda and Toeplitz lattices, *Comm. Pure Appl. Math.* **54**, pp. 153–205, (arXiv: math.CO/9912143).
3. Gessel, I. M. (1990) Symmetric functions and P-recursiveness, *J. of Comb. Theory*, Ser. A, **53**, pp. 257–285.
4. Tracy, C. A. and Widom, H. (2001) On the distributions of the lengths of the longest monotone subsequences in random words. *Probab. Theory Related Fields,* **119**(3), pp. 350–380. (arXiv:math.CO/9904042).
5. Rains, E. M. (1998) *Increasing subsequences and the classical groups*, Elect. J. of Combinatorics, **5**, R12.
6. Adler, M. and van Moerbeke, P. (2004) Integrals over Grassmannians and Random permutations, *Adv. in Math.* **181**, pp. 190–249. (arXiv: math.CO/0110281).

7. MacDonald, I. G. (1995) "Symmetric functions and Hall polynomials", Clarendon Press.

8. Sagan, B. E. (1991) The Symmetry Group, Wadsworth & Brooks, Pacific Grove, California.

9. Stanley, R. S. (1999) *Enumerative Combinatorics*, Vol. 2, Cambridge studies in advanced mathematics **62**, Cambridge University Press.

10. Stanton, D. and White, D. (1986) Constructive Combinatorics, Springer-Verlag, NY.

11. Borodin, A. and Olshanski, G. (2001) *Z-measures on partitions, Robinson-Shensted-Knuth correspondence, and $\beta = 2$ random matrix ensembles*, Random matrix models and their applications, pp. 71–94, *Math. Sci. Res. Inst. Publ.*, **40**, Cambridge University Press, Cambridge.

12. Okounkov, A. (2001) Infinite wedge and measures on partitions, *Selecta Math.* **7**, pp. 57–81. (arXiv: math.RT/9907127)

13. Adler, M. and van Moerbeke, P. (2004) Virasoro action on Schur function expansions, skew Young tableaux and random walks, *Comm. Pure Appl. Math.* (ArXiv: math.PR/0309202)

14. Forrester, P. J. (2001) Random walks and random permutations, *J. Phys.* A **34**(31), pp. L417–L423. (arXiv:math.CO/9907037).

15. Grabiner, D. J. and Magyar, Peter (1993) Random walks in Weyl chambers and the decomposition of tensor powers. *J. Algebraic Combin.* **2**(3), pp. 239–260.

16. Grabiner, D. J. (2002) Random walk in an alcove of an affine Weyl group, and non-colliding random walks on an interval, *J. Combin. Theory,* Ser. A **97**(2), pp. 285–306.

17. Baik, J. (2000) Random vicious walks and random matrices. *Comm. Pure Appl. Math.* **53**, pp. 1385–1410.

18. McMillan, E. M. (1971) *A problem in the stability of periodic systems*, Topics in Modern Physics, A tribute to E. U. Condon, eds. W. E. Brittin and H. Odabasi (Colorado Ass. Univ. Press, Boulder), pp. 219–244.

19. Borodin, A. (2003) Discrete gap probabilities and discrete Painlevé equations, *Duke Math. J.* **117**, pp. 489–542. (arXiv:math-ph/0111008).

20. Baik, J. (2003) Riemann-Hilbert problems for last passage percolation. Recent developments in integrable systems and Riemann-Hilbert problems (Birmingham, AL, 2000), pp. 1–21, *Contemp. Math.*, 326, *Amer. Math. Soc.*, Providence, RI. (arXiv:math.PR/0107079).

21. Suris, Yu. B. (1987) Integrable mappings of standard type, *Funct. Anal. Appl.* **23**, pp. 74–76.

22. Grammaticos, B. Nijhoff, F., and Ramani, A. (1999) Discrete Painlevé equations, The Painlevé property, *CRM series in Math. Phys.*, Chapter 7, Springer, New York, 1999, pp. 413–516.

23. Tracy, C. A. and Widom, H. (1999) Random unitary matrices, permutations and Painlevé, *Commun. Math. Phys.* **207**, pp. 665–685.

24. Adler, M. and van Moerbeke, P. (2003) Recursion relations for Unitary integrals of Combinatorial nature and the Toeplitz lattice, *Comm. Math. Phys.* **237**, pp. 397–440. (arXiv: math-ph/0201063).

25. Borodin, A. and Okounkov, A. (2000) A Fredholm determinant formula for Toeplitz determinants, *Integral Equations Operator Theory* **37**, pp. 386–396. (math.CA/9907165).
26. Borodin, A. and Olshanski, G. (2000) Distributions on partitions, point processes, and the hypergeometric kernel, (*Comm. Math. Phys.* **211**(2), pp. 335–358. (math.RT/9904010).
27. Borodin, A. and Deift, P. (2002) Fredholm determinants, Jimbo-Miwa-Ueno tau-functions, and representation theory, *Comm. Pure Appl. Math.* **55**, pp. 1160–1230. (arXiv:math-ph/0111007).
28. Gessel, I. M. (1990) Symmetric functions and P-recursiveness, *J. of Comb. Theory*, Ser. A, **53**, pp. 257–285.
29. Gessel, I. M. and Zeilberger, Doron (1992) Random walk in a Weyl chamber, *Proc. Amer. Math. Soc.* **115**(1), pp. 27–31.

ON REDUCTIONS OF SOME KdV-TYPE SYSTEMS AND THEIR LINK TO THE QUARTIC HÉNON-HEILES HAMILTONIAN

C. Verhoeven and M. Musette
Dienst Theoretische Natuurkunde, Vrije Universiteit Brussel
Pleinlaan 2, B-1050 Brussels, Belgium

R. Conte
Service de physique de l' état condensé (URA 2464)
CEA–Saclay, F-91191 Gif-sur-Yvette Cedex, France

Abstract   A few $2 + 1$-dimensional equations belonging to the KP and modified KP hierarchies are shown to be sufficient to provide a unified picture of all the integrable cases of the cubic and quartic Hénon–Heiles Hamiltonians.

1   INTRODUCTION

The Hénon–Heiles (HH) Hamiltonian [1] with a generalized cubic potential is defined as

$$\text{HH3}: H = \frac{1}{2}(p_1^2 + p_2^2 + c_1 q_1^2 + c_2 q_2^2) + \alpha q_1 q_2^2 - \frac{\beta}{3} q_1^3 + \frac{c_3}{q_2^2}, \quad (1)$$

in which α, β, c_1, c_2, c_3 are constants.

The corresponding equations of motion pass the Painlevé test for only three sets of values of the ratio β/α, which are also the only three cases for which an additional first integral K has been found [2–4]. These three cases have been integrated [5, 6] with genus-two hyperelliptic functions. Moreover, they are equivalent [7] to the stationary reduction of three fifth-order soliton equations, called fifth-order Korteweg de Vries (KdV$_5$), Sawada–Kotera (SK), and Kaup–Kupershmidt (KK) equations, belonging respectively to the KP, BKP, and CKP hierarchies whose Hirota bilinear forms can be found in [8].

L. Faddeev et al. (eds.),
Bilinear Integrable Systems: From Classical to Quantum, Continuous to Discrete, 363–374.
© 2006 Springer. Printed in the Netherlands.

If the potential is taken as the most general cubic polynomial in (q_1, q_2), there exists a fourth Liouville integrable case,

$$V = q_1^3 + \frac{1}{2}q_2^2 q_1 + \frac{i}{6\sqrt{3}}q_2^3, \tag{2}$$

detected by Ramani et al. [9], but up to now its general solution is unknown.

Another Hénon–Heiles-type Hamiltonian with an extended quartic potential has been considered,

$$\text{HH4}: H = \frac{1}{2}(P_1^2 + P_2^2 + aQ_1^2 + bQ_2^2) + CQ_1^4 + BQ_1^2 Q_2^2 + AQ_2^4$$
$$+ \frac{1}{2}\left(\frac{\alpha}{Q_1^2} + \frac{\beta}{Q_2^2}\right) + \mu Q_1, \tag{3}$$

in which $A, B, C, \alpha, \beta, \mu, a, b$ are constants. Again, the equations of motion pass the Painlevé test for only four values of the ratios $A : B : C$ [9–11], which happen to be the only known cases of Liouville integrability. However, it is not yet completely settled whether, in all four cases, the quartic Hamiltonian (3) displays the same pattern as the cubic Hamiltonian (1), i. e.

the equations of motion can be integrated with hyperelliptic functions of genus two,

there exists an equivalence with the stationary reduction of some partial differential equation (PDE) belonging to the KP, BKP, and CKP hierarchies.

In this paper, we first summarize the results already established for the systems (1) and (3). We then establish new links between the coupled KdV (c-KdV) systems considered in [12] and some other ones [8, 13, 14] belonging to the BKP and CKP hierarchies. These links could be useful to find the explicit general solution without any restriction on the parameters other than those generated by the Painlevé test.

2 ALREADY INTEGRATED CASES

The four cases for which the quartic Hamiltonian passes the Painlevé test are,

1. $A : B : C = 1 : 2 : 1$, $\mu = 0$. The system is then equivalent to the stationary reduction of the Manakov system [15] of two coupled nonlinear Schrödinger (NLS) equations and has been integrated [16] with genus two hyperelliptic functions.
2. $A : B : C = 1 : 6 : 1$, $a = b$, $\mu = 0$,
3. $A : B : C = 1 : 6 : 8$, $a = 4b$, $\alpha = 0$,
4. $A : B : C = 1 : 12 : 16$, $a = 4b$, $\mu = 0$.

Table 1. All the cases of HH3 and HH4 which pass the Painlevé test, with the extra terms c_3 or α, β, μ. First column indicates the cubic or quartic case. Second column is the value of β/α (if cubic) or the ratio $A : B : C$ (quartic), followed by the values selected by the Painlevé test. Third column indicates the polynomial degree of the additional constant of the motion K in the momenta (p_1, p_2). Next column displays the PDE system connected to the HH case. Last column shows the reference to the general solution and the not yet integrated cases. When the general solution is known, it is a single-valued rational function of genus-two hyperelliptic functions

HH	case	deg K	PDE	General solution
3	$-1, c_1 = c_2$	4	SK	[6]
3	$-6, c_1, c_2$ arb.	2	KdV$_5$	[5]
3	$-16, c_1 = 16c_2$	4	KK	[6]
4	$1:2:1$ $\mu = 0$	2	c-NLS	[16]
4	$1:6:1$ $a = b, \mu = 0$	4	c-KdV2, Lax order 4	$\alpha = \beta$ [18], $\alpha \neq \beta$?
4	$1:6:8$ $a = 4b, \alpha = 0$	4	c-KdV1, Lax order 4	$\beta_\mu = 0$ [18], $\beta_\mu \neq 0$?
4	$1:12:16$ $a = 4b, \mu = 0$	4	c-KdVb Lax order 5	$\alpha\beta = 0$ [19], $\alpha\beta \neq 0$?

Each of the last three cases is equivalent [12] to the stationary reduction of a coupled KdV system possessing a fourth or fifth order Lax pair. Canonical transformations have been found [12,17] which allow us [18,19] to define the separating variables of the Hamilton–Jacobi equation, however with additional restrictions on α, β, μ, as showed in Table 1.

3 LINK BETWEEN KP HIERARCHIES AND INTEGRABLE HH CASES

Let us consider the following three systems of the KP and modified KP hierarchies [8],

$$\begin{cases} \left(D_1^4 - 4D_1 D_3 + 3D_2^2\right) (\tau_0 \cdot \tau_0) = 0, \\ \left((D_1^3 + 2D_3) D_2 - 3D_1 D_4\right) (\tau_0 \cdot \tau_0) = 0, \end{cases} \tag{4}$$

$$\begin{cases} \left(D_1^4 - 4D_1 D_3 + 3D_2^2\right) (\tau_0 \cdot \tau_0) = 0, \\ \left(D_1^6 - 20D_1^3 D_3 - 80D_3^2 + 144D_1 D_5 - 45D_1^2 D_2^2\right) (\tau_0 \cdot \tau_0) = 0, \end{cases} \tag{5}$$

$$\begin{cases} \left(D_1^2 + D_2\right) (\tau_0 \cdot \tau_1) = 0, \\ \left(D_1^6 - 20D_1^3 D_3 - 80D_3^2 + 144D_1 D_5 \right. \\ \left. \quad + 15 \left(4D_1 D_3 - D_1^4\right) D_2\right) (\tau_0 \cdot \tau_1) = 0, \end{cases} \tag{6}$$

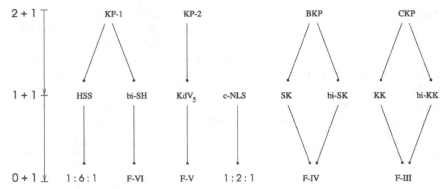

Figure 1. Reductions from (2+1)-dimensional PDEs to (1+1)-dimensional PDEs, then to ODEs (the notation *F-xxx* denotes the autonomous case of the ODE denoted *F-xxx* in citeCos2000a) or to Hamiltonian systems. The symbol c-NLS represents the Manakov cite Manakov 1973 system of two coupled NLS equations

in which the subscripts of the bilinear operators correspond to the components of the vector $\vec{x} = (x_1, x_2, \ldots, x_n)$, while τ_0 and τ_1 are functions of $\vec{x}$. By further putting some symmetry constraint on τ_0 and τ_1, let us define as follows four (2+1)-dimensional PDEs (see line "2+1" in Figure 1).

1. With the system (4), one defines by $D_4 = 0$ [20] the (2+1)-dim PDE labeled "KP-1" in Figure 1.
2. With the system (5), one defines by $D_2 = 0$ the (2+1)-dim PDE-labeled "KP-2" in Figure 1.
3. With the system (6) and the B_∞ symmetry constraint [8, p. 968]

$$\begin{cases} \tau_0(x) = f(x_{\text{odd}}) + x_2\, g(x_{\text{odd}}) + \frac{1}{2}x_2^2 h_1(x_{\text{odd}}) + x_4 h_2(x_{\text{odd}}) + \cdots, \\[2mm] \tau_1(x) = f(x_{\text{odd}}) - x_2\, g(x_{\text{odd}}) + \frac{1}{2}x_2^2 h_1(x_{\text{odd}}) - x_4 h_2(x_{\text{odd}}) + \cdots, \end{cases} \tag{7}$$

one defines the (2+1)-dim BKP equation

$$9z_{x_1,x_5} - 5z_{2x_3}$$
$$+ \left(z_{5x_1} + 15z_{x_1} z_{3x_1} + 15(z_{x_1})^3 - 5z_{2x_1,x_3} - 15z_{x_1} z_{x_3}\right)_{x_1} = 0, \tag{8}$$

in which $z = \partial_{x_1} \log \tau_0(\vec{x})|_{x_2=x_4=\cdots=0}$ and $z_{2x_3} \equiv z_{x_3x_3\cdots}$.

4. With (5) and the C_∞ symmetry constraint [8, p. 968]

$$\tau_0(x) = f(x_{\text{odd}}) + \frac{1}{2}x_2^2\, g(x_{\text{odd}}) + \frac{1}{2}x_4^2\, h(x_{\text{odd}}) + \ldots, \tag{9}$$

one defines the (2+1)-dim CKP equation

$$9z_{x_1,x_5} - 5z_{2x_3} + \left(z_{5x_1} + 15z_{x_1}z_{3x_1} \right.$$

$$\left. + 15(z_{x_1})^3 - 5z_{2x_1,x_3} - 15z_{x_1}z_{x_3} + \frac{45}{4}(z_{2x_1})^2 \right)_{x_1} = 0, \qquad (10)$$

in which $z = \partial_{x_1} \operatorname{Log} \tau_0(\vec{x})|_{x_2=x_4=\cdots=0}$.

Next, from these $(2+1)$-dimensional PDEs, one performs the following natural reductions to $(1+1)$-dimensional PDEs (see line "1 + 1" in Figure 1).

1. In KP-1, the C_∞ symmetry constraint (9) defines

$$\begin{cases} (D_1^4 - 4D_1D_3)(f \cdot f) + 6fg = 0, \\ (D_1^3 + 2D_3)(f \cdot g) = 0, \end{cases} \qquad (11)$$

which we call bi-SH [20] for reasons explained in next section.
2. In KP-1, the constraint

$$\tau_0(x) = f(x_{\text{odd}}) + x_2 g(x_{\text{odd}}) + \frac{1}{2}x_2^2 h_1(x_{\text{odd}}) + x_4 h_2(x_{\text{odd}}) + \cdots, \qquad (12)$$

defines

$$\begin{cases} (D_1^4 - 4D_1D_3)(f \cdot f) - 6g^2 = 0, \\ (D_1^3 + 2D_3)(f \cdot g) = 0, \end{cases} \qquad (13)$$

which is called coupled KdV system of Hirota–Satsuma (HSS) [20].
3. In KP-2, the elimination of x_3 [8, p. 962] yields the potential KdV$_5$ equation

$$z_t + z_{xxxxx} + 5z_{xx}^2 + 10z_x z_{xxx} + 10z_x^3 = 0, \qquad (14)$$

with the notation $x \equiv x_1, t \equiv -x_5/16, z = 2\partial_x \log \tau_0$.
4. In BKP (8), the reduction $z_{x_3} = 0$ defines the potential SK equation [21]

$$z_t + z_{xxxxx} + 15z_x z_{xxx} + 15(z_x)^3 = 0, \qquad (15)$$

with the notation $x_5 \equiv 9t, x_1 \equiv x$.
5. In BKP (8), the reduction $z_{x_5} = 0$ defines the 1+1-dimensional bi-SK or Ramani equation [22]

$$\left(z_{xxxxx} + 15z_x z_{xxx} + 15(z_x)^3 - 15z_x z_t - 5z_{xxt} \right)_x - 5z_{tt} = 0, \qquad (16)$$

with the notation $x_3 \equiv t, x_1 \equiv x$.

6. In CKP (10), the reduction $\partial_{x_3}\tau_0 = 0$ defines the fifth-order potential KK equation [23]

$$z_t + z_{xxxxx} + 15z_x z_{xxx} + 15(z_x)^3 + \frac{45}{4}(z_{xx})^2 = 0, \qquad (17)$$

with the notation $x_1 \equiv x$, $x_5 \equiv 9t$.

7. In CKP (10), the reduction $\partial_{x_5}\tau_0 = 0$ defines the sixth-order bi-KK equation [24]

$$\left(z_{xxxxx} + 15z_x z_{xxx} + 15(z_x)^3 - 15z_x z_t - 5z_{xxt} + \frac{45}{4}(z_{xx})^2 \right)_x - 5z_{tt} = 0, \qquad (18)$$

with the notation $x_1 \equiv x$, $x_3 \equiv t$.

Finally, the stationary reduction $(x, t) \to x - ct$ of these $(1+1)$-dimensional PDEs leads directly to the Hamiltonian systems or the ODE listed in the line "$0 + 1$" of Figure 1.

The four ODEs F-III, F-IV, F-V, F-VI have a single-valued general solution, obtained by the Jacobi postmultiplier method [24], which is expressed with genus-two hyperelliptic functions. Three of them (F-III, F-IV, F-V), which are the stationary reductions of respectively (17), (15), and (14), have been shown [7] to have a one-to-one correspondence with the q_1 component of the three integrable cases of HH3. Therefore the chain of reductions generated from the systems (5) and (6) contains the full information for the integration of HH3.

Let us now show that Figure 1 also contains the full information for the integration of HH4. This will involve two kinds of coupled KdV (c-KdV) systems: some with a fourth-order Lax pair, some with a fifth-order Lax pair.

4   LINK OF COUPLED KdV SYSTEMS WITH HH4

In the variables $u = \partial_x^2 \log f$, $v = 4g/f$, the bilinear system (11) is rewritten as the c-KdV system [8, 13, 20, 26]

$$\begin{cases} -4u_t + \left(6u^2 + u_{xx} + 3v \right)_x = 0, \\ 2v_t + 6uv_x + v_{xxx} = 0, \end{cases} \qquad (19)$$

with the notation $x_3 \equiv t$, $x_1 \equiv x$. This system possesses the fourth-order Lax pair [20]

$$\begin{cases} \left(\partial_x^4 + 4u\,\partial_x^2 + 4u_x\,\partial_x + 2u_{xx} + 4u^2 + v \right)\psi = \lambda\psi, \\ \left(\partial_x^3 + 3u\,\partial_x + \frac{3}{2}u_x \right)\psi = \partial_t\psi, \end{cases} \qquad (20)$$

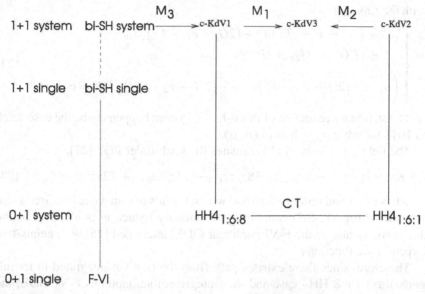

Figure 2. Path from an already integrated ODE (autonomous F-VI) to the quartic cases 1:6:1 and 1:6:8. All $1 + 1$-dimensional systems involved (on the top line) have fourth-order Lax pairs. The dashed vertical line from the level "$1 + 1$-system" to the level "$1 + 1$-single" represents the elimination of one dependent variable. All the other vertical lines represent the stationary reduction. The horizontal lines represent Miura transformations at the level "$1 + 1$-system" and canonical transformations at the Hamiltonian level "$0 + 1$-system." The systems are defined as (22) for c-KdV$_1$, (13) for c-KdV$_2$, cite [p. 79] Baker Thesis for c-KdV$_3$, (19) for the bi-SH system. The Miura maps M$_1$, M$_2$ can be found in cite(Eq. (5.3)) Baker Thesis and cite (Eq. (5.8)) Baker Thesis

Under the Miura transformation denoted M_3 in Figure 2

$$\begin{cases} 4u = 2G - F_x - F^2, \\ 2v = 2F_{xxx} + 4FF_{xx} + 8GF_x + 4FG_x + 3F_x^2 - 2F^2F_x - F^4 + 4GF^2, \end{cases}$$

(21)

the system (19) is mapped to the following c-KdV system (denoted c-KdV$_1$ in Figure 2) given in [12,17]:

$$\begin{cases} 4F_t = \left(-2F_{xx} - 3FF_x + F^3 - 6FG \right)_x, \\ 8G_t = 2G_{xxx} + 12GG_x + 6FG_{xx} + 12GF_{xx} + 18F_xG_x - 6F^2G_x \\ \qquad + 3F_{xxxx} + 3FF_{xxx} + 18F_xF_{xx} - 6F^2F_{xx} - 6FF_x^2, \end{cases}$$

(22)

with the Lax pair

$$
\begin{cases}
\big(\partial_x^4 + (2G - F_x - F^2)\partial_x^2 + (2G - F_x - F^2)_x \partial_x \\
\quad + (FG)_x + G_{xx} + G^2\big)\psi = \lambda\psi, \\
\Big(\partial_x^3 + \dfrac{3}{4}(2G - F_x - F^2)\partial_x + \dfrac{3}{8}(2G - F_x - F^2)_x\Big)\psi = \partial_t\psi,
\end{cases}
\tag{23}
$$

The stationary reduction of this c-KdV$_1$ system happens to be the case 1:6:8 of HH4 for arbitrary values of (β, μ).

The field $z = \int u\,dx$ of (19) satisfies the sixth-order PDE [27],

$$
-8z_{tt} + z_{xxxxxx} - 2z_{xxxt} + 18z_x z_{xxxx} + 36z_{xx} z_{xxx} + 72z_x^2 z_{xx} = 0,
\tag{24}
$$

which is of second order in time and which for this reason we call bidirectional Satsuma-Hirota (bi-SH) equation. Its stationary reduction is identical to the autonomous case of the F-VI nonlinear ODE, integrated [25] with genus-two hyperelliptic functions.

Therefore, since there exists a path from the (not yet integrated in its full generality) 1:6:8 HH4 case and the (integrated) autonomous F-VI ODE, the general solution of the 1:6:8 can in principle be obtained, this will be addressed in future work.

All the links between the system (19) and other c-KdV systems considered by S. Baker and which reduce to the integrable cases 1:6:1 and 1:6:8 of HH4 are displayed in Figure 2.

Finally, let us explain the link between the 1:12:16 integrable case of HH4 and two c-KdV systems possessing a fifth-order Lax pair, systems respectively equivalent to the bi-SK equation (16) and the bi-KK equation (18).

The following coupled system [13]:

$$
\begin{cases}
u_t = \Big(-2au_{xx} - bu^2 + \dfrac{9a^2}{5b}v\Big)_x, \\
v_t = av_{xxx} - bu_{xxxxx} - \dfrac{5b^2}{3a}uu_{xxx} - \dfrac{5b^2}{3a}u_x u_{xx} + buv_x - bu_x v,
\end{cases}
\tag{25}
$$

where a, b are nonzero constants, arises from the compatibility condition of the fifth-order Lax pair

$$
\begin{cases}
\Big(\partial_x^5 + \dfrac{5b}{3a}u\partial_x^3 + \dfrac{5b}{3a}u_x\partial_x^2 + v\partial_x\Big)\varphi = \lambda\varphi, \\
(a\partial_x^3 + bu\partial_x)\varphi = \partial_t\varphi.
\end{cases}
\tag{26}
$$

The field $z = \int u\,dx$ of (25) satisfies the sixth-order PDE

$$
5z_{tt} + \Big(5z_{xxt} + 5\dfrac{b}{a}z_t z_x - az_{xxxxx} - 5bz_x z_{xxx} - \dfrac{5b^2}{3a}z_x^3\Big)_x = 0,
\tag{27}
$$

identical to the bi-SK equation (16) for $a = 1, b = 3$.

Similarly, the coupled system [13]

$$
\begin{cases}
u_t = \left(-\dfrac{7}{2}au_{xx} - bu^2 + \dfrac{9a^2}{5b}v\right)_x, \\[2mm]
v_t = \dfrac{5}{2}av_{xxx} - \dfrac{19}{4}bu_{xxxxx} - \dfrac{25b^2}{6a}uu_{xxx} - \dfrac{5b^2}{a}u_x u_{xx} \\[2mm]
\qquad + buv_x - bu_x v,
\end{cases}
\tag{28}
$$

arises from the compatibility condition of the other fifth-order Lax pair

$$
\begin{cases}
\left(\partial_x^5 + \dfrac{5b}{3a}u\,\partial_x^3 + \dfrac{5b}{2a}u_x\,\partial_x^2 + v\,\partial_x + \dfrac{1}{2}v_x - \dfrac{5b}{12a}u_{xxx}\right)\varphi = \lambda\varphi, \\[3mm]
\left(a\,\partial_x^3 + bu\,\partial_x + \dfrac{b}{2}u_x\right)\varphi = \partial_t\varphi.
\end{cases}
\tag{29}
$$

The field $z = \int u\,dx$ satisfies the sixth-order PDE

$$
5z_{tt} + a\left(5z_{xxxt} + 5\dfrac{b}{a}z_t z_x - az_{xxxxxx} - 5bz_x z_{xxx} - \dfrac{5b^2}{3a}z_x^3 - \dfrac{15b}{4}z_{xx}^2\right)_x = 0,
\tag{30}
$$

identical to the potential bi-KK equation (18) for $a = 1, b = 3$. The property of these two systems which is of interest to us is the existence of two mappings, respectively (setting $a = 5$), for the system (25) the Miura transformation denoted M_b

$$
\begin{cases}
u_{\text{bi-SK}} = \dfrac{3}{b}\left(2G + 3F_x - F^2\right), \\[3mm]
v_{\text{bi-SK}} = F_{xxx} + G_{xx} - FF_{xx} + GF_x - FG_x + G^2,
\end{cases}
\tag{31}
$$

and, for the system (28), the transformation denoted M_a

$$
\begin{cases}
u_{\text{bi-KK}} = \dfrac{3}{b}\left(2G - 2F_x - F^2\right), \\[3mm]
v_{\text{bi-KK}} = -F_{xxx} + 3G_{xx} - FF_{xx} + 2FG_x - F_x^2 + G^2,
\end{cases}
\tag{32}
$$

to a common coupled KdV-type system [17, p. 65] (denoted c-KdV$_a$ in Figure 3)

$$
\begin{cases}
F_t = \left(-7F_{xx} - 3G_x - 3FF_x - 9FG + 2F^3\right)_x, \\[2mm]
G_t = 3F_{xxxx} + 2G_{xxx} + 3FG_{xx} - 3F^2F_{xx} - 3F^2G_x - 3FF_x^2 \\[2mm]
\qquad + 3FGF_x + 9F_xF_{xx} + 9F_xG_x + 3GF_{xx} + 3GG_x.
\end{cases}
\tag{33}
$$

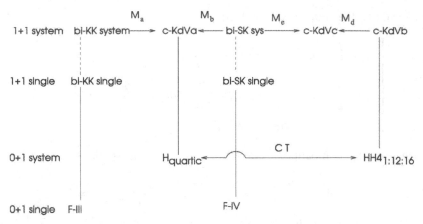

Figure 3. Path from an already integrated ODE (autonomous F-III or F-IV, which are ODEs for q_1 in the HH3-KK and HH3-SK cases) to the quartic case 1:12:16. All 1 + 1-dimensional systems involved (on the top line) have fifth-order Lax pairs. The dashed vertical line from the level "1 + 1-system" to the level "1 + 1-single" represents the elimination of one dependent variable. All the other vertical lines represent the stationary reduction. The horizontal lines represent Miura transformations at the level "1 + 1-system" and birational canonical transformations at the Hamiltonian level "0 + 1-system." The Miura maps M_a, M_b, are given in the text, M_d is given in cite [p. 95] Baker Thesis, $M_e = M_b M_c$, in which M_c is the Miura transformation from c-KdVa to c-KdVc given in cite [p. 95] Baker Thesis. The systems are defined as cite (Eq. (6.9)) Baker Thesis for c-KdVb, and cite [p. 95] Baker Thesis for c-KdVc

This system also possesses a fifth-order Lax pair, which can be written in two different ways, either

$$\begin{cases} \left(\partial_x^2 + F\partial_x + F_x + G\right)\partial_x \left(\partial_x^2 - F\partial_x + G\right)\varphi = \lambda\varphi, \\ \left(5\partial_x^3 + 3\left(2G - 2F_x - F^2\right)\partial_x + 3\left(G_x - F_{xx} - FF_x\right)\right)\varphi = \partial_t\varphi \end{cases} \tag{34}$$

or

$$\begin{cases} \left(\partial_x^2 - F\partial_x + G\right)\left(\partial_x^2 + F\partial_x + F_x + G\right)\partial_x\varphi = \lambda\varphi, \\ \left(5\partial_x^3 + 3\left(2G + 3F_x - F^2\right)\partial_x\right)\varphi = \partial_t\varphi. \end{cases} \tag{35}$$

It happens that the stationary reduction of (33), which is an unphysical Hamiltonian system [17, pp. 98, 103], is mapped by a canonical transformation to the 1:12:16 case of HH4.

In Figure 3, we display the link between c-KdV systems possessing a fifth-order Lax pair and the 1:12:16 integrable Hamiltonian.

5 CONCLUSION

We have linked each of the three not yet integrated quartic Hénon–Heiles cases to fourth-order ODEs recently integrated by Cosgrove, via a path involving, on one hand canonical transformations between Hamiltonian systems, and on the other hand Bäcklund transformations between coupled KdV systems. This proves that these three cases have a general solution expressed with hyperelliptic functions of genus two. Their explicit closed form expression will be given in future work.

ACKNOWLEDGMENTS

The authors acknowledge the financial support of the Tournesol grants T99/040 and T2003.09. MM and RC thank the organizers of this ARW for invitation. CV is a research assistant of the FWO.

REFERENCES

1. Hénon, M. and Heiles, C. (1964) The applicability of the third integral of motion: some numerical experiments, *Astron. J.* **69**, pp. 73–79.
2. Bountis, T. Segur, H., and Vivaldi, F. (1982) Integrable Hamiltonian systems and the Painlevé property, *Phys. Rev. A.* **25**, pp. 1257–1264.
3. Chang, Y. F., Tabor, M., and Weiss, J. (1982) Analytic structure of the Hénon–Heiles Hamiltonian in integrable and nonintegrable regimes, *J. Math. Phys.* **23**, pp. 531–538.
4. Grammaticos, B., Dorizzi, B., and Padjen, R. (1982) Painlevé property and integrals of motion for the Hénon–Heiles system, *Phys. Lett. A.* **89**, pp. 111–113.
5. Drach, J. (1919) Sur l'intégration par quadratures de l'équation $\frac{d^2y}{dx^2} = [\varphi(x) + h]y$, *C.R. Acad. Sc. Paris* **168** pp. 337–340.
6. Verhoeven, C., Musette, M., and Conte, R. (2002) Integration of a generalized Hénon-Heiles Hamiltonian, *J. Math. Phys.* **43**, pp. 1906–1915. http://arXiv.org/abs/nlin.SI/0112030.
7. Fordy, A. P. (1991) The Hénon–Heiles system revisited, *Physica. D.* **52**, pp. 204–210.
8. Jimbo, M. and Miwa, T. (1983) Solitons and infinite dimensional Lie algebras, *Publ. RIMS, Kyoto* **19**, pp. 943–1001.
9. Ramani, A., Dorizzi, B., and Grammaticos, B. (1982) Painlevé conjecture revisited, *Phys. Rev. Lett.* **49**, pp. 1539–1541.
10. Grammaticos, B., Dorizzi, B., and Ramani, A. (1983) Integrability of Hamiltonians with third- and fourth-degree polynomial potentials, *J. Math. Phys.* **24**, pp. 2289–2295.

11. Hietarinta, J. (1987) Direct method for the search of the second invariant, *Phys. Rep.* **147**, pp. 87–154.
12. Baker, S., Enol'skii, V. Z., and Fordy, A. P. (1995) Integrable quartic potentials and coupled KdV equations, *Phys. Lett. A.* **201** pp. 167–174.
13. Drinfel'd, V. G. and Sokolov, V. V. (1984) Lie algebras and equations of Korteweg-de Vries type, Itogi Nauki i Tekhniki, Seriya Sovremennye Problemy Matematiki **24**, pp. 81–180 [English: (1985) *J. Soviet Math.* **30**, pp. 1975–2036].
14. Verhoeven, C. and Musette, M. (2003) Soliton solutions of two bidirectional sixth order partial differential equations belonging to the KP hierarchy, *J. Phys. A.* **36**, pp. L133–L143.
15. Manakov, S. V. (1973) On the theory of two-dimensional stationary self-focusing of electromagnetic waves, *Zh. Eksp. Teor. Fiz.* **65**, pp. 505–516 [(1974) JETP **38**, pp. 248–253].
16. Wojciechowski, S. (1985) Integrability of one particle in a perturbed central quartic potential, *Physica. Scripta.* **31**, pp. 433–438.
17. Baker, S. (1995) Squared eigenfunction representations of integrable hierarchies, PhD Thesis, University of Leeds.
18. Verhoeven, C. Musette, M., and Conte, R. (2003) General solution for Hamiltonians with extended cubic and quartic potentials, *Theor. Math. Phys.* **134**, pp. 128–138. http://arXiv.org/abs/nlin.SI/0301011.
19. Verhoeven, C. (2003) Integration of Hamiltonian systems of Hénon-Heiles type and their associated soliton equations, PhD thesis, Vrije Universiteit Brussel.
20. Satsuma, J. and Hirota, R. (1982) A coupled KdV equation is one case of the four-reduction of the KP hierarchy, *J. Phys. Soc. Japan* **51**, pp. 3390–3397.
21. Sawada, K. and Kotera, T. (1974) A method for finding N-soliton solutions of the K.d.V. equation and K.d.V.-like equation, *Prog. Theor. Phys.* **51**, pp. 1355–1367.
22. Ramani, A. (1981) in: *Fourth International Conference on Collective Phenomena*, Annals of the New York Academy of Sciences, Vol. 373, ed. J. L.Lebowitz, New York Academy of Sciences, New York, pp. 54–67.
23. Kaup, D. J. (1980) On the inverse scattering problem for cubic eigenvalue problems of the class $\psi_{xxx} + 6Q\psi_x + 6R\psi = \lambda\psi$, *Stud. Appl. Math.* **62**, pp. 189–216.
24. Dye, J. M. and Parker, A. (2001) On bidirectional fifth-order nonlinear evolution equations, Lax pairs, and directionally dependent solitary waves, *J. Math. Phys.* **42**, pp. 2567–2589.
25. Cosgrove, C. M. (2000) Higher order Painlevé equations in the polynomial class: I. Bureau symbol *P2*, *Stud. Appl. Math.* **104**, pp. 1–65.
26. Drinfel'd, V. G. and Sokolov, V. V. (1981) Equations of Korteweg-de Vries type and simple Lie Algebras, *Soviet Math. Dokl.* **23**, pp. 457–462.
27. Bogoyavlensky, O. I. (1990), Breaking solitons in 2 + 1-dimensional integrable equations, Usp. Matem. Nauk **45** pp. 17–77 [English: (1990) *Russ. Math. Surveys* **45**, pp. 1–86].

ON THE BILINEAR FORMS OF PAINLEVÉ'S 4TH EQUATION

R. Willox
Graduate School of Mathematical Sciences, The University of Tokyo,
3-8-1 Komaba, Meguro-ku, 153-8914 Tokyo, Japan

J. Hietarinta
Department of Physics, University of Turku, FIN-20014 Turku, Finland

1 INTRODUCTION

Equations in Hirota's bilinear form appear naturally whenever one needs to write integrable differential equations in terms of "nice" functions. In the case of soliton equations nice typically means polynomials of exponentials with linear exponents, whereas in the context of Painlevé equations nice often means entire functions. The natural or original dependent variables of the equation are usually not the best in this respect and a change of the dependent variables is necessary. This dependent variable transformation can be somewhat involved.

Since the solutions of Painlevé equations are meromorphic by definition (the movable singularities can be at worst poles) it is natural to express them as ratios of entire functions, and this leads to homogeneous equations and often to equations in Hirota's bilinear form. Indeed, bilinear forms were already derived by Painlevé himself [1]. It should be noted, however, that the converse of the above is not necessarily true: a bilinear form does not by itself imply that the independent functions are regular in any sense, in fact bilinear forms exist even for non integrable equations.

In the field of integrable partial differential equations, the so-called τ-functions play a major role. These functions actually provide a huge pool of such "nice" functions, in both of the aforementioned interpretations, and they also possess an extremely rich algebraic and algebro-geometric structure. This is the main theme of Sato-theory, one of the great unifying theories for the description of integrable systems.

The relevance of τ-functions for the study of the Painlevé equations has been recognized ever since the seminal work of Jimbo and Miwa on isomonodromy deformations [2] and in particular since the results obtained by Okamoto, regarding the algebro-geometric properties of the Painlevé equations [3].

L. Faddeev et al. (eds.),
Bilinear Integrable Systems: From Classical to Quantum, Continuous to Discrete, 375–390.
© 2006 *Springer. Printed in the Netherlands.*

Recently [4], Noumi and Yamada provided a nearly complete description of the symmetry properties of the P_{II}, P_{IV}, and P_V equations and their higher order generalizations [5], in terms of their τ-function descriptions, i.e., in terms of their bilinear representations.

Here we want to place the above bilinear descriptions, and especially the Okamoto, Noumi–Yamada-type description of the symmetry properties of the Painlevé equations, in a more general setting: that of the Kadomtsev–Petviashvili (KP) hierarchy. For brevity we shall restrict our discussion to the Painlevé IV equation. First we shall explain how a bilinearization of this equation, based on an expression of its solutions in terms of entire functions, can be related to a bilinearization in terms of genuine τ-functions (in the sense of Sato-theory). We then explain how, yet another, bilinearization of this Painlevé equation is related to a reduction of a Darboux chain for KP τ-functions. This approach is inspired by Adler's results on reductions of dressing chains [6].

It will also be shown that in this reduction the fundamental Bäcklund transformations for KP τ-functions give rise to Bäcklund transformations for the Painlevé IV equation, which correspond to the full automorphism group of the $A_2^{(1)}$ root lattice [7]. These transformations are well known [6–8] but apart from the Weyl group, extended with rotations of the Dynkin diagram [9], a description of such transformations in terms of τ-functions was still lacking. All results pertaining to periodic reductions of Darboux chains, including the discussion of their Bäcklund transformations, can be extended to the case of general periods. This will be the topic of a future publication.

2   BILINEARIZING THE PAINLEVÉ IV EQUATION

2.1   Solution in Terms of Entire Functions and the Implied Bilinearization

We will start by describing a bilinearization of the Painlevé IV equation (P_{IV}) in terms of entire functions. Its connection to τ-functions will be described later.

P_{IV} in its canonical form is given by

$$\frac{d^2y}{dz^2} = \frac{1}{2y}\left(\frac{dy}{dz}\right)^2 + \frac{3}{2}y^3 + 4zy^2 + 2(z^2 - a)y + \frac{b}{y}. \tag{1}$$

It is well known that its solution can have movable poles, and around such poles the expansion is [10]

$$y = \frac{\pm 1}{z - z_0} - z_0 + \frac{1}{3}(-4 \pm (2a + z_0^2))(z - z_0) + d(z - z_0)^2 + \dots, \tag{2}$$

where z_0 and d are the two required free parameters.

The question we want to address is how to express y in terms of entire functions, or from a different point of view, how to construct entire functions from y. This problem was studied in [10] with an algorithmic answer for most Painlevé equations. (For a similar approach that works for all Painlevé equations, see [11].) The idea is to first construct an expression which has a double pole with coefficient one and no single poles or other singularities, and then by integrating twice and by exponentiating the result, to construct an entire function. In the case of P_{IV} this approach leads to the definition

$$F(z) := \exp\left\{-\int dz \int dz[y(z^2) + 2zy(z) + \gamma]\right\}. \qquad (3)$$

It is now easy to verify that the function F indeed has a simple zero at the singularities of y, given in (2). Furthermore, this observation implies that

$$G(z) := y(z)F(z). \qquad (4)$$

is another entire function. Thus we have constructed two entire functions from y, but at the same time we have obtained an expression for y in terms of entire functions: $y = G/F$.

The definitions (3, 4) imply the bilinear relation

$$(D_z^2 - 2\gamma)F \cdot F + 2(G^2 + 2zFG) = 0, \qquad (5)$$

where we have used the bilinear derivative operator D_z, defined by

$$D_z^n A(z) \cdot B(z) = (\partial_{z_1} - \partial_{z_2})^n A(z_1)A(z_2)\big|_{z=z_1=z_2}.$$

In order to obtain a second bilinear equation, thus fully determining the two functions F and G, one can substitute $y = G/F$ into P_{IV}. The result is not bilinear—it is in fact quadrilinear—but by suitable application of (5) one can derive [10]

$$(D_z^4 - 4(3\gamma^2 - 4b))F \cdot F + 4(zD_z^2 + 2D_z + 10\gamma z)G \cdot F$$
$$- 2(3D_z^2 + 16a - 18\gamma)G \cdot G = 0. \qquad (6)$$

Equations (5, 6) then offer a bilinear description of the P_{IV} equation. Note that the parameter γ is arbitrary (in [10] $\gamma = 0$).

2.2 P_{IV} as a Similarity Reduction of the Modified Classical Boussinesq Equation

In relation to the above bilinear formulation there exists an alternative bilinearization of the P_{IV} equation, where the τ-functions that appear in it can be explicitly related to the τ-functions of the 2-component KP hierarchy (see e.g., [12] for a definition of this hierarchy and for examples of the equations contained in it). To see how this comes about, let us note that the similarity reduction (for arbitrary constants α_1, α_2 and with $\varepsilon \neq 0$)

$$d_y = \alpha_1 - \varepsilon x d_x, \quad v_y = -\alpha_2 - \varepsilon x v_x, \tag{7}$$

of the so-called modified classical Boussinesq system [13],

$$\begin{cases} d_y = d_{2x} + d_x(d_x + 2v_x), \\ v_y = -v_{2x} + v_x(v_x + 2d_x), \end{cases} \tag{8}$$

yields (with $g_1 = d_x$, $g_2 = v_x$, $' = \frac{d}{dx}$) a Hamiltonian formulation of P_{IV}:

$$g_1' = \frac{\partial H}{\partial g_2}, \quad g_2' = \frac{\partial H}{\partial g_1}, \tag{9}$$

$$H = g_1 g_2^2 + g_2 g_1^2 + \varepsilon x g_1 g_2 + \alpha_2 g_1 - \alpha_1 g_2. \tag{10}$$

Eliminating g_2 from (9, 10) and introducing

$$y(z) = \kappa g_1(x), \quad x = \kappa z, \quad \kappa^2 = 2/\varepsilon, \tag{11}$$

one obtains (1) with parameters $a = 1 + (\alpha_1 + 2\alpha_2)/\varepsilon$ and $b = -2\alpha_1^2/\varepsilon^2$.

Modified classical Boussinesq (8) is in fact an alternative version of the well known Chen–Lee–Liu system [13] and it is known to bilinearize as [14]

$$(D_y + D_x^2)f \cdot g = 0, \quad D_x^2 g \cdot \tilde{g} + D_x f \cdot \tilde{f} = 0, \tag{12}$$

$$(D_y + D_x^2)f \cdot \tilde{g} = 0, \quad D_x g \cdot \tilde{g} + f\tilde{f} = 0, \tag{13}$$

where the new variables f, g, and $\tilde{g}$ are connected to d and v by

$$v = \log \frac{g}{f}, \quad d = \log \frac{\tilde{g}}{g}. \tag{14}$$

It is also well known that this bilinear system is a so-called (1, 1)-reduction of the 2-component KP hierarchy [12]. One may therefore conclude that the τ-functions f, $\tilde{f}$, g, and $\tilde{g}$ that appear in it are indeed "genuine" (in the sense of Sato-theory) reduced, 2-component KP τ-functions. Implementing the similarity reduction (7) on the τ-functions ($\alpha_2 \equiv c_1 - c_2$, $\alpha_1 \equiv c_4 - c_2$)

$$f_y = c_1 f - \varepsilon x f_x, \quad g_y = c_2 g - \varepsilon x g_x, \tag{15}$$

$$\tilde{f}_y = c_3 \tilde{f} - \varepsilon x \tilde{f}x, \quad \tilde{g}_y = c_4 \tilde{g} - \varepsilon x \tilde{g}_x, \tag{16}$$

we obtain the following bilinear system for the Hamilton equations (9, 10) and hence for the Painlevé IV equation:

$$(D_x^2 - \varepsilon x D_x + \alpha_2)f \cdot g = 0, \tag{17}$$

$$(D_x^2 - \varepsilon x D_x + \alpha_2 - \alpha_1)f \cdot \tilde{g} = 0, \tag{18}$$

$$D_x^2 g \cdot \tilde{g} + D_x f \cdot \tilde{f} = 0, \tag{19}$$

$$D_x g \cdot \tilde{g} + f\tilde{f} = 0, \tag{20}$$

with the connection to g_1, g_2 given by

$$g_2 = \partial_x \log(g/f), \quad g_1 = \partial_x \log(\tilde{g}/g) \equiv \frac{D_x \tilde{g} \cdot g}{\tilde{g}g}. \tag{21}$$

How are the bilinearizations (5, 6) and (17–20) related? The answer lies in the form of g_1, which is basically y. Indeed, if we define

$$F = \tilde{g}g, \quad G = \kappa D_x \tilde{g} \cdot g \tag{22}$$

we obtain the desired correspondence. In detail: Solve $\tilde{f}$ and f_x from Eqs. (20) and (19), respectively, and then $\tilde{g}$ and g_x from (22). Then a suitable linear combination of Eqs. (18, 17) leads to (5). Using (5) and its derivatives we finally get from either one—after changing coordinates as in (11)—the other Eq. (6) for the choice $\gamma = 2\alpha_1/\epsilon$, and with the previous parameter identifications.

3 PAINLEVÉ EQUATIONS AND THE KP HIERARCHY

3.1 The Dressing Chain and P_{IV}

Since Adler's seminal paper [6] it is known that the Painlevé equations (P_{II} through P_{VI}) can be obtained from periodic reductions of chain equations which result from repeated application of particular Darboux transformations. A well known result is that the P_{IV} equation (1) is obtained as a period 3 reduction of the dressing chain [15]. Here we shall first recall Adler's construction of P_{IV}, which we shall then implement on the level of KP τ-functions. A systematic approach to the problem of constructing chain equations related to Schrödinger-type linear problems has been developed in [16].

3.1.1 The Dressing Chain

The following infinite chain of equations for the functions $F_j(j = 0, 1, \ldots)$

$$F_j(x)' + F_{j+1}(x') = F_j(x)^2 - F_{j+1}(x)^2 + v_{j+1} - v_j. \tag{23}$$

is called the *dressing chain* (the v_j are arbitrary constants). It is associated with the spectral problem for the Schrödinger operator

$$L_j(u, \lambda)\psi_j(\lambda, x) = 0, \quad \text{where} \quad L_j(u, \lambda) := \partial_x^2 + u_j(x) - \lambda, \tag{24}$$

(j indexes a sequence of eigenproblems with eigenvalue λ).

The operator

$$G_j(x) := (\partial_x - F_j(x)), \tag{25}$$

can be used to define new functions

$$\psi_{j+1}(\lambda, x) := G_j(x)\psi_j(\lambda, x), \tag{26}$$

for each eigenfunction ψ_j of the original problem. The new functions ψ_{j+1} are eigenfunctions of a new operator L_{j+1} that satisfies

$$L_{j+1}G_j(x) = G_j(x)L_j(u, \lambda). \tag{27}$$

In fact, one has

$$L_{j+1} = L_j(u_{j+1}, \lambda), \tag{28}$$
$$u_{j+1}(x) = u_j(x) + [2F_j(x)]', \tag{29}$$
$$F_j(x)' + F_j(x)^2 + u_j(x) - \lambda = \mu_j(\lambda), \tag{30}$$

where μ_j is an integration constant (as before, $'$ denotes $\frac{d}{dx}$).

The transformation from ψ_j to ψ_{j+1} is called a *Darboux transformation* iff the operator G_j is such that it annihilates some chosen eigenfunction φ_j of (24) with eigenvalue v_j. This implies that $F_j(x) = (\log \varphi_j)_x$ and that it satisfies the equation

$$F_j(x)' + F_j(x)^2 + u_j(x) - v_j = 0. \tag{31}$$

Hence we have $\mu_j(\lambda) = v_j - \lambda$ in (30) and subsequent elimination of the potentials u from (30) yields the dressing chain (23).

3.1.2 P_{IV} as a Reduction of the Dressing Chain

Closing this chain periodically (with period N) by imposing

$$\varphi_{j+N}(v_{j+N}) = \varphi_j(v_j), \quad v_{j+N} = v_j - \varepsilon, \tag{32}$$

for some non zero constant ε, yields trivial systems at $N = 1$ and 2. At $N = 3$ however, introducing the variables

$$\begin{aligned}
g_1 &= (\log \varphi_1\varphi_2)_x, \\
g_2 &= (\log \varphi_2\varphi_3)_x, \\
g_3 &= (\log \varphi_3\varphi_1)_x,
\end{aligned} \tag{33}$$

we obtain

$$\begin{cases}
g_1' = g_1(g_3 - g_2) + \alpha_1, \\
g_2' = g_2(g_1 - g_3) + \alpha_2, \\
g_3' = g_3(g_2 - g_1) + \alpha_3,
\end{cases} \tag{34}$$

with $\alpha_1 = v_2 - v_1$, $\alpha_2 = v_3 - v_2$, and $\alpha_3 = v_1 - v_3 - \varepsilon$. Obviously we have a conserved quantity $(g_1 + g_2 + g_3)' = -\varepsilon$ and this allows us to reduce the order of (34), e.g., by eliminating g_3 in terms of the other variables. This gives rise to the Hamiltonian form (9, 10) of the P_{IV} equation. The system (34) (which first appeared in [17]) is often referred to as the symmetric form of the P_{IV} equation [9]. We shall now proceed to show that this symmetric form for P_{IV} possesses a bilinear formulation in terms of KP τ-functions (see [16] for a direct bilinearization of the dressing chain).

3.2 KP τ-Functions

We shall describe the P_{IV} equation in terms of (this time, single component) KP τ-functions. This approach can be generalized to include all the higher order Painlevé equations introduced in [5].

3.2.1 The KP Hierarchy and Its Symmetries

Denote the infinite set of coordinates $x_1, x_2, \ldots$ by the vector $\mathbf{x} = (x_1, x_2, \ldots)$, then $\tau(\mathbf{x}) \in \mathbb{C}[x_1, x_2, \ldots]$ (the space of formal power series in $x_1, x_2, \ldots$) is called a KP τ-function iff it satisfies

$$\text{Res}_\lambda \left[\tau(\mathbf{x} - \epsilon_\lambda)\tau(\mathbf{x}' + \epsilon_\lambda)\, e^{\xi_\lambda(\mathbf{x}-\mathbf{x}')} \right] = 0 \quad \forall \mathbf{x}, \mathbf{x}'. \tag{35}$$

The operation Res_λ is defined by $\text{Res}_\lambda[\sum_{n=-\infty}^{+\infty} a_n \lambda^n] = a_{-1}$, $\xi_\lambda(\mathbf{x})$ denotes the formal power series $\xi_\lambda(\mathbf{x}) = \sum_{n=1}^{\infty} x_n \lambda^n$ and ϵ_λ denotes the infinite vector $\epsilon_\lambda = (1/\lambda, 1/(2\lambda^2), 1/(3\lambda^3), \ldots)$. Relation (35) actually encodes all the equations in the KP hierarchy, expressed in bilinear form. An extensive list of such equations can be found in [12], to which we also refer for a detailed account of the algebraic machinery underlying the KP hierarchy and its solutions.

We also define the KP vertex operators

$$\Gamma_\lambda^\pm := e^{\pm \xi_\lambda(\mathbf{x})}\, e^{\mp \xi_\lambda(\tilde{\partial})}, \tag{36}$$

where $\tilde{\partial}$ denotes the vector $\tilde{\partial} = (\frac{\partial}{\partial_{x_1}}, \frac{1}{2}\frac{\partial}{\partial_{x_2}}, \ldots)$ as well as the so-called *solitonic* vertex operator:

$$\Gamma_{\lambda\mu} := e^{\xi_\lambda(\mathbf{x})-\xi_\mu(\mathbf{x})}\, e^{-\xi_\lambda(\tilde{\partial})+\xi_\mu(\tilde{\partial})}. \tag{37}$$

Note that $\Gamma_\lambda^\pm[\tau(\mathbf{x})] = \tau(\mathbf{x} \mp \epsilon_\lambda)e^{\pm\xi_\lambda(\mathbf{x})}$ and also that the solitonic vertex operator is intimately related to the fundamental symmetry of the equations in the KP hierarchy. Expanding

$$\frac{\Gamma_{\lambda\mu} - 1}{\lambda - \mu} = \sum_{i,j=-\infty}^{+\infty} Z_{ij} \lambda^i \mu^{-j-1}, \tag{38}$$

then $\{\sum_{i,j} a_{ij} Z_{ij} \mid a_{ij} = 0 \text{ for } |i - j| \gg 0\}$ is nothing but the well-known vertex representation of the Lie algebra $gl(\infty)$ (see e.g., [12, 18, 19]). The bilinear relation (35) expresses the fact that the τ-function $\tau(\mathbf{x})$ lies on the orbit $GL(\infty) \cdot 1$ of the Lie group $GL(\infty)$ generated by $gl(\infty)$.

Of particular interest are certain elements of this symmetry group, expressible in terms of solutions to the linear problem that underlies the KP hierarchy. Consider the following formulation of the Zakharov–Shabat equations for the KP hierarchy (and their formal adjoints):

$$\forall n \geq 2 : p_n(-\tilde{\partial})\Phi^+ = \Phi^+[p_{n-1}(-\tilde{\partial})(\log \tau)_{x_1}], \tag{39}$$

$$\forall n \geq 2 : p_n(\tilde{\partial})\Phi^- = -\Phi^-[p_{n-1}(\tilde{\partial})(\log \tau)_{x_1}], \tag{40}$$

where the $p_n(\mathbf{z})$ in some general argument $\mathbf{z} = (z_1, z_2, \ldots)$ denote the Schur polynomials generated by $\exp\left[\sum_{n=1}^{\infty} z_n \lambda^n\right] = \sum_{n=0}^{\infty} p_n(\mathbf{z})\lambda^n$. We shall call the solutions Φ^+ (Φ^-) to these linear equations *KP eigenfunctions* (or *adjoint eigenfunctions*) for the τ-function $\tau(\mathbf{x})$. The equations in the KP hierarchy are of course obtained as the compatibility conditions of the above linear equations (expressed in terms of the field $(\log \tau)_{2x}$). As for the eigenfunctions and adjoint eigenfunctions for a particular $\tau(\mathbf{x})$, they can be expressed in terms of the vertex operators (36):

$$\Phi^{\pm}(\mathbf{x}) = \int_{C_\lambda} \frac{d\lambda}{2\pi i} h^{\pm}(\lambda) \frac{\Gamma_\lambda^{\pm}[\tau(\mathbf{x})]}{\tau(\mathbf{x})}, \tag{41}$$

for spectral densities $h^{\pm}(\lambda) = \frac{1}{\lambda}\Phi^{\pm}(\mathbf{x}' \pm \epsilon_\lambda)\frac{\Gamma_\lambda^{\pm}[\tau(\mathbf{X}')]}{\tau(\mathbf{X}')}$ (for some arbitrary $\mathbf{x}'$) and a contour C_λ around $\lambda = \infty$ that does not enclose any of the other singularities of the density $h^{\pm}(\lambda)$ [20].

Whereas $gl(\infty)$ is the fundamental symmetry algebra for the KP hierarchy, this symmetry is restricted to the affine algebra $A_{\ell-1}^{(1)}$ for the so-called ℓ-reduced KP hierarchies. We define an ℓ-reduced τ-function as a KP τ-function that, besides relation (35), also satisfies

$$\text{Res}_\lambda\left[\lambda^\ell \tau(\mathbf{x} - \epsilon_\lambda)\tau(\mathbf{x}' + \epsilon_\lambda)e^{\xi_\lambda(\mathbf{x}-\mathbf{x}')}\right] = 0 \quad \forall \mathbf{x}, \mathbf{x}'. \tag{42}$$

These τ-functions are all $x_{j\ell}$-independent ($j = 1, 2, \ldots$) and $gl(\infty)$ is reduced to the direct sum of a Heisenberg algebra $\mathcal{H}_\ell$, formed by $jx_{j\ell}$ and $\partial/\partial x_{j\ell}(j = 1, 2, \ldots)$, and an algebra isomorphic to $A_{\ell-1}^{(1)}$. When dealing with ℓ-reduced τ-functions we shall always restrict the coordinate set to $\{x_n | n \neq 0 \mod \ell\}$, whereby effectively eliminating $\mathcal{H}_\ell$.

3.2.2 Darboux Transformations for KP τ-Functions

Let $\tau(\mathbf{x})$ be a KP τ-function. We can then define a Darboux transformation for $\tau(\mathbf{x})$ as a mere multiplication by an eigenfunction Φ^+ (for that τ):

$$\tau \longrightarrow \tilde{\tau} = \tau \times \Phi^+. \tag{43}$$

As this transformation has its origin in a particular action of $GL(\infty)$, the new function $\tilde{\tau}(\mathbf{x})$ is also a KP τ-function [20]. Eigenfunctions for this $\tilde{\tau}$ are produced from those for τ by means of the operator (25) defined in terms of $F = (\log \Phi^+)_{x_1}$ (as explained in Section 3.1.1).

Repeated application of such a Darboux transformation to a particular $\tau(\mathbf{x})$ yields KP τ-functions at every step and moreover the, say, kth Darboux iterate $\tau_{[k]}$ and the "seed" τ-function $\tau(\mathbf{x})$ satisfy the defining relation for the so-called kth-modified KP hierarchy:

$$\text{Res}_\lambda \left[\lambda^k \tau_{[k]}(\mathbf{x} - \epsilon_\lambda) \tau(\mathbf{x'} + \epsilon_\lambda) e^{\xi_\lambda(\mathbf{x} - \mathbf{x'})} \right] = 0 \quad \forall \mathbf{x}, \mathbf{x'}. \tag{44}$$

The basic member of the 1st-modified KP hierarchy (obtained from (44) at $k = 1$) is the following very simple bilinear equation, which will be used extensively in the next sections:

$$(D_{x_2} - D_{x_1}^2) \tau_{[1]} \cdot \tau = 0. \tag{45}$$

In fact, as this equation expresses a relation between τ and its Darboux transformation $\tilde{\tau} \equiv \tau_{[1]}$, i.e., between τ and an eigenfunction $\Phi^+ = \tau_{[1]}/\tau$ associated to it, (45) is nothing but the Hirota form of the Zakharov–Shabat equation for this particular Φ^+, obtained from (39) at $n = 2$.

Similarly, we define an adjoint Darboux transformation $\tau \to \overline{\tau} = \tau \times \Phi^-$ in terms of an adjoint eigenfunction $\Phi^- = \overline{\tau}/\tau$ for τ. This adjoint eigenfunction will satisfy the adjoint Zakharov–Shabat equations (40), and in particular at $n = 2$ one has

$$(D_{x_2} + D_{x_1}^2) \overline{\tau} \cdot \tau = 0. \tag{46}$$

Note that due to the symmetry properties of the Hirota operators, Eq. (46) can also be written as

$$(D_{x_2} - D_{x_1}^2) \tau \cdot \overline{\tau} = 0, \tag{47}$$

suggesting that the ratio $\tau/\overline{\tau}$ is in fact an eigenfunction for $\overline{\tau}$. This is indeed the case: In general we have that if Φ^- is an adjoint eigenfunction for a τ-function τ, then its reciprocal $1/\Phi^-$ is an eigenfunction for the τ-function $\overline{\tau} \equiv \tau \times \Phi^-$. It is self-evident that the converse statement (i.e., for eigenfunctions) also holds.

We can also define a so-called *binary* Darboux transformation [20]

$$\tau \longrightarrow \hat{\tau} = \tau \times \Omega(\Phi^+, \Phi^-), \tag{48}$$

as multiplication by an *eigenfunction potential* $\Omega(\Phi^+, \Phi^-)$. This potential is defined in terms of an eigenfunction and an adjoint eigenfunction for $\tau(\mathbf{x})$, through the exact differential

$$d\Omega(\Phi^+, \Phi^-) := \sum_{n=1}^{\infty} A_n dx_n, \quad (A_n)_{x_m} = (A_m)_{x_n}, \tag{49}$$

$$A_n := n\Phi^- p_{n-1}(\tilde{\partial})\Phi^+ - \sum_{k=1}^{n-1} \left(\Phi^- p_{n-k-1}(\tilde{\partial})\Phi^+\right)_{x_k}. \tag{50}$$

Its x_1 and x_2 derivatives are: $\Omega_{x_1} \equiv A_1 = \Phi^+\Phi^-$ and $\Omega_{x_2} \equiv A_2 = \Phi_x^+\Phi^- - \Phi^+\Phi_x^-$.

A binary Darboux transformation generalizes the action of the solitonic vertex operator (37) [20]. More precisely:

$$\Omega(\Phi^+, \Phi^-) = \frac{1}{\tau} \int_{C_\lambda} \frac{d\lambda}{2\pi i} \int_{C_\mu} \frac{d\mu}{2\pi i} h^+(\lambda)h^-(\mu)\frac{\Gamma_{\lambda\mu}[\tau]}{\lambda - \mu}, \tag{51}$$

for contours and spectral densities as defined in connection to (41). A binary Darboux transformation can therefore be thought of as the equivalent of a general action of $GL(\infty)$ (remember relation (38)).

Furthermore, a binary Darboux transformation $\tau \to \hat{\tau} \equiv \tau \times \Omega(\Phi^+, \Phi^-)$ is always associated with a "Bianchi diagram" for Darboux transformations

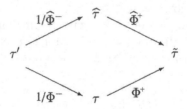

and vice versa (arrows indicate Darboux transformations involving the eigenfunctions indicated). This diagram is unique for given Φ^+ and Φ^-, up to multiplication of the τ-functions by a constant:

$$\hat{\Phi}^- \equiv \frac{\Phi^-}{\Omega(\Phi^+, \Phi^-)}, \quad \hat{\Phi}^+ \equiv \frac{\Phi^+}{\Omega(\Phi^+, \Phi^-)}. \tag{52}$$

Note that this property of binary Darboux transformations generalizes the well known Bianchi permutation theorem for Bäcklund transformations for $1+1$ dimensional integrable systems.

3.2.3   Periodic Darboux Chains and Self Similarity

Consider a chain of KP τ-functions generated by successive Darboux transformations:

$$\tau_{[0]} = \tau, \quad \tau_{[n+1]} = \tau_{[n]} \times \Phi_n^+, \quad n = 0, 1, \ldots \tag{53}$$

where Φ_n^+ denotes an eigenfunction for $\tau_{[n]}$. We wish to discuss the case where such a *Darboux chain* closes periodically. However, we shall only treat the period 3 case (an extension of the analysis below to general periods is possible and will be reported in a forthcoming publication), i.e., we shall consider the periodic Darboux chain:

$$\tau_{[0]} \xrightarrow{\Phi_0^+} \tau_{[1]} \xrightarrow{\Phi_1^+} \tau_{[2]} \xrightarrow{\Phi_2^+} \tau_{[3]} \equiv \tau_{[0]}. \tag{54}$$

From now on indices referring to particular τ-functions or eigenfunctions etc. are to be understood modulo 3.

It can be shown that the τ-functions in the chain (54) are all 3-reduced τ-functions and vice versa that, starting from an arbitrary 3-reduced τ-function, one can always construct such a periodic Darboux chain.

As to a description of such chains by means of specific equations, since every pair $(\tau_{[n+1]}, \tau_{[n]})$ satisfies the 1st-modified KP hierarchy, each such pair satisfies (45) and we find that the following system of Hirota equations

$$\begin{cases} (D_{x_2} - D_{x_1}^2) \, \tau_{[1]} \cdot \tau_{[0]} = 0 \\ (D_{x_2} - D_{x_1}^2) \, \tau_{[2]} \cdot \tau_{[1]} = 0 \\ (D_{x_2} - D_{x_1}^2) \, \tau_{[0]} \cdot \tau_{[2]} = 0 \end{cases} \tag{55}$$

completely characterizes the Darboux chain (54).

If we now require the τ-functions (which do not depend on the x_{3j} variables, $j = 1, 2, \ldots$) in the chain to be self-similar, i.e., to be "eigenvectors" for the operator L:

$$L := \sum_{\substack{k=1 \\ k \neq 0 \bmod 3}}^{\infty} k x_k \frac{\partial}{\partial x_k}, \tag{56}$$

$$\forall n = 0, 1, 2, \quad \exists c_n : L[\tau_{[n]}(\mathbf{x})] = c_n \tau_{[n]}(\mathbf{x}), \tag{57}$$

we can reduce (55) to a system of ordinary differential equations. It suffices to restrict the coordinates by means of

$$\mathcal{P} : x_1 \to x, \quad x_2 \to -\frac{3}{2\varepsilon}, \quad x_{n>3} \to 0, \tag{58}$$

for an arbitrary non zero constant ε, to obtain:

$$\begin{cases} \left(D_x^2 - \frac{\varepsilon x}{3} D_x - \kappa_0 \right) \tau_1 \cdot \tau_0 = 0 \\ \left(D_x^2 - \frac{\varepsilon x}{3} D_x - \kappa_1 \right) \tau_2 \cdot \tau_1 = 0 \\ \left(D_x^2 - \frac{\varepsilon x}{3} D_x - \kappa_2 \right) \tau_0 \cdot \tau_2 = 0 \end{cases} \tag{59}$$

In this system, the τ-functions $\tau_n := \tau_{[n]}(\mathcal{P}[\mathbf{x}])$ only depend on a single variable (x, ε being treated as a mere parameter). The constants κ_n are expressed in terms of the weights c_n associated to each $\tau_{[n]}$:

$$\kappa_n = \frac{\varepsilon}{3}(c_n - c_{n+1}). \tag{60}$$

If we now introduce the variables

$$g_n := ([\log \frac{\tau_{[n+1]}}{\tau_{[n-1]}})_x - \frac{\varepsilon x}{3} \tag{61}$$

and parameters

$$\alpha_n := \kappa_n - \kappa_{n-1} - \frac{\varepsilon}{3} \equiv \frac{\varepsilon}{3}(2c_n - c_{n+1} - c_{n-1} - 1), \tag{62}$$

we recover the symmetric form (34) of the P_{IV} equation from (59). Thus, system (59) is nothing but yet another Hirota bilinear form of the P_{IV} equation [21, 9, 16], this time expressed in terms of 3-reduced, self similar (singlecomponent) KP τ-functions. Note that the dependent variable transformation (61) is again a logarithmic derivative of the type (21). Note also that the operator L used to select self similar τ-functions among the 3-reduced ones is quite a well known object: it is related to the Virasoro energy operator $\sum_{k=1}^{\infty} kx_k \partial_{x_k}$ and is in fact located in a Cartan subalgebra of $A_2^{(1)}$ [19].

3.3 Bäcklund Transformations for P_{IV}

From the Bianchi diagram in Section 3.2.2 it is easily seen [22] that a binary Darboux transformation applied to one of the τ-functions in the periodic Darboux chain (54), using its neighbours to construct appropriate eigenfunctions and adjoint eigenfunctions, produces again a periodic Darboux chain. Furthermore, it can be shown that there always exists an eigenfunction potential such that the self similarity of the τ-functions in the Darboux chains is preserved by such a transformation. The following diagram summarizes this construction:

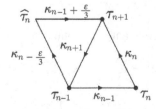

This is essentially the Bianchi diagram of Section 3.2.2, with $\tau = \tau_n$, $\tau' = \tau_{n-1}$, $\tilde{\tau} = \tau_{n+1}$, and $\hat{\tau} = \tau_n \times \Omega(\Phi_n^+, \Phi_n^-)$, with $\Phi_n^+ = \frac{\tau_{n+1}}{\tau_n}$ and $\Phi_n^- = \frac{\tau_{n-1}}{\tau_n}$ but for an additional arrow as the chains $\{\tau_{n-1}, \tau_n, \tau_{n+1}\}$ and $\{\tau_{n-1}, \hat{\tau}_n, \tau_{n+1}\}$ close periodically. Next to the arrows, the "eigenvalues" κ_n in (59) are indicated instead of the relevant eigenfunctions.

In the generic case, i.e., when α_n (62) is non zero, the eigenfunction potential used in the above diagram can be written explicitly:

$$\Omega(\Phi_n^+, \Phi_n^-) \equiv \frac{1}{\alpha_n} \left[(\Phi_n^+)_x \Phi_n^- - \Phi_n^+ (\Phi_n^-)_x - \frac{\varepsilon x}{3} \Phi_n^+ \Phi_n^- \right]. \tag{63}$$

Using this expression, we then define the following 3 transformations, denoted as $\mathbf{B}_n$, that map the chain (59) onto a new chain ($n = 0, 1, 2$):

$$\mathbf{B}_n : \mathbf{B}_n(\tau_n) = \tau_n \times \Omega(\frac{\tau_{n+1}}{\tau_n}, \frac{\tau_{n-1}}{\tau_n}), \quad \mathbf{B}_n(\tau_{n\pm1}) = \tau_{n\pm1}. \tag{64}$$

The resulting $\hat{\tau}_n = \mathbf{B}(\tau_n)$ are self similar (or rather, their preimages $\hat{\tau}_{[n]}$ w.r.t., the projection $\mathcal{P}$ are) with weights

$$\hat{c}_n = c_{n+1} + c_{n-1} - c_n + 1. \tag{65}$$

This allows us to lift the action of $\mathbf{B}_n$ to the weights c_n

$$\mathbf{B}_n(c_n) = c_{n+1} + c_{n-1} - c_n + 1, \quad \mathbf{B}_n(c_{n\pm1}) = c_{n\pm1}. \tag{66}$$

It should be clear that a binary Darboux transformation which maps $\tau_{[n]}$ to $\hat{\tau}_{[n]}$ (and which, after restricting the coordinates by $\mathcal{P}$, gives rise to a particular $\mathbf{B}_n$) actually corresponds to an element of $A_2^{(1)}$, in the sense of (51). We can therefore represent $\hat{\tau}_{[n]}$ as $\hat{\tau}_{[n]} = \Gamma_{[n]}[\tau_{[n]}]$, for some element $\Gamma_{[n]}$ in the vertex representation of $A_2^{(1)}$. In terms of this operator, relation (63) can be shown to be equivalent to

$$\left([L, \Gamma_{[n]}]_- + \frac{3\alpha_n}{\varepsilon} \Gamma_{[n]} \right) [\tau_{[n]}] = 0. \tag{67}$$

As L is located in the Cartan subalgebra of $A_2^{(1)}$, this last relation suggests that the α_n are proportional to the L-components of certain roots of $A_2^{(1)}$. Indeed, using (62, 66), the action of the $\mathbf{B}_n$ on the α_n's

$$\mathbf{B}_n(\alpha_n) = -\alpha_n, \quad \mathbf{B}_n(\alpha_{n\pm1}) = \alpha_{n\pm1} + \alpha_n, \tag{68}$$

is seen to be that of the (affine) Weyl group $W(A_2^{(1)})$ on the simple roots of $A_2^{(1)}$.

When α_n happens to be zero, i.e., when $\kappa_n - \kappa_{n-1} = \frac{\varepsilon}{3}$, one immediately notices that the eigenvalues in the two chains in the above diagram coincide pair-wise, such that the two Darboux chains can actually be taken to be identical. Hence, when $\alpha_n = 0$ we define the transformation $\mathbf{B}_n$ to be the identity, compatible with the Weyl action (68).

The action of these transformations on the fields g_n (61) that appear in the symmetric form (34) of P_{IV} is

$$\mathbf{B}(g_n) = g_n, \quad \mathbf{B}_n(g_{n\pm1}) = g_{n\pm1} \mp \frac{\alpha_n}{g_n}. \tag{69}$$

There exist also symmetry transformations expressible in terms of mere Darboux or adjoint Darboux transformations. These can be seen to correspond to automorphisms of the Dynkin diagram of $A_2^{(1)}$.

First of all there is the transformation

$$\mathbf{S} : \mathbf{S}(\tau_n) = \tau_{n+1} \quad (\equiv \tau_n \times \Phi_n^+) \tag{70}$$

which merely "rotates" the Darboux chain and which can be thought of as a succession of Darboux transformations, as can be seen from the definition (70). Obviously, $\mathbf{S}(c_n) = c_{n+1}$ and hence we have

$$\mathbf{S}(g_n) = g_{n+1}, \quad \mathbf{S}(\alpha_n) = \alpha_{n+1}. \tag{71}$$

This last relation tells us that $\mathbf{S}$ acts as a cyclic permutation on the roots α_n and thus corresponds to a rotation of the Dynkin diagram for $A_2^{(1)}$.

A second type of transformation is slightly more involved ($n = 0, 1, 2$):

$$\mathbf{R}_n : \tau_j(x) \longrightarrow \tau_{2n-j}(ix), \quad \kappa_j \longrightarrow -\kappa_{2n-j-1}. \tag{72}$$

It arises from the fundamental symmetry of the 1st-modified KP equations (in bilinear form), already commented upon in Section 3.2.2: Eq. (47) can obviously be cast into the form (46) if we transform $x \to \pm ix$. Under this transformation any eigenfunction for a particular τ-function is transformed into an adjoint eigenfunction for that τ and vice versa. Hence, any local permutation of two τ-functions in the Darboux chain can be "compensated" by changing the x-variable to ix and hence system (59) is covariant under $\mathbf{R}_n$. On the symmetric form of P_{IV} this transformation acts as:

$$\mathbf{R}_n : g_j(x) \longrightarrow -ig_{2n-j}(ix), \quad \alpha_j \longrightarrow \alpha_{2n-j}. \tag{73}$$

The transformations $\mathbf{R}_n$, together with the $\mathbf{S}^j (j = 1, 2, 3)$, form the dihedral group D_3 which exhausts the automorphisms of the Dynkin diagram for $A_2^{(1)}$. Hence, together with the $\mathbf{B}_n$, these transformations generate the full automorphism group of the root lattice for $A_2^{(1)}$.

4 SUMMARY

We have explained, on the example of the Painlevé IV equation, how different bilinearizations give insight into different aspects of the "integrability" of this equation. In particular it was shown how a bilinearization in terms of entire functions relates to integrable equations obtained from Sato-theory and, vice-versa, how a systematic study of particular reductions of equations in the KP-hierarchy yields insight into the structure of the Bäcklund transformations for the P_{IV} equation. Extending the above scheme to the P_{III} and P_{VI} equations remains a challenging problem.

ACKNOWLEDGMENTS

R.W. would like to acknowledge the support of the Fund for Scientific Research Flanders (F.W.O.), with which he was affiliated at the time of the NATO Advanced Research Workshop.

REFERENCES

1. Painlevé, P. (1902) Sur les équations différentielles du second ordre et d'ordre supérieur dont l'intégrale générale est uniforme, *Acta Math.* **25**, pp. 1–85.
2. Jimbo, M. and Miwa, T. (1981) Monodromy preserving deformation of linear ordinary differential equations with rational coefficients. **II**, *Physica D* **2**, pp. 407–448.
3. Okamoto, K. (1981) On the τ-function of the Painlevé equations, *Physica D* **2**, pp. 525–535.
4. Noumi, M. and Yamada, Y. (1998) Affine Weyl groups, discrete dynamical systems and Painlevé equations, *Comm. Math. Phys.* **199**, pp. 281–295.
5. Noumi, M. and Yamada, Y. (1998) Higher order Painlevé equations of type $A_\ell^{(1)}$, *Funkcialaj Ekvacioj* **41**, pp. 483–503.
6. Adler, V. E. (1994) Nonlinear chains and Painlevé equations, *Physica D* **73**, pp. 335–351.
7. Okamoto, K. (1986) Studies on the Painlevé equations III. Second and Fourth Painlevé equations, P_{II} and P_{IV}, *Math. Ann.* **275**, pp. 221–255.
8. Okamoto, K. (1992) The Painlevé equations and the Dynkin diagrams, in: eds. *Painlevé Transcendents*, D. Levi and P. Winternitz, Plenum, pp. 299–313.
9. Noumi, M. (2000) *Painlevé Equations—An Introduction via Symmetry*, Asakura Shoten, Tokyo [in Japanese].
10. Hietarinta, J. and Kruskal, M. (1992) Hirota forms for the six Painlevé equations from singularity analysis, in: eds. *Painlevé Transcendents*, Plenum, D. Levi and P. Winternitz, pp. 175–185.
11. Hietarinta, J. (2002) Taming the movable singularities, *ANZIAM J.* **44**, pp. 1–9.
12. Jimbo, M. and Miwa, T. (1983) Solitons and infinite dimensional Lie algebras, *Pub. RIMS* **19**, pp. 943–1001.
13. Chen, H. H., Lee, Y. C., and Liu, C. S. (1979) Integrability of nonlinear hamiltonian systems by inverse scattering method, *Phys. Scripta* **20**, pp. 490–492.

14. Pelinovsky, D., Springael, J., Lambert, F., and Loris, I. (1997) On modified NLS, Kaup and NLBq equations: differential transformations and bilinearization, *J. Phys. A* **30**, pp. 8705–8717.

15. Shabat, A. (1992) The infinite-dimensional dressing dynamical system, *Inv. Prob.* **8**, pp. 303–308.

16. Willox, R. and Hietarinta, J. (2003) Painlevé equations from Darboux chains, Part 1: P_{III}–P_V, *J. Phys. A: Math. Gen.* **36**, pp. 10615–10635.

17. Bureau, F. J. (1980) Sur un système d'équations différentielles non linéaires, *Bulletin de la Classe des Sciences de l'Académie Royale de Belgique* **66**, pp. 280–284 [in French].

18. Date, E., Jimbo, M., Kashiwara, M., and Miwa, T. (1982) Transformation groups for soliton equations—Euclidean Lie algebras and reduction of the KP hierarchy, *Publications of RIMS* **18**, pp. 1077–1110.

19. Kac, V. G. (1990) *Infinite Dimensional Lie Algebras*, 3rd. ed., Cambridge University Press.

20. Willox, R., Tokihiro, T., Loris, I., and Satsuma, J. (1998) The fermionic approach to Darboux transformations, *Inv. Prob.* **14**, pp. 745–762.

21. Noumi, M. and Yamada, Y. (1999) Symmetries in the fourth Painlevé equation and Okamoto polynomials, *Nagoya Math. J.* **153**, pp. 53–86.

22. Willox, R. (2003) On Darboux chains and Painlevé equations, *RIMS Kōkyūroku* **1302**, pp. 21–37 [in Japanese].